Advanced Lectures in Mathematics (ALM)

ALM 1: Superstring Theory
ALM 2: Asymptotic Theory in Probability and Statistics with Applications
ALM 3: Computational Conformal Geometry
ALM 4: Variational Principles for Discrete Surfaces
ALM 6: Geometry, Analysis and Topology of Discrete Groups
ALM 7: Handbook of Geometric Analysis, No. 1
ALM 8: Recent Developments in Algebra and Related Areas
ALM 9: Automorphic Forms and the Langlands Program
ALM 10: Trends in Partial Differential Equations
ALM 11: Recent Advances in Geometric Analysis
ALM 12: Cohomology of Groups and Algebraic K-theory
ALM 13: Handbook of Geometric Analysis, No. 2
ALM 14: Handbook of Geometric Analysis, No. 3
ALM 15: An Introduction to Groups and Lattices: Finite Groups and Positive Definite Rational Lattices
ALM 16: Transformation Groups and Moduli Spaces of Curves
ALM 17: Geometry and Analysis, No. 1
ALM 18: Geometry and Analysis, No. 2
ALM 19: Arithmetic Geometry and Automorphic Forms
ALM 20: Surveys in Geometric Analysis and Relativity
ALM 21: Advances in Geometric Analysis
ALM 22: Differential Geometry: Under the Influence of S.-S. Chern
ALM 23: Recent Developments in Geometry and Analysis
ALM 24: Handbook of Moduli, Volume I
ALM 25: Handbook of Moduli, Volume II
ALM 26: Handbook of Moduli, Volume III
ALM 27: Number Theory and Related Areas
ALM 28: Selected Expository Works of Shing-Tung Yau with Commentary, Volume I
ALM 29: Selected Expository Works of Shing-Tung Yau with Commentary, Volume II
ALM 30: Automorphic Forms and *L*-functions
ALM 31: Handbook of Group Actions, Volume I
ALM 32: Handbook of Group Actions, Volume II
ALM 33: Introduction to Modern Mathematics

Advanced Lectures in Mathematics
Volume 33

Introduction to Modern Mathematics

edited by

Shiu-Yuen Cheng · Lizhen Ji · Yat-Sun Poon

Jie Xiao · Lo Yang · Shing-Tung Yau

International Press
www.intlpress.com

HIGHER EDUCATION PRESS

Advanced Lectures in Mathematics, Volume 33
Introduction to Modern Mathematics

Volume Editors:
Shiu-Yuen Cheng (Hong Kong University of Science and Technology)
Lizhen Ji (University of Michigan, Ann Arbor)
Yat-Sun Poon (University of California at Riverside)
Jie Xiao (Tsinghua University)
Lo Yang (Institute of Mathematics, Chinese Academy of Sciences)
Shing-Tung Yau (Harvard University)

ISBN: 978-1-57146-305-0

Printed in the United States of America.

19 18 17 16 15 1 2 3 4 5 6 7 8 9

ADVANCED LECTURES IN MATHEMATICS

Contents

Course Notes

Course Notes

*Introduction to
Modern Mathematics*
ALM 33, pp. 3–91

Introductory Lectures on Surgery Theory

Thomas Farrell* and Somnath Basu†

To Marlene

Preface

This article is closely based on a graduate level topics course in topology given in the fall of 2012. The course had a large and enthusiastic audience which consisted of mathematicians with many varying interests ranging from graduate students looking for or working on theses to researchers in dynamical systems, topology and geometry. Because of this the lectures were designed to accommodate the backgrounds and interests of the audience. These lectures eventually focussed on the early work of Kervaire and Milnor [30] and ended with a long discussion of Kervaire's paper [29] which constructed the first example of a topological manifold that does not support a smooth structure. Thus these lectures are only meant to be an introduction to *Surgery Theory*, not a systematic account of this field. Hopefully they will whet the reader's appetite to delve deeper into this fascinating subject. Such a reader can pursue it further in the two classic textbooks [9] by W. Browder and [56] by C.T.C. Wall. The monograph [31] by R. Kirby and L. Siebenmann extends the reach of surgery theory to encompass topological manifolds while that by M. Freedman and F. Quinn [17] extends it to include 4-dimensional manifolds. For the more recent algebraic point of view on this subject, the reader should see A. Ranicki's monograph [45]. Finally, to obtain a taste of the current research in the area, we recommend the following sources [15], [11], [12] and [14].

Acknowledgements

We would like to thank M. Bustamante, A. Gogolev, G. Sorcar, and X. Wu for their help in preparing this article. This work was supported in part by NSF grant DMS-1206622 and a Simons Foundation grant.

*Department of Mathematical Sciences, Binghamton University, NY 13902-6000, USA. Email: farrell@math.binghamton.edu

†Department of Mathematics, Rkm Vivekananda University, Howrah, WB 711202, India. Email: somnath@maths.rkmvu.ac.in

List of Lectures

List of Figures

1 The surgery theory program

Let us start by talking about this theory that tries to answer the basic question:

How to classify manifolds by numerical invariants?

We should begin by discussing what *numerical*, *manifolds* and *classify* mean. A manifold will be generally assumed to be *closed*, i.e., compact without boundary. Usually, there are two types of manifolds: topological and smooth. A hundred years ago, people did not make the distinction. It was assumed that there was no difference. Poincaré started looking at this problem. The first invariant is the dimension of the manifold. An n-dimensional manifold is locally homeomorphic to $\mathbb{R}^n$. For $n = 1$ we have the circle as the only closed one while objects like $[0, 1]$ have a boundary. When we go to $n = 2$ we get various manifolds-sphere, torus, two-holed torus, and more generally, a torus with n-holes. This is the full list for orientable, closed surfaces. Poincaré [43] defined homology groups and showed that the first homology group of a closed, oriented surface is the free abelian group of rank twice the number of holes. The notion of groups only came about later. Poincaré never really talked about homology groups but rather about Betti numbers. When we go to $n = 3$, we have

$$S^3 := \{\mathbf{x} = (x_1, x_2, x_3, x_4) \in \mathbb{R}^4 | \|\mathbf{x}\|^2 := x_1^2 + x_2^2 + x_3^2 + x_4^2 = 1\}.$$

The first homology group $H_1(S^3; \mathbb{Z})$ is zero. Poincaré had said that any closed 3-manifold that has this property is S^3. He even published a paper on that but within a year he had found a counter-example [43]. He found a group G, of order 120, acting freely on S^3 such that the quotient manifold $M := S^3/G$ has the requisite properties. The reason he could distinguish the two manifolds is that he came up with a definition of the *fundamental group*. Since $\pi_1(M) = G$, once we replace the first homology being zero by the fundamental group being zero, we get the Poincaré conjecture.

Let us now jump fifty years to the notion of *homotopy equivalence*. Given a pair of spaces X and Y, not only do they have homology groups or homotopy groups, but there is a relevant notion of maps between them.

Definition 1. We say X is *homotopic* to Y if there exist continuous maps $f : X \to Y$ and $g : Y \to X$ such that $f \circ g \sim \mathrm{id}_Y$ and $g \circ f \sim \mathrm{id}_X$. In other words, there exists $h : [0, 1] \times X \to Y$ such that $h(0, \cdot) \equiv \mathrm{id}_X$ and $h(1, \cdot) \equiv g \circ f$. A similar homotopy exists between $f \circ g$ and the identity map of Y.

To make this definition useful, we need to know how to interpret homotopy equivalence. The word *numerical* is very closely related to this notion.

Definition 2 (CW complex)**.** A CW complex X is a Hausdorff space equipped with a partition of X into open cells that satisfies two additional properties:

(1) For each n-dimensional open cell e (of dimension n) in the partition of X, there exists a continuous map $\varphi_e : \mathbb{D}^n \to X$ such that the restriction to the interior of the closed ball is a homeomorphism onto the cell e, and the image of the boundary of the closed ball is contained in the union of a finite number of elements of the partition, each having cell dimension less than n.

(2) A subset of X is closed if and only if it meets the closure of each cell in a closed set.

The importance of working with CW complexes is due to the following result [59], [60] of J.H.C. Whitehead.

Theorem 3 (J.H.C. Whitehead '49). Let X and Y be two path connected, topological spaces homotopy equivalent to CW complexes. A continuous map $f: X \to Y$ is a *homotopy equivalence* if and only if it induces an isomorphism on homotopy groups, i.e.,

$$f_*: \pi_i(X) \xrightarrow{\cong} \pi_i(Y)$$

is an isomorphism for any $i \geqslant 0$. Moreover, if $\pi_1(X) = 0 = \pi_1(Y)$ then f is a homotopy equivalence if and only if f_* is an isomorphism on homology groups.

Note that closed manifolds (in dimension 4) need not be CW complexes but this theorem does hold for closed manifolds as compact manifolds have the homotopy type of CW complexes. If we knew that simply connected homotopy equivalent spaces (closed manifolds) are homeomorphic then that would be pretty good! But this is where surgery theory comes in. The surgery program has the following problem that it seeks to address:

Given a homotopy equivalence $f: M \to N$ of closed manifolds, decide whether f is homotopic to a homeomorphism.

For a long time, when restricted to simply connected manifolds, this was considered to be true. Until the thirties this was believed to be true in all cases. Later on the simply connected case was left open.

Hurewicz Conjecture. Given a homotopy equivalence $f: M \to N$ of simply connected, closed manifolds, the map f is homotopic to a homeomorphism.

This was solved negatively after the Kervaire-Milnor paper [30] was written. It requires machinery developed by various people, notably by Novikov [38], [39] in 1965. There are three steps to this surgery program:

Step 1: Find a *membrane* (the technical term is *cobordism*) connecting M to N. In other words, find a compact $(n+1)$-dimensional manifold W with two boundary components such that

$$\partial W := \partial^+ W \sqcup \partial^- W = N \sqcup M.$$

We would then want a map F between W and $N \times [0,1]$, extending f on $\partial^+ W$ and is identity on $\partial^- W$. Such a map would be called a *tangential* if there is an isomorphism between TW and the pullback bundle $F^*(T(N \times [0,1]))$.

Thom, in his paper [54] on cobordism, defined these objects. In fact, a smooth manifold has many more numerical invariants other than the usual homology, homotopy groups. Characteristic classes such as Stiefel-Whitney classes, Chern classes and Pontrjagin classes are some examples. Given a smooth manifold we have a notion of tangent space at each point which gives us the tangent bundle.

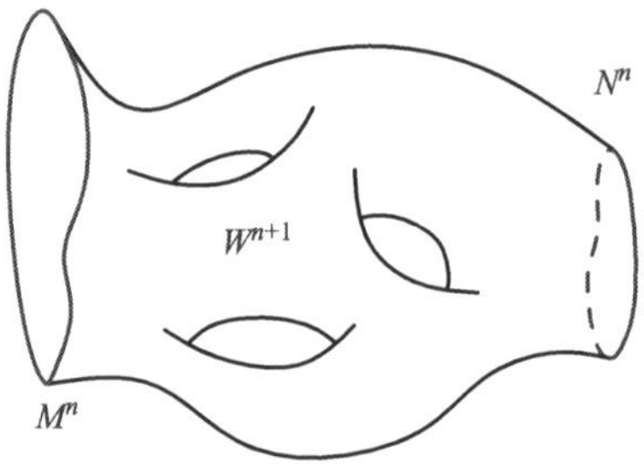

Figure 1. A cobordism between M and N.

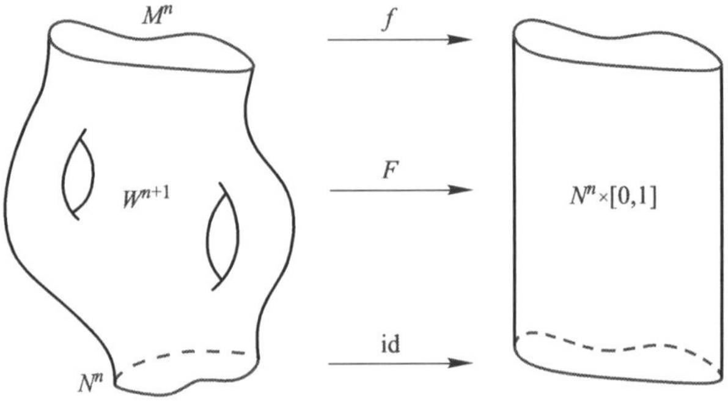

Figure 2. An extension of a map via a cobordism.

This is usually not present for arbitrary topological spaces. Embed M^n into some big Euclidean space $\mathbb{R}^N$ via $\tau : M^n \hookrightarrow \mathbb{R}^N$ and translate the tangent space at any point in M^n to a vector subspace of $\mathbb{R}^N$. We get a continuous function

$$\tau_M : M^n \longrightarrow G_n(\mathbb{R}^N),$$

where $G_n(\mathbb{R}^N)$ is the space of n-dimensional subspaces of $\mathbb{R}^N$. It is an example of Grassmann manifolds. When $n = 1$ we get real projective spaces. This map τ_M is well defined up to homotopy. The Whitney embedding theorem tells us that two embeddings τ, τ' are isotopic and this provides a homotopy between the two maps τ_M and τ'_M. We are implicitly talking about *stable* tangent bundles. Therefore, we have a list of numerical invariants and we would like to say that this list classifies closed manifolds. Well, it does not but it comes pretty close for simply connected manifolds.

By Thom's work [54], if two manifolds have the same stable tangent bundle then they are cobordant. Consider a map $f : M \to N$ between closed n-dimensional manifolds. We want f to satisfy a homotopy triangle

$$\begin{array}{ccc} M & \xrightarrow{\quad f \quad} & N \\ & {\scriptstyle \tau_M} \searrow \quad \swarrow {\scriptstyle \tau_N} & \\ & G_n(\mathbb{R}^{n+k}) & \end{array}$$

Theorem 4 (Thom '54). If $f^*(TN) = TM$ then M is cobordant to N.

We can ask the following: find F such that the following diagram commutes, up to homotopy

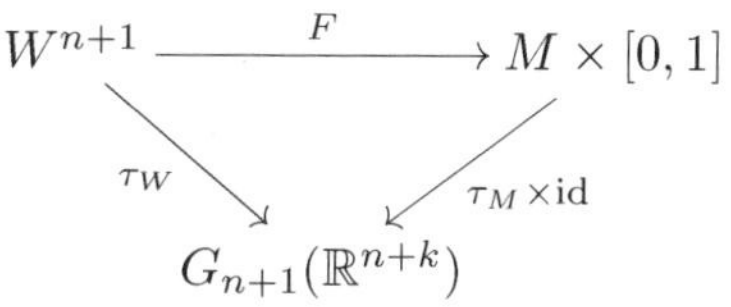

For the Hurewicz conjecture, it is the first step that goes wrong! If we have a tangential[1] cobordism W (between M and N) then M and N have the same stable tangent bundle. Hurewicz conjecture deals with homeomorphisms while we are talking about tangent bundles now. It is known that there is a pair of simply connected, smooth, closed manifolds which are homotopy equivalent but not diffeomorphic. For instance, we can take $S^4 \times S^5$ and the unit sphere bundle N of some appropriate $\mathbb{R}^6$-bundle over S^4. These oriented rank 6 bundles are classified by $\pi_3(\mathrm{SO}_6) \cong \mathbb{Z}$. The first Pontrjagin class of N is non-zero while it is zero for $S^4 \times S^5$. If these were diffeomorphic then they would have the same Pontrjagin class. We will discuss how Novikov got around this issue. His big result was that even if two manifolds were just homeomorphic, the Pontrjagin classes have to be the same. So, the pair in question cannot even be homeomorphic!

Let us modify the program a bit.

Question. *When is a homotopy equivalence $f : M \to N$ between a pair of closed, smooth manifolds homotopic to a diffeomorphism?*

Step 2: This is the surgery step. If we looked at the homology groups then

$$F_* : H_i(W, \partial W) \longrightarrow H_i(M \times [0,1], \partial)$$

is an epimorphism-this follows from Poincaré duality. We want to do a surgery on W without touching ∂W so that F becomes a homotopy equivalence.

Once we have cut out enough, we will show that this is an isomorphism on homology, and by Whitehead's result (cf. Theorem 3), this is a homotopy equivalence in the simply connected case. We may not be able to always do this step unless we allow tacking on the bottom self-tangential cobordisms. We may now rewrite Step 2.

[1] A *tangential cobordism* is a cobordism W^{n+1} between M^n and N^n which is covered by a map of tangent bundles. The (stable) tangent bundle on M is classified by a map $\tau_M : M \to BO_N \simeq G_n(\mathbb{R}^N)$. If $\iota_M : M \hookrightarrow W$ be the inclusion then we would want $\tau_W \circ \iota = \tau_M$. A similar equality holds for N. Finally, a tangential cobordism is *self-tangential* if $M = N$.

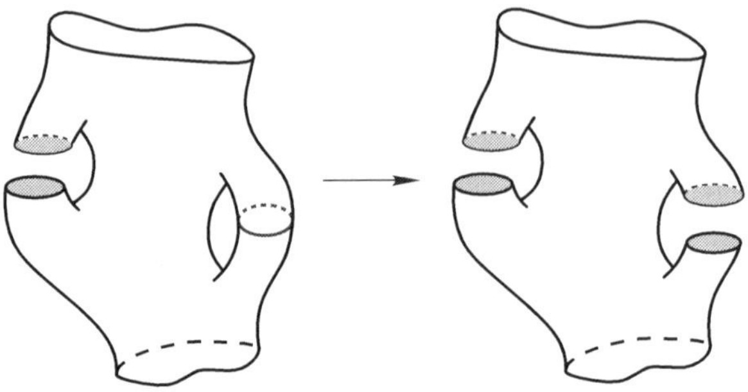

Figure 3. A surgery avoiding the boundary.

Step 2: After concatenating the given tangential cobordism W with a self-tangential cobordism $\mathcal{K}$ of M, do a surgery on $W \cup \mathcal{K}$ to get an h-cobordism between M and N.

Definition 5 (h-cobordism)**.** A h-cobordism W^{n+1} between M^n and N^n is a cobordism such that the two inclusion maps of the two boundaries are homotopy equivalences.

Step 3: Show that this h-cobordism is a *cylinder*, i.e., diffeomorphic to $M \times [0,1]$. We can codify the problem in the *structure set*.

Definition 6 (Structure set)**.** Given a closed, smooth manifold N^n define a set $\mathcal{S}(N)$, called the *structure set*, as follows: an element of $\mathcal{S}(N)$ is an equivalence class of pairs (M^n, f), where M^n is a smooth manifold and $f : M \to N$ is a homotopy equivalence. Two pairs (M_1, f_1) and (M_2, f_2) are equivalent if there exists a diffeomorphism $F : M_1 \to M_2$ such that $f_2 \circ F$ is homotopic to f_1.

If the set has only one element then $[(M, f)] = [(N, \mathrm{id})]$, i.e., M is diffeomorphic to N. But this is very rare! We know from Kervaire-Milnor's paper [30] that

$$\mathcal{S}(S^6) = 1 = \mathcal{S}(S^{12}).$$

They did not know it at that time but we now know that $\mathcal{S}(S^5) = 1$. In general, for $n \geqslant 1$ we have

$$|\mathcal{S}(S^{4n+3})| > 1.$$

Remark 7. For the rest of the lectures, whenever spaces are concerned, we shall use $\simeq$ to denote homotopy equivalences and $\cong$ to denote homeomorphisms. We will continue to use $\cong$ to denote isomorphisms in the natural contexts of vector spaces or groups.

2 Homotopy equivalent manifolds that are not homeomorphic

Previously we had an overview about the surgery theory program. Before getting into the Kervaire-Milnor paper [30], we would like to discuss a couple of examples:

(1) There is a pair of homotopy equivalent, simply connected, closed, smooth manifolds $f : M \to N$ such that f is not homotopic to a diffeomorphism. It was later shown that these manifolds were non-homeomorphic. Take $M = S^4 \times S^5$ and N as the total space of a fibre bundle over S^4 with fibre S^5. More explicitly, N arises as the sphere bundle of an oriented rank 6 bundle over S^4 (cf. discussion to follow later).

(2) There is a pair of homeomorphic, closed, smooth manifolds M and N which are not diffeomorphic. We set $M = S^7$ and let $N = S(\xi)$, the sphere bundle of ξ, where $\xi \to S^4$ is a bundle, suitably chosen, with fibre $\mathbb{R}^4$. If we take the disk bundle $\mathbb{D}(\xi)$ and glue on $\mathbb{D}^8$ by a homeomorphism $f : M \to N$, we get a closed 8-manifold K^8 which does not support a smooth structure.

The dates for when these happened are worth mentioning. The first example above is in Milnor-Stasheff's book [37] on characteristic classes and they attribute this to three people—Thom (1955—1956), Shimada-Tamura (1957). These extra results, that happened later, needed Novikov's theorem. The second example is by Milnor [33] in 1956.

Pontrjagin classes associate to each real vector bundle $\xi \to B$ with fibre $\mathbb{R}^n$ a sequence of integral cohomology classes $p_0(\xi), p_1(\xi), \ldots$, with $p_i(\xi) \in H^{4i}(B;\mathbb{Z})$, satisfying some properties:

(1) (Normalization) $p_0(\xi) = 1 \in H^0(B;\mathbb{Z})$.

(2) (Naturality) If we have a bundle map $\mathcal{F} : \xi_1 \to \xi_2$ covering a map of bases $f : B_1 \to B_2$ which is an isomorphism on fibres then $f^*(p_i(\xi_2)) = p_i(\xi_1)$.

(3) The *total Pontrjagin class* is defined to be

$$p(\xi) := (1 + p_1(\xi) + p_2(\xi) + \cdots) \in \oplus_{i=0}^{\infty} H^{4i}(B;\mathbb{Z}).$$

For a sum $\xi_1 \oplus \xi_2$ of two bundles $p(\xi_1 \oplus \xi_2) = p(\xi_1) \cup p(\xi_2)$ up to 2-torsion. More precisely,

$$2\big[p(\xi_1 \oplus \xi_2) - p(\xi_1) \cup p(\xi_2)\big] = 0.$$

(4) (Stability) Let ε be the trivial rank one bundle (over B). Then

$$p_i(\xi \oplus \varepsilon) = p_i(\xi).$$

A smooth manifold M has a tangent bundle TM. By definition, $p_i(M) := p_i(TM)$. If $f : M \to N$ is a smooth map between smooth manifolds then there is an induced map of bundles $Df : TM \to TN$ which, by naturality of Pontrjagin classes, imply

$$f^*(p_i(N)) = p_i(M).$$

We will prove this result (cf. [39]) later but here it is:

Theorem 8 (Novikov '65). If $f : M \to N$ is a homeomorphism of smooth manifolds then
$$f^*(p_i(N)) = p_i(M) \in H^{4i}(M;\mathbb{Q}).$$

Recall that two vector bundles ξ_1, ξ_2 over B are called *stably equivalent* if there exists an isomorphism of bundles $\xi_1 \oplus \varepsilon^{n_1} \cong \xi_2 \oplus \varepsilon^{n_2}$. A bundle $\xi \to B$, when thought of as a stable bundle, simply means the equivalence class of ξ under this equivalence relation. The key lemma for Novikov's theorem is the following:

Lemma 9. Let $E_i \to S^n$, $i = 1, 2$, be $\mathbb{R}^m$-bundles for $m > n$. Let E_1 be topologically equivalent to E_2 as spaces over S^n, i.e., there exists a homeomorphism $f : E_1 \to E_2$ covering the identity map. Then E_1 and E_2 are equivalent as vector bundles.

We will try to get to the easier material from ten years earlier. Recall that we are considering bundles of the form
$$\mathbb{R}^6 \longrightarrow E \xrightarrow{q} S^4.$$

Think of the 4-sphere as a union of two hemispheres $\mathbb{D}^4_\pm$. Let $E^\pm := q^{-1}(\mathbb{D}^4_\pm)$; these are both trivial bundles over $\mathbb{D}^4$. For each $x \in S^3 = \partial\mathbb{D}^4$ we have a map $A_x : \mathbb{R}^6 \to \mathbb{R}^6$ that identifies the fibre over $\{x\}$ in the two copies of $\mathbb{D}^4$. In other words, we have a map
$$A : S^3 \longrightarrow \mathrm{GL}_6(\mathbb{R}).$$

This is known as the *clutching construction.* It turns out that the isomorphism class of the bundle depends only on the homotopy class of the clutching map. Therefore, isomorphism classes of real vector bundles (of rank 6) over S^4 are in bijective correspondence with $[S^3, \mathrm{GL}_6(\mathbb{R})]$, the set of homotopy classes of maps from S^3 to $\mathrm{GL}_6(\mathbb{R})$. It follows from basic linear algebra that $\mathrm{GL}_n(\mathbb{R})$ is homotopy equivalent to O_n. Therefore,
$$[S^3, \mathrm{GL}_6(\mathbb{R})] = [S^3, \mathrm{O}_6] = \pi_3(\mathrm{O}_6) = \pi_3(\mathrm{SO}_6) \cong \mathbb{Z}.$$

This is perhaps easy for this dimension but later on we will need Bott's result [7]. Let us write down two very important homotopy theorems (refer [49], [50] and [7]).

Theorem 10 (Serre '51). The homotopy groups of spheres are given by
$$\pi_i(S^n) \cong \begin{cases} \text{finite}, & \text{if } i \neq n, 2n-1 \\ \mathbb{Z}, & \text{if } i = n \\ \text{finite}, & \text{if } i = 2n-1 \text{ and } n \text{ is odd} \\ \mathbb{Z} \oplus \text{finite}, & \text{if } i = 2n-1 \text{ and } n \text{ is even.} \end{cases}$$

Theorem 11 (Bott '56)**.** We have a natural embedding

$$O_1 \subset O_2 \subset O_3 \subset \cdots$$

and denote the union by O. The homotopy groups are given by

$$\begin{array}{llll} \pi_0(O) = \mathbb{Z}_2, & \pi_1(O) = \mathbb{Z}_2, & \pi_2(O) = 0, & \pi_3(O) \cong \mathbb{Z}, \\ \pi_4(O) = 0, & \pi_5(O) = 0, & \pi_6(O) = 0, & \pi_7(O) \cong \mathbb{Z}, \end{array}$$

and repeated thereafter. Moreover, $\pi_i(O_n) = \pi_i(O)$ if $i < n - 1$.

Let us switch our consideration to

$$O_n \subseteq \text{Diff}(S^{n-1}) \subseteq \text{Top}(S^{n-1}) \subseteq G_n,$$

where G_n is the H-space of all self-homotopy equivalences of S^{n-1}. We have a sequence of groups:

$$\pi_3(O_6) \longrightarrow \pi_3(\text{Diff}(S^5)) \longrightarrow \pi_3(\text{Top}(S^5)) \longrightarrow \pi_3(G_6).$$

Lemma 12. If $E \in \pi_3(O_6)$ goes to zero in $\pi_3(G_6)$ then $S(E) \simeq S^4 \times S^5$.

Proof. We need to define a map $f : S(E) \to S^5$ such that restricted to any fibre it is a homotopy equivalence. Thus, if it is degree one for one fibre then it is degree one for each fibre. Define f on the southern hemisphere by projection and on the northern hemisphere by the null-homotopy. Using $q : S(E) \to S^4$ we get a map

$$q \times f : S(E) \longrightarrow S^4 \times S^5.$$

We can invoke Whitehead's theorem and show that this induces an isomorphism in homology. There are various things one can invoke at this point but since we are going to need this later on, we will use a Serre spectral sequence (with integer coefficients). There exists a spectral sequence such that

$$E^2_{p,q} = H_p(S^4; H_q(S^5; \mathbb{Z})) = H_p(S^4; \mathbb{Z}) \otimes H_q(S^5; \mathbb{Z})$$

and this is supposed to converge to $H_{p+q}(S(E); \mathbb{Z})$. We then use the fact that both bundles have cross-sections to get an isomorphism on homology, i.e., that both spectral sequences collapse. □

3 Freudenthal's suspension & the J-homomorphism

Let us start by giving another proof of the result that we had stated last time.

Lemma 13 (cf. Lemma 12). Let E be a 6-dimensional real vector bundle over S^4 corresponding to $\alpha \in \pi_3(\mathrm{O}_6)$ such that the image of α is zero in $\pi_3(\mathrm{G}_6)$. Then $S(E)$ is homotopy equivalent to $S^4 \times S^5$.

Proof. Recall that we had constructed a map $f : S(E) \to S^5$ which when restricted to each fibre is a degree one map. This gave us a map $q \times f : S(E) \to S^4 \times S^5$. We will invoke Whitehead's theorem on homotopy equivalence in terms of homotopy groups. We have the diagram

$$\begin{array}{ccc} S^5 & \xrightarrow{f|_{\text{fibre}}} & S^5 \\ \downarrow & & \downarrow \\ S(E) & \xrightarrow{q\times f} & S^4 \times S^5 \\ \downarrow{\scriptstyle q} & & \downarrow{\scriptstyle \text{projection}} \\ S^4 & \xrightarrow{\text{id}} & S^4. \end{array}$$

There is a long exact sequence of homotopy groups and maps between them:

$$\begin{array}{ccccccccccc} \cdots \to & \pi_{n+1}(S^4) & \to & \pi_n(S^5) & \to & \pi_n(S(E)) & \to & \pi_n(S^4) & \to & \pi_{n-1}(S^5) & \to \cdots \\ & \downarrow\cong & & \downarrow\cong & & \downarrow & & \downarrow\cong & & \downarrow\cong & \\ \cdots \to & \pi_{n+1}(S^4) & \to & \pi_n(S^5) & \to & \pi_n(S^4\times S^5) & \to & \pi_n(S^4) & \to & \pi_{n-1}(S^5) & \to \cdots \end{array}$$

By the five lemma, the middle vertical arrow is an isomorphism. □

Let us consider the J-homomorphism. It was defined by the other Whitehead-G.W. Whitehead[2]. Let $*$ be a base point for S^{n-1} and define F_n to be all self-homotopy equivalences of S^{n-1} which fix the base point $*$. The suspension map $\Sigma : \mathrm{G}_n \to \mathrm{G}_{n+1}$ actually lands in F_{n+1} as the suspension of any map fixes the north pole.

Lemma 14. The homotopy groups of F_{n+1} are given by

$$\pi_k(\mathrm{F}_{n+1}) \cong \pi_{k+n}(S^n).$$

Proof. We claim that F_{n+1} is the nth based loop space of S^n. We are looking at maps $(S^n, *) \to (S^n, *)$ which can also be thought of as maps from $(\mathbb{D}^n, \partial\mathbb{D}^n)$; this

[2] They are not related by blood. One is British while the other is American. Although they never wrote papers together, they wrote on each others' papers.

is precisely $\Omega^n S^n$. This is a H-space and all components are homotopy equivalent. Picking a component, we see that

$$\pi_k(\Omega^n S^n, 0) \cong \pi_{n+k}(S^n)$$

and this completes the proof. □

The inclusion $\mathrm{O}_n \subseteq \mathrm{G}_n \subseteq \mathrm{F}_{n+1}$ induces a map

$$J : \pi_k(\mathrm{O}_n) \longrightarrow \pi_k(\mathrm{F}_{n+1}) \cong \pi_{n+k}(S^n).$$

This is the *J-homomorphism.* In our case, we have

$$J : \mathbb{Z} \cong \pi_3(\mathrm{O}_6) \longrightarrow \pi_9(S^6) \cong \mathbb{Z}_{24}.$$

This map factors through $\pi_3(\mathrm{G}_6)$. For a long time people used to think that higher homotopy groups are zero for spheres. Hopf came along and showed [24] that $\pi_3(S^2) \cong \mathbb{Z}$.

We look at the fibration $\mathrm{F}_n \to \mathrm{G}_n \to S^{n-1}$. We have the long exact sequence

$$\cdots \longrightarrow \pi_{k+1}(S^{n-1}) \longrightarrow \pi_k(\mathrm{F}_n) \longrightarrow \pi_k(\mathrm{G}_n) \longrightarrow \pi_k(S^{n-1}) \longrightarrow \cdots$$

and when $k \leqslant n-3$ we have an isomorphism between the middle groups. Moreover, we also have the diagram:

$$\begin{array}{ccccccccc}
\cdots \longrightarrow & \pi_{k+1}(S^n) & \longrightarrow & \pi_k(\mathrm{F}_{n+1}) & \xrightarrow{i_*} & \pi_k(\mathrm{G}_{n+1}) & \longrightarrow & \pi_k(S^n) & \longrightarrow \cdots \\
 & & & \Sigma\uparrow & \nwarrow^{\boldsymbol{\Sigma}} & \uparrow\Sigma & & & \\
\cdots \longrightarrow & \pi_{k+1}(S^{n-1}) & \longrightarrow & \pi_k(\mathrm{F}_n) & \xrightarrow{i_*} & \pi_k(\mathrm{G}_n) & \longrightarrow & \pi_k(S^{n-1}) & \longrightarrow \cdots .
\end{array}$$

We have to recall another important theorem [18].

Theorem 15 (Freudenthal '38)**.** The suspension map E (or Σ)

$$\Sigma : \pi_k(S^m) \longrightarrow \pi_{k+1}(S^{m+1})$$

is an isomorphism if $k \leqslant 2m - 2$.

The letter 'E' in German was used to denote 'einhängung' meaning 'suspension'.

Corollary 16. If $k \leqslant n - 3$ then $\boldsymbol{\Sigma} : \pi_k(\mathrm{G}_n) \longrightarrow \pi_k(\mathrm{F}_{n+1})$ is an isomorphism.

Proof. For k and n as in the statement, the maps i_* (cf. the grid above) are isomorphisms. Now consider the square

$$\begin{array}{ccc}
\pi_k(\mathrm{F}_{n+1}) & \xrightarrow{i_*} & \pi_k(\mathrm{G}_{n+1}) \\
\Sigma\uparrow & \nwarrow^{\boldsymbol{\Sigma}} & \uparrow\Sigma \\
\pi_k(\mathrm{F}_n) & \xrightarrow{i_*} & \pi_k(\mathrm{G}_n)
\end{array}$$

The left vertical map is an isomorphism by Theorem 15, whence the diagonal map is also an isomorphism. □

It follows from the above that

$$\pi_3(\mathrm{G}_6) \cong \pi_3(\mathrm{F}_7) \cong \pi_9(S^6) \cong \mathbb{Z}_{24}.$$

Since $\pi_3(\mathrm{O}_6) \cong \mathbb{Z}$ and $J : \pi_3(\mathrm{O}_6) \to \pi_3(\mathrm{G}_6) \cong \mathbb{Z}_{24}$, if Θ generates $\pi_3(\mathrm{O}_6)$ then

$$S(E_{n\Theta}) \simeq S^4 \times S^5$$

provided 24 divides n.

Theorem 17. Let $n \in \mathbb{Z}$. The smooth manifolds $M_{24n} := S(E_{24n})$ are pairwise non-diffeomorphic.

The method of proof is to show that the first Pontrjagin class of M_{24n} is $2n$, once we identify $H^4(M_{24n};\mathbb{Z})$ with $H^4(S^4 \times S^5;\mathbb{Z})$.

Let us recall some facts about canonical line bundles. For $\mathbb{RP}^1$ we get the Möbius line bundle while for $\mathbb{CP}^1$, the associated circle bundle to the canonical line bundle is the Hopf fibration. We can do this with the quarternions too. We are denoting the tautological (canonical) line bundle over $S^4 = \mathbb{HP}^1$ by $\mathbb{H}\eta$. The associated sphere bundle is given by

$$S^3 \longrightarrow S(\mathbb{H}\eta) = S^7 \longrightarrow S^4 = \mathbb{HP}^1.$$

In [37] they calculate the various characteristic classes for this bundle.

Lemma 18 (Lemma 20.9 [37]). The first Pontrjagin class is given by

$$p_1(\mathbb{H}\eta) = -2 \in H^4(\mathbb{HP}^1;\mathbb{Z}).$$

This class is the negative of the Euler class of the bundle. This is naturally defined as the fibres are quarternionic spaces which have a natural orientation, and over a sphere such bundles have a natural Euler class. We should keep in mind that the theorems come after the examples which cover a large percentage of the material. A worthwhile endeavour (for this course) is to understand the examples really well. Finally, there is a map

$$\mathscr{P} : \pi_3(\mathrm{O}_6) \longrightarrow \mathbb{Z},\ \alpha \mapsto \langle p_1(E_\alpha), [S^4] \rangle .$$

We will talk more about this map in subsequent discussions.

4 Bundles of rank six on the 4-sphere

Let us recall what we did last time. We are trying to construct manifolds which are homotopy equivalent but not homeomorphic to each other. We are studying real vector bundles (of rank 6) over S^4; such bundles are classified by $\pi_3(\mathrm{O}_6)$. Recall that $\pi_3(\mathrm{O}_6) \cong \mathbb{Z}$ and $\pi_9(S^6) \cong \mathbb{Z}_{24}$. There is the J-homomorphism

$$J : \pi_3(\mathrm{O}_6) \longrightarrow \pi_9(S^6).$$

If $\alpha \in \ker J$ then $S(E_\alpha) \simeq S^4 \times S^5$. These are smooth, closed manifolds.

Lemma 19. The stable canonical line bundle $\mathbb{H}\eta \oplus \varepsilon^2$ over $\mathbb{HP}^1 = S^4$ generates $\pi_3(\mathrm{O}_6)$.

Well, how do we prove something like this? We claim that SO_4 is topologically $S^3 \times \mathbb{RP}^3$. We have a principal fibre bundle

$$\mathrm{SO}_3 \longrightarrow \mathrm{SO}_4 \longrightarrow S^3$$

by mapping a matrix $A \in \mathrm{SO}_4$ to its first column. We think of S^3 as the unit quarternions. Fix $\mathbf{a} \in S^3$ and define

$$f_{\mathbf{a}} : \mathbb{H} \longrightarrow \mathbb{H}, \quad \mathbf{x} \mapsto \mathbf{a} \cdot \mathbf{x}.$$

This is a linear transformation and, moreover, it is orthogonal. A cross section $s : S^3 \to \mathrm{SO}_4$ is given by mapping $\mathbf{a} \in S^3$ to $f_{\mathbf{a}} \in \mathrm{SO}_4$. The principal bundle, therefore, is trivial. Since $\mathbb{RP}^3 \cong \mathrm{SO}_3$, we know that

$$\pi_3(\mathrm{SO}_4) \cong \pi_3(S^3 \times \mathbb{RP}^3) \cong \mathbb{Z} \oplus \mathbb{Z}.$$

By Bott periodicity,

$$\pi_3(\mathrm{SO}_5) \cong \mathbb{Z} \cong \pi_3(\mathrm{SO}_6).$$

Since $\mathbb{Z} \cong \pi_3(\mathrm{O}_6)$ is generated by $\pi_3(S^3)$ that generates the stable π_3 (cf. Lemma 23). This proves the lemma contingent on the proof of Lemma 23 given in the next lecture.

Let us define another homomorphism $\mathscr{P} : \pi_3(\mathrm{O}_6) \longrightarrow \mathbb{Z}$. Given $\alpha \in \pi_3(\mathrm{O}_6)$ we take the first Pontrjagin class $p_1(E_\alpha) \in H^4(S^4; \mathbb{Z})$, where E_α is the bundle obtained via the clutching construction. Fix a generator $[S^4] \in H_4(S^4; \mathbb{Z})$.

Proposition 20. The map $\mathscr{P} : \pi_3(\mathrm{O}_6) \longrightarrow \mathbb{Z}$ defined by

$$\mathscr{P}(\alpha) := \langle p_1(E_\alpha), [S^4] \rangle \in \mathbb{Z}.$$

is a group homomorphism.

We still have to verify that $\mathscr{P}$ is a group homomorphism. We are going to use $G_n(\mathbb{R}^\infty)$ as the Grassmann manifold of n-planes in $\mathbb{R}^\infty$. Let $n = 6$ and we can also look at 6-*frames*, i.e., 6-tuples $(v_1, \ldots, v_6) \in (\mathbb{R}^\infty)^6$ such that $v_i \cdot v_j = 0$ if $i \neq j$

and 1 otherwise. Let $V_6(\mathbb{R}^\infty)$ be the *Stiefel manifold* consisting of all 6-frames. We have a map

$$V_6(\mathbb{R}^\infty) \longrightarrow G_6(\mathbb{R}^\infty), \quad (v_1, \ldots, v_6) \mapsto \mathrm{span}\{v_1, \ldots, v_6\}.$$

The map above defines a Serre fibration and the fibre is O_6.

Lemma 21. The space $V_6(\mathbb{R}^\infty)$ is contractible.

Proof. If we look at unit length vectors in $\mathbb{R}^\infty$ then we see that S^∞ is contractible. If we look at 2-frames then we have a map

$$V_2(\mathbb{R}^\infty) \longrightarrow V_1(\mathbb{R}^\infty)$$

with fibre S^∞. Therefore, $V_2(\mathbb{R}^\infty)$ is also contractible. In general, we boot-strap our way to the claim which holds for $V_n(\mathbb{R}^\infty)$. □

Proof of Proposition. Recall the long exact sequence of homotopy groups induced by the fibration

$$\mathrm{O}_6 \longrightarrow V_6(\mathbb{R}^\infty) \longrightarrow G_6(\mathbb{R}^\infty).$$

In particular, we have an isomorphism

$$\delta : \pi_4(G_6(\mathbb{R}^\infty)) \longrightarrow \pi_3(\mathrm{O}_6).$$

Therefore, $\delta^{-1}(\alpha)$ defines a map $S^4 \to G_6(\mathbb{R}^\infty)$. Now consider the canonical bundle over $G_6(\mathbb{R}^\infty)$ given by

$$E_{\xi^6} := \{(V, v) \mid V \in G_6(\mathbb{R}^\infty), v \in V\}.$$

This is called the *universal* 6-dimensional vector bundle. In particular, there is a bijection between isomorphism classes of 6-dimensional vector bundles over B and homotopy classes of maps $[B, G_6(\mathbb{R}^\infty)]$. The correspondence is as follows: if $f : B \to G_6(\mathbb{R}^\infty)$ is continuous then the pullback $f^*(E_{\xi^6})$ is a 6-dimensional bundle over B.

Let us see what we have until now. Given $\alpha \in \pi_3(\mathrm{O}_6)$ we get

$$\delta^{-1}(\alpha) \in \pi_4(G_6(\mathbb{R}^\infty)) = [S^4, G_6(\mathbb{R}^\infty)].$$

The exercise is to show that the bundle induced by $\delta^{-1}(\alpha)$ and E_α are isomorphic. Under this assumption,

$$p_1(E_\alpha) = p_1(E_{\delta^{-1}(\alpha)}) = p_1(f^*(E_{\xi^6})) = f^*(p_1(E_{\xi^6})).$$

Therefore, we get

$$\langle p_1(E_\alpha), [S^4] \rangle = \langle f^*(p_1(E_{\xi^6})), [S^4] \rangle = \langle p_1(E_{\xi^6}), f_*[S^4] \rangle = \langle p_1(E_{\xi^6}), \delta^{-1}(\alpha) \rangle.$$

The right hand side of the right-most term is a composite of two homomorphisms - δ^{-1} and Hurewicz-whence is itself a homomorphism. □

What we want, while we are at it, is that the very same argument will prove that something else is a homomorphism. Let us talk about the Euler class of an oriented bundle. But before that let us go back and modify the definition of Grassmann manifold slightly. We can talk about Grassmann manifold $\widetilde{G}_6(\mathbb{R}^\infty)$ of oriented 6-frames in $\mathbb{R}^\infty$. This fibres over $G_6(\mathbb{R}^\infty)$ with fibre $\mathbb{Z}_2$. We also know that homotopy classes of maps $[B, \widetilde{G}_6(\mathbb{R}^\infty)]$ are in bijective correspondence with isomorphism classes of oriented 6-dimensional vector bundles over B. Now we can talk about the Euler class.

Given a bundle $\mathbb{R}^6 \to E \to B$ we want to define $e(E) \in H^6(B;\mathbb{Z})$. This is defined by obstruction theory. We need E to be oriented as a bundle to work over $\mathbb{Z}$; otherwise we work over $\mathbb{Z}_2$ coefficients. We shall define another number:

$$\mathscr{E} : \pi_3(\mathrm{SO}_6) \longrightarrow \mathbb{Z}, \quad \alpha \mapsto \langle e(E_\alpha), [S^4] \rangle \in \mathbb{Z}.$$

At the end, we will arrive at Milnor's result [33] where he studies 4-dimensional bundles over S^4. We have two maps from $\pi_3(\mathrm{SO}_6)$ given by $\mathscr{E}$ and $\mathscr{P}$. Any bundle with the same pair of numbers will be isomorphic but not any pair is realizable. The numerics will play a crucial role in Milnor's paper. Let us get back to what we started with. Let $\alpha \in \pi_3(\mathrm{O}_6)$ and consider $M^9 = S(E_\alpha)$. We want to calculate $|p_1(E_\alpha)|$ and show that it equals $|\mathscr{P}(\alpha)|$. Since $H^4(M^9;\mathbb{Z}) \cong \mathbb{Z}$ has no canonical identification, we cannot get rid of the absolute value.

Let $E^6 \to S^4$ be a bundle equipped with a section $\sigma : S^4 \to E^6$. It is a tautology when one says that

$$TE|_{0-\text{section}} = TS^4 \oplus E^6.$$

More precisely,

$$\sigma^*(TE) = TS^4 \oplus E$$

and, therefore,

$$p_1(\sigma^*(TE)) = p_1(TS^4) + p_1(E) = p_1(E)$$

as S^4 is stably parallelizable. By naturality, we get

$$p_1(E) = \sigma^*(p_1(TE)).$$

We want to look at the sphere bundle $\iota : S(E) \to E$. We have

$$TE|_{S(E)} = TS(E) \oplus \varepsilon.$$

Therefore,

$$p_1(TS(E)) = \iota_*(p_1(TE)).$$

Notice that σ^* is an isomorphism on the fourth cohomology. We claim that under

$$\iota_* \circ q_* : H^4(S^4;\mathbb{Z}) \longrightarrow H^4(S(E);\mathbb{Z}), \quad p_1(E) \mapsto p_1(S(E)).$$

5 Rank six bundles & Pontrjagin classes

Let us recall the setup: let η be a 6-dimensional real vector bundle with base S^4. Let E denote the total space of η; it is a 10-dimensional smooth manifold and we denote $M^9 := S(\eta)$, the sphere bundle associated with η. We have also been talking about Pontrjagin classes of bundles and manifolds. Whenever we write $p_i(X)$ we mean the ith Pontrjagin class of the tangent bundle of X unless X is a vector bundles over some base. In such a case X will be denoted appropriately. We have the following diagram:

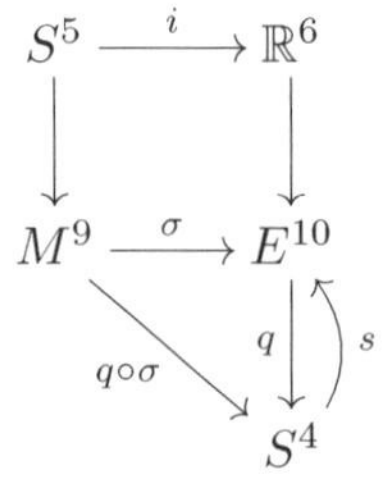

We have the following isomorphisms

$$\mathbb{Z} \cong H^4(S^4;\mathbb{Z}) \xrightarrow{q^*} H^4(E^{10};\mathbb{Z}) \xrightarrow{\sigma^*} H^4(M^9;\mathbb{Z}).$$

Last time we showed that

$$\sigma^*(p_1(E^{10})) = p_1(M^9), \quad s^*(p_1(E^{10})) = p_1(\eta).$$

Applying q^* to the second equality we get

$$p_1(E^{10}) = q^*(p_1(\eta)).$$

A substitution gives us

$$\sigma^* \circ q^*(p_1(\eta)) = p_1(M^9).$$

Let a be *the* generator of $H^4(M^9;\mathbb{Z})$ such that

$$q_*\sigma_*(a) = [S^4] \in H_4(S^4;\mathbb{Z}).$$

Therefore,

$$\langle p_1(M), a\rangle = \langle \sigma^* q^*(p_1(\eta)), a\rangle = \langle p_1(\eta), q_*\sigma_*(a)\rangle = \left\langle p_1(\eta), [S^4]\right\rangle.$$

Lemma 22. The following equality holds

$$|p_1(M)| = |p_1(\eta)|.$$

We apply the lemma to $\eta = E_\alpha$, where $\alpha \in \pi_3(\mathrm{O}_6) \cong \mathbb{Z}$, we get

$$|p_1(S(E_\alpha))| = |\mathscr{P}(\alpha)|. \tag{1}$$

We have already shown that $\mathscr{P}$ is a homomorphism. By Lemma 18

$$\langle p_1(\mathbb{H}\eta), [S^4]\rangle = -2.$$

Since Pontrjagin classes are stable invariants,

$$\langle p_1(\mathbb{H}\eta \oplus \varepsilon^2), [S^4]\rangle = -2.$$

If we denote $E_\omega := \mathbb{H}\eta \oplus \varepsilon^2$ then $\mathscr{P}(\omega) = -2$.

Define $\alpha_n \in \pi_3(\mathrm{O}_6)$ as $\alpha_n := (24n)\omega$. There are two homomorphisms $\mathscr{P}, J$ from $\pi_3(\mathrm{O}_6)$ to $\mathbb{Z}$ and $\pi_9(S^6) \cong \mathbb{Z}_{24}$ respectively. Then we have

$$J(\alpha_n) = (24n)J(\omega) = 0, \tag{2}$$

$$\mathscr{P}(\alpha_n) = (24n)(-2) = -48n. \tag{3}$$

Let $M_n := S(E_{\alpha_n})$. By Lemma 12 and (2) we know that $M \simeq S^4 \times S^5$. But by (3) and (1) we get $|\mathscr{P}(M_n)| = 48|n|$. We claim that ω generates $\pi_3(\mathrm{O}_6)$. This may not seem like much but it gives examples of manifolds that cannot be triangulated. There are two main theorems here—that of Novikov and Rokhlin.

Lemma 23. The element ω generates $\pi_3(\mathrm{O}_6)$. Moreover, this is equivalent to the following: the minimum positive value of $|p_1(\xi)|$ is 2 as ξ varies over all real vector bundles with base S^4.

Proof. If ω generates the group then by the fact that $\mathscr{P}$ is a group homomorphism and $\mathscr{P}(\omega) = -2$, the minimum value is as stated. On the other hand, if ω is not a generator then write $\omega = n\alpha$ for some $n > 1$. Therefore,

$$|\mathscr{P}(\alpha)| = \frac{|\mathscr{P}(\omega)|}{n} = \frac{2}{n} < 2$$

and it suffices to prove the second claim.

We need to study the Stiefel manifolds. In particular, we need $V_2(\mathbb{R}^5)$, the space of unit 2-frames in $\mathbb{R}^5$. This is also the total space of the sphere bundle of TS^4. We will digress by noting that any SO_4-bundle over S^4 can be turned into an SO_6-bundle by stabilization. There is a natural map $\pi_3(\mathrm{SO}_4) \to \pi_3(\mathrm{SO}_6)$. Since $\mathrm{SO}_4 \cong \mathrm{SO}_3 \times S^3$, it follows that $\pi_3(\mathrm{SO}_4) \cong \mathbb{Z} \oplus \mathbb{Z}$ and one of the summands is generated by $\mathbb{H}\eta$, which comes from the S^3 factor of SO_4. We claim that this maps to ω under the natural map. Since SO_5 is in the stable range, consider

$$\mathrm{SO}_3 \xrightarrow{\sigma} \mathrm{SO}_5 \longrightarrow V_2(\mathbb{R}^5)$$

given by mapping a matrix to its first two columns. The stabilizer is given by SO_3. This induces a long exact sequence of homotopty groups:

$$\cdots \longrightarrow \pi_3(\mathrm{SO}_3) \xrightarrow{\sigma_*} \pi_3(\mathrm{SO}_5) \longrightarrow \pi_3(V_2(\mathbb{R}^5)) \longrightarrow \pi_2(\mathrm{SO}_3) \longrightarrow \cdots.$$

As the second homotopy group of SO_3 is zero we get

$$\cdots \longrightarrow \mathbb{Z} \xrightarrow{\sigma_*} \mathbb{Z} \longrightarrow \pi_3(V_2(\mathbb{R}^5)) \longrightarrow 0.$$

We claim that the rightmost group is $\mathbb{Z}_2$. There are many ways to do this-we will use the Gysin sequence. Consider the disk bundle $\mathbb{D}S^4$ sitting inside TS^4. The unit tangent bundle is $\partial\mathbb{D}S^4$. We have a long exact sequence:

$$\cdots \longrightarrow H^4(\mathbb{D}S^4, \partial\mathbb{D}S^4; \mathbb{Z}) \longrightarrow H^4(\mathbb{D}S^4; \mathbb{Z}) \longrightarrow H^4(\partial\mathbb{D}S^4; \mathbb{Z}) \longrightarrow 0.$$

The leftmost map above is multiplication by the Euler class (which is multiplication by the Euler characteristic of S^4) which is 2. By Poincaré duality, $H_3(V_2(\mathbb{R}^5); \mathbb{Z}) = \mathbb{Z}_2$ and by Hurewicz isomorphism, $\pi_3(V_2(\mathbb{R}^5)) = H_3(V_2(\mathbb{R}^5); \mathbb{Z})$. Therefore, the image of σ_* is the set of even integers in $\pi_3(SO_5)$. Consequently, the image of $(\mathscr{P} \circ \sigma)_*$ is contained in the even integers. We want to check that the $\mathbb{Z}$ generated by S^3 inside SO_4 also maps to even integers under $\mathscr{P}$. In other words, the map

$$\pi_3(S^3) \longrightarrow \pi_3(SO_4) \longrightarrow \pi_3(SO_6) \xrightarrow{\mathscr{P}} \mathbb{Z}$$

is contained in $2\mathbb{Z}$. Doing the clutching construction to the identity map of S^3 we get $\mathbb{H}\eta$ over S^4. This is true and one of those little things that we have to sit down and convince ourselves of. But by Lemma 18 we have

$$p_1(\mathbb{H}\eta) = -2$$

which proves the lemma. □

We need to talk about Rokhlin's theorem and this itself will require some rather useful notions.

Definition 24 (Parallelizability of manifolds). Let M^n be a connected, smooth manifold.
(i) The manifold M^n is *parallelizable* if $TM \cong \varepsilon^n$.
(ii) The manifold M^n is *stably parallelizable* if $TM \oplus \varepsilon^k \cong \varepsilon^{n+k}$.
(iii) The manifold M^n is *almost parallelizable* if $M - \{\text{point}\}$ is parallelizable.

Note that (ii) is equivalent to requiring $TM \oplus \varepsilon^1 \cong \varepsilon^{n+1}$ (cf. Remark 67). It can be shown that (i) implies (ii) which in turn implies (iii). The two-sphere S^2 is an example of (ii) which is not (i). An example of (iii) which is not (ii) is a little more non-trivial and we will see one as we go through Rokhlin's theorem. We need an invariant from algebraic topology-the *index* of a manifold.

Definition 25 (Index of a manifold). Let M^{4n} be a closed, oriented manifold with an orientation class $[M^{4n}] \in H_{4n}(M;\mathbb{Z})$. We have an intersection form

$$\mathcal{B} : H^{2n}(M^{4n};\mathbb{R}) \times H^{2n}(M^{4n};\mathbb{R}) \xrightarrow{\cup} H^{4n}(M^{4n};\mathbb{R}) \xrightarrow{\langle \cdot, [M] \rangle} \mathbb{R}.$$

It is non-degenerate by Poincaré duality. The *index* is defined to be the difference of the number of $+1$'s and -1's, once $\mathcal{B}$ is diagonalized.

Having defined the index of a manifold, we will state Rokhlin's result [48].

Theorem 26 (Rokhlin '52). Let M^4 be a connected, closed, smooth 4-manifold which is almost parallelizable. Then the index of M is divisible by 16.

The almost parallelizable condition is there to rule out objects like $\mathbb{CP}^2$. In modern terms, it is the same as a *spin* manifold, i.e., a choice of a trivialization of TM over the 2-skeleton of M.

6 Rokhlin's theorem & Thom complexes

Last time we had talked about Rokhlin's theorem. Historically, what Rokhlin had proved is not what we stated. What he had actually proved is the following:

Theorem 27 (cf. Theorem 26). If M^4 is a closed, orientable, almost parallelizable, smooth 4-manifold then 48 divides $p_1(M)$.

A couple of comments are in order.

Remark 28. An orientable manifold M is almost parallelizable if and only if $w_2(M)$ is zero, i.e., M is spin. Rokhlin also proved the four dimensional version of what later in 1956 became the Hirzebruch Index formula, i.e., the index

$$I(M^4) = \langle \tfrac{1}{3}\, p_1(M), [M] \rangle.$$

Combining this and Rokhlin's theorem we get what has now become Rokhlin's theorem.

In proving Rokhlin's theorem we need to revisit the J-homomorphism defined by G.W. Whitehead [57] in 1942.

Let $\alpha : S^k \to \mathrm{SO}_n$ be a map. We define the function

$$\alpha : S^k \times S^{n-1} \longrightarrow S^{n-1}, (x, \mathbf{v}) \mapsto \alpha(x)(\mathbf{v}).$$

Whenever we have a map between spaces we can suspend the function, i.e.,

$$\Sigma\alpha : \Sigma(S^k \times S^{n-1}) \longrightarrow \Sigma S^{n-1} = S^n.$$

Let us see what the domain is. We could collapse the two ends of $S^k \times S^{n-1} \times [0,1]$ to S^k and S^{n-1} respectively, and then collapse the ends to points. Therefore, the domain, being the suspension of the quotient of the join of S^k and S^{n-1}, is S^{k+n}. This induces a map

$$J : \pi_k(\mathrm{SO}_n) \longrightarrow \pi_{k+n}(S^n)$$

called the *J-homomorphism.* It is worthwhile to chase through the maps for id : $S^1 \to \mathrm{SO}_2$ and check that $J(\mathrm{id}) : S^3 \to S^2$ is the *Hopf map.*

We have to talk about the *Thom complex.* Let $\xi = \xi^n$ be a real n-dimensional vector bundle. If $\mathbb{D}(\xi)$ is the unit disk bundle then let $\mathrm{Th}(\xi) = \mathbb{D}(\xi)/S(\xi)$ be the quotient space. If the base B is compact then $\mathrm{Th}(\xi)$ is the one-point compactification of the total space of the bundle. We are interested in this when the base space is a manifold. Let us assume that $B = S^k$. In this situation, the Thom complex is closely related to J-homomorphism.

Proposition 29. Let $\alpha \in \pi_{k-1}(\mathrm{SO}_n)$. If the bundle ξ_α is constructed with base S^k via the clutching construction then the Thom complex $\mathrm{Th}(\xi_\alpha)$ is homotopy equivalent to S^n with a $(k+n)$-cell attached via the J-homomorphism.

The proof is an exercise using Whitehead's definition of $J(\alpha)$. Look at the disk bundle over the southern hemisphere. This is homeomorphic to $\mathbb{D}^{k+n}$. Now the complement of the interior of $\mathbb{D}^{k+n}$ in $\mathrm{Th}(\xi_\alpha)$ is a thickening of S^n.

Example 30. Let $\mathbb{C}\eta$ be the canonical complex line bundle over S^2, now thought of as a real dimensional bundle. We have seen that this bundle is just ξ_{id}. On the other hand,

$$\mathrm{Th}(\mathbb{C}\eta) = \mathrm{Th}(\xi_{\mathrm{id}}) = S^2 \cup_{J(\mathrm{id}_{S^1})} \mathbb{D}^4 = \mathbb{CP}^2.$$

The proposition is not hard to prove and the example above shows that everything agrees.

Remark 31. Let $q : \xi \to S^k$ be a bundle projection with fibre $\mathbb{R}^n$ and $* \in S^k$ be the north pole. If E is the total space then

$$\mathrm{Th}(\xi) = E \cup \{\infty\}$$

can be broken up into three cells: $\{\infty\}$ as the 0-cell, $q^{-1}(*)$ as the n-cell, and $q^{-1}(S^k - \{*\}) = \mathbb{R}^n \times \mathbb{R}^k$ is the $(n+k)$-cell.

It is interesting to look up Rokhlin's paper and see how he did it. Quite presumably, Serre's calculation of homotopy groups of spheres was done at around the same time.

Proposition 32. The space $S^n \cup_\gamma \mathbb{D}^{n+k+1}$ is homotopy equivalent to $S^n \vee S^{n+k+1}$ for $\gamma : S^{n+k} \to S^n$ if and only if γ is homotopy equivalent to the constant map.

Proof. One way is easy. For the other direction, let $\sigma : S^n \to A := S^n \cup_\gamma \mathbb{D}^{n+k+1}$ be the inclusion map. Assume that $f : A \xrightarrow{\simeq} S^n \vee S^{n+k+1}$ is a homotopy equivalence. By a cellular approximation we may assume that $f \circ \sigma$ is the identity map. Let us look at the composition

$$S^{n+k} \xrightarrow{\gamma} S^n \xrightarrow{\sigma} A \xrightarrow{f} S^n \vee S^{n+k+1} \xrightarrow{\text{projection}} S^n.$$

We can verify that the above composition is γ. Moreover, $\sigma \circ \gamma$ is homotopic to the constant map; call such a homotopy h. This induces a homotopy between γ and the constant map. □

Let us now recall the Thom isomorphism theorem.

Theorem 33 (Thom Isomorphism). Let ξ^n be an n-dimensional oriented bundle over an oriented base B. Then for any k

$$\widetilde{H}_{k+n}(\mathrm{Th}(\xi); \mathbb{Z}) \cong H_k(B; \mathbb{Z}).$$

Proof. Let us prove this for the case when B is a connected, orientable, closed manifold. Let us look at $\widetilde{H}_i(\mathrm{Th}(\xi); \mathbb{Z})$. This is the relative homology $H_i(\mathbb{D}(\xi), S(\xi); \mathbb{Z})$. Notice that $\mathbb{D}(\xi)$ is an oriented manifold of dimension $(n+m)$ with boundary $S(\xi)$. By Lefschetz duality,

$$H_i(\mathbb{D}(\xi), S(\xi); \mathbb{Z}) \cong H^{n+m-i}(\mathbb{D}(\xi); \mathbb{Z}) \cong H^{n+m-i}(B; \mathbb{Z}),$$

where we have collapsed the disks fibrewise. Now the base admits Poincaré duality and we have

$$H^{n+m-i}(B; \mathbb{Z}) \cong H_{i-n}(B; \mathbb{Z}).$$

If we shift the indices we recover the version we have stated. □

We also have the fundamental class $[B] \in H_m(B; \mathbb{Z})$ and applying the isomorphism to this class we get a class $\mathcal{U} \in H_{n+m}(\mathrm{Th}(\xi); \mathbb{Z})$. We can now talk about whether a map $S^{n+m} \to \mathrm{Th}(\xi)$ generates this element $\mathcal{U}$. This is the same thing as looking at the Hurewicz homomorphism and asking if it is surjective.

Definition 34 (Reducible Thom complexes). When the base B^m is a connected, closed manifold and ξ^n is a bundle over it, the Thom complex $\mathrm{Th}(\xi)$ is called *reducible* if and only if the Hurewicz homomorphism is surjective, i.e., there exists a degree 1 map $f : S^{n+m} \to \mathrm{Th}(\xi)$ which generates $H_{n+m}(\mathrm{Th}(\xi); \mathbb{Z}) \cong \mathbb{Z}$ in the top dimension $(n+m)$.

Proposition 35. If the base space of ξ^n is S^m then $\mathrm{Th}(\xi)$ is reducible if and only if $J(\alpha) = 0$, where $\xi = \xi_\alpha$ for $\alpha \in \pi_{m-1}(\mathrm{SO}_n)$.

Note that if $m > n$ then $\mathrm{Th}(\xi)$ is reducible if and only if the bundle is fibre homotopically trivial. Recall that when we proved that the bundle is fibre homotopically trivial if and only if J-homomorphism is zero, we were working with stable bundles. This suits us fine as we will be working with such bundles in general.

Proof. By Proposition 29

$$\mathrm{Th}(\xi_\alpha) \simeq S^n \cup_{J(\alpha)} \mathbb{D}^{n+m}.$$

If $J(\alpha) = 0$ then the Thom complex is $S^n \vee S^{m+n}$ and the the inclusion $S^{m+n} \hookrightarrow S^n \vee S^{m+n}$ induces the isomorphism we seek. Conversely, if $\mathrm{Th}(\xi_\alpha)$ is reducible, we have a map

$$S^{m+n} \vee S^n \xrightarrow{f \vee \sigma} \mathrm{Th}(\xi).$$

This is an isomorphism on homology and by J.H.C. Whitehead's theorem this is a homotopy equivalence. By Proposition 32, $J(\alpha) = 0$. □

Proposition 36. Let ψ^n be the normal bundle of a smooth embedding of a closed oriented manifold $M^m \subset \mathbb{R}^{n+m} \subset S^{m+n}$. Then $\mathrm{Th}(\psi)$ is reducible.

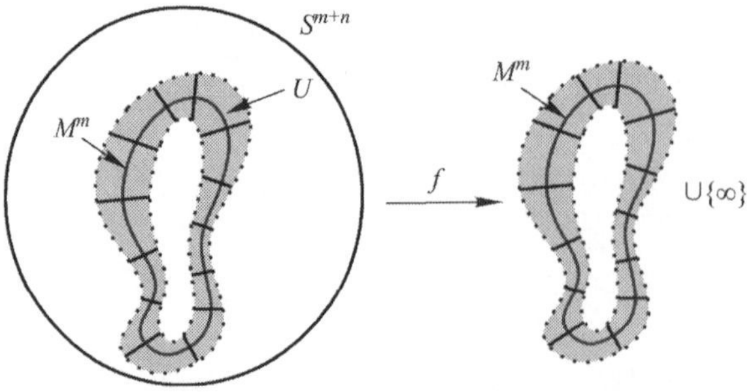

Figure 4. A picture of the map f.

It is known that any manifold embeds into Euclidean space and we get a reducible bundle by taking the normal bundle of such an embedding. Given such an embedding we take the tubular neighbourhood U which can be identified with the disk normal bundle $\mathbb{D}(\psi)$. Define a map $f : S^{n+m} \to \mathrm{Th}(\psi)$ as follows:

$$f(x) = \begin{cases} x \text{ if } x \in \mathbb{D}(\psi) - S(\psi) \\ \infty \text{ otherwise.} \end{cases}$$

This shows that $\mathrm{Th}(\psi)$ is reducible.

We now start the proof of Rokhlin's theorem. The references for it are two papers of Rokhlin: [46] and [48]. Let $M^4 \hookrightarrow \mathbb{R}^{10}$ be an embedding. Consider the normal bundle ψ of M. There is a natural degree one map $\varphi : M^4 \to S^4$. We need the following proposition:

Proposition 37. If M^4 is almost parallelizable then there exists a bundle ξ^6 over S^4 and a bundle map $\widehat{\varphi} : \psi^6 \to \xi^6$ covering $\varphi : M^4 \to S^4$.

7 The Thom isomorphism & Rokhlin's theorem revisited

Last time we had sketched some material but let us revisit it.

Proposition 38 (cf. Proposition 37). Let ψ^6 be the normal bundle of $M^4 \subset \mathbb{R}^{10} \subset S^{10}$, where M^4 is almost parallelizable. Then there exists a bundle ξ^6 over S^4 such that $\varphi^*(\xi^6) \cong \psi^6$, where $\varphi : M^4 \to S^4$ is a degree one map.

Proof. The strategy is to show that there is a bundle $\eta^4 \to S^4$ such that $\varphi^*(\eta^4) \cong TM^4$.

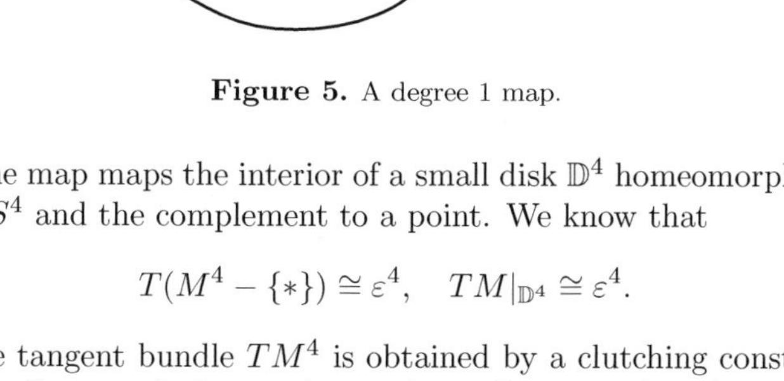

Figure 5. A degree 1 map.

The degree one map maps the interior of a small disk $\mathbb{D}^4$ homeomorphically to the open disk in S^4 and the complement to a point. We know that

$$T(M^4 - \{*\}) \cong \varepsilon^4, \quad TM|_{\mathbb{D}^4} \cong \varepsilon^4.$$

Therefore, the tangent bundle TM^4 is obtained by a clutching construction, i.e., two trivial bundles attached over the 3-sphere. In particular, there exists a map

$\alpha : S^3 \to \mathrm{SO}_4$ with which we identify the two trivial bundles above. This map can also be used to form a bundle over S^4. Now we claim that we can put a bundle map $\widehat{\varphi} : TM^4 \to \eta^4$ covering $\varphi : M^4 \to S^4$. This is the first part of the proof. For the second part, we use the fact that given η^4 one can find a bundle ξ^6 (a unique one at that) such that $\xi^6 \oplus \eta^4 \cong \varepsilon^{10}$ over S^4. We know that

$$\varepsilon^{10} \cong \varphi^*(\xi^6 \oplus \eta^4) \cong \psi^6 \oplus TM^4$$

and $\varphi^*(\xi^6) \cong \psi^6$. □

We have a covering map

$$\begin{array}{ccc} \psi^6 & \xrightarrow{\widehat{\varphi}} & \xi^6 \\ \downarrow & & \downarrow \\ M^4 & \xrightarrow{\varphi} & S^4. \end{array}$$

We claim that $\widehat{\varphi}$ is a degree 1 map on the Thom complex. To verify this, let us go back to a certain formulation of the Thom isomorphism theorem.

Theorem 39 (Thom Isomorphism). Let $q : \eta^n \to B$ be an oriented n-dimensional real vector bundle over a connected base B. Then there exists a cohomology class $\mathcal{U} \in H^n(\mathbb{D}(\eta^n), S(\eta^n); \mathbb{Z})$, unique up to a sign, called the *Thom class* such that for each point $x \in B$ the following holds: let $\mathbb{D}^n_x := q^{-1}(x)$ and $S^{n-1}_x = \partial \mathbb{D}^n_x$. Let

$$\sigma_x : (\mathbb{D}^n_x, S^{n-1}_x) \longrightarrow (\mathbb{D}(\eta^n), S(\eta^n))$$

denote the inclusion. Then

$$\sigma^*_x : H^n(\mathbb{D}(\eta^n), S(\eta^n); \mathbb{Z}) \longrightarrow H^n(\mathbb{D}^n_x, S^{n-1}_x; \mathbb{Z}) \cong \mathbb{Z}$$

is surjective and $\sigma_x(\mathcal{U})$ generates the range. Moreover,

$$\mathrm{T} : H^i(B; \mathbb{Z}) \longrightarrow H^{n+i}(\mathbb{D}(\eta^n), S(\eta^n); \mathbb{Z})$$

is an isomorphism for all $i \geqslant 0$, where $\mathrm{T}(\theta) := q^*(\theta) \cup \mathcal{U}$.

All of this is very geometric and relies on some hard stuff coming from intersection theory and cohomology theory. It is good to have a solid algebraic foundation for it. We now go back to Proposition 38 and consider ψ^6 over M^4. By the Thom isomorphism

$$\mathrm{T} : H^4(S^4; \mathbb{Z}) \xrightarrow{\cong} H^{10}(\mathrm{Th}(\xi^6); \mathbb{Z}).$$

Since φ is covered by a bundle map, we get

$$\begin{array}{ccc} H^{10}(\mathrm{Th}(\xi^6);\mathbb{Z}) & \xrightarrow{\widehat{\varphi}^*} & H^{10}(\mathrm{Th}(\psi^6);\mathbb{Z}) \\ {\scriptstyle \mathrm{T}}\uparrow{\scriptstyle \cong} & & {\scriptstyle \cong}\uparrow{\scriptstyle \mathrm{T}} \\ H^4(S^4;\mathbb{Z}) & \xrightarrow{\varphi^*} & H^4(M^4;\mathbb{Z}). \end{array}$$

If the above is a commutative diagram, since the bottom row is an isomorphism, we would conclude that φ is of degree 1. In one direction,

$$\theta \mapsto q_\xi^*(\theta) \mapsto q_\xi^*(\theta) \cup \mathcal{U}_\xi \mapsto \widehat{\varphi}^*(q_\xi^*(\theta)) \cup \widehat{\varphi}^*(\mathcal{U}_\xi).$$

In the other direction,

$$\theta \mapsto \varphi^*(\theta) \mapsto q_\psi^*(\varphi^*(\theta)) \mapsto q_\psi^*(\varphi^*(\theta)) \cup \mathcal{U}_\psi.$$

Note that $\widehat{\varphi}^*(q_\xi^*(\theta)) = q_\psi^*(\varphi^*(\theta))$ by naturality while $\widehat{\varphi}^*(\mathcal{U}_\xi) = \mathcal{U}_\psi$ by the uniqueness of the Thom class. To be more precise, consider

$$\begin{array}{ccc} \mathbb{D}(\psi) & \xrightarrow{\widehat{\varphi}} & \mathbb{D}(\xi) \\ {\scriptstyle q_\psi}\downarrow & & \downarrow{\scriptstyle q_\xi} \\ M^4 & \xrightarrow{\varphi} & S^4. \end{array}$$

Let $x \in M^4$. We have the following diagram

$$\begin{array}{ccc} H^6(\mathbb{D}(\psi), S(\psi)) & \xleftarrow{\widehat{\varphi}^*} & H^6(\mathbb{D}(\xi), S(\xi)) \\ {\scriptstyle \sigma_x^*}\downarrow & & \downarrow{\scriptstyle \sigma_{\varphi(x)}^*} \\ H^6(\mathbb{D}_x^6(\psi), S_x^5(\psi)) & \xleftarrow{\widehat{\varphi}^*} & H^6(\mathbb{D}_{\varphi(x)}^6(\xi), S_{\varphi(x)}^5(\xi)), \end{array}$$

which proves the naturality of the Thom class.

We have the map

$$S^{10} \longrightarrow \mathrm{Th}(\psi^6) \xrightarrow{\widehat{\varphi}} \mathrm{Th}(\xi^6),$$

where both the maps are of degree 1. Therefore, $\mathrm{Th}(\xi^6)$ is reducible and $\xi^6 = \xi_\alpha$ for $\alpha \in \pi_3(\mathrm{SO}_6)$. But Serre showed that the J-homomorphism $J : \pi_3(\mathrm{SO}_6) \to \pi_9(S^6) \cong \mathbb{Z}_{24}$ is surjective. Therefore, $\xi = 24sE_\omega$, where $E_\omega = \mathbb{H}\eta \oplus \varepsilon^2$ and ω generates $\pi_4(B\mathrm{SO}_6)$. In particular,

$$|p_1(\xi^6)| := \left| \langle p_1(\xi^6), [S^4] \rangle \right| = |\mathscr{P}(24sE_\omega)| = 24s \cdot |\mathscr{P}(E_\omega)| = 48s$$

since $\langle \mathscr{P}(E_\omega), [S^4] \rangle = -2$. But $p_1(\psi^6) = \varphi^*(p_1(\xi^6))$ whence 48 divides $p_1(\psi^6)$. This completes the proof of Theorem 27 since $p_1(M^4) = -p_1(\psi^6)$. In view of Theorem 41, which we will prove in the subsequent lecture, this also proves Theorem 26.

Remark 40. Rokhlin wrote four papers [46], [47] in a few years. The first one was where he claimed $\pi_9(S^6) \cong \mathbb{Z}_{12}$. This was wrong and then he had a second paper where he fixed it to $\mathbb{Z}_{24}$. But this was never published. In 1951, Massey and G.W. Whitehead wrote two short notes [32] in the Bulletin of AMS claiming that the group is indeed $\mathbb{Z}_{24}$. Later that year Serre [49] wrote about the various results he had about homotopy groups of spheres.

Let us connect this to the case of spin manifolds. An oriented, almost parallelizable 4-manifold is the same thing as an oriented, spin manifold, i.e., $w_1(M) = 0$ and $w_2(M) = 0$. Let us think a little bit about this. If M is almost parallelizable then $T(M - \{*\}) \cong \varepsilon^4$. Moreover, if we consider the inclusion

$$\sigma : M - \{*\} \longrightarrow M$$

then this induces an isomorphism on dimensions 1 and 2. Since the classes are trivial for $M - \{*\}$, it is for M. Now to go the other way, consider the Stiefel manifold $V_3(\mathbb{R}^4)$ which is homeomorphic to SO_4. The second Stiefel-Whitney class $w_2(M^4)$ is the first obstruction to getting a cross section to $V_3(TM) \to M$ with fibre $S^3 \times \mathbb{RP}^3$. Therefore, we have an obstruction in $H^2(M^4; \pi_1(\mathrm{SO}_4))$. Since we are working with smooth manifolds, we may assume that the manifold is triangulable. This is not true in general. We have a cross-section over the 2-skeleton as $w_2(M^4) = 0$. This means we can extend it to the 3-skleton as $H^3(M; \pi_2(\mathrm{SO}_4)) = 0$. We may not be able to extend this to the whole manifold as $\pi_3(\mathrm{SO}_4) \neq 0$. We claim that $M - \{*\}$ deformation retracts to the 3-skeleton. Pick a point $*$ in M in the centre of a top cell and start pushing the 3-dimensional walls until we get to the 3-skeleton. We can do this is because $M - \{*\}$ has no 3-dimensional homology.

We are going to connect this to the four dimensional case of what is now known as *Hirzebruch index formula* (cf. Theorem 46).

Theorem 41 (Rokhlin's Index Theorem)**.** If M^4 is an oriented, closed, smooth 4-manifold then

$$I(M^4) = \langle \tfrac{1}{3} p_1(M), [M] \rangle.$$

We now indicate how to prove this. Rokhlin pretty much had the first formula of this kind and all the higher dimension ones are essentially done the same way. The key is to calculate the *oriented bordism group* Ω_n. We say that two closed, oriented manifolds M_1 and M_2 are *cobordant* if there exists a compact, smooth, oriented manifold W^{n+1} such that ∂W^{n+1} is the disjoint union of M_1 and M_2. There is an issue with orientation, i.e., we want the orientation of W near the boundary to have one normal direction pointing inside and the rest to agree with M_1 and disagree with M_2. As a set, Ω_n consists of equivalence classes of closed, oriented, smooth n-manifolds. The *sum* of M_1 and M_2 is given by the disjoint union. The standard n-sphere, for example, will be equivalent to $\varnothing$ as spheres bound disks. It is also the identity of the group.

8 Cobordism groups

Let us revisit what we did last time, clarify some things and then talk about the index theorem. We need the notion of cobordism group.

Definition 42 (Cobordism group). Let M^n, N^n be a pair of closed, oriented, smooth manifolds. Then M^n is *cobordant* to N^n if there exists a compact, oriented smooth manifold W^{n+1} such that ∂W^{n+1} is diffeomorphic, preserving orientation, to $M^n - N^n$.

Let Ω_n be the set of *cobordism classes* of closed, smooth n-manifolds. The additive structure is given by defining $[M^n] + [N^n] := [M^n \sqcup N^n]$.

We are using the convention that the induced orientation on the boundary comes from the inward pointing normal. Note that Ω_n is an abelian group. We claim that $[S^n] = 0$. Take a cylinder $M^n \times [0,1]$ and remove a ball from the interior, i.e.,

$$W^{n+1} := M^n \times [0,1] - \text{int}(\mathbb{D}^{n+1}).$$

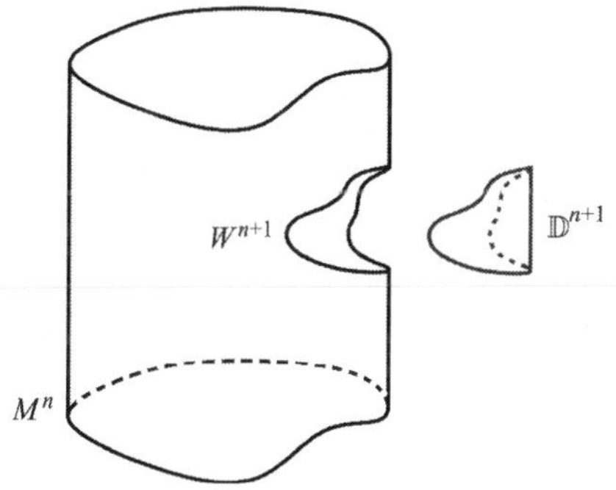

Figure 6. A cobordism between M and $M \sqcup S^n$.

It follows from the construction that

$$\partial W^{n+1} = M^n \sqcup S^n \sqcup -M^n = (M^n \sqcup S^n) \sqcup -M^n.$$

Therefore, as $M^n \sqcup S^n$ is cobordant to M^n, we have

$$[M^n] = [M^n \sqcup S^n] = [M^n] + [S^n].$$

The inverse of $[M^n]$ is given by $[-M^n]$.

Theorem 43 (Thom '53). Let Ω_4 be the four dimensional cobordism group. Then $\Omega_4 \otimes \mathbb{Q} \cong \mathbb{Q}$.

Apparently this fact was known to Rokhlin earlier. Let us see how Rokhlin's index theorem (Theorem 41) follows from this.

Proof of Theorem 41. Let us assume Thom's theorem [54] as stated. We want to look at two functions from Ω_4 to $\mathbb{Z}$ given by index and $\langle p_1(M^4), [M^4]\rangle$. Why are these two functions well defined? If M^4 is cobordant to N^4 then we need to show that $\text{index}(M^4) = \text{index}(N^4)$; this follows from Poincaré duality and it will be left as an exercise. We also need to show that

$$\langle p_1(M^4), [M^4]\rangle = \langle p_1(N^4), [N^4]\rangle .$$

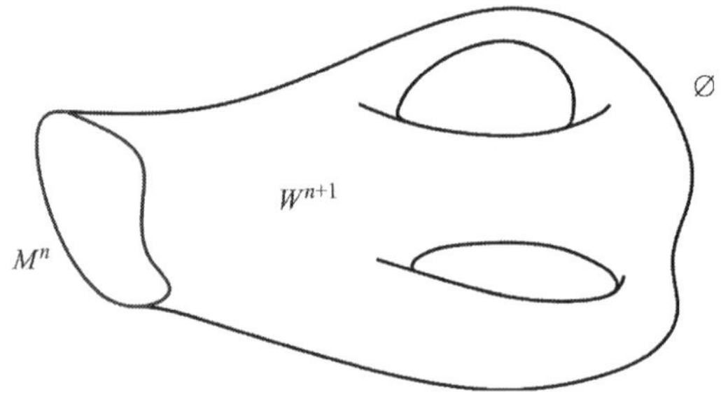

Figure 7. A cobordism between M and $\varnothing$.

Let us see how to prove the last equality. If $M^4 = \partial W^5$ then we need to show that $p_1(M^4) = 0$. Let $\sigma : M^4 \hookrightarrow W^5$ be the inclusion. The normal bundle of M^4 is trivial and $\varepsilon^1 \oplus TM^4 = \sigma^*(TW^5)$. In particular,

$$p_1(M^4) = \sigma^*(p_1(W^5))$$

and when we evaluate this we have

$$\langle p_1(M^4), [M^4]\rangle = \langle \sigma^*(p_1(W^5)), [M^4]\rangle = \langle p_1(W^5), \sigma_*[M^4]\rangle = 0.$$

What we really have is that these two maps from Ω_4 are actually group homomorphisms. When we tensor this with $\mathbb{Q}$, we have homomorphisms $\Omega_4 \otimes \mathbb{Q} \to \mathbb{Q}$. By Thoms theorem, $\langle p_1(\cdot), [\cdot]\rangle$ and $\text{index}(\cdot)$ are both elements in $\mathbb{Q}^* \cong \mathbb{Q}$. Thus, there exists $a, b \in \mathbb{Q}$ such that

$$a \cdot \text{index}(M^4) + b\,\langle p_1(M^4), [M^4]\rangle = 0$$

for any M^4. We take $M^4 = \mathbb{CP}^2$. This has an index of 1. This also shows that $\mathbb{CP}^2$ does not bound. We get

$$a \cdot \text{index}(\mathbb{CP}^2) + b\,\langle p_1(\mathbb{CP}^2), [\mathbb{CP}^2]\rangle = 0.$$

We also have the geometric fact that

$$T\mathbb{CP}^2 \oplus \varepsilon^1_{\mathbb{C}} = \mathbb{C}\eta \oplus \mathbb{C}\eta \oplus \mathbb{C}\eta.$$

Therefore,

$$1 + p_1(\mathbb{CP}^2) = p(T\mathbb{CP}^2) = p(\mathbb{C}\eta) \cup p(\mathbb{C}\eta) \cup p(\mathbb{C}\eta).$$

Going through Milnor-Stasheff's book [37] we will find that $p_1(\mathbb{C}\eta) = \alpha$ such that $\langle \alpha, [\mathbb{CP}^2] \rangle = 1$. In conclusion, $p_1(\mathbb{CP}^2) = 3\alpha$, whence $a = 3b$; we may divide out by b and conclude

$$I(M^4) := \text{index}(M^4) = \tfrac{1}{3} \langle p_1(M^4), [M^4] \rangle .$$

This completes the proof. □

How does one calculate $\Omega_4 \otimes \mathbb{Q}$? This computation was a big stumbling block. Take a four dimensional manifold M^4 and smoothly embed it in S^{10} with ν^6 being its normal bundle. Any bundle is classified by a map into a Grassmann manifold just like the Gauss map. Therefore,

$$\begin{array}{ccccc} \nu^6 & \xrightarrow{\widehat{\varphi}} & i^*\boldsymbol{\xi}^6 & \longrightarrow & \boldsymbol{\xi}^6 \\ \downarrow & & \downarrow & & \downarrow \\ M^4 & \xrightarrow{\varphi} & \widetilde{G}_6(\mathbb{R}^{10}) & \xrightarrow{i} & \widetilde{G}_6(\mathbb{R}^{\infty}). \end{array}$$

This determines a map $\overline{\varphi} : \text{Th}(\nu^6) \to \text{Th}(\boldsymbol{\xi}^6)$ of Thom spaces. Since $\text{Th}(\nu^6)$ is reducible (by Proposition 36), there is a degree-one map $\psi : S^{10} \to \text{Th}(\nu^6)$. The composition of these two maps $\overline{\varphi} \circ \psi : S^{10} \to \text{Th}(\boldsymbol{\xi}^6)$ determines an element in $\pi_{10}(\text{Th}(\boldsymbol{\xi}^6))$.

Theorem 44 (Pontrjagin-Thom construction; cf. [37], pp. 215-216)**.** This correspondence determines an isomorphism $\Omega_4 \cong \pi_{10}(\text{Th}(\boldsymbol{\xi}^6))$.

The canonical bundles arise from the principal fibration

$$\text{SO}_6 \hookrightarrow V_6(\mathbb{R}^{\infty}) \longrightarrow \widetilde{G}_6(\mathbb{R}^{\infty}),$$

where $\widetilde{G}_6(\mathbb{R}^{\infty}) \to G_6(\mathbb{R}^{\infty})$ is a double cover. This is the classifying space for oriented rank 6 bundles.

The Thom space $\text{Th}(\boldsymbol{\xi}^6)$ is often denoted by $M\text{SO}_6$, the Thom space of the total space over the classifying space $B\text{SO}_6$. Note that $\text{Th}(\boldsymbol{\xi}^6)$ is simply connected and the first 5 homology groups vanish due to Thom isomorphism. This implies, by the Hurewicz theorem, that $\text{Th}(\boldsymbol{\xi}^6)$ is 5-connected. The rational Hurewicz theorem of Serre [50] (cf. Theorem 18.3 [37], p. 207)now implies that

$$\pi_j(\text{Th}(\boldsymbol{\xi}^6)) \otimes \mathbb{Q} \cong H_j(\text{Th}(\boldsymbol{\xi}^6); \mathbb{Q})$$

for $1 \leqslant j \leqslant 10$. In particular, $\Omega_4 \otimes \mathbb{Q}$ is isomorphic to the tenth homology group of the Thom complex with rational coefficients. The latter is isomorphic to $H_4(\widetilde{G}_6(\mathbb{R}^{\infty}); \mathbb{Q})$ via the Thom isomorphism. Now note that the Grassmann manifold is homotopy equivalent to BSO_6 as these are both models for classifying orientable vector bundles of rank 6. Let us apply the homotopy exact sequence for the fibration

$$SO_6 \hookrightarrow ESO_6 \longrightarrow BSO_6 \simeq \widetilde{G}_6(\mathbb{R}^\infty).$$

Therefore, $\pi_j(BSO_6) \cong \pi_{j-1}(SO_6)$ as ESO_6 is contractible. The Lie group SO_6 has the following associated groups:

$$\pi_1(SO_6) = \mathbb{Z}_2, \quad \pi_2(SO_6) = 0, \quad \pi_3(SO_6) \cong \mathbb{Z}.$$

This implies that BSO_6 is simply connected and

$$\pi_2(BSO_6) \otimes \mathbb{Q} = 0, \quad \pi_3(BSO_6) \otimes \mathbb{Q} = 0, \quad \pi_4(BSO_6) \otimes \mathbb{Q} \cong \mathbb{Q}.$$

Serre's mod $\mathcal{C}$ Hurewicz theorem (cf. Theorem 15 [52], p. 508) now provides the isomorphism

$$H_4(BSO_6; \mathbb{Q}) \cong \pi_4(BSO_6) \otimes \mathbb{Q} \cong \mathbb{Q}.$$

In particular, we see that $\Omega_4 \otimes \mathbb{Q} \cong \mathbb{Q}$ and this proves Thom's result. Let us recall an important fact; another result of Thom [54].

Theorem 45 (Thom '53). The eight dimensional cobordism group satisfies

$$\Omega_8 \otimes \mathbb{Q} \cong \mathbb{Q} \oplus \mathbb{Q}.$$

This implies another index formula, i.e.,

$$\operatorname{index}(M^8) = \tfrac{7}{45}\left\langle p_2(M^8), [M^8]\right\rangle - \tfrac{1}{45}\left\langle p_1^2(M^8), [M^8]\right\rangle.$$

The proof for the index theorem is precisely the same as last time. Consider the three linear functionals-index, $\langle p_2(\cdot), [\cdot]\rangle$ and $\langle p_1^2(\cdot), [\cdot]\rangle$. Since we have a two dimensional vector space there are rational numbers $a, b, c \in \mathbb{Q}$ such that

$$a \cdot \operatorname{index}(M^8) + b\left\langle p_2(M^8), [M^8]\right\rangle + c\left\langle p_1^2(M^8), [M^8]\right\rangle \equiv 0.$$

We try $\mathbb{CP}^4$ and $\mathbb{CP}^2 \times \mathbb{CP}^2$ and solve for a, b and c. Essentially there are two constants as we may assume that $a = 1$ by scaling. Milnor used this in his famous result [33] on exotic spheres. In dimension 12 we get

$$\Omega_{12} \otimes \mathbb{Q} \cong \mathbb{Q} \oplus \mathbb{Q} \oplus \mathbb{Q}$$

and there are four such functionals; we determine the constants by evaluating it on $\mathbb{CP}^6, \mathbb{CP}^4 \times \mathbb{CP}^2$ and $(\mathbb{CP}^2)^3$. We will just state the general result [23] for now.

Theorem 46 (Hirzebruch Index Formula '56). For any $4n$-dimensional closed, oriented, smooth manifold M^{4n},

$$\operatorname{index}(M^{4n}) = a_n\left\langle p_n(M^{4n}), [M^{4n}]\right\rangle + \cdots,$$

where $a_n \neq 0$ and the a_i's can be computed in terms of Bernoulli numbers.

9 Milnor's exotic spheres

Let us review some things from last time and also get to Milnor's result on exotic spheres. If we look at Rokhlin's paper then it seems a bit sketchy. It is hard to find a complete reference by any one to it. In this paper he shows that $\Omega_4 \cong \mathbb{Z}$ but it is hard to see how his (very geometric) argument generalizes to higher dimensions. Rokhlin's Theorem will be used again later when we discuss Freedman's work [16]. But let us shift over now to Milnor's work [33].

Theorem 47 (Milnor's 56)**.** There exists a 7-dimensional smooth manifold Σ^7 which is homeomorphic to S^7 but not diffeomorphic to S^7.

In fact, we will find a 4-dimensional real vector bundle ξ^4 with base S^4 such that $\Sigma^7 = S(\xi^4)$.

Recall, via the clutching construction, that oriented $\mathbb{R}^4$-bundles over S^4 are in one-to-one correspondence with $\pi_3(\mathrm{SO}_4) \cong \mathbb{Z}\oplus\mathbb{Z}$. This provides a group structure on the space of such bundles. We will also need to know the first Pontrjagin class and the Euler class of these bundles. We can verify that $T := TS^4$ and $\eta := \mathbb{H}\eta$ freely generate a subgroup G of finite index in this group $\pi_3(\mathrm{SO}_4)$ of bundles. Therefore, any bundle in G is of the form $\xi^4 = aT + b\eta$ for some $a, b \in \mathbb{Z}$. Notice that there are two maps

$$\xi^4 \mapsto \langle e(\xi^4), [S^4]\rangle, \quad \xi^4 \mapsto \langle p_1(\xi^4), [S^4]\rangle .$$

Therefore, to know the values of these two maps on any bundle ξ^4 in G we need to know

$$\langle e(T), [S^4]\rangle, \quad \langle e(\eta), [S^4]\rangle, \quad \langle p_1(T), [S^4]\rangle, \quad \text{and} \quad \langle p_1(\eta), [S^4]\rangle .$$

Recall that $\langle e(T), [S^4]\rangle = 2$ no matter how we orient S^4. Notice that η is an oriented bundle and, in fact, $\langle e(\eta), [S^4]\rangle = \pm 1$. We put the orientation on S^4 such that we get $+1$. This does not effect $\langle e(T), [S^4]\rangle = 2$. The first Pontrjagin class of T is zero as T is stably trivial. We may decide not to be exactly specific for the last one but we have seen that $\langle p_1(\eta), [S^4]\rangle = -2$. Therefore, if $\xi^4 = aT + b\eta$ then

$$\xi^4 \mapsto 2a + b = \langle e(\xi), [S^4]\rangle, \quad \xi^4 \mapsto -2b = \langle p_1(\xi), [S^4]\rangle$$

under the Euler class and the first Pontrjagin class respectively. We are going to insist that

$$\langle e(\xi^4), [S^4]\rangle = 1.$$

This implies that $2a + b = 1$ and $b = 1 - 2a$. For a fixed $a \in \mathbb{Z}$ denote by $\xi^4_a = aT + (1-2a)\eta$ and $M^7_a := S(\xi^4_a)$.

Now let us go back to the Gysin sequence to argue some more. We want to show that M^7_a is homeomorphic to S^7. We know that M^7_a fibres over S^4 with fibre S^3. It is easy to see from the long exact sequence of homotopy groups that

$$\pi_1(M^7_a) = 0 = \pi_2(M^7_a).$$

This implies what the homology is in these low dimensions. By Poincaré duality, we know $H_5(M_a^7;\mathbb{Z})$ and $H_6(M_a^7;\mathbb{Z})$ are both zero. Now we look at the Gysin sequence. Consider a bundle $S^3 \hookrightarrow E = S(\xi^4) \to B$. We have a long exact sequence

$$H^0(B;\mathbb{Z}) \xrightarrow{\cup e(\xi^4)} H^4(B;\mathbb{Z}) \xrightarrow{q^*} H^4(E;\mathbb{Z}) \longrightarrow H^1(B;\mathbb{Z}) \longrightarrow \cdots.$$

In our case, the Gysin sequence tells us

$$\mathbb{Z} \xrightarrow{\cong} \mathbb{Z} \xrightarrow{q^*} H^4(M_a^7;\mathbb{Z}) \to 0.$$

Therefore, $H^4(M_a^7;\mathbb{Z}) = 0$ and by duality we have $H_3(M_a^7;\mathbb{Z}) = 0$. Since M_a^7 is also simply connected, by Hurewicz isomorphism,

$$\pi_7(M_a^7) \xrightarrow{\cong} H_7(M_a^7;\mathbb{Z}).$$

We choose a generator of the seventh homotopy group and get a map $f : S^7 \to M_a^7$. This will induce a homotopy equivalence by Whitehead's theorem. Now we will invoke Smale's result and claim that this will imply that S^7 and M_a^7 are homeomorphic. Since these bundles are explicit, Milnor actually constructs Morse functions with two critical points to conclude the claim. This was important since Smale's result was not known in 1956.

Now let us analyze when M_a^7 is diffeomorphic to S^7. Assume that M_a^7 is diffeomorphic to S^7. We construct

$$N_a^8 := \mathbb{D}(\xi_a) \cup_{M_a^7} \mathbb{D}^8.$$

Then N_a^8 is a closed, smooth, oriented 8-manifold with index given by

$$\operatorname{index}(N_a^8) = \tfrac{7}{45}\left\langle p_2(N_a^8), [N_a^8]\right\rangle - \tfrac{1}{45}\left\langle p_1^2(N_a^8), [N_a^8]\right\rangle.$$

Let us get rid of the denominators:

$$45 \cdot (\operatorname{index}(N_a^8)) = 7\left\langle p_2(N_a^8), [N_a^8]\right\rangle - \left\langle p_1^2(N_a^8), [N_a^8]\right\rangle.$$

The above is an equation in integers. We hardly know anything about the second Pontrjagin class; so we reduce modulo 7:

$$4 \cdot (\operatorname{index}(N_a^8)) \equiv \left\langle p_1^2(N_a^8), [N_a^8]\right\rangle \bmod 7.$$

Therefore, we need to calculate $p_1(N_a^8)$.

Let us look at the inclusion maps

$$S^4 \xrightarrow{\tau} \mathbb{D}(\xi_a^4) \xrightarrow{\sigma} N_a^8,$$

where τ is the zero section of ξ_a^4. The first claim is that σ^* is an isomorphism in dimension 4. We also have, by naturality,

$$\sigma^*(p_1(N_a^8)) = p_1(\mathbb{D}(\xi_a^4)).$$

We have

$$TS^4 \oplus (\text{normal bundle to } \tau) = TS^4 \oplus \xi_a^4 = \tau^*(T\mathbb{D}(\xi_a^4)).$$

Putting the above together we get

$$\tau^*\sigma^*(p_1(N_a^8)) = p_1(\xi_a^4)$$

and we know that $\tau^*\sigma^*$ is an isomorphism in dimension 4. We know that

$$\langle p_1(\xi_a^4), [S^4] \rangle = -2b = -2(1-2a).$$

We are going to switch to absolute values, i.e.,

$$|p_1(N_a^8)| = |2(1-2a)|.$$

We need to find $p_1^2(N_a^8)$. Notice that the homology of N_a^8 is $\mathbb{Z}$ in dimensions 0, 4 and 8. By Poincaré duality, the cup product of a generator in dimension 4 has to generate the top class. Let Θ be a generator in dimension 4. We may write

$$p_1(N_a^8) = \pm 2(1-2a)\Theta.$$

Therefore,

$$\pm 4 \equiv 4 \cdot \operatorname{index}(N_a^8) \equiv \langle p_1^2(N_a^8), [N_a^8] \rangle \equiv 4(1-2a)^2 \bmod 7.$$

We have an ambiguity in the index of N_a^8 but it is rather elementary arithmetic at this point. Take $a = 2$ and check that

$$\pm 4 \not\equiv 4(1 - 2 \times 2)^2 \bmod 7.$$

Therefore, $S(\xi_a^4)$ is homeomorphic but *not* diffeomorphic to S^7.

10 Exotic spheres are stably parallelizable

What we will focus on is to get a little bit into the paper [30].

Definition 48 (Homotopy spheres). A *homotopy n-sphere* is a smooth, n-dimensional manifold Σ^n which is homotopy equivalent to S^n.

Remark 49. By Smale's work [51], when $n \neq 3, 4$ we know that Σ^n is homeomorphic to S^n. When Milnor's work [33] came out this was not known. And now we know that this is true for all n: the case $n = 4$ is due to Freedman [16] in 1982 and the case $n = 3$ is by Perelman [40], [42], [41] in 2003.

Theorem 50 (Kervaire-Milnor '62). Every homotopy sphere Σ^m is stably parallelizable, i.e., $T\Sigma^m \oplus \varepsilon^1 \cong \varepsilon^{m+1}$.

The *standard* sphere is thought of as $S^m \subset \mathbb{R}^{m+1}$. Therefore,

$$TS^m \oplus \varepsilon^1 \cong T\mathbb{R}^{m+1}\big|_{S^m} = \varepsilon^{m+1}$$

and the result holds for the standard sphere.

Proof. We are going to use some heavy-duty machinery. Recall that by Bott periodicity

$$\pi_i(\mathrm{O}) = \begin{cases} \mathbb{Z}_2, & \text{if } i \equiv 0, 1 \bmod 8 \\ \mathbb{Z}, & \text{if } i \equiv 3, 7 \bmod 8 \\ 0, & \text{if } i \equiv 2, 4, 5, 6 \bmod 8. \end{cases}$$

Recall that the space O is the union of O_n's. If ξ^r is any bundle over S^m, with $r > m$, and $m \equiv 3, 5, 6, 7 \bmod 8$ then by the clutching construction, ξ^r is classified (stably) by an element $\alpha : S^{m-1} \to \mathrm{O}$, which is zero. Therefore, ξ^r is stably trivial. Now we simply choose $\xi^r = T\Sigma^m \oplus \varepsilon^1$ to prove the theorem when $m \equiv 3, 5, 6, 7 \bmod 8$.

Now let $m \equiv 1, 2 \bmod 8$. As a bundle, ξ^r is classified by a clutching map $\alpha : S^{m-1} \to \mathrm{O}_r$. By Bott periodicity we know that $\pi_{m-1}(\mathrm{O}_r) = \mathbb{Z}_2$ in the stable range. We will need another deep result due to Frank Adams [3].

Theorem 51 (Adams '61)**.** The J-homomorphism

$$J : \pi_{m-1}(\mathrm{O}_r) \longrightarrow \pi_{m-1+r}(S^r)$$

is monic when $m \equiv 1, 2 \bmod 8$.

This is a hard result and we just take it for what it is. In our case, $\xi^r = T\Sigma^m \oplus \varepsilon^1$ and we need to show that $J(\alpha) = 0$ for the clutching map α. Recall that if $\Sigma^m \subset \mathbb{R}^{m+k}$ then the (stable) normal bundle ν is reducible, i.e., $S^{m+k} \to T(\nu)$ is of degree 1. But ν is reducible implies that $J(\nu) = 0$ and hence $\nu \cong \varepsilon^k$. Since

$$T\Sigma^m \oplus \nu \cong \varepsilon^{m+k},$$

we conclude that $T\Sigma^m$ is stably trivial.

We need to take care of $m \equiv 0, 4 \bmod 8$. Instead of getting bogged down with numbers let us do an example.

Example 52. Let $m = 8$ and $\varphi : \Sigma^8 \xrightarrow{\simeq} S^8$ and we know all about the homology of S^8. In particular, the index is zero. Therefore,

$$0 = \operatorname{index}(\Sigma^8) = \tfrac{7}{45}\left\langle p_2(\Sigma^8), [\Sigma^8]\right\rangle - \tfrac{1}{45}\left\langle p_1^2(\Sigma^8), [\Sigma^8]\right\rangle.$$

Since $p_1(\Sigma^8) = 0$ we conclude that

$$\left\langle p_2(\Sigma^8), [\Sigma^8]\right\rangle = 0.$$

Stable bundles over S^8 are classified by $\pi_7(\mathrm{O}_9) \cong \pi_7(\mathrm{O}) \cong \mathbb{Z}$. There is a homomorphism

$$\pi_7(\mathrm{O}_9) \xrightarrow{\langle p_2(\xi),[S^8]\rangle} \mathbb{Z}.$$

If we can show that this is a monomorphism then we are done.

Lemma 53. For each $n \geqslant 1$ there exists a bundle ξ^{4n} over S^{4n} such that its nth Pontrjagin class $p_n(\xi^{4n}) \neq 0$.

When $m \equiv 0, 4 \bmod 8$ then $m = 4n$ and $\pi_{4n-1}(\mathrm{O}_{4n+1})$ is in the stable range, whence

$$\pi_{4n-1}(\mathrm{O}_{4n+1}) \cong \pi_{4n-1}(\mathrm{O}) \cong \mathbb{Z}.$$

Therefore, it follows from the lemma that the homomorphism

$$\pi_{4n-1}(\mathrm{O}_{4n+1}) \xrightarrow{\langle p_n(\xi^{4n}),[S^{4n}]\rangle} \mathbb{Z} \tag{4}$$

is a monomorphism. Now let us prove the remainder of the theorem using this. We are going to need another big result. We do not need the full Hirzebruch index formula in dimension $4n$ but only the weaker form stated in Theorem 46.

Theorem 54 (Hirzebruch). If M^{4n} is an oriented, smooth manifold then its index is a rational linear combination of its Pontrjagin numbers where the coefficient a_n of $\langle p_n(M^{4n}), [M^{4n}]\rangle$ is non-zero.

So what do we mean by these Pontrjagin numbers in general?

Definition 55 (Pontrjagin numbers). A *Pontrjagin number* of M^{4n} is a monomial in cohomology classes $p_1, p_2, \ldots, p_n$ of degree $4n$ evaluated on $[M^{4n}]$.

The full Hirzebruch's result is more algebraic and is done in [37] and [23]. Let us prove the remainder of Theorem 50 assuming Hirzebruch's result. Therefore,

$$0 = \operatorname{index}(\Sigma^{4n}) = a_n \langle p_n(\Sigma^{4n}), [\Sigma^{4n}]\rangle$$

as there are no cohomology classes in the intermediate dimensions. By Theorem 54, $a_n \neq 0$ and hence

$$\langle p_n(\Sigma^{4n}), [\Sigma^{4n}]\rangle = 0.$$

Therefore, $T\Sigma^{4n} \oplus \varepsilon$ is trivial because of the fact that (4) is a monomorphism. □

Let BO be the classifying space of stable bundles. Elements of $[X, BO]$ are homotopy classes of maps and these are in one-to-one correspondence with isomorphism classes of stable real bundles over X. In [37] they use $BO_n = G_n(\mathbb{R}^\infty)$ as a model. They calculate the rational cohomology of these spaces. It is known that

$$H^\bullet(BO;\mathbb{Q}) := \oplus_{i\geqslant 0} H^i(BO;\mathbb{Q}) \cong \mathbb{Q}[p_1, p_2, \ldots],$$

where p_i's are the universal Pontrjagin classes. We really want to understand the space BO. Everything we are talking about here had been continuously developed since the 40's into the 60's.

Definition 56 (Eilenberg-MacLane space)**.** An *Eilenberg-MacLane space* of type (n, A), for $n \in \mathbb{Z}^+$ and an abelian group A, is a topological space $K(A, n)$ satisfying

$$\pi_i(K(n,A)) = \begin{cases} 0, & \text{if } i \neq n \\ A, & \text{if } i = n. \end{cases}$$

We have a one-to-one correspondence between $[X, K(n,A)]$ and $H^n(X;A)$. The correspondence works as follows: a map $f : X \to K(n,A)$ induces a map

$$f^* : H^n(K(n,A);A) \longrightarrow H^n(X;A).$$

There is a natural class $\Theta \in H^n(K(n,A);A)$ corresponding to the identity map (via universal coefficient theorem) and the class in $H^n(X;A)$ is $f^*(\Theta)$.

For each generator $p_i \in H^{4i}(BO;\mathbb{Q}) = [BO, K(4i,\mathbb{Q})]$ we thus have a continuous map $\mathcal{P}_i : BO \to K(4i,\mathbb{Q})$. Let

$$\mathcal{P} := \times \mathcal{P}_i : BO \longrightarrow \mathcal{K} := \textstyle\prod_{i\geqslant 1} K(4i,\mathbb{Q}).$$

This map is a rational homotopy equivalence. Dennis Sullivan looked at this way of thinking about BO but we will go back to a more primitive way of looking at things à la Serre. We claim that

$$\mathcal{P}^* : H^j(\mathcal{K};\mathbb{Q}) \longrightarrow H^j(BO;\mathbb{Q})$$

is an isomorphism. Let us switch back to $B\mathrm{SO}$ as opposed to BO. This will turn out not to be a serious concern as everything works out. By Serre's work [50]

$$H^\bullet(K(4i,\mathbb{Q});\mathbb{Q}) = \oplus_{j\geqslant 0} H^j(K(4i,\mathbb{Q});\mathbb{Q}) \cong \mathbb{Q}[\theta_i],\ |\theta_i| = 4i$$

and generators are sent to generators. By duality we have an isomorphism

$$\mathcal{P}_* : H_\bullet(B\mathrm{SO};\mathbb{Q}) \longrightarrow H_\bullet(\mathcal{K};\mathbb{Q}).$$

We finally need to conclude that

$$\mathcal{P}_* : \pi_*(B\mathrm{SO}) \otimes \mathbb{Q} \longrightarrow \pi_*(\mathcal{K}) \otimes \mathbb{Q}$$

is an isomorphism. This can be deduced from the following (cf. [52], p. 512):

Theorem 57 (Serre-Whitehead). *If $f : A \to B$ is a continuous map with $f_* : \pi_1(A) \xrightarrow{\cong} \pi_1(B)$ and f induces an isomorphism on homology with rational coefficients then f induces an isomorphism on higher rational homotopy groups.*

Proof of Lemma 53. Let ξ^{4n} generate $\pi_{4n}(BO) = \pi_{4n-1}(O)$. We have the composite

$$S^{4n} \xrightarrow{\xi^{4n}} BO \xrightarrow{\mathcal{P}} \mathcal{K} \xrightarrow{\text{proj}} K(4n, \mathbb{Q}).$$

This map generates $\pi_{4n}(K(4n,\mathbb{Q})) \otimes \mathbb{Q}$. Moreover, $\mathcal{P}_n$ is the class given by composing the two rightmost arrows. We have

$$H^{4n}(K(4n,\mathbb{Q});\mathbb{Q}) \xrightarrow{\mathcal{P}_n^*} H^{4n}(BO;\mathbb{Q}) \xrightarrow{\xi^*} H^{4n}(S^{4n};\mathbb{Q}).$$

In particular, a generator Θ of $H^{4n}(K(4n,\mathbb{Q});\mathbb{Q})$ goes to $p_n(\xi^{4n})$ and this is nonzero by construction. □

Corollary 58. *Let B be a finite complex and let ξ^m be a real vector bundle with base B. If all of the higher Pontrjagin classes are zero then there exists $n > 0$ such that*

$$\xi^m \oplus \xi^m \oplus \cdots \oplus \xi^m \cong \varepsilon^{nm}.$$

We are not going to go much into this but rather get back into the J-homomorphism. Let us talk about framed bordism theory.

Definition 59 (Framed manifold). A framed n-manifold in $\mathbb{R}^{n+k}$ is a closed, smooth n-dimensional submanifold M^n together with a framing $\sigma_1, \ldots, \sigma_k$ of its normal bundle.

A lot of manifolds do not admit a framing. Any embedding of $\mathbb{CP}^2$ in an Euclidean space does not have a trivial normal bundle. But $S^n \subset \mathbb{R}^{n+1}$ is framed. If we include $\mathbb{R}^{n+1}$ inside a higher Euclidean space then S^n is still framed. It turns out that if we consider $S^1 \subset \mathbb{R}^2 \subset \mathbb{R}^3$ we can either have a trivial 2-framing of S^1 or a twisted 2-framing.

11 Framed cobordism & the J-homomorphism

We were talking about framed manifolds last time. A framed n-manifold in $\mathbb{R}^{n+k}$ is a closed, smooth submanifold M^n together with a framing[3] $\overline{\sigma} = (\sigma_1, \ldots, \sigma_k)$ for the

[3]When we talk about a framing, unless otherwise specified, we always think of an orthonormal framing.

normal bundle ν^k of the embedding. We also have a diffeomorphism of $M^n \times \mathbb{D}^k$ with $\mathbb{D}(\nu^k) \subset \mathbb{R}^{n+k}$ given by σ. The tubular neighbourhood is appropriately chosen to be small enough.

Definition 60 (Framed cobordism)**.** Two framed submanifolds $(M_0, \overline{\sigma}_0)$ and $(M_1, \overline{\sigma}_1)$ are *framed cobordant* if there exists a compact, smooth, framed $(n+1)$-submanifold $W^{n+1} \subseteq \mathbb{R}^{n+k} \times [0,1]$ such that

$$\partial W = \partial_1 W \cup \partial_0 W, \quad \partial_1 W = M_1 \subset \mathbb{R}^{n+k} \times \{1\}, \quad \partial_0 W = M_0 \subset \mathbb{R}^{n+k} \times \{0\}.$$

The framing $\overline{\sigma} = (\sigma_1, \sigma_2, \ldots, \sigma_k)$ restricts to the boundary to the prescribed framing of M_0 and M_1.

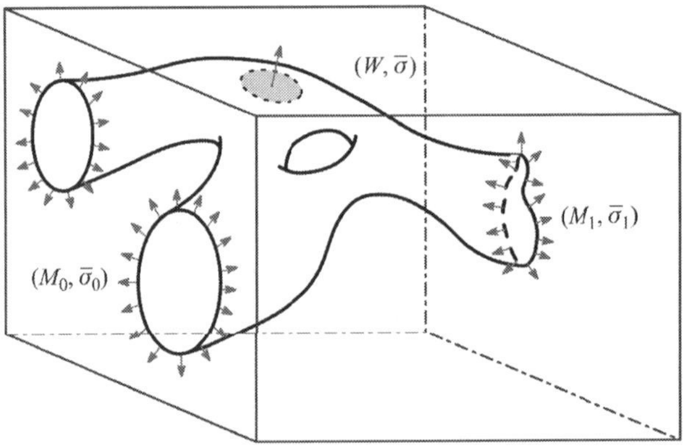

Figure 8. A framed cobordism.

Let $\Omega^{\text{fr}}_{n,k}$ be the set of equivalence classes of framed n-manifolds in $\mathbb{R}^{n+k}$. Given two elements $[(M, \sigma)]$ and $[(N, \tau)]$ we define

$$[(M, \overline{\sigma})] + [(N, \overline{\tau})] := [(M \sqcup N, \overline{\sigma} \sqcup \overline{\tau})].$$

We know that the sphere S^n sits inside $\mathbb{R}^{n+1}$ with a framing of outward pointing normal vectors. This induces a framing e of S^n inside $\mathbb{R}^{n+k}$. One can use $W^{n+1} = (M^n \times [0,1]) \cup \mathbb{D}^{n+1}$ to conclude that $[(S^n, e)]$ is the identity element in $\Omega^{\text{fr}}_{n,k}$. Let $(M^n, \overline{\sigma})$ be a framed submanifold in $\mathbb{R}^{n+k}$. Let $\mathcal{R} : \mathbb{R}^{n+k} \to \mathbb{R}^{n+k}$ be the reflection map that changes the sign of the last coordinate. Then $(\mathcal{R}(M^n), \mathcal{R}{\cdot}\overline{\sigma})$ is the inverse of $(M^n, \overline{\sigma})$. Consider an embedding of M^n in $\mathbb{D}^{n+k}$ and use a cylinder $\mathbb{D}^{n+k} \times [0,1]$ and make a U-turn to map it inside $\mathbb{R}^{n+k} \times [0,1]$ which interpolates between M^n and $\mathcal{R}(M^n)$.

In fact, $\Omega^{\text{fr}}_{n,k}$ is an abelian group. And in the stable range $n \leqslant k+2$, $-[(M^n, \overline{\sigma})]$ is represented simply by changing the framing $\overline{\sigma} = (\sigma_1, \ldots, \sigma_k)$ on the normal bundle ν^k of the embedding $M^n \subset \mathbb{R}^{n+k}$ to $\widetilde{\sigma} := (-\sigma_1, \sigma_2, \ldots, \sigma_k)$, i.e.,

$$-[(M^n, \overline{\sigma})] = [(M^n, \widetilde{\sigma})].$$

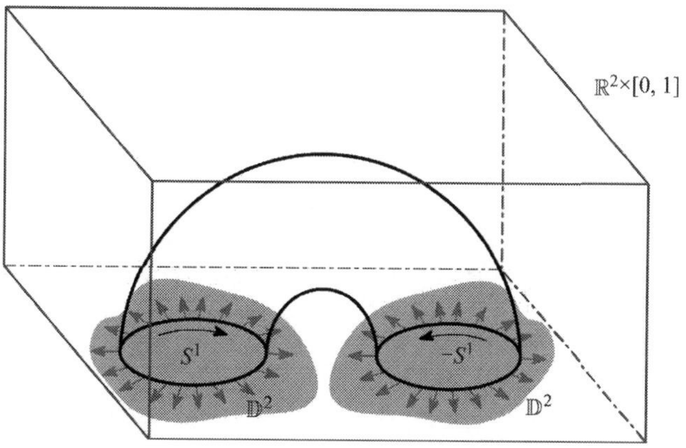

Figure 9. A framed cobordism between a framed S^1 in $\mathbb{R}^2$ and its inverse.

An important result [44] is the following.

Theorem 61 (Pontrjagin '50)**.** There is an isomorphism of groups

$$\Omega^{\mathrm{fr}}_{n,k} \cong \pi_{n+k}(S^k).$$

These kind of results are a major theme in algebraic topology. There are different ways of looking at things - algebraic and geometric - and it is good to have both.

Proof. Let $(M^n, \overline{\sigma}) \subset \mathbb{R}^{n+k}$ be a framed manifold. Let ν^k be an appropriate tubular neighbourhood. We have a map

$$\mathrm{proj} : \nu^k \xrightarrow{\cong} M^n \times \mathbb{D}^k \xrightarrow{\mathrm{proj}} \mathbb{D}^k \longrightarrow \mathbb{D}^k/S^{k-1} \cong S^k.$$

We define a map $f : S^{n+k} := \mathbb{R}^{n+k} \cup \{\infty\} \to S^k$ by sending a point $x \in \nu^k$ to $\mathrm{proj}(x)$ and everything else to the point at infinity.

Conversely, given a map $f : S^{n+k} \to S^k$ we can always perturb it to get a smooth map homotopic to f. By Sard's theorem we choose a regular value and the inverse image under f is a smooth submanifold M^n of S^{n+k}. Moreover, the pullback of the tangent bundle of a regular value in $\mathbb{R}^k$ gives a framing of M^n. This depends on a lot of choices and Milnor proves all of this in his book [35]. □

Now let us discuss some applications. Using that $\Omega^{\mathrm{fr}}_{1,2} \cong \pi_3(S^2)$, we can get an upper bound for $\pi_3(S^2)$. It suffices to look at framings of the standard circle in $\mathbb{R}^3$ (or S^3). This is because there are no non-trivial knots in $\mathbb{R}^4$; this provides a (framed) bordism between a knot and the unknot. It suffices to look at framings of the standard circle and its copies. Since one normal direction on the circle at a point determines the other, we conclude that for each map of the circle to itself we have a framing. Such self-maps are classifed by its degree; consequently the integers surject to $\pi_3(S^2)$. If we can show that these framings are non-equivalent then we would have Hopf's result. One can do something similar for $\Omega^{\mathrm{fr}}_{1,3} \cong \pi_4(S^3)$.

We consider the isomorphism

$$\Omega^{\mathrm{fr}}_{2,k} \cong \pi_{k+2}(S^k)$$

and let $k \geqslant 4$. Notice that if $M^2 := S^2 \subset \mathbb{R}^6$ then with any framing of S^2 we have the same element; these are classified by maps from S^2 to SO_4 and any such map is nullhomotopic. Let us look at the torus. The torus can be thought of as a cylinder and when we glue the two ends of the cylinder we require a map $S^1 \to \mathrm{SO}_4$. This is $\mathbb{Z}_2$ and this is essential and crucial. We actually know that

$$\Omega^{\mathrm{fr}}_{2,4} \cong \pi_6(S^4) = \mathbb{Z}_2.$$

Notice that without framing the torus is bordant to sphere by thickening the torus with a disk attached. That is where Pontrjagin went wrong when he was claiming that $\pi_{k+2}(S^k) = 0$ which is not the case.

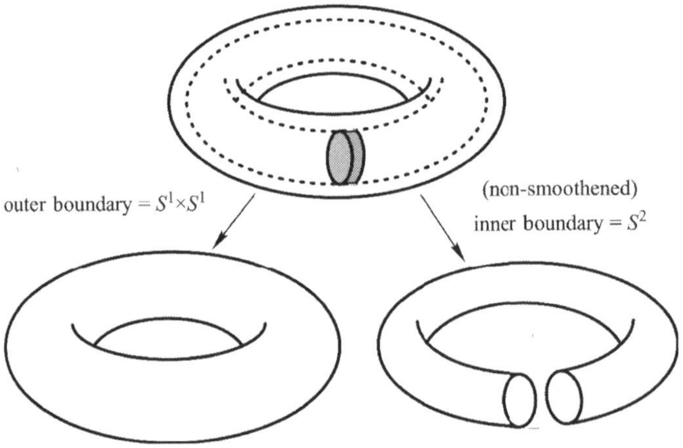

Figure 10. A cobordism between a torus and the 2-sphere.

Let us look at the Freudenthal suspension theorem [18].

Theorem 62 (cf. Theorem 15). The suspension map sending $f : S^k \to S^l$ to $\Sigma f : S^{k+1} \to S^{l+1}$ induces a homomorphism

$$\Sigma : \pi_{n+k}(S^k) \longrightarrow \pi_{n+k+1}(S^{k+1})$$

which is an isomorphism when $n \leqslant k - 2$.

Therefore, we can define the *stable* n^{th} *homotopy group* of spheres as

$$\pi^s_n := \pi_{n+k}(S^k)$$

for large enough k. Basically this theorem is the Whitney embedding theorem [61] in disguise.

Proof. We know that

$$\pi_{n+k}(S^k) \cong \Omega^{\text{fr}}_{n,k}, \quad \pi_{n+k+1}(S^{k+1}) \cong \Omega^{\text{fr}}_{n,k+1}.$$

A framed manifold $(M^n, \overline{\sigma})$ in $\mathbb{R}^{n+k}$ is also a framed submanifold of $\mathbb{R}^{n+k+1}$ by including the new direction. This map is the geometric realization of the suspension map under consideration. To prove that Σ is surjective, let $(N^n, \overline{\omega})$ be a framed submanifold in $\mathbb{R}^{n+k+1}$. We need to find a framed submanifold in $\mathbb{R}^{n+k}$ which is framed cobordant to N^n in one dimension higher. By Whitney's result there is an isotopy

$$\varphi_t : N \longrightarrow \mathbb{R}^{n+k+1}$$

indexed by $t \in [0,1]$ such that $\varphi_0 = \varphi$ and $\varphi_1 = \psi$, where ψ is an embedding of N^n in $\mathbb{R}^{n+k}$. This embedding exists as $k+1 \geqslant n+3$ by choice. We can extend the framing of $N^n \times \{1\}$ to $N^n \times [0,1]$. The framing on $\psi(N^n)$ is given by elements of SO_{k+1}. But since N^n maps to SO_{k+1} and composing this with $q : \mathrm{SO}_{k+1} \to S^k$ is null homotopic (as $k > n$), we may lift the map to a map $N^n \to \mathrm{SO}_k$. This proves surjectivity. The argument for injectivity is left as an exercise. □

There is a nice paper [28] by Kervaire (in English) where he talks about all this and the J-homomorphism. Recall that the J-homomorphism (by G.W. Whitehead) is the map

$$J : \pi_{n-1}(\mathrm{SO}_k) \longrightarrow \pi_{n+k-1}(S^k).$$

It turns out that in a lot of geometric problems we have to understand this map. We want to define the map

$$J : \pi_{n-1}(\mathrm{SO}_k) \longrightarrow \Omega^{\text{fr}}_{n-1,k}.$$

Let $\alpha \in \pi_{n-1}(\mathrm{SO}_k)$ be given by a map $\alpha : S^{n-1} \to \mathrm{SO}_k$. For the standard embedding $S^{n-1} \hookrightarrow \mathbb{R}^{k+n-1}$ we may change the framing $(\sigma_1, \ldots, \sigma_k)$ of the normal bundle at $\mathbf{x} \in S^{n-1}$ by the element $\alpha(\mathbf{x})$ acting on $(\sigma_1, \ldots, \sigma_k)$. This produces another framed sphere. Even in the case $n-1 = 1$ this is non-trivial. Consider $S^1 \subset \mathbb{R}^2 \subset \mathbb{R}^3$ with S^1 is framed by (n, e_3), where n is the outward normal framing of the unit circle in the plane. We have the map

$$\mathscr{R} : S^1 \longrightarrow \mathrm{SO}_2, \quad \theta \mapsto \begin{pmatrix} \cos\theta & \sin\theta \\ -\sin\theta & \cos\theta \end{pmatrix}$$

and we get a new framing applying the matrix at θ to the framing (n, e_3).

In general, we do the same thing to higher spheres. Moreover, we see, as a corollary, the following:

Corollary 63. The J-homomorphism is onto if and only if every framed $(n-1)$-manifold is framed cobordant to the standard $(n-1)$-sphere.

In the stable range we can take any embedding of the sphere S^{n-1} with any framing.

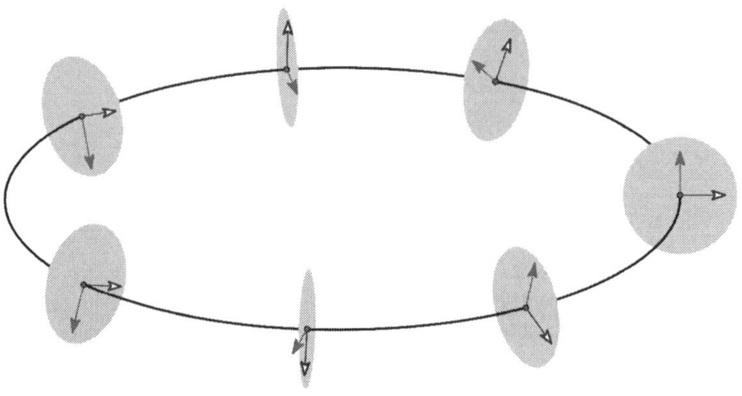

Figure 11. A non-trivial framing of the unknot in $\mathbb{R}^3$.

12 Understanding the group of exotic spheres

We shall use the following notation. Let Θ_n denote the set of all oriented diffeomorphism classes of homotopy n-spheres. An oriented, closed, smooth n-manifold homotopic to S^n is called a *homotopy* n-sphere. We are going to talk about high dimensional manifolds, i.e., $n \geqslant 5$. By Smale's result [51], in this range of dimensions, we know that if Σ^n is a homotopy n-sphere then Σ^n is homeomorphic to S^n.

Theorem 64. The set Θ_n is an abelian group using connected sum as the group operation.

Let us discuss a bit more about connected sums. If M_1 and M_2 are oriented smooth n-manifolds then we want to make sense of $M_1 \# M_2$. Take $M_1 \times [0,1]$ and $M_2 \times [0,1]$ and attach a cylinder $\mathbb{D}^n \times [0,1]$ such that $\mathbb{D}^n \times \{0\}$ attaches to $M_1 \times \{1\}$ and $\mathbb{D}^n \times \{1\}$ attaches to $M_2 \times \{1\}$. The top boundary is just $M_1 \# M_2$, if done correctly, while the bottom boundary is $M_1 \times \{0\} \sqcup M_2 \times \{0\}$. Note that we have a lot of choices - attach $\mathbb{D}_n \times \{0\}$ to $M_1 \times [0,1]$ by an orientation preserving map while the other end gets attached by an orientation reversing map.
These things are done so that we get something consistent at the end. With this notion in place we want to show that Θ_n is an abelian group.

Definition 65 (Addition in Θ_n). Given two homotopy n-spheres Σ_1, Σ_2 we define

$$[\Sigma_1] + [\Sigma_2] := [\Sigma_1 \# \Sigma_2].$$

We claim that

$$[\Sigma^n] + [-\Sigma^n] = [S^n].$$

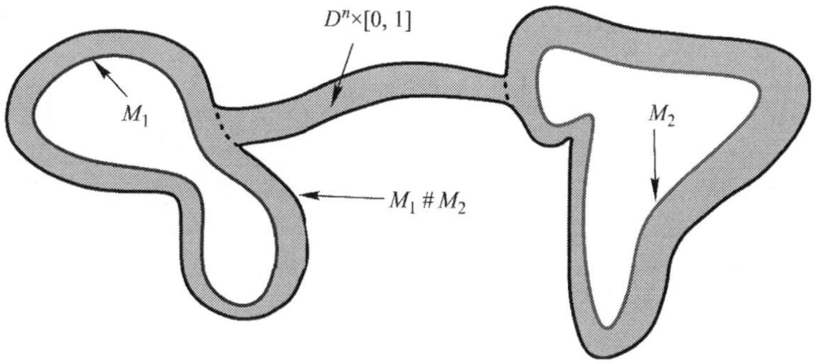

Figure 12. A cobordism between connected sum and disjoint union.

Consider the cylinder $\Sigma^n \times [0, 1]$. Let

$$W^{n+1} = \Sigma^n \times [0, 1] - \text{int}(\mathbb{D}^n) \times [0, 1],$$

the cylinder minus a tunnel.

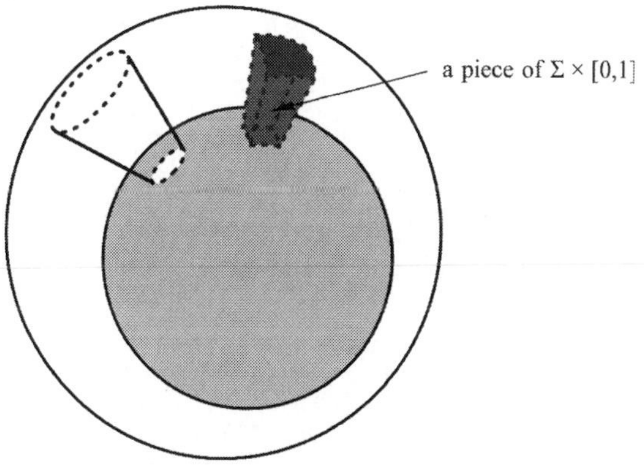

Figure 13. Tunneling through a cylinder.

We orient W^{n+1} with an orientation ω which restricts to the preferred orientation on Σ^n at $t = 0$ and the opposite orientation on Σ^n at $t = 1$. It is easy to see that

$$\partial W^{n+1} = \Sigma^n \#(-\Sigma^n).$$

We also remove a ball from the cylinder, i.e.,

$$\mathcal{W}^{n+1} := W^{n+1} - \text{int}(\mathbb{D}^{n+1}), \quad \partial \mathcal{W}^{n+1} = (\Sigma^n \# - \Sigma^n) \sqcup S^n.$$

We claim that $\mathcal{W}^{n+1}$ is an h-cobordism between the respective manifolds. Once we have this, by Smale's result on h-cobordism we will conclude that these two manifolds are diffeomorphic. We know that

$$W^{n+1} \sim (\Sigma^n - \text{int}(\mathbb{D}^n)) = \mathbb{D}^n \sim \{*\}.$$

We also know that
$$\mathcal{W}^{n+1} \cup_{S^n} \mathbb{D}^{n+1} = W^{n+1} \sim \{*\}.$$
By Van-Kampen's theorem we conclude that $\pi_1(\mathcal{W}^{n+1}) = 0$. To finish the proof we need to show that the inclusion $S^n \hookrightarrow \mathcal{W}^{n+1}$ is a homotopy equivalence. By Whitehead's theorem, it suffices to prove that the map is an isomorphism on homology as both spaces are simply connected. We may use Mayer-Vietoris sequence to conclude that
$$\widetilde{H}_*(S^n; \mathbb{Z}) \longrightarrow \widetilde{H}_*(W^{n+1}; \mathbb{Z})$$
is an isomorphism as $W^{n+1} \sim \{*\} \sim \mathbb{D}^{n+1}$. Now we are going to need another subgroup.

Definition 66. Let bP_{n+1} be the subset of Θ_n consisting of all homotopy n-spheres which bound a π-manifold, i.e., $[\Sigma^n] \in bP_{n+1}$ if and only if there exists a compact, smooth manifold W^{n+1} such that $\partial W^{n+1} = \Sigma^n$ and $TW^{n+1} \oplus \varepsilon^1 \cong \varepsilon^{n+2}$.

Remark 67. A connected, smooth manifold W^{n+1} with non-empty boundary is a π-manifold if and only if it is parallelizable. The reason for this is the following fact: if ξ^n is a real vector bundle over a base B such that $\dim B < n$ then $\xi^n \oplus \varepsilon^1 \cong \varepsilon^{n+1}$ implies that $\xi^n \cong \varepsilon^n$. To prove this fact, suppose we have $\xi^n \oplus \varepsilon^1 \cong \varepsilon^{n+1}$. This means, in particular, that we have a continuous map
$$f : \xi^n \oplus \varepsilon^1 \longrightarrow \mathbb{R}^{n+1}$$
such that f restricted to each fibre is a linear isomorphism. If we apply f to the fibre ξ^n_x over $x \in B$ arising from ξ^n then we get a codimension 1 subspace of $\mathbb{R}^{n+1}$. Moreover, as $x \in B$ varies we get one such codimension 1 subspace. In this way we have a map
$$\varphi : B \longrightarrow G_n(\mathbb{R}^{n+1}), \quad \varphi(x) := f(\xi^n_x).$$
By duality, $G_n(\mathbb{R}^{n+1})$ is isomorphic to $G_1(\mathbb{R}^{n+1}) \simeq \mathbb{RP}^n$ which is a little complicated. Let us work with oriented bundles instead. In fact, we we know that ξ^n has a natural pointwise orientation arising from the vector bundle isomorphism. So we really have a map
$$\varphi : B \longrightarrow \widetilde{G}_n(\mathbb{R}^{n+1}) \cong \widetilde{G}_1(\mathbb{R}^{n+1}) \simeq S^n.$$
Therefore, $\varphi : B \to S^n$ is homotopic to a constant map. On the other hand, by construction, ξ^n is isomorphic to the pullback of the canonical bundle over S^n:
$$\begin{array}{ccc} \xi^n & \longrightarrow & \eta^n \\ {\scriptstyle q}\downarrow & & \downarrow \\ B & \xrightarrow{\varphi} & S^n. \end{array}$$
In other words, $\xi^n = \varphi^*(\eta^n)$ which is the same as the pullback under the constant map.

Remark 68. In the paper [30], π-manifolds are called S-parallelizable or stably parallelizable.

Lemma 69. The subset bP_{n+1} is a subgroup of Θ_n.

Proof. It is clear that $0 = [S^n] \in bP_{n+1}$. If $[\Sigma^n] \in bP_{n+1}$ then its inverse also belongs to bP_{n+1}. So it suffices to show closure under addition. Let $[\Sigma_i^n] \in bP_{n+1}$ for $i = 0, 1$. We need the notion of boundary connected sum. Let W_i^{n+1} be the manifolds bounding Σ_i^n. We join W_0^{n+1} with W_1^{n+1} by a line and thicken it with a trough:

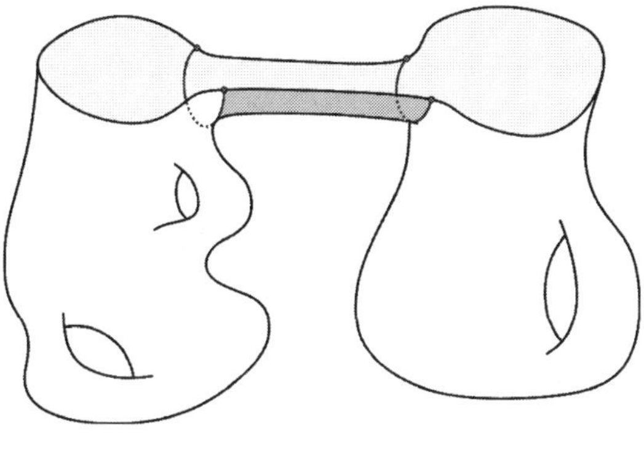

Figure 14. A boundary connected sum.

This results in

$$\partial(W_0^{n+1} \#_b W_1^{n+1}) = \partial W_0^{n+1} \# \partial W_1^{n+1}.$$

It remains to prove that $W_0^{n+1} \#_b W_1^{n+1}$ is parallelizable. This is not hard to show and is left as an exercise. □

The aim here is to show that the quotient group is finite.

Theorem 70 (Theorem 4.1 [30]). The group Θ_n / bP_{n+1} is finite. In fact, this is isomorphic to a subgroup of π_n^s modulo the image of the J-homomorphism, i.e., to a subgroup of $\pi_{n+k}(S^k)/J(\pi_n(\mathrm{SO}_k))$ for $k \gg n$.

What is bP_{n+1}? It is actually finite cyclic with concrete generators coming from plumbing constructions. Moreover, if n is even then $bP_{n+1} = 0$. If n is congruent to 1 mod 4 then either bP_{n+1} is 0 or $\mathbb{Z}_2$. Due to recent work [21] by Hill, Hopkins, and Ravenel it follows that for $n > 125$ everything about bP_{n+1} is known. Of course, we also need to figure out

$$\frac{\pi_n^s/\text{image } J}{\Theta_n/bP_{n+1}}.$$

To prove the theorem we need framed cobordism. What we want to do is to define for each homotopy sphere Σ^n a subset $P(\Sigma^n) \subseteq \pi_n^s$. In other words, we want to define a map

$$P : \Theta_n \longrightarrow \mathcal{P}(\pi_n^s) = \text{power set of } \pi_n^s.$$

We first embed Σ^n in $\mathbb{R}^{n+k}$ by Whitney embedding theorem. This is unique up to isotopy for suitable choice of $k \geqslant n+2$. Let ν^k be the normal bundle. We claim that $\nu^k \cong \varepsilon^k$. We know that the tangent bundle of a homotopy n-sphere is stably parallelizable, i.e.,

$$T\Sigma^n \oplus \nu^k \cong \varepsilon^{n+k}$$

and adding a trivial bundle we get

$$\varepsilon^{n+1} \oplus \nu^k = \varepsilon^1 \oplus T\Sigma^n \oplus \nu^k \cong \varepsilon^{n+k+1}.$$

By the arguments in Remark 67 we know that $\nu^k \cong \varepsilon^k$. We frame Σ^n by choosing a framing $\overline{\sigma}$:

$$[(\Sigma^n, \overline{\sigma})] \in \Omega^{\text{fr}}_{n,k} = \pi_{n+k}(S^k) = \pi_n^s.$$

This defines the map P which sends Σ^n to the collection $\{[(\Sigma^n, \overline{\sigma})]\}$ as the choice of frames vary.

Remark 71. We only use those framings $\overline{\sigma}$ whose orientation is consistent with the orientation on Σ^n, i.e., such that the induced orientation on $T\Sigma^n \oplus \nu^k$ agrees with the standard orientation on $\mathbb{R}^{n+k}$. Hence, $P(-\Sigma^n) = -P(\Sigma^n)$.

If we carefully follow how the J-homomorphism is defined then it follows that $P(S^n)$ is the image of the (stable) J-homomorphism. We need to prove:

(1) $P(\Sigma^n)$ is a coset of $P(S^n)$;
(2) $P(\Sigma_1^n \# \Sigma_2^n) = P(\Sigma_1^n) + P(\Sigma_2^n)$.

13 Kervaire's example & plumbing

Let us begin from where we were last time. We are looking at the group Θ_n and want to get good qualitative information about them. Inside we have a subgroup $bP_{n+1} \subseteq \Theta_n$ and we want to show that Θ_n / bP_{n+1} is finite. In fact, it is isomorphic to a subgroup of $\pi_n^s / \text{image}\, J$. We had defined a map

$$P : \Theta_n \longrightarrow \{\text{subsets of } \pi_n^s\} = \{\text{subsets of } \Omega^{\text{fr}}_{n,k}\}.$$

Recall that

$$P(\Sigma^n) = \{[(\Sigma^n, \overline{\sigma})] \,|\, \overline{\sigma} \text{ varies over all orientation consistent framings of the stable normal bundle of } \Sigma^n \subset \mathbb{R}^{n+k}\}.$$

Lemma 72 (Lemma 4.4 [30]). If Σ_0^n, Σ_1^n are a pair of homotopy n-spheres then

$$P(\Sigma_0^n) + P(\Sigma_1^n) \subseteq P(\Sigma_0^n \# \Sigma_1^n).$$

As a consequence of the lemma, with $\Sigma_0^n = \Sigma^n$ and $\Sigma_1^n = S^n$,

$$\bigsqcup_{x \in P(\Sigma^n)} (x + P(S^n)) = P(\Sigma^n) + P(S^n) \subseteq P(\Sigma^n \# S^n) = P(\Sigma^n).$$

On the other hand, $0 \in P(S^n)$ and therefore,

$$P(\Sigma) \subseteq \bigsqcup_{x \in P(\Sigma^n)} (x + P(S^n)).$$

Therefore, $P(\Sigma^n)$ is a union of cosets. To show that $P(\Sigma^n)$ is a single coset, we need to show that if $x, y \in P(\Sigma^n)$ then $x - y \in P(S^n)$. We will use the lemma again; take $\Sigma_0^n = \Sigma^n$ and $\Sigma_1^n = -\Sigma^n$, then

$$P(\Sigma^n) + P(-\Sigma^n) \subseteq P(\Sigma^n \# - \Sigma^n) = P(S^n).$$

Recall from Remark 71 that $P(-\Sigma^n) = -P(\Sigma^n)$ and therefore, $x - y \in P(S^n)$. In other words, P determines a homomorphism

$$P : \Theta_n \longrightarrow \pi_n^s / \operatorname{image} J.$$

If we analyze the kernel then we can ask if the kernel is bP_{n+1}. Note that $\Sigma^n \in \ker P$ if and only if $P(\Sigma^n) = P(S^n)$ and this is equivalent to $0 \in P(\Sigma^n)$. This is tantamount to having a framed submanifold $(W^{n+1}, \overline{\sigma}) \subset \mathbb{R}^{n+k} \times [0, \infty)$ such that $\partial W^{n+1} = \Sigma^n$ and $[(\Sigma^n, \overline{\sigma}|_{\Sigma^n})] = 0$.

Take Σ^n in the kernel of P. Then $\Sigma^n = \partial W^{n+1}$ and $TW^{n+1} \oplus \varepsilon^1 \cong \varepsilon^2$. This implies that $TW^{n+1} \cong \varepsilon^{n+1}$ and $\Sigma^n \in bP_{n+1}$. For the other inclusion, take $\Sigma^n = \partial W^{n+1}$ such that W^{n+1} is parallelizable. Choose an embedding of W^{n+1} such that the sum of the tangent bundle and the normal bundle is trivial. Since the tangent bundle is trivial, this implies that the normal bundle is stably trivial but we are in a stable range; therefore, the normal bundle is trivial and this induces a framing on Σ^n which represents the zero element. Therefore,

$$\ker P = bP_{n+1}.$$

The proof of the lemma is similar and it is done in the paper [30].

One of the main objectives of [30] is the following result.

Theorem 73. The group Θ_n is finite.

Because of what we have just seen in Theorem 70, this is equivalent to the proving the following:

Theorem 74. The group bP_{n+1} is finite.

More precisely, we will show that bP_{n+1} is a finite cyclic group; in fact,

(i) $bP_{n+1} = 0$ when n is even;
(ii) bP_{n+1} is (cyclic) of order 1 or 2 when $n \equiv 1 \mod 4$.

We will identify an explicit generator Σ of bP_{n+1} in all cases and show that it has finite order. For this let us introduce the plumbing construction; starting with *basic* plumbing and then proceed to *advanced* plumbing in the next lecture. Take two copies of $\mathbb{D}S^n$, the unit disk bundle of TS^n. More formally,

$$\mathbb{D}S^n := \left\{(\mathbf{x}, \mathbf{y}) \in \mathbb{R}^{n+1} \times \mathbb{R}^{n+1} \mid \|\mathbf{x}\| = 1, \|\mathbf{y}\| \leqslant 1, \mathbf{x} \cdot \mathbf{y} = 0\right\}.$$

This is a compact, smooth $2n$-dimensional manifold with boundary which fibres over S^n with fibre $\mathbb{D}^n$. We can glue two copies according to the picture:

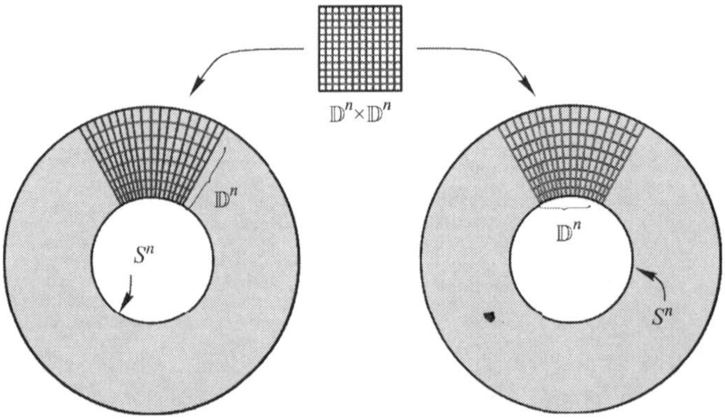

Figure 15. Plumbing two unit disk bundles over S^n.

Let P^{2n} be the resulting manifold (cf. Definition 77). For instance, in the actual picture above we would get a torus minus an open disk.
When $n = 1$ we have $\partial P^2 \cong S^1$ and when we attach a disk $\mathbb{D}^2$ we get $S^1 \times S^1$. We want to do this in all dimensions and when $n = 1, 3, 7$ the tangent bundle is trivializable. In these dimensions $\mathbb{D}S^n \cong S^n \times \mathbb{D}^n$ and

$$\partial P^{2n} \cong S^n, \quad P^{2n} \cup \mathbb{D}^{2n} \cong S^n \times S^n.$$

It is imperative that we need to smoothen things appropriately. We usually get manifolds with corners and there is some work required to smoothen things out. For those who have not seen these things before, it is worthwhile working out the details. A good reference is Connor & Floyd's book (see [13], pp. 9-10) called

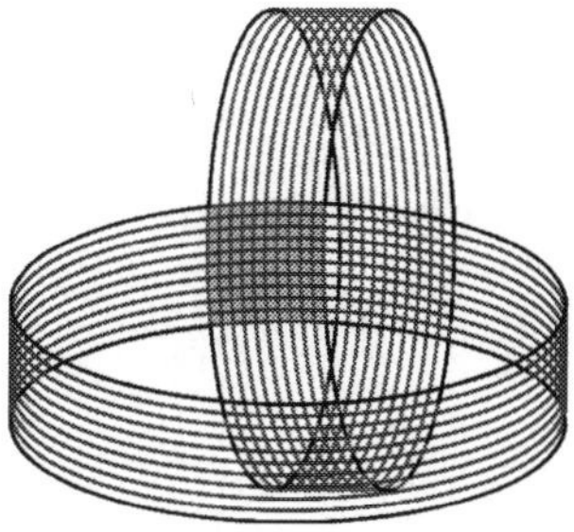

Figure 16. The result of plumbing two unit disk bundles over S^1.

Differentiable Period Maps. Another relevant thing is a paper [22] by Hirsch on how we can put a smooth structure on M (canonically) if we knew that $M \times \mathbb{R}$ has a smooth structure. We also note that the plumbing does not always give a homotopy sphere.

Theorem 75. Let $P^{2n} = \mathbb{D}S^n \wp \mathbb{D}S^n$ be the plumbed manifold. Then ∂P^{2n} is a homotopy $(2n-1)$-sphere if and only if n is odd.

Milnor invented plumbing [34] in 1958 but Kervaire used it to find a 10-manifold that did not have a smooth structure. We have a big result [29] of Kervaire.

Theorem 76 (Kervaire '60)**.** The 10-manifold

$$K^{10} = (\mathbb{D}S^5 \wp \mathbb{D}S^5) \cup_{\Sigma^9} \mathbb{D}^{10}$$

has no smooth structure. In fact, this does not even have the homotopy type of a smooth, closed 10-manifold.

Even more is true: any manifold of the form $K^{10} \times M^n$ is not a smooth manifold. Let us have a specific definition written down and then we will prove Theorem 75.

Definition 77. Let $\mathbb{D}^n \hookrightarrow \mathbb{D}S^n \xrightarrow{q} S^n$ be the unit disk bundle associated to the tangent bundle. Since a bundle over any disk is trivializable, i.e., for any $\mathbb{D}^n \subset S^n$ we may choose a trivialization

$$q^{-1}(\mathbb{D}^n) \cong \mathbb{D}^n \times \mathbb{D}^n.$$

We define

$$P^{2n} := \mathbb{D}S^n \wp \mathbb{D}S^n = (\mathbb{D}S^n \sqcup \mathbb{D}S^n)/\sim,$$

where $(x,y) \sim (y,x) \in \mathbb{D}^n \times \mathbb{D}^n$.

Proof of Theorem 75. Let us look at homology:

$$\cdots \longrightarrow H_i(\partial P^{2n};\mathbb{Z}) \longrightarrow H_i(P^{2n};\mathbb{Z}) \longrightarrow H_i(P^{2n},\partial P^{2n};\mathbb{Z}) \\ \longrightarrow H_{i-1}(\partial P^{2n};\mathbb{Z}) \longrightarrow H_{i-1}(P^{2n};\mathbb{Z}) \longrightarrow \cdots .$$

We know that P^{2n} is homotopic to $S^n \vee S^n$. Therefore, we know the (co)homology of P^{2n}. By Lefschetz duality

$$H^{2n-i}(P^{2n};\mathbb{Z}) \cong H_i(P^{2n},\partial P^{2n};\mathbb{Z}).$$

When $n > 1$ we can show that $\pi_1(\partial P^{2n}) = 0$. Therefore, it suffices to show that the boundary is a homology sphere. We know that

$$H_i(\partial P^{2n};\mathbb{Z}) = \begin{cases} \mathbb{Z}, & \text{if } i = 0, 2n-1 \\ ?, & \text{if } i = n, n-1 \\ 0, & \text{otherwise.} \end{cases}$$

Consider the following:

$$0 \longrightarrow H_n(\partial P^{2n};\mathbb{Z}) \longrightarrow H_n(P^{2n};\mathbb{Z}) \xrightarrow{\varphi} H_n(P^{2n},\partial P^{2n};\mathbb{Z}) \\ \longrightarrow H_{n-1}(\partial P^{2n};\mathbb{Z}) \longrightarrow 0.$$

We claim that the map $H_n(P^{2n};\mathbb{Z}) \to H^n(P^{2n};\mathbb{Z})$, defined by the composite of φ followed by Lefschetz duality, is the adjoint map to the intersection bilinear form. Let $a, b \in H_n(P^{2n};\mathbb{Z})$. If n is even the intersection matrix is

$$\begin{pmatrix} 2 & 1 \\ 1 & 2 \end{pmatrix}$$

while when n is odd we get

$$\begin{pmatrix} 0 & -1 \\ 1 & 0 \end{pmatrix}.$$

The last matrix is invertible while the first one is not! Therefore, when n is odd, ∂P^{2n} is a homology $(2n-1)$-sphere, which by Smale's result, is homeomorphic to S^{2n-1}. □

When n is odd

$$[\partial P^{2n}] \in bP_{2n+1}.$$

as P^{2n} is a π-manifold. To prove this, note that it suffices to show that TP^{2n} restricted to $S^n \vee S^n$ is trivial. Equivalently, it suffices to check that TP^{2n} restricted to any one of S^n is trivial. We know that $TS^n \oplus \varepsilon^1 \cong \varepsilon^{n+1}$ and therefore

$$TS^n \oplus TS^n \oplus \varepsilon^1 \cong TS^n \oplus \varepsilon^{n+1} = (TS^n \oplus \varepsilon^1) \oplus \varepsilon^n \cong \varepsilon^{2n} \oplus \varepsilon^1.$$

By one of the earlier lemma we know that TP^{2n} restricted to each S^n is trivial.

Theorem 78. The group bP_{4n+2} is finite and cyclic of order 1 or 2 generated by $\partial(\mathbb{D}S^{2n+1} \wp \mathbb{D}S^{2n+1})$.

Finally,

$$(\mathbb{D}S^{2n+1} \wp\, \mathbb{D}S^{2n+1}) \cup_\partial \mathbb{D}^{4n+2}$$

is not smoothable if and only if $\partial(\mathbb{D}S^{2n+1} \wp\, \mathbb{D}S^{2n+1})$ is not diffeomorphic to S^{4n+1}.

In the remaining time, let us talk about advanced plumbing. We would like to do a plumbing for a given graph. For each edge we take a pipe which actually goes around in a circle. We would like this to be such that the boundary is a homotopy sphere. Let n be even and we consider the diagram:

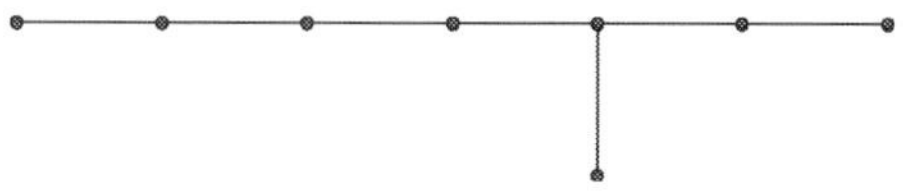

Figure 17. The Dynkin diagram for E_8.

If we plumb using this by the method described then we will get a manifold. This will be relevant when we discuss Freedman's work on four dimensional manifolds. Notice that the diagram is the Dynkin diagram of the exceptional Lie group E_8. The boundary of this plumbed manifold is a homotopy $(2n-1)$-sphere (as long as $n > 2$ is even) which generates bP_{2n+1} and is not diffeomorphic to the standard sphere.

14 Advanced plumbing & Freedman's result

The topic for today is advanced plumbing. Plumbing with the interval as a blueprint will result in the Kervaire manifold $\mathbb{D}S^{2n+1} \wp\, \mathbb{D}S^{2n+1}$. How do we see that the boundary is connected or is a homology sphere? For that we have to do some computations. We may plumb along a graph with cycles. For example in the following cyclic graph on 3 vertices we get an oriented surface[4] (of genus two) minus an open disk.

Figure 18. The graph for the Kervaire manifold.

Theorem 79. Let $P^{2n}_{E_8}$ be the smooth, compact $2n$-manifold with boundary obtained by gluing together eight copies of $\mathbb{D}S^n$ via the Dynkin diagram of E_8. Then $P^{2n}_{E_8}$ is a π-manifold and $\partial P^{2n}_{E_8}$ is a homology $(2n-1)$-sphere. It is a homotopy sphere except when $n = 2$.

Proof. Recall that $\mathbb{D}S^n$ is parallelizable. Any diagram of a tree T leads to a manifold P^{2n}_T which is parallelizable. Let us denote the manifold $P^{2n}_{E_8}$ by P. We

[4]The coloured patches denote the site of the three plumbing operations. There are no other intersections between the three bands.

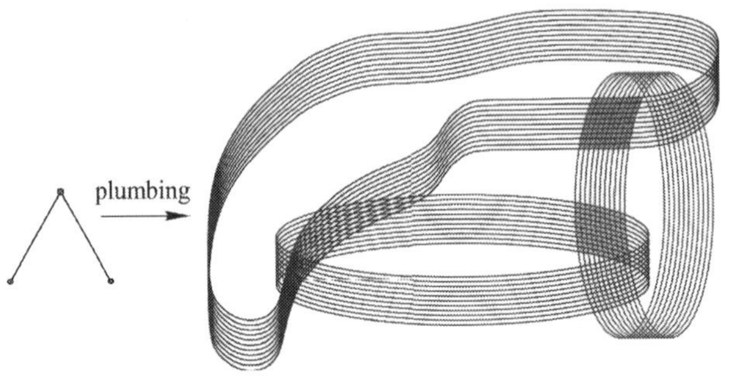

Figure 19. Plumbing along a cycle.

use the homology exact sequence

$$\cdots \longrightarrow H_i(\partial P;\mathbb{Z}) \longrightarrow H_i(P;\mathbb{Z}) \longrightarrow H_i(P,\partial P;\mathbb{Z}) \longrightarrow H_{i-1}(\partial P;\mathbb{Z}) \longrightarrow \cdots.$$

Notice that P is homotopic to a bouquet of eight copies of S^n, whence

$$H_i(P;\mathbb{Z}) = \begin{cases} \mathbb{Z}, & \text{if } i = 0 \\ \mathbb{Z}^8, & \text{if } i = n \\ 0, & \text{otherwise.} \end{cases}$$

By the universal coefficient theorem and Lefschetz duality we conclude that

$$H_i(P,\partial P;\mathbb{Z}) = \begin{cases} \mathbb{Z}, & \text{if } i = 2n \\ \mathbb{Z}^8, & \text{if } i = n \\ 0, & \text{otherwise.} \end{cases}$$

This tells us that

$$H_i(\partial P;\mathbb{Z}) = \begin{cases} \mathbb{Z}, & \text{if } i = 0, 2n-1 \\ ?, & \text{if } i = n, n-1 \\ 0, & \text{otherwise.} \end{cases}$$

We need to investigate the sequence when $i = n$:

$$\begin{array}{ccccccccc} 0 \longrightarrow & H_n(\partial P;\mathbb{Z}) & \longrightarrow & H_n(P;\mathbb{Z}) & \xrightarrow{\varphi} & H_n(P,\partial P;\mathbb{Z}) & \longrightarrow & H_{n-1}(\partial P;\mathbb{Z}) & \longrightarrow 0 \\ & & & & \searrow^{\psi} & \downarrow \cong & & & \\ & & & & & H^n(P;\mathbb{Z}). & & & \end{array}$$

We know that ψ is adjoint to the intersection pairing

$$H_n(P;\mathbb{Z}) \times H_n(P;\mathbb{Z}) \longrightarrow \mathbb{Z}.$$

The generators of $H_n(P;\mathbb{Z})$ are geometric and are given by the zero sections of $\mathbb{D}S^n$. We thus have an intersection matrix A. Recall that we can read off the matrix from the diagram:

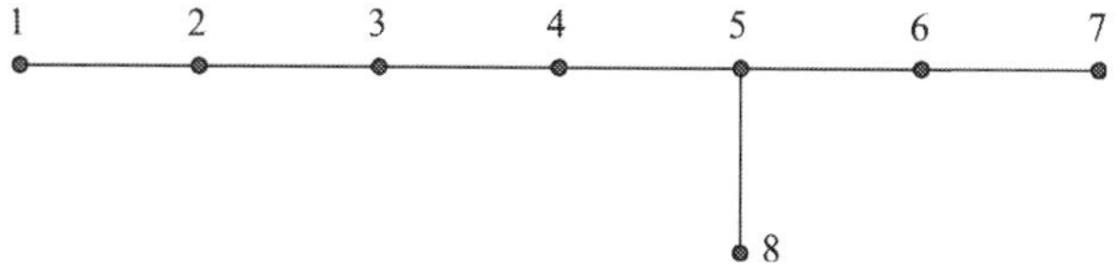

Figure 20. The labelled Dynkin diagram for E_8.

The intersection matrix Q looks different depending on the parity of n:

$$Q_{\text{even}} = \left(\begin{array}{cccc|cccc} 2 & 1 & & & & & & \\ 1 & 2 & 1 & & & & & \\ & 1 & 2 & 1 & & & & \\ & & 1 & 2 & 1 & & & \\ \hline & & & 1 & 2 & 1 & & 1 \\ & & & & 1 & 2 & 1 & \\ & & & & & 1 & 2 & \\ & & & & 1 & & & 2 \end{array}\right),$$

$$Q_{\text{odd}} = \left(\begin{array}{cccc|cccc} 0 & -1 & & & & & & \\ 1 & 0 & -1 & & & & & \\ & 1 & 0 & -1 & & & & \\ & & 1 & 0 & -1 & & & \\ \hline & & & 1 & 0 & -1 & & -1 \\ & & & & 1 & 0 & -1 & \\ & & & & & 1 & 0 & \\ & & & & 1 & & & 0 \end{array}\right)$$

One can do Gaussian elimination to find out the determinant of Q_{odd}. For symmetric matrices we have its signature and it is related to the index theory. For n odd, we can take the matrix mod 2 and get what is called a *symplectic matrix.* Arf defined a $\mathbb{Z}_2$-invariant from it and we will talk about it later. In both cases, the determinant is 1 and the map ψ is an isomorphism.

In general, for $n \geqslant 3$ we can show that ∂P is simply connected and then we deduce that ∂P is a homotopy $(2n-1)$-sphere. This homotopy sphere is a nontrivial generator of the subgroup of homotopy spheres which bound a π-manifold when $n \neq 2$ is even. When n is odd it is always diffeomorphic to S^{2n-1}. When $n = 2$ we have

$$\partial P^4_{E^8} = \Sigma^3, \quad |\pi_1(\Sigma^3)| = 120.$$

Moreover, $\pi_1(\Sigma^3)$[5] is a central extension of the alternating group A_5 by $\mathbb{Z}_2$. In fact, $\Sigma^3 = \mathrm{SO}_3/A_5$. We can take P^4 and glue along its boundary the cone on Σ^3. However, this is not a manifold. What Freedman did [16] was to construct a compact, contractible 4-manifold $\mathcal{D}^4$ such that $\partial\mathcal{D}^4 = \Sigma^3$. Now we can talk about

$$F^4 := P^4 \cup_{\Sigma^3} \mathcal{D}^4.$$

[5]This is called the *binary icosahedral group.*

Theorem 80 (Freedman '82). The closed 4-manifold F^4 does not support a smooth structure.

We may compare this with the following:

Theorem 81 (Kervaire '60). The manifold

$$K^{10} = (\mathbb{D}S^5 \wp \mathbb{D}S^5) \cup_{S^9} \mathbb{D}^{10}$$

does not support a smooth structure.

We can now discuss the argument of Freedman's result. The hard thing is to construct $\mathcal{D}^4$.

Proof of Freedman's theorem. Recall by Rokhlin's theorem that if M^4 is a closed, smooth 4-manifold with $w_1(M^4) = 0 = w_2(M^4)$ then 16 divides index(M^4). It is an easy exercise to show that the index of F^4 is 8. The key point is that the signature of the matrix Q_{even} is 8. If we assume F^4 supports a differential structure then by Rokhlin's theorem, one of the Stiefel-Whitney classes has to be non-zero. In other words, either F^4 is not orientable or F^4 is not spin. Since F^4 is simply connected, $w_1(F^4) = 0$. We have a smooth structure on P^4 by restriction from F^4. But this may not be the smooth structure connected to the intersection matrix. Let us talk about how one can read off $w_2(M)$ [62].

Theorem 82 (Wu's formula '48). Let M^n be a closed, smooth manifold. Then the total Stiefel-Whitney class of the manifold

$$w(M^n) = 1 + w_1(M^n) + w_2(M^n) + \cdots + w_n(M^n) \in \oplus_{i=0}^{n} H^i(M^n; \mathbb{Z}_2)$$

is the Steenrod square of the total *Wu class* $v(M^n)$.

The *Wu class* is defined algebraically as follows: by Poincaré duality, any linear map

$$\alpha : H^i(M^n; \mathbb{Z}_2) \longrightarrow \mathbb{Z}_2$$

can be identified with a unique element $a \in H^{n-i}(M; \mathbb{Z}_2)$ via $\alpha(x) = \langle a \cup x, [M^n] \rangle$. Then $v_i(M^n)$ is defined by

$$\langle v_i \cup x, [M^n] \rangle = \left\langle \mathrm{Sq}^i(x), [M^n] \right\rangle.$$

We shall talk a little bit about Steenrod squares. This is actually a course in itself. The Steenrod square

$$\mathrm{Sq} = \mathrm{Sq}^0 + \mathrm{Sq}^1 + \cdots,$$

where $\mathrm{Sq}^i : H^n(X; \mathbb{Z}_2) \to H^{n+i}(X; \mathbb{Z}_2)$ are natural transformations. A better way to understand this is to think about

$$\mathrm{Sq}^i \in [K(n,\mathbb{Z}_2), K(n+i,\mathbb{Z}_2)].$$

One can give an axiomatic treatment by saying that

$$\begin{aligned} \mathrm{Sq}^0 &= \text{identity}, \\ \mathrm{Sq}^n(x) &= x \bigcup x, \quad |x| = n, \\ \mathrm{Sq}^i(x) &= 0, \quad i > |x|. \end{aligned}$$

This is key to a lot of things actually. With these axioms (one needs Adem relations) there is unique such natural transformation that exists.

By Wu's formula, if $v = 1 + v_2 + v_4$ then

$$1 + w_2 + w_4 = w(F^4) = \mathrm{Sq}(v) = \mathrm{Sq}(1) + \mathrm{Sq}(v_2) + \mathrm{Sq}(v_4).$$

In particular, $w_2(F^4) = v_2(F^4)$. Now if we go back to the definition of the second Wu class,

$$\langle v_2 \cup x, [F^4] \rangle = \langle x^2, [F^4] \rangle .$$

Therefore, $v_2 = w_2 = 0$ if and only if $x^2 = 0$ for any $x \in H^2(F^4;\mathbb{Z}_2)$. Since we are using the cup product, which is a homotopy invariant, we can calculate it with respect to the ambient structure. It is obvious in this setting, the key fact being that the diagonal entries in the matrix Q_{even} are all even integers. This proves the theorem. □

In fact, if the Poincaré conjecture is true (which is true now after Perelman's work) then F^4 cannot be triangulated. Casson [4] showed that this is true without going through the Poincaré conjecture.

15 Towards the Kervaire invariant

The topic that we will eventually get to is Kervaire and Arf invariants. For now we want to look at the eight dimensional analogue of Freedman's result. Recall that we have this Dynkin diagram of E_8 and and we can plumb eight copies of $\mathbb{D}S^4$ using this as a blueprint. Let P^8 be the resulting manifold. Last time we had showed that

(1) P^8 is parallelizable;
(2) The boundary ∂P^8 is homeomorphic to S^7.

Theorem 83 (Kervaire-Milnor). The boundary ∂P^8 is not diffeomorphic to S^7.

The strategy is to use a proof by contradiction; we show that this is a homotopy 7-sphere and then use Smale's result.

Proof. We assume that ∂P^8 is diffeomorphic to S^7. Form a smooth 8-manifold M^8 by gluing $\mathbb{D}^8$ to P^8 along the common boundary S^7. The index formula applied to M^8 yields

$$\operatorname{index}(M^8) = \left\langle \tfrac{1}{45}\left(7p_2 - p_1^2\right), [M^8]\right\rangle$$

on the one hand. On the other hand, the intersection pairing

$$H_4(M^8;\mathbb{Z}) \times H_4(M^8;\mathbb{Z}) \longrightarrow \mathbb{Z}$$

is the one given by that of P^8 via Mayer-Vietoris. We know that the matrix A, associated to E_8, is of index 8. Therefore,

$$45 \times 8 = 45 \cdot \operatorname{index}(M^8) = \left\langle 7p_2(M^8), [M^8]\right\rangle - \left\langle p_1^2(M^8), [M^8]\right\rangle.$$

We know the structure of $\sigma : P^8 \hookrightarrow M^8$, i.e., $\sigma^*(p_1(M^8)) = p_1(P^8)$. Since P^8 is parallelizable,

$$\sigma^*(p_1(M^8)) = p_1(P^8) = 0.$$

But σ^* is an isomorphism on the fourth cohomology group. Therefore, $p_1^2(M^8) = 0$ and we get

$$45 \times 8 = \left\langle 7p_2(M^8), [M^8]\right\rangle.$$

This tells us that

$$5 \times 3^2 \times 2^3 = 7 \times \text{integer},$$

which is a contradiction. □

The result above happened after Milnor's paper [34] on plumbing in 1959.

Remark 84. Using the explicit formula for coefficients of $p_n(M^{4n})$ obtained by the Hirzebruch index formula (cf. [37], p 230, problem 19-c) we get

$$8 = \operatorname{index}(M^{4n}) = a_n \left\langle p_n(M^{4n}), [M^{4n}]\right\rangle.$$

But a number-theoretic contradiction to this equation results from Kervaire-Milnor [36] extension of Rokhlin's theorem (Theorem 41) to higher dimensions. Therefore, ∂P^{4n} is homeomorphic but not diffeomorphic to S^{4n-1}.

In the big conference in Seattle in 1963, Milnor listed 7 problems (cf. [1]) and a lot of those involved these ideas. What we are left with here is to construct a non-smoothable manifold. We have seen an example twenty years ahead but now let us get back to Kervaire's example [29].

Theorem 85 (Kervaire '60). Take two copies of $\mathbb{D}S^5$ and plumb using to get

Figure 21. The tree on two vertices.

$\mathbb{D}S^5 \wp \mathbb{D}S^5$. The following holds:

1. The manifold Σ^9 is homeomorphic to S^9.
2. Glue $\mathbb{D}^{10}$ to $\mathbb{D}S^5 \wp \mathbb{D}S^5$ along $\partial\mathbb{D}^{10} = S^9 \sim \Sigma^9$ to get K^{10}. Then K^{10} is a closed, topological manifold that does not support a smooth structure.

Corollary 86. The manifold Σ^9 is not diffeomorphic to S^9.

For a discussion of the above, we need to digress into analyzing bundles over spheres. What we want is the following:

Proposition 87. Let ξ^n be an n-dimensional real vector bundle over S^n where n is odd and $\xi^n \oplus \varepsilon^1 \cong \varepsilon^{n+1}$. Then either $\xi^n \cong \varepsilon^n$ or $\xi^n \cong TS^n$.

Observe that both the conclusions of the above result could hold simultaneously as it does for $n = 1, 3, 7$. It is worth mentioning, at this point, a deep result [2] that is relevant:

Theorem 88 (Adams '61)**.** The tangent bundle $TS^n \cong \varepsilon^n$ implies that $n = 1, 3, 7$.

Proof of proposition. Let us look at the fibration[6]

$$\mathrm{SO}_n \hookrightarrow \mathrm{SO}_{n+1} \longrightarrow S^n.$$

An element in the fibre over $x \in S^n$ (in the associated principal bundle) is given by an orthonormal framing $(v_1, \ldots, v_n)$ of the tangent space at x. This gives us an element

$$A := (x, v_1, \ldots, v_n) \in \mathrm{SO}_{n+1}.$$

Moreover, this gives us a framing of $\mathbb{R}^{n+1}$. If we look at the homotopy exact sequence of the bundle we get

$$\cdots \to \pi_n(\mathrm{SO}_n) \to \pi_n(\mathrm{SO}_{n+1}) \to \pi_n(S^n) \xrightarrow{\delta} \pi_{n-1}(\mathrm{SO}_n) \to \pi_{n-1}(\mathrm{SO}_{n+1}) \to \cdots.$$

We know that $\pi_n(S^n) \cong \mathbb{Z}$ and $\pi_{n-1}(\mathrm{SO}_n)$ is the equivalence classes of oriented $\mathbb{R}^n$-bundles over S^n.
Under the map

$$\pi_{n-1}(\mathrm{SO}_n) \longrightarrow \pi_{n-1}(\mathrm{SO}_{n+1})$$

the bundle ξ^n maps to zero as it is stably trivial. Therefore, ξ^n is in the image of δ. One can chase through this and check that $\delta(1) = TS^n$ generates the image of δ. We claim that the map

$$\chi : \pi_n(\mathrm{SO}_{n+1}) \longrightarrow \pi_n(S^n)$$

is given by the Euler characteristic. Recall that $\pi_n(\mathrm{SO}_{n+1})$ is the set of equivalence classes of (oriented) $\mathbb{R}^{n+1}$-bundles over S^{n+1}. Let E be such a bundle over S^{n+1}.

[6]This is, actually, a principal bundle associated to the tangent bundle.

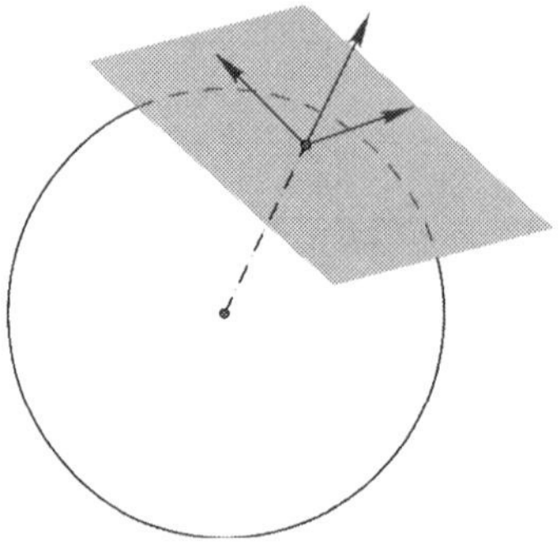

Figure 22. An orthonormal framing of TS^2 at a point.

We are going to associate $\chi(E) \in H^{n+1}(S^{n+1};\mathbb{Z}) \cong \mathbb{Z}$. One can trivialize the bundle over each hemisphere but the clutching construction captures the difference of this trivialization over the equator which is S^n. In other words, 2 lies in the image of χ. From the long exact sequence, image δ equals 0 or $\mathbb{Z}_2$. This finishes the proof. □

Why is this useful? We may want to distinguish TS^5 from ε^5. The Euler class is of no help as it is zero for both bundles. However, we know from Adams' work that the bundles are different. Let

$$P^{10} = \mathbb{D}S^5 \wp \mathbb{D}S^5 \sim S^5 \vee S^5.$$

By the Hurewicz theorem, the map

$$\pi_5(P^{10}) \longrightarrow H_5(P^{10};\mathbb{Z}) \cong \mathbb{Z} \oplus \mathbb{Z}$$

is an isomorphism. We want to have an embedded S^5 with well defined normal bundles for any homology class $\alpha : S^5 \to P^{10}$. We need the hard Whitney theorem [61] and this is what Haefliger [20] did. Let us try to see what is going on. If we have a double point on the plane, then we can get rid of it in $\mathbb{R}^3$. If we have a triple point then we shift to get three double points.

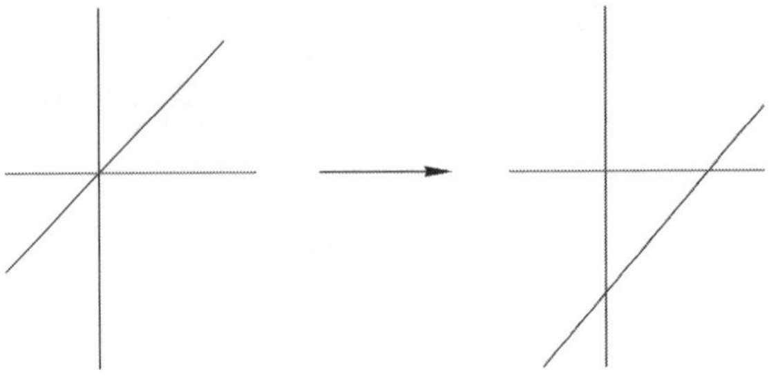

Figure 23. Resolving a triple point in $\mathbb{R}^2$.

The picture gets more and more complicated with triple points and higher as the dimension goes up.

Theorem 89 (Haefliger's Embedding Theorem '61)**.** If $f : N^s \to M^m$ is $(k+1)$-connected, where $s > 2k+2$ and $m \geqslant 2s-k$ then f is homotopic to an embedding. Moreover, any two such embeddings are isotopic if $m > 2s-k$.

Let us check it for homotopy spheres; Σ^n embeds in $\mathbb{R}^{2n}$ by Whitney's result and by Haefliger's result it embeds in $\mathbb{R}^m$ where $m = 3n/2$. So, where do we want to use this? We say M^{2n} satisfies [$*$] if $n > 3$ is odd and

(i) The manifold M^{2n} is compact and smooth;
(ii) The manifold M^{2n} is a π-manifold;
(iii) Either $\partial M^{2n} = \varnothing$ or ∂M^{2n} is a homotopy $(2n-1)$-sphere;
(iv) The manifold M^{2n} is $(n-1)$-connected.

Assume that M^{2n} satisfies property [$*$]. Then $H_n(M^{2n};\mathbb{Z})$ is a finitely generated free abelian group. We use Lefschetz duality and the universal coefficient theorem to get this. We want to define a function

$$\Psi : H_n(M^{2n};\mathbb{Z}) \longrightarrow \mathbb{Z}_2.$$

For each $\alpha : S^n \to M^{2n}$ we homotope it into an embedding, which is unique up to a smooth isotopy, and assign a value of 0 or 1 according to its normal bundle being trivial or not.

16 The Arf invariant

We will come back to Haefliger's theorem but let us first talk about the Arf invariant [5].

Definition 90 (Quadratic form)**.** Let V be a finite vector space over $\mathbb{Z}_2$. A function $q : V \to \mathbb{Z}_2$ is a *quadratic form* if the associated function

$$(\cdot,\cdot) : V \times V \longrightarrow \mathbb{Z}_2, \quad (x,y) \mapsto x \cdot y := q(x) + q(y) + q(x+y)$$

is a bilinear form. It is called non-degenerate if the associated bilinear form is non-degenerate.

Note that $(\cdot,\cdot)$ is alternating, i.e., $(x,x) = 0$. To check this note that

$$x \cdot x = q(x) + q(x) + q(x+x) = q(0).$$

Moreover,

$$q(0) = 0 \cdot 0 = (0+0) \cdot 0 = 0$$

by bilinearity.

Theorem 91. There are exactly two non-degenerate quadratic forms up to isomorphism on any even dimensional vector space. There are none on odd dimensional vector spaces.

We will have to define isomorphism classes of forms. Two forms (V, φ) and (W, ψ) will be called *isomorphic* if there is a vector space isomorphism $f : V \xrightarrow{\cong} W$ such that $\psi \circ f = \varphi$.

Definition 92 (Symplectic basis)**.** Given a quadratic form on V, a symplectic basis for V is a basis $e_1, \sigma_1, \ldots, e_s, \sigma_s$ such that

$$\begin{aligned} e_i \cdot \sigma_j &= \delta_{ij} = \sigma_j \cdot e_i \\ e_i \cdot e_j &= 0 = \sigma_i \cdot \sigma_j. \end{aligned}$$

The matrix with respect to a symplectic basis looks like

$$\left(\begin{array}{c|c|c|c} \begin{matrix} 0 & 1 \\ 1 & 0 \end{matrix} & & & \\ \hline & \begin{matrix} 0 & 1 \\ 1 & 0 \end{matrix} & & \\ \hline & & \ddots \quad \ddots & \\ \hline & & & \begin{matrix} 0 & 1 \\ 1 & 0 \end{matrix} \end{array}\right)$$

Let us do the example for dimension 2, i.e., $V = \mathbb{Z}_2 \oplus \mathbb{Z}_2$ with $e = (1, 0)$ and $\sigma = (0, 1)$. Define

$$q_0(x) := \begin{cases} 0, & \text{if } x \neq e + \sigma \\ 1, & \text{if } x = e + \sigma \end{cases}, \quad q_1(x) := \begin{cases} 0, & \text{if } x = 0 \\ 1, & \text{if } x \neq 0. \end{cases}$$

The matrices associated to the quadratic form for both these look like

$$\begin{pmatrix} 0 & 1 \\ 1 & 0 \end{pmatrix}.$$

Lemma 93. If (V, q) is a non-degenerate quadratic form then V has dimension $2s$ and the associated bilinear form has a symplectic basis $\{e_i, \sigma_i \mid i = 1, \cdots, s\}$.

Proof. Pick a non-zero vector $e_1 \in V$ and extend it to a basis. As q is non-degenerate, there is an isomorphism between V and its dual $V^* := \mathrm{Hom}_{\mathbb{Z}_2}(V, \mathbb{Z}_2)$. Pick $\varphi_1 : V \to \mathbb{Z}_2$ given by mapping e_1 to 1 and the rest of the basis elements to zero. There exists $\sigma_1 \in V$ such that

$$\sigma_1 \cdot x = \varphi_1(x)$$

for any $x \in V$. Moreover, $\sigma_1 \cdot \sigma_1 = 0$ and $\sigma_1 \cdot e_1 = 1$. Now we basically use induction. Look at

$$V_1^{\perp} := \{x \in V \mid x \cdot e_1 = x \cdot \sigma_1 = 0\},$$

where V_1 is the span of e_1 and σ_1. The dimension of $V_1^{\perp}$ is $2(s-1)$ and q restricted to this subspace is a non-degenerate quadratic form. By induction, we get our symplectic basis. □

Definition 94 (Arf invariant). Let (V, q) be a non-degenerate quadratic form and $\mathfrak{B} = \{e_i, \sigma_i \mid i = 1, \ldots, s\}$ be a symplectic basis. Then

$$c_{\mathfrak{B}}(q) := \sum_{i=1}^{s} q(e_i)q(\sigma_i)$$

is the Arf invariant.

We are going to look at this geometrically where we think of the middle dimensional homology classes as embedded spheres. We then use its normal bundle and calculate intersection numbers. Anyway, the definition implicitly assumes that the Arf invariant is independent of the symplectic basis. Actually, the proof of the previous lemma along with induction implies this.

Lemma 95. Up to isomorphism there are at most 2 non-degenerate quadratic forms on $(\mathbb{Z}_2)^{2s}$. In fact, if (V, q) is a non-degenerate quadratic form and dimension of V is $2s$ then (V, q) is isomorphic to either $q_0 + \cdots + q_0$ or $q_1 + q_0 + \cdots + q_0$.

Definition 96. If $\varphi_0 = (V_0, q_0)$ and $\varphi_1 = (V_1, q_1)$ are a pair of non-degenerate quadratic forms then we define their sum to be

$$\varphi_0 + \varphi_1 := (V_0 \oplus V_1, q_0 \oplus q_1).$$

It is a routine check that this definition gives a non-degenerate quadratic form. To prove the lemma, we need another fact. We need to check that

$$q_0 + q_0 \simeq q_1 + q_1. \tag{5}$$

Proof. Pick a symplectic basis $\{e_i, \sigma_i \mid i = 1, \ldots, s\}$ for V. Let V_i be the subspace generated by e_i and σ_i and define $\overline{q}_i$ to be the restriction of q to V_i. Therefore, q

is determined by $\overline{q}_i$'s in a block diagonal form. Either $\overline{q}_i$ is q_0 or q_1 as these are the only 2-dimensional quadratic forms up to isomorphism. Therefore,

$$q \simeq \underbrace{(q_0 + \cdots + q_0)}_{a \text{ copies}} + \underbrace{(q_1 + \cdots + q_1)}_{s-a \text{ copies}}.$$

If $s-a$ is even then one can get rid of all the q_1's and replace it by q_0's. Otherwise, we are left with one copy of q_1 with the rest being q_0. □

Observe that this argument yields the following important extra fact. We started off by picking a symplectic basis $\mathfrak{B}$. If $c_{\mathfrak{B}}(q) = 0$ then

$$q = q_0 + q_0 + \cdots + q_0.$$

In the decomposition, we would get an even number of q_1's and we can get rid of all of them. On the other hand

$$q = q_1 + q_0 + \cdots + q_0$$

if $c_{\mathfrak{B}}(q) = 1$.

Let (V, q) be a non-degenerate quadratic form. We will define $p(q), n(q), r(q) \in \mathbb{Z}$. Think of these as the number of positive vectors, negative vectors and their difference. Let

$$p(q) := \big|\{x \in V \,|\, q(x) = 1\}\big|, \quad n(q) := \big|\{x \in V \,|\, q(x) = 0\}\big|.$$

We set $r(q) = p(q) - n(q)$ and note that $p(q) + n(q) = 4^s$.

Proposition 97. Let q_0 and q_1 be as before.
(1) If $q = sq_0 := q_0 + \cdots + q_0$ then $r(q) = -2^s$.
(2) If $q = q_1 + (s-1)q_0 := q_1 + q_0 + \cdots + q_0$ then $r(q) = 2^s$.

Corollary 98. The quadratic forms sq_0 and $q_1 + (s-1)q_0$ are not isomorphic. Moreover, this also shows that the Arf invariant is independent of the symplectic basis.

Before we prove the fact (5) let us do the calculation.

Proof of Proposition 97. We will prove this by induction. We need a lemma in order to prove this.

Lemma 99. Let (V, q) be a non-degenerate quadratic form. Then

$$p(q + q_0) = 3\,p(q) + n(q),$$
$$n(q + q_0) = 3\,n(q) + p(q).$$

Proof of Lemma. Let $\Phi := q + q_0$ be defined on $V \oplus (\mathbb{Z}_2 \oplus \mathbb{Z}_2)$. We need to solve

$$\Phi(a,b) = q(a) + q_0(b) = 1.$$

Either $q(a) = 0$ and $q_0(b) = 1$ which happens $n(q)$ times while if $q(a) = 1$ and $q_0(b) = 0$ then this happens $3\,p(q)$ times. The second identity is similar. □

Therefore, using this intermediate lemma, we get

$$r(q + q_0) = r(\Phi) = p(\Phi) - n(\Phi) = 2\,p(q) - 2\,n(q) = 2\,r(q).$$

Therefore,

$$r(sq_0) = r((s-1)q_0 + q_0) = 2\,r((s-1)q_0) = 2^{s-1}r(q_0) = -2^s.$$

On the other hand

$$r(q_1 + (s-1)q_0) = 2^{s-1}r(q_1) = 2^s.$$

This finishes the proof. □

The Arf invariant is often called the *democratic invariant* - if there are more 0's then the 0's win while if there are more 1's then the 1's win.

Proof of fact (5). With respect to the symplectic basis $e_1, \sigma_1, e_2, \sigma_2$ of $\mathbb{Z}_2^4$ we have the usual block diagonal matrix. Now consider the basis

$$\begin{aligned} e_1 &= e_1 + e_2, \\ \sigma_1 &= \sigma_1 + e_2, \\ e_2 &= e_2 + \sigma_2 + e_1 + \sigma_1, \\ \sigma_2 &= \sigma_2 + e_1 + \sigma_1. \end{aligned}$$

We claim that this is a symplectic basis and we define the linear map $f : \mathbb{Z}_2^4 \to \mathbb{Z}_2^4$ by the equations above. Let us do one of them - they are all pretty much the same.

$$\begin{aligned} [(q_1 + q_1) \circ f](e_1) &= (q_1 + q_1)(e_1 + e_2) = q_1(e_1) + q_1(e_2) + e_1 \cdot e_2 \\ &= 1 + 1 + 0 = 0 = (q_0 + q_0)(e_1). \end{aligned}$$

One can similarly check the remaining. □

At this point we want to go back to the following (cf. Theorem 89).

Theorem 100 (Haefliger's Embedding Theorem). Let $f : N^s \to M^m$ be a continuous map between smooth manifolds (where N^s is closed) which is $(k+1)$-connected.

(1) If $m \geqslant 2s - k$ and $s > 2k + 2$ then f is homotopic to an embedding.
(2) Any two such embeddings are isotopic if $m > 2s - k$.

Note that the two inequalities required in (1) force us to be in the meta-stable range $2m > 3s$ where there are no triple points for f. Recall that if we consider the relative groups in

$$\cdots \longrightarrow \pi_i(N^s) \longrightarrow \pi_i(M^m) \longrightarrow \pi_i(M^m, N^s) \longrightarrow \pi_{i-1}(N^s) \longrightarrow \cdots$$

then we say f is $(k+1)$-connected if and only if $\pi_i(M,N)=0$ for $i \leqslant k$. This is equivalent to saying that $f_* : \pi_i(N^s) \to \pi_i(M^m)$ is surjective for $i \leqslant k+1$ and isomorphism for $i \leqslant k$.

So, where do we really want to use this? Let M^{2s} be a manifold with $\pi_i(M^{2s}) = 0$ for $i < s$. If we have any continuous map $f : S^s \to M^{2s}$ (we also assume $s \geqslant 5$) then we let $k=1$. So, $s > 2k+2$ and $2s = m \geqslant 2s-k$ are automatically satisfied.

17 The Kervaire invariant for highly connected π-manifolds

Last time we had defined the Arf invariant. Now we want to lead up to Kervaire invariant and give Kervaire's original proof of his Theorem 85.

Definition 101. A $2n$-dimensional smooth manifold M^{2n} satisfies [$*$] provided the following conditions are satisfied:

(i) M^{2n} is compact and either $\partial M^{2n} = \varnothing$ or ∂M^{2n} is homeomorphic to S^{2n-1};
(ii) M^{2n} is a π-manifold, i.e., $TM^{2n} \oplus \varepsilon^1 \cong \varepsilon^{2n+1}$;
(iii) M^{2n} is $(n-1)$-connected, i.e., $\pi_i(M^{2n}) = 0$ if $i < n$;
(iv) n is odd and $n \neq 1, 3, 7$.

We would like to define an invariant $c(M^{2n}) \in \mathbb{Z}_2$ called its *Kervaire invariant*. Recall that if M^{2n} satisfies [$*$] then

$$H_i(M^{2n};\mathbb{Z}) = 0 \quad \text{if} \quad i \in \{1, 2, \ldots, n-1, n+1, n+2, \ldots, 2n-1\}$$

by Poincaré-Lefschetz duality. Moreover, $H_n(M^{2n};\mathbb{Z}) \cong \mathbb{Z}^s$ and $H_{2n}(M^{2n};\mathbb{Z})$ is 0 or $\mathbb{Z}$. By Hurewicz theorem

$$\pi_n(M^{2n}) \longrightarrow H_n(M^{2n};\mathbb{Z})$$

is an isomorphism. Haefliger's embedding theorem says that every element $x \in H_n(M^{2n};\mathbb{Z})$ can be represented by an embedded n-sphere $S^n_x \subset M^{2n}$ which is unique up to isotopy. Hence, the normal bundle ξ_x is well defined. This bundle is also stably trivial and therefore ξ_x is either ε^n or TS^n. Moreover, $\varepsilon^n \ncong TS^n$ as $n \neq 1, 3, 7$. We can define a function

$$q : H_n(M^{2n};\mathbb{Z}) \longrightarrow \mathbb{Z}_2, \quad q(x) = \begin{cases} 1, & \text{if } \xi_e = \varepsilon^n \\ 0, & \text{if } \xi_x = TS^n. \end{cases}$$

Theorem 102. The function q is a non-degenerate quadratic form on $H_n(M^{2n};\mathbb{Z})$, i.e.,

$$q(x+y) = q(x) + q(y) + x \cdot y$$

whose associated alternating bilinear form is the intersection pairing on $H_n(M^{2n};\mathbb{Z})$ mapped into $H_n(M^{2n};\mathbb{Z}_2)$ via the natural coefficient homomorphism $\mathbb{Z} \to \mathbb{Z}_2$.

Corollary 103. The function q induces a non-degenerate quadratic form $\overline{q}$ in the sense of Arf, i.e.,

$$\overline{q} : H_n(M^{2n};\mathbb{Z}_2) \longrightarrow \mathbb{Z}_2$$

by setting $\overline{q}(\overline{x}) := q(x)$, where x is an element mapping to $\overline{x}$ under

$$0 \longrightarrow H_n(M^{2n};\mathbb{Z}) \longrightarrow H_n(M^{2n};\mathbb{Z}) \longrightarrow H_n(M^{2n};\mathbb{Z}_2) \longrightarrow 0.$$

We leave the proof of the corollary to the reader. Once we have the theorem, we will define the *Kervaire invariant* to be

$$c(M^{2n}) := \text{Arf invariant of } \overline{q}.$$

The proof of the theorem is going to take some work. We need to introduce cohomology operations. We are going to make these things less geometric and more algebraic. So let us write down another theorem [29] of Kervaire.

Theorem 104 (Kervaire '60)**.** If M^{10} is closed, i.e., $\partial M^{10} = \varnothing$ and satisfies [*] (cf. Definition 101) then $c(M^{10}) = 0$.

Corollary 105. The Kervaire manifold $K^{10} := (\mathbb{D}S^5 \wp \mathbb{D}S^5) \cup \mathbb{D}^{10}$, obtained by gluing a disk to the plumbed manifold, does not have a smooth structure. In fact, there is no smooth manifold which is homotopy equivalent to K^{10}.

We want to go back and review some facts about Whitehead products.

Definition 106 (J.H.C. Whitehead product)**.** Let $\alpha \in \pi_m(X)$ and $\beta \in \pi_n(X)$ then their *Whitehead product* $[\alpha, \beta] \in \pi_{m+n-1}(X)$ is defined by the composite

$$S^{m+n-1} \xrightarrow{\gamma} S^m \vee S^n \xrightarrow{\alpha\vee\beta} X.$$

There is a natural cell structure on $S^m \times S^n$ with four cells - a base point, e^m, e^n and e^{m+n}. Now e^{m+n} is attached to $S^m \vee S^n$ by some attaching map γ.

In the special case when $m = n = 1$ we get the torus $S^1 \times S^1$. If we have maps $\alpha, \beta : S^1 \to X$ then $[\alpha, \beta] = \alpha\beta\alpha^{-1}\beta^{-1}$. Let us list some properties of these products:

(1) The map $[\cdot,\cdot]$ is bilinear when $m, n > 1$ and

$$[\alpha, \beta] = (-1)^{|\alpha||\beta|}[\beta, \alpha].$$

Hence, $[\alpha, \alpha]$ has order 1, or possibly 2 if $|\alpha|$ is odd.

(2) Let $\iota_n := \mathrm{id}_{S^n} : S^n \to S^n$. Then $[\iota_n, \iota_n] : S^{2n-1} \to S^n$ is the attaching map for the Thom complex of TS^n, i.e.,

$$\mathrm{Th}(TS^n) = S^n \cup_{[\iota_n, \iota_n]} \mathbb{D}^{2n}.$$

[To prove (2) use the fact that the tangent bundle of S^n is the normal bundle ξ_Δ of the diagonal map $\Delta : S^n \to S^n \times S^n$. This is true for any manifold.]

(3) The Whitehead product $[\iota_n, \iota_n] = 0$ if and only if $n = 1, 3, 7$. For this, one uses G.W. Whitehead's EHP[7] sequence. This is a hard result [2] by Adams.

The material we are covering now is basically §8 from the paper [30] by Kervaire and Milnor. We will talk about cohomology operations. This is heavily based on the last part of Steenrod's book [53] on fibre bundles. Let (K, L) be a CW pair, i.e., $L \subseteq K$, satisfying the following hypothesis

$$H^i(K, L; G) = 0 \quad \text{if} \quad n < i < 2n \quad \text{for any } G. \qquad \text{(hypothesis)}$$

Here G is abelian and we are working with untwisted coefficients. The main example to keep in mind is $K = M^{2n}$ and $L = \partial M^{2n}$. Recall that a homology class is called *spherical* if it is in the image of the Hurewicz homomorphism. There is a dual notion in cohomology: we say $x \in H^i(X)$ is called *co-spherical* if there exists a map $\varphi : X \to S^i$ such that $\varphi^*([S^i]) = x$. There are notions of *reducible* and *co-reducible* classes and often these are delicate questions.

Question *Assume that (K, L) satisfies the hypothesis and dim $K = 2n$. Which $x \in H^n(K, L)$ are co-spherical, i.e., does there exists $\varphi : K/L \to S^n$ such that $\varphi^*([S^n]) = x$?*

There are some issues with what happens when we change triangulations if K has dimension higher than $2n$ but let us not get too concerned about that.

Let us study the special case when $K = M^{2n}$ and $L = \partial M^{2n}$. We assume from now that such pairs satisfy [$*$] (cf. Definition 101). By Lefschetz duality, we have an isomorphism

$$\mathcal{D} : H_n(M^{2n}; \mathbb{Z}) \xrightarrow{\cong} H^n(M^{2n}, \partial M^{2n}; \mathbb{Z}).$$

We have $q : H_n(M^{2n}; \mathbb{Z}) \to \mathbb{Z}_2$ and we may want to define it on cohomology. Given $y \in H_n(M^{2n}; \mathbb{Z})$ let ν_y be the normal bundle of $S^n_y \subseteq M^{2n}$. Let U_y be the Thom class in $H^n(\mathrm{Th}(\nu_y); \mathbb{Z})$. Recall that the attaching map of the Thom space is given by $[\iota_n, \iota_n]$ or zero. We want to apply it to the pair $(\mathrm{Th}(\nu_y), \varnothing)$. But let us go back and define the cohomology operations.

[7] E stands for *suspension*, H stands for *Hopf* and P stands for *Whitehead products*.

Consider the Eilenberg-Maclane spaces $K(\mathbb{Z}, n)$. By Hurewicz, we have a map $p : S^n \to K(\mathbb{Z}, n)$ which induces an isomorphism on π_n. We can change S^n a little bit to make this a Serre fibration. We thus have a fibration

$$F \hookrightarrow S^n \to K(\mathbb{Z}, n).$$

The long exact sequence of homotopy groups tell us that

$$\pi_i(F) = \begin{cases} 0, & \text{if } i \leqslant n \\ \pi_i(S^n), & \text{if } i > n. \end{cases}$$

Given an element $x \in H^n(K/L; \mathbb{Z})$ we have a map $x : K/L \to K(\mathbb{Z}, n)$. To make this class co-spherical we need to lift x to S^n. This is obstruction theory and we are going to get a sequence of obstructions living in $H_{i+1}(K/L; \pi_i(F))$. By the hypothesis and the homotopy groups of the fibre, we will have an obstruction $H^{2n}(K/L; \pi_{2n-1}(S^n))$. This is what Steenrod calls the *primary obstruction* in his book. This defines a map

$$\varphi : H^n(K/L; \mathbb{Z}) \longrightarrow H^{2n}(K/L; \pi_{2n-1}(S^n)).$$

We note that this obstruction is well defined and independent of the cell structure by Corollary 35.8 [53].

18 Manifolds that do not support a smooth structure

We are trying to put together a fairly complete proof that there are manifolds that do not support a smooth structure. Let us recall where we were last time. Let n be a fixed positive odd integer different from $1, 3, 7$. Let A be a CW complex which

(1) is simply connected;
(2) has only finitely many cells in each dimension;
(3) has cohomology groups $H^i(A; G) = 0$ when $n < i < 2n$ and G is any abelian group [cf. (hypothesis)].

If A satisfies the above hypothesis then given $x \in H^n(A; \mathbb{Z})$ there is a well defined obstruction

$$\Psi(x) \in H^{2n}(A; \pi_{2n-1}(S^n))$$

to finding a continuous map $f : A^{2n} \to S^n$ such that $f^*(\Theta) = x$. Recall that $\iota_n \in \pi_n(S^n)$ is the canonical generator and $\Theta \in H^n(S^n; \mathbb{Z})$ be the corresponding generator. Moreover, we have an isomorphism between the nth cohomology of A^{2n} and A. The reference for this is [53], where $\Psi(x)$ is what Steenrod calls the *primary obstruction* to lifting the map.

Given $x \in H^n(A;\mathbb{Z})$ there is a homotopy class of maps from $A \to K(\mathbb{Z},n)$, call it φ_x. We would want to lift φ_x:

$$\begin{array}{ccc} & & F \\ & & \downarrow \\ & \overset{\widehat{\varphi}_x}{\nearrow} & S^n \\ & & \downarrow \\ A & \xrightarrow[\varphi_x]{} & K(\mathbb{Z},n). \end{array}$$

Here F is the homotopy fibre of the map $S^n \to K(\mathbb{Z},n)$. In his book ([53], p. 179), Steenrod carefully uses these rules to define the primary obstruction and to check that it is well defined:

(1) $H^{i+1}(A;\pi_i(F)) = 0$ if $i = 0,1,\ldots,2n-2$;
(2) $H^i(A;\pi_i(F)) = 0$ if $i = 0,1,\ldots,2n-2$.

According to Corollary 35.8 ([53], p. 179) the primary obstruction is a topological invariant, i.e., it does not depend on the CW structure. We want to relate these things to manifolds that satisfy condition [$*$] and show that these are the same things.

Suppose M^{2n} satisfies condition [$*$]. Recall that we have a function

$$q : H_n(M;\mathbb{Z}) \longrightarrow \mathbb{Z}_2$$

defined geometrically and $q(y)$ is either 0 or 1 depending on whether the normal bundle of the embedding S^n_y of the sphere (representing y) is trivial or not.

Lemma 107. We have the identity $q = \Psi \circ \mathcal{D}$.

Notice that $\mathcal{D}$ is Poincaré-Lefschetz duality and

$$\Psi \circ \mathcal{D} : H_n(M^{2n};\mathbb{Z}) \longrightarrow H^{2n}(M^{2n};\pi_{2n-1}(S^n)).$$

However, $\pi_{2n-1}(S^n) \neq \mathbb{Z}_2$ in general. On the other hand, a subgroup $\mathbb{Z}_2$ is generated by $[\iota_n,\iota_n]$. For manifolds satisfying [$*$] the image lands in this subgroup.

Proof. For $y \in H_n(M^{2n};\mathbb{Z})$ we take an embedded representative $S^n_y \subset M^{2n}$ with normal bundle ν_y. There is the Thom collapse map $\varphi : M \to \mathrm{Th}(\nu_y)$ by sending everything outside the disk bundle to the point at infinity. Moreover, there is an isomorphism

$$\varphi^* : H^{2n}(\mathrm{Th}(\nu_y);\pi_{2n-1}(S^n)) \xrightarrow{\cong} H^{2n}(M^{2n};\pi_{2n-1}(S^n)).$$

We also know that Ψ is a natural cohomology operation, i.e., we have the commuting diagram:

$$\begin{array}{ccc} H^{2n}(\mathrm{Th}(\nu_y);\pi_{2n-1}(S^n)) & \xrightarrow{\varphi^*} & H^{2n}(M^{2n};\pi_{2n-1}(S^n)) \\ \uparrow{\scriptstyle\Psi} & & {\scriptstyle\Psi}\uparrow \\ H^n(\mathrm{Th}(\nu_y);\mathbb{Z}) & \xrightarrow{\varphi^*} & H^n(M^{2n};\mathbb{Z}). \end{array}$$

We also can check that if $\mathcal{U}_y \in H^n(\mathrm{Th}(\nu_y);\mathbb{Z})$ is the Thom class then $\varphi^*(\mathcal{U}_y) = \mathcal{D}(y)$. Therefore, $\Psi(\mathcal{U}_y) = 0$ if and only if $\Psi(\mathcal{D}(x)) = 0$. Notice that the image under $\Psi \circ \mathcal{D}$ lands in $\mathbb{Z}_2$. Moreover, $\Psi(\mathcal{D}(y)) = 0$ if and only if $q(y) = 0$. But then $\Psi(\mathcal{U}_y) = 0$ which implies that there exists a map

$$\varphi_y : \mathrm{Th}(\nu_y) \longrightarrow S^n, \quad \varphi_y^*(\Theta) = \mathcal{D}(y).$$

The Thom complex is easy to describe. It has three cells: a point, an n-cell and then a $2n$-cell attached via $\alpha : S^{2n-1} \to S^n$ to S^n. In fact, if $\eta = \varepsilon^n$ then $\alpha = 0$ and if $\eta = TS^n$ then $\alpha = [\iota_n, \iota_n]$. One can check that $\Psi(\mathcal{U}_\eta) = \alpha$. If $\Psi(\mathcal{U}_y) = 0$ then $\mathrm{Th}(\mathcal{U}_y) \simeq S^n \vee S^{2n}$. On the other hand, if $\Psi(\mathcal{U}_y) = [\iota_n, \iota_n]$ then $\mathrm{Th}(\mathcal{U}_y) \simeq S^n \cup_{[\iota_n,\iota_n]} \mathbb{D}^{2n}$. It is clear that the homotopy type of the two Thom complexes are different. $\square$

Corollary 108. Let M^{2n} and N^{2n} be smooth manifolds satisfying [$*$] and $f : (M, \partial M) \to (N, \partial N)$ be a homotopy equivalence of pairs. Then $q_N \circ f_* = q_M$.

We have to use Lefschetz duality and the fact that Ψ is a cohomology operation.

We would like to prove two theorems - the proof of the first is long and that of the second is hard.

Theorem 109 (Lemma 8.2 [30])**.** The function q_M is a quadratic form. Alternatively, if A satisfies the hypothesis then the cohomology operation Ψ satisfies

$$\Psi(x + y) = \Psi(x) + \Psi(x) + [\iota_n, \iota_n](x \cup y).$$

When we reduce it mod 2 then $\overline{q}_M$ is a quadratic form in the sense of Arf and this is defined to be the Kervaire invariant $c(M^{2n}) := c(\overline{q}_M)$. Recall Theorem 104, i.e., the Kervaire invariant of a closed manifold M^{10} satisfying [$*$] vanishes.

Corollary 110. Suppose M^{10} is a closed manifold homotopy equivalent to the Kervaire manifold

$$K^{10} := (\mathbb{D}S^5 \wp \mathbb{D}S^5) \cup \mathbb{D}^{10}.$$

Then M^{10} cannot support a smooth structure.

To prove this we first need the following.

Proposition 111. Let M^{10} be a closed, smooth manifold homotopy equivalent to $K^{10} = (\mathbb{D}S^5 \wp \mathbb{D}S^5) \cup \mathbb{D}^{10}$. Then M^{10} satisfies condition [$*$], i.e., 4-connected and is a π-manifold.

Proof. Take the Kervaire manifold and subtract off the interior of a 10-disk, i.e.,

$$K^{10} - \mathrm{Int}(\mathbb{D}^{10}) \simeq S^5 \vee S^5.$$

The same is true for M^{10} minus an open disk. Let $\mathcal{M}^{10}$ be the resulting manifold. Let

$$\varphi : \mathcal{M}^{10} \simeq K^{10} - \mathrm{Int}(\mathbb{D}^{10}) \simeq S^5 \vee S^5$$

be the (restricted) homotopy equivalence. Then $T\mathcal{M}^{10} = \varphi^*(\xi^{10})$, where ξ^{10} is a bundle over $S^5 \vee S^5$. By Bott periodicity, we know that $\pi_4(\mathrm{O}) = 0$ and, therefore, $\xi^{10} = \varepsilon^{10}$. As a consequence, $T\mathcal{M}^{10} = \varepsilon^{10}$. Notice that TM^{10} is the pull-back of η^{10}, a bundle over S^{10}. The tangent bundle TM^{10} is stably trivial if and only if $\nu^{12} = \varepsilon^{12}$, where we embed M^{10} in $\mathbb{R}^{22}$ and ν is the normal bundle. Moreover, $\nu^{12} = \rho^*(\psi^{12})$, where ρ is a classifying map. Recall that ψ^{12} is obtained by the clutching construction applied to $\alpha \in \pi_9(\mathrm{O}_{12})$.

We also have the fibration

$$S^{22} \hookrightarrow \mathrm{Th}(\nu^{12}) \longrightarrow \mathrm{Th}(\psi^{12}).$$

Therefore, $\mathrm{Th}(\psi^{12})$ is reducible and ψ^{12} is fibre-homotopically trivial. This implies that $J(\alpha) = 0$. But by Adams' result, the J-homomorphism

$$J : \pi_9(\mathrm{O}_{12}) \longrightarrow \pi_{22}(S^{12})$$

is injective and hence $\alpha = 0$. □

Proof of Corollary. We assume that M^{10} has a smooth structure. Let $f : K^{10} \to M^{10}$ be a homotopy equivalence. We can perturb the map f such that f is identity on appropriate disks and maps the exterior to the exterior. We have a homotopy equivalence

$$f : (K^{10} - \mathbb{D}^{10}, \Sigma^9) \longrightarrow (M^{10} - \mathbb{D}^{10}, \partial)$$

of pairs, both satisfying [*] by Proposition 111. We know that

$$H_5(K^{10};\mathbb{Z}) \cong \mathbb{Z}x \oplus \mathbb{Z}y$$

and $q_K(x) = q_K(y) = 1$. By Corollary 108,

$$H_5(M^{10};\mathbb{Z}) \cong \mathbb{Z}\bar{x} \oplus \mathbb{Z}\bar{y}$$

such that $q_M(\bar{x}) = q_M(\bar{y}) = 1$. This means that

$$\bar{q}_M : H_5(M^{10};\mathbb{Z}_2) \longrightarrow \mathbb{Z}_2$$

has to have Arf invariant 1 as the images of x and y take the value 1. But this violates Theorem 104. □

In his paper [29] Kervaire uses some homotopy computations done by Toda [55] which is specific to dimension 10. Apparently, he had heard of these computations and a large part of his paper covers his proof of these computations.

19 Kervaire manifold revisited

We will continue with this construction of Kervaire of a manifold that is not smoothable.

Proof of Theorem 109. We turn the natural map $S^n \hookrightarrow K(\mathbb{Z}, n)$ into a fibration. We construct a space U equipped with an inclusion map $\imath : S^n \hookrightarrow U$ satisfying the property

$$\pi_j(U) = \begin{cases} 0, & \text{if } i \geqslant 2n-1 \\ \pi_i(S^n), & \text{if } i < 2n-1. \end{cases}$$

We will kill the homotopy groups of S^n in degree $(2n-1)$ and higher. Serre proved that any simply connected, finite CW complex has finitely generated homotopy groups in each dimension (cf. Corollary 16 [52], p. 509). Using this result, U can be constructed to have finitely many cells in each dimension. In fact, we may take

$$U = S^n \cup (2n - \text{cells}) \cup ((2n+1) - \text{cells}) \cup \cdots$$

such that $\pi_i(U) = 0$ for $i \geqslant 2n-1$. Now recall the hypothesis we had:

(1) The integer n is odd and not equal to 1, 3 or 7;
(2) The space A is simply connected, i.e., $\pi_1(A) = 0$;
(3) The space A is a CW complex with finitely many cells in each dimension;
(4) [cf. (hypothesis)] For any G and $n < i < 2n$, we have $H_i(A; G) = 0$.

It is clear that U satisfies the hypothesis. Check that

$$U \times U = (S^n \times \{*\}) \cup (\{*\} \times S^n) \cup (S^n \times S^n) \cup (\text{cells of dimension } 2n \text{ or higher}).$$

satisfies the hypothesis. Let $u \in H^n(U;\mathbb{Z})$ be an element which gets mapped to a fixed generator $\Theta \in H^n(S^n;\mathbb{Z})$.

Given an element $x \in H^n(A;\ Z)$, we choose a classifying map $\varphi_x : A \to K(\mathbb{Z}, n)$. We also have a classifying map $\varphi_u : U \to K(\mathbb{Z}, n)$. Is there a lift?

$$\begin{array}{ccc} & & U \\ & \overset{\widehat{\varphi}_x}{\nearrow} & \Big\downarrow \varphi_u \\ A & \xrightarrow[\varphi_x]{} & K(\mathbb{Z}, n). \end{array}$$

We can always find a lift as the obstructions lie in $H^{i+1}(A; \pi_i(F'))$, where F' is the fibre of the vertical map in the previous diagram. By construction of U, the appropriate homotopy groups of F' vanish and there are no obstructions. We define the composite

$$\varphi : A \xrightarrow{\Delta} A \times A \xrightarrow{\widehat{\varphi}_x \times \widehat{\varphi}_y} U \times U.$$

Assuming Proposition 112, we can verify that

$$\varphi^*(\Psi(u \times 1 + 1 \times u)) = \varphi^*(\Psi(u) \times 1) + \varphi^*(1 \times \Psi(u)) + [\iota_n, \iota_n]\,\varphi^*(u \times u).$$

Notice that the last term can be written as

$$\varphi^*(u \times u) = \Delta^*(\widehat{\varphi}_x^*(u) \times \widehat{\varphi}_x^*(u)) = [\iota_n, \iota_n](x \cup y).$$

For the other term, observe that

$$\varphi^*(\Psi(u) \times 1) = \Delta^*(\widehat{\varphi}_x^*(\Psi(u)) \times \widehat{\varphi}_y^*(1)) = \widehat{\varphi}_x^*(\Psi(u)) \cup 1 = \Psi(x).$$

This implies that

$$\varphi^*(\Psi(u \times 1 + 1 \times u)) = \varphi^*(\Psi(u \times 1)) + \varphi^*(\Psi(1 \times u)) = \Psi(x) + \Psi(y).$$

Therefore, it suffices to prove the proposition. □

Proposition 112. Let $u \in H^n(U; \mathbb{Z})$ be as before. Then

$$\Psi(u \times 1 + 1 \times u) = \Psi(u) \times 1 + 1 \times \Psi(u) + [\iota_n, \iota_n](u \times u).$$

Proof. It follows from Künneth formula that

$$H^{2n}(U \times U; \mathbb{Z}) = (H^{2n}(U; \mathbb{Z}) \otimes 1) \oplus (1 \otimes H^{2n}(U; \mathbb{Z})) \oplus (H^n(U; \mathbb{Z}) \otimes H^n(U; \mathbb{Z}))$$

and there is a map

$$H^{2n}(U \times U; \mathbb{Z}) \longrightarrow H^{2n}(U \times U; \pi_{2n-1}(S^n)).$$

Therefore,

$$\Psi(u \times 1 + 1 \times u) = a \times 1 + 1 \times b + \gamma(u \times u).$$

We just need to show that $a = b = \Psi(u)$ and $\gamma = [\iota_n, \iota_n]$. We define three inclusion maps

$$\begin{aligned} f : U &\longrightarrow U \times \{*\} \subset U \times U \\ g : U &\longrightarrow \{*\} \times U \subset U \times U \\ h : S^n \times S^n &\longrightarrow U \times U. \end{aligned}$$

Notice that $f = (\mathrm{id} \times \{*\}) \circ \Delta$. Therefore,

$$\begin{aligned} \Psi \circ f^*(u \times 1 + 1 \times u) &= f^*(a \times 1 + 1 \times b + \gamma(u \times u)) \\ &= f^*(a \times 1) + f^*(1 \times b) + \gamma f^*(u \times u) \\ &= a \cup 1 = a. \end{aligned}$$

on the one hand while on the other

$$\Psi \circ f^*(u \times 1 + 1 \times u) = \Psi \circ f^*(u \times 1) + \Psi \circ f^*(1 \times u) = \Psi(u).$$

A similar argument shows that $b = \Psi(u)$. Now look at

$$\begin{aligned} \Psi \circ h^*(u \times 1 + 1 \times u) &= h^*(\Psi(u) \times 1 + 1 \times \Psi(u) + \gamma(u \times u)) \\ &= h^*(\Psi(u) \times 1) + h^*(1 \times \Psi(u)) + h^*(\gamma(u \times u)) \\ &= \sigma^*(\Psi(u)) \times 1 + 1 \times \sigma^*(\Psi(u)) + \gamma\, h^*(u \times u) = \gamma\, h^*(u \times u). \end{aligned}$$

On the other hand we have

$$\Psi(\Theta \times 1 + 1 \times \Theta) = \gamma(\Theta \times \Theta).$$

This is a formula about $S^n \times S^n$. How did we define Ψ? Given the cohomology class $\Theta \times 1 + 1 \times \Theta$ we have a classifying map

$$\varphi_{\Theta \times 1 + 1 \times \Theta} : S^n \times S^n \longrightarrow K(\mathbb{Z}, n).$$

We define Ψ by the first obstruction to lifting this map to S^n which fibres over $K(\mathbb{Z}, n)$. Notice that $S^n \times S^n$ is homotopy equivalent to $S^n \vee S^n$ with a $2n$-disk attached via the Whitehead product $[\iota_n, \iota_n]$. Notice that there is a lift on the n-skeleton $S^n \vee S^n$ by $\iota_n \vee \iota_n$. By the definition of the obstruction class, $\gamma = [\iota_n, \iota_n]$. □

This was slightly tedious but that is the proof. Now we are left with the hard theorem. Let us recall what it is.

Theorem 113 (cf. Theorem 104)**.** Let M^{10} be a closed, 4-connected, smooth manifold satisfying property [$*$], i.e., $TM^{10} \oplus \varepsilon^1 \cong \varepsilon^{11}$. Then the Kervaire invariant vanishes, i.e., $c(M^{10}) = 0$.

We will need results about the stable 10-stem as well as surgery theory which is from the paper [30]. It is worth separating out the two things:

(1) (Toda '61 [55]) The stable 10-stem is

$$\pi_{10}^s \cong \mathbb{Z}_2 \oplus \mathbb{Z}_3.$$

A generator $\gamma \in \mathbb{Z}_2 \subset \pi_{10}^s$ is homotopic to $\alpha \circ \Sigma^{19} h$, where $h : S^3 \to S^2$ is the Hopf map and $\alpha : S^{21} \to S^{12}$ is an element in π_9^s.

(2) (Easy) Given any framed, smooth compact manifold M^m, it is framed cobordant (relative boundary) to a highly connected, framed manifold N^m. In particular, N^m is $([m/2]-1)$-connected.
(Hard) If $m = 2n+1$ and ∂M^m is empty or a homotopy sphere then N^m is n-connected.

20 Surgery on manifolds

Today we are finally going to delve into a discussion of the word *surgery.* This is a course on an introduction to surgery theory and it has been a long introduction. We have talked about some easy and some hard theorems last time.

Theorem 114 (Easy Surgery). Let M^n be a compact, smooth framed manifold. Then M^n is framed cobordant (relative boundary) to a "highly connected" framed manifold N^n, i.e., $\pi_i(N^n) = 0$ if $i < [n/2]$.

The proof uses something called *spherical modification* which people eventually ended up calling *surgery.* We shall start with an example.

Example 115. We can think of S^2 as the join of S^1 with S^0. Now

$$\partial(\mathbb{D}^2 \times \mathbb{D}^1) = (\partial\mathbb{D}^2 \times \mathbb{D}^1) \cup (\mathbb{D}^2 \times \partial\mathbb{D}^1) = S^1 \times \mathbb{D}^1 \cup \mathbb{D}^2 \times S^0.$$

What is a 0-surgery on a 2-manifold? We have an embedding $\Phi : S^0 \times \mathbb{D}^2 \to M^2$ and then we form

$$(M^2 - \operatorname{image}\Phi) \cup_\varphi S^1 \times \mathbb{D}^1.$$

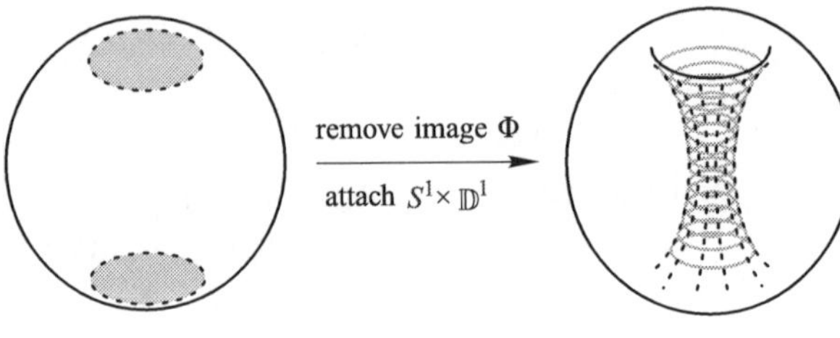

Figure 24. A surgery on S^2.

Depending on φ we may get the torus (as above) or the Klein bottle from $M^2 = S^2$.

Definition 116 (Surgery). A p-surgery on a $n = (p+q+1)$-dimensional manifold M^n is done as follows. Find an embedding $f : S^p \times \mathbb{D}^{q+1} \to M$ and attach a copy of $\mathbb{D}^{p+1} \times \mathbb{D}^{q+1}$ to $M^n \times \{1\} \subset M^n \times [0,1]$ along $S^p \times \mathbb{D}^{q+1}$ using f. Then the resulting manifold is a cobordism W^{n+1} such that

$$\partial W^{n+1} = M^n \times \{0\} \sqcup N^n.$$

The resulting manifold N^n is going to depend on f. In fact, the basic driving notion is the decomposition

$$S^p * S^q \cong S^n = \partial(\mathbb{D}^{p+1} \times \mathbb{D}^{q+1}) = (S^p \times \mathbb{D}^{q+1}) \cup (\mathbb{D}^{p+1} \times S^q).$$

In other words, $S^p \times S^q$ is the boundary of two different manifolds. The manifold N^n can be obtained from M^n by removing $f(S^p \times \mathbb{D}^{q+1})$ and attaching $\mathbb{D}^{p+1} \times S^q$ along the boundary $S^p \times S^q$.

Lemma 117. If $p < [n/2]$ then $H_i(N^n;\mathbb{Z}) = H_i(M^n;\mathbb{Z})$ if $i < p$ and $H_p(N^n;\mathbb{Z}) = H_p(M^n;\mathbb{Z}) / \langle f_*[S^p] \rangle$.

Proof. Notice that $W^{n+1} \sim M^n \cup (p+1)$-cell. It follows that $H_i(W^{n+1};\mathbb{Z}) = H_i(M^n;\mathbb{Z})$ for $i < p$ and we also have

$$H_p(W^{n+1};\mathbb{Z}) = H_p(M^n;\mathbb{Z}) / \langle f_*[S^p] \rangle .$$

We also know that

$$-W^{n+1} \sim N^n \cup (q+1)\text{-cell}.$$

Therefore, we conclude that $H_i(W^{n+1};\mathbb{Z}) = H_i(N^n;\mathbb{Z})$ for $i < q$. But $p < q$ implies the first statement as well as the second. □

This is a nice method in that we know that the homology groups are finitely generated. We do surgery and kill off the second homology and so on. But there is a problem with this method.

Example 118. Let us assume that we can apply surgery to $\mathbb{CP}^6$ to obtain N^{12}, where $\pi_i(N^{12}) = 0$ if $i < 6$. This also gives us a sequence of cobordisms and, in particular, $\mathbb{CP}^6$ is cobordant to N^{12}. Since Pontrjagin numbers are cobordism invariant, it follows that

$$\langle p_1^3(\mathbb{CP}^6), [\mathbb{CP}^6] \rangle = \langle p_1^3(N^{12}), [N^{12}] \rangle = 0.$$

But we know that ([37], p. 178) that $p_1(\mathbb{CP}^6) = 5a^2$. Therefore, $p_1^3(\mathbb{CP}^6) = 125a^6$. In general, we cannot do surgery on any arbitrary manifold to make it highly connected. Another reason why this example fails is the fact that the normal bundle of $S^2 = \mathbb{CP}^1$ in $\mathbb{CP}^6$ is non-trivial.

Suppose we have a simply connected, framed manifold M^n. We are trying to kill $H_2(M^n;\mathbb{Z})$. By Hurewicz theorem we know that generators of $H_2(M^n;\mathbb{Z})$ can be represented by continuous maps of 2-spheres. By the easy version of Whitney theorem (as $4 = 2 \cdot 2 < n$), we can actually choose embeddings $f : S^2 \to M^n$ of spheres instead. We know that

$$TS^2 \oplus \nu = f^*(TM^n),$$

where ν is the normal bundle of S^2 in M^n. Therefore, ν is stably trivial. However, we have to take some care as the resulting manifold may not be framed.

We will be analyzing bundles over spheres. We start with a framed manifold and do a surgery using $f : S^p \times \mathbb{D}^{q+1} \to M^n$, where $n = p + q + 1$. We have a framing on $M^n \times [0,1]$ and we also have a framing on $S^p \times \mathbb{D}^{q+1}$. The framings have to match up and, in general, the obstruction lives in $\pi_p(\mathrm{O}_{n+1})$. We change f to $f \circ \varphi_\alpha$, where $\alpha \in \pi_p(\mathrm{O}_{q+1})$ and

$$\varphi_\alpha : S^p \times \mathbb{D}^{q+1} \longrightarrow S^p \times \mathbb{D}^{q+1}, \quad \varphi_\alpha(x,y) := (x, \alpha(x)(y)).$$

If the map $\pi_p(\mathrm{O}_{q+1}) \to \pi_p(\mathrm{O}_{n+1})$ is surjective then we are done. But this is surjective as $p < q$ and can be seen via the long exact sequence of homotopy groups associated to the fibration $\mathrm{O}_{q+1} \hookrightarrow \mathrm{O}_{q+2} \to S^{q+1}$. For a detailed discussion of framed surgery which includes a complete proof of Theorem 114 see §5 and §6 of [30].

We intend to talk about Kervaire's program. His invariant is defined for highly connected π-manifolds which are framable manifolds. We would like to extend the definition of Kervaire invariant from manifolds satisfying [*] to all framed manifolds which are closed or with boundary homeomorphic to a sphere. We know that M^n is framed cobordant to N^n satisfying [*]. We may define the invariant of M^n as that of N^n provided it does not depend on the choice of N^n. The proof of Theorem 104 depends on defining the Kervaire invariant for this larger class of manifolds.

Proposition 119. Let $n = 4m + 2$ be such that $n + 2$ is not a power of 2. If W^{n+1} is a compact, framed manifold whose boundary $M^n := \partial W^{n+1}$ satisfies condition [*] then the Kervaire invariant vanishes, i.e., $c(M^n) = 0$.

By the easy surgery theorem, we can do a framed surgery on W^{n+1} without changing its boundary such that we have a highly connected, framed manifold $\mathcal{W}^{n+1}$ with boundary M^n. Let us assume $n = 10$, the case in which Kervaire was really interested. Let $\sigma : M^{10} \hookrightarrow \mathcal{W}^{11}$. We have

$$\begin{aligned} 0 \longrightarrow H^5(\mathcal{W}, M;\mathbb{Z}) \longrightarrow H^5(\mathcal{W};\mathbb{Z}) \xrightarrow{\sigma^*} H^5(M;\mathbb{Z}) \\ \longrightarrow H^6(\mathcal{W}, M;\mathbb{Z}) \longrightarrow H^6(\mathcal{W};\mathbb{Z}) \longrightarrow 0. \end{aligned}$$

Let $\mathscr{I}$ be the image of σ^*. We claim that $q : H_5(M^{10};\mathbb{Z}) \to \mathbb{Z}_2$ restricted to $\mathcal{D}^{-1}(\mathscr{I})$ is zero. Recall that we identify $H^5(M;\mathbb{Z})$ with $H_5(M;\mathbb{Z})$ via the Poincaré duality isomorphism $\mathcal{D}$ and that $q = \Psi \circ \mathcal{D}$. We prove the claim by showing that the cohomology operation $\Psi : H^5(M^{10}; \mathbb{Z}) \to \mathbb{Z}_2$ is zero on $\mathscr{I}$. In order to show that $\Psi(\sigma^*(x)) = 0$ we choose a representative of x, i.e.,

$$\begin{array}{ccccc} & & & & S^5 \\ & & & \nearrow & \downarrow \\ M^{10} & \xrightarrow[\sigma]{} & W^{11} & \xrightarrow{x} & K(\mathbb{Z}, 5). \end{array}$$

We show that $\Psi(\sigma^*(x)) = 0$ by proving that there is a lift from M^{10} to S^5. Let F be the homotopy fibre of $S^n \to K(\mathbb{Z}, 5)$. Then F is 5-connected and the primary obstruction to finding a lift is an element of $H^{10}(M^{10}; \pi_9(F)) \cong H^{10}(M^{10}; \pi_9(S^5))$.

21 The Kervaire invariant

The main topic today is to define the *Kervaire invariant*, i.e., the map

$$c : \pi^s_{4m+2} \longrightarrow \mathbb{Z}_2.$$

Recall that previously we had defined the Kervaire invariant for highly connected π-manifolds. We will concentrate on $m = 2$ and see that $c : \pi^s_{10} \to \mathbb{Z}_2$ is the zero map. We did so much work to define the zero map but what this implies is the existence of a manifold, of dimension 10, which does not admit any smooth structure.

Kervaire Invariant One Problem. The invariant $c = 0$ except for $m = 0, 1, 3, 7, 15$.

We take

$$K^{4m+2} := (\mathbb{D}S^{2m+1} \wp \mathbb{D}S^{2m+1}) \cup \mathbb{D}^{4m+2}$$

as the Kervaire manifold. The Kervaire problem is equivalent to saying that K^{4m+2} has a smooth structure only when $m = 0, 1, 3, 7, 15$ except $m = 31$. In [29] Kervaire had proved that K^{10} does not have a smooth structure. Brown and Peterson [10] had proved that the Kervaire invariant for $(8m+2)$-manifolds vanish. Browder [8] then extended the definition of the Kervaire invariant to smooth, stably parallelizable manifolds of dimension $2k$ (with k odd) and proved that it vanishes if $k \neq 2^i - 1$. He also connected this question to deep questions in the stable homotopy groups of spheres (cf. §1, §2 of [26] for a review of the history). Jones [27] worked on the case $m = 7$ while Barratt, Jones and Mahowald [6] answered the case $m = 15$. By a recent result [21] of Hill, Hopkins, and Ravenel, we know that the above is true except possibly $m = 31$.

Let us recall the result stated without proof (cf . Proposition 119) from last time.

Proposition 120. Let M^n be a closed, smooth manifold satisfying condition [$*$] and $n = 4m+2$. If $n \neq 2^r - 2$ for any integer r and W^{n+1} is a framed manifold with boundary M^n then $c(M^n) = 0$.

Consider the $(2m+1)^{\text{th}}$ homology. We have to show that more than "half" the embedded spheres have trivial normal bundles. We look at it more algebraically. For the ease of discussion we prove it for $n = 10$ or $m = 2$ although the general proof is similar.

Proof. We restrict ourselves to $n = 2$ or $m = 10$. Let $\sigma : M^{10} \hookrightarrow W^{11}$ be the inclusion map. Consider the cohomology exact sequence (with integer coefficients) of (W^{11}, M^{10}):

$$\begin{aligned} 0 = H^4(M^{10}) &\to H^5(W^{11}, M^{10}) \to H^5(W^{11}) \to H^5(M^{10}) \\ &\to H^6(W^{11}, M^{10}) \to H^6(W^{11}) \to H^6(M^{10}) = 0, \end{aligned}$$

as M^{10} is 4-connected. Although W^{11} is not necessarily 4-connected, we had shown last time that we can do a surgery and make W^{11} a 4-connected manifold. We claim that

$$\Psi : H^5(M^{10}; \mathbb{Z}) \longrightarrow H^{10}(M^{10}; \pi_9(S^5)) = H^{10}(M^{10}; \mathbb{Z}_2)$$[8]

is identically zero when restricted to $\mathscr{I} := \operatorname{image} \sigma^*$. Let $y \in H^5(M^{10}; \mathbb{Z})$ and we choose a representative of y, i.e.,

$$\begin{array}{ccc} & & S^5 \\ & \nearrow & \downarrow \\ M^{10} & \xrightarrow[\varphi_y]{} & K(\mathbb{Z}, 5). \end{array}$$

We know that $\Psi(y) = 0$ if and only if φ_y lifts to S^5. But $y = \sigma^*(x)$ implies we have the following diagram:

$$\begin{array}{ccccc} & & & & S^5 \\ & & & \nearrow & \downarrow \\ M^{10} & \xrightarrow[\sigma]{} & W^{11} & \xrightarrow[\varphi_x]{} & K(\mathbb{Z}, 5). \end{array}$$

We also have $\varphi_y = \varphi_x \circ \sigma$. Notice that it suffices to lift φ_x over the 10-skeleton of W^{11}. By obstruction theory, the obstructions to lifting φ_x lie in $H^{i+1}(W^{11}; \pi_i(F))$, where F is the fibre of $S^5 \to K(\mathbb{Z}, 5)$. Observe that Θ_{i+1} is defined if there exists a lift $\widehat{\varphi}_x$ over $W^{11}(i)$, the ith skeleton of W^{11}. Moreover, $\Theta_{i+1} = 0$ if and only if there exists a lift $\widetilde{\varphi}_x$ over $W^{11}(i+1)$ such that

$$\widetilde{\varphi}_x\big|_{W^{11}(i-1)} \equiv \widehat{\varphi}_x\big|_{W^{11}(i-1)}.$$

[8]When $m \neq 0, 1, 3$ and $\iota := \mathrm{id} : S^{2m+1} \to S^{2m+1}$, the element $[\iota, \iota] \in \pi_{4m+1}(S^{2m+1})$ generates a subgroup of order 2. If $m = 2$ then there is an isomorphism $\pi_9(S^5) \cong \langle [\iota, \iota] \rangle$.

We will show that φ_x lifts over $W^{11}(10)$ for all $x \in H^5(W^{11};\mathbb{Z})$. Since $\pi_j(F) = 0$ if $j \leqslant 5$ it suffices to show that

$$H^{j+1}(W;\pi_j(F)) = 0, \quad 6 \leqslant j \leqslant 9.$$

By Poincaré-Lefschetz duality

$$H^{j+1}(W^{11};G) \cong H_{10-j}(W^{11},M^{10};G).$$

Via the long exact sequence in (reduced) homology for (W^{11},M^{10}), the last group in the isomorphism above is between $H_{10-j}(W^{11};\mathbb{Z})$ and $H_{9-j}(M^{10};\mathbb{Z})$. Since W^{11} and M^{10} are both 4-connected, both of these groups vanish for $6 \leqslant j \leqslant 9$. This implies the claim.

To show that the Kervaire invariant is zero, we have to make use of the just verified claim. Let us write down the corresponding sequence in cohomology with $\mathbb{Z}_2$ coefficients:

$$\begin{aligned} 0 \to H^5(W^{11},M^{10};\mathbb{Z}_2) \to H^5(W^{11};\mathbb{Z}_2) \xrightarrow{\sigma^*} H^5(M^{10};\mathbb{Z}_2) \\ \to H^6(W^{11},M^{10};\mathbb{Z}_2) \to H^6(W^{11};\mathbb{Z}_2) \to 0. \end{aligned}$$

The first and the fifth groups are isomorphic while the second and the fourth groups are also isomorphic. This follows from Poincaré-Lefschetz duality and the universal coefficient theorem. Let $\bar{\mathscr{I}}$ be the image of σ^*. We claim that

$$2\dim_{\mathbb{Z}_2}\bar{\mathscr{I}} = \dim_{\mathbb{Z}_2} H^5(M^{10};\mathbb{Z}_2). \tag{6}$$

We have three short exact sequences. The first one

$$0 \longrightarrow H^5(W^{11},M^{10};\mathbb{Z}_2) \longrightarrow H^5(W^{11};\mathbb{Z}_2) \longrightarrow \bar{\mathscr{I}} \longrightarrow 0$$

implies

$$\dim_{\mathbb{Z}_2} H^5(W^{11};\mathbb{Z}_2) = \dim_{\mathbb{Z}_2}\bar{\mathscr{I}} + \dim_{\mathbb{Z}_2} H^5(W^{11},M^{10};\mathbb{Z}_2).$$

The second short exact sequence is

$$0 \longrightarrow \operatorname{coker}\sigma^* \longrightarrow H^6(W^{11},M^{10};\mathbb{Z}_2) \longrightarrow H^6(W^{11};\mathbb{Z}_2) \longrightarrow 0$$

and implies

$$\dim_{\mathbb{Z}_2} H^6(W^{11},M^{10};\mathbb{Z}_2) = \dim_{\mathbb{Z}_2}\operatorname{coker}\sigma^* + \dim_{\mathbb{Z}_2} H^6(W^{11};\mathbb{Z}_2)$$

while the third

$$0 \longrightarrow \bar{\mathscr{I}} \longrightarrow H^5(M^{10};\mathbb{Z}_2) \longrightarrow \operatorname{coker}\sigma^* \longrightarrow 0$$

implies that

$$\dim_{\mathbb{Z}_2} H^5(M^{10};\mathbb{Z}_2) = \dim_{\mathbb{Z}_2}\bar{\mathscr{I}} + \dim_{\mathbb{Z}_2}\operatorname{coker}\sigma^*.$$

If we put all of these together with universal coefficient theorem and duality then we may conclude that (6) holds.

We have the following commutative diagram:

$$\begin{array}{ccc} H^5(W^{11};\mathbb{Z}) & \xrightarrow{\sigma^*} & H^5(M^{10};\mathbb{Z}) \\ \downarrow & & \downarrow\!\!\downarrow \eta \\ H^5(W^{11};\mathbb{Z}_2) & \xrightarrow{\sigma^*} & H^5(M^{10};\mathbb{Z}_2). \end{array}$$

We know that $H^5(W^{11};\mathbb{Z})$ is free abelian (by universal coefficient theorem and the fact that W^{11} has no 4th homology). Let us assume that

$$\bar{\mathscr{I}} \subseteq \eta(\mathscr{I}). \tag{7}$$

Notice that Ψ induces a quadratic form $\overline{\Psi}$ on $V := H^5(M^{10};\mathbb{Z})\otimes\mathbb{Z}_2$. This is a non-degenarate quadratic form and isomorphic to $\overline{q}$ on $H^5(M^{10};\mathbb{Z}_2)$. By definition, $c(M^{10}) = \text{Arf}(\overline{q})$ which is equal to $\text{Arf}(\overline{\Psi})$. We know that $\overline{\Psi}$ is zero on $\bar{\mathscr{I}}$ and this subspace is half the dimension of $H^5(M^{10};\mathbb{Z}_2)$. Since $\overline{\Psi}$ is a *democratic invariant* (see the comment following the proof of Lemma 99) we conclude that$\text{Arf}(\overline{\Psi}) = 0$ when assumption 7 is satisfied. in our next lecture we will show how to remove the assumption. □

Corollary 121. The Kervaire invariant of a framed manifold M^{4m+2} is well defined when $m \neq 2^l - 1$.

Proof. Let N_i^{4m+2} be a pair $(i = 1, 2)$ of highly connected framed manifolds, framed cobordant to M^{4m+2}. We need to show that $c(N_1) = c(N_2)$. We can concatenate the two cobordisms to get a framed cobordism W with boundary $N_1 \sqcup N_2$. This disjoint union does not satisfy condition [*]. We can rectify that by taking little disks $\mathbb{D}^{4m+2}$ from N_1 and N_2 and drill a tunnel connecting them.

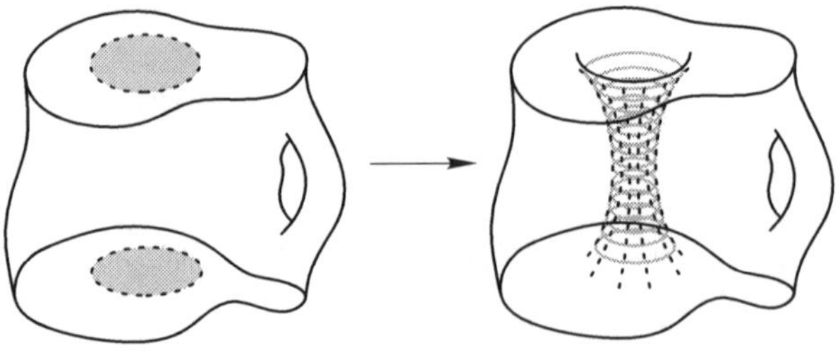

Figure 25. A surgery replacing $N_1 \sqcup N_2$ with $N_1 \# N_2$.

We get a new framed manifold $\mathcal{W}$ with boundary $N_1\#N_2$, which satisfy condition [*]. Therefore, $c(N_1\#N_2) = 0$. The homology classes can be represented by spheres avoiding the disks and, therefore,

$$c(N_1\#N_2) = c(N_1) + c(N_2).$$

Since this is an equation in $\mathbb{Z}_2$, we conclude that $c(N_1) = c(N_2)$. □

In fact, the same argument shows that the Kervaire invariant c that maps framed cobordism classes of framed manifolds of dimension $(4m+2)$ to $\mathbb{Z}_2$ is a group homomorphism. We will show in the next lecture how to remove the assumption (7). This is important since the assumption is probably not true in general as the map $H^5(W^{11};\mathbb{Z}) \to H^5(W^{11}, \mathbb{Z}_2)$ need not be onto.

22 Vanishing of the Kervaire invariant

Let $M^{10} = \partial W^{11}$ be such that the inclusion $\sigma : M^{10} \hookrightarrow W^{11}$ is that of 4-connected manifolds. We had seen that

$$\overline{\Psi} : H^5(M;\mathbb{Z}_2) \longrightarrow H^{10}(M;\mathbb{Z}_2) = \mathbb{Z}_2$$

was zero on $\bar{\mathscr{I}}$, the image of σ^* with $\mathbb{Z}_2$-coefficients, under the assumption (7), i.e., $\bar{\mathscr{I}} \subseteq \eta(\mathscr{I})$. This follows if the natural map $H^5(W^{11};\mathbb{Z}) \to H^5(W^{11}, \mathbb{Z}_2)$ is surjective. However, not every class in $H^5(W;\mathbb{Z}_2)$ lifts to $H^5(W;\mathbb{Z})$. We will discuss how to remove the assumption. Let $y = \sigma^*(x) \in H^5(M;\mathbb{Z}_2)$ and consider the following diagram:

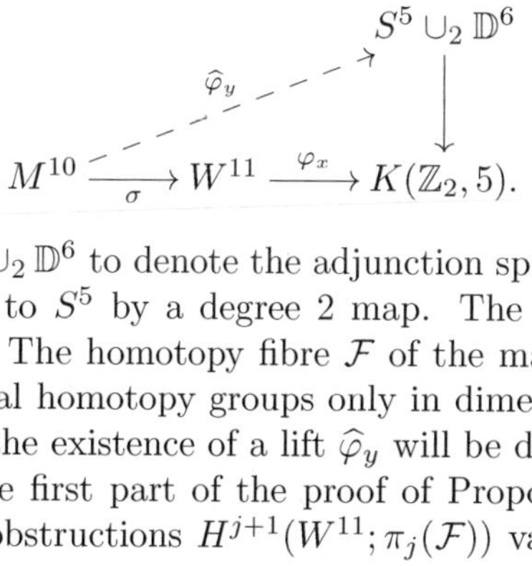

We shall use $\mathcal{M}_6 := S^5 \cup_2 \mathbb{D}^6$ to denote the adjunction space formed by attaching $\mathbb{D}^6$ along the boundary to S^5 by a degree 2 map. The resulting space is often called the *Moore space*. The homotopy fibre $\mathcal{F}$ of the map $\mathcal{M}_6 \to K(\mathbb{Z}_2, 5)$ is a space that has non-trivial homotopy groups only in dimensions 6 or higher. The primary obstruction to the existence of a lift $\widehat{\varphi}_y$ will be denoted by $\Phi(y)$. By the same reasoning as in the first part of the proof of Proposition 120 we conclude that a lift exists as all obstructions $H^{j+1}(W^{11}; \pi_j(\mathcal{F}))$ vanish for $6 \leqslant j \leqslant 9$. In particular, $\Phi(y) = 0$.

Now consider any element $\widehat{y} \in H^5(M;\mathbb{Z})$ such that $\widehat{y}$ is mapped to y under the natural map. Consider the following diagram:

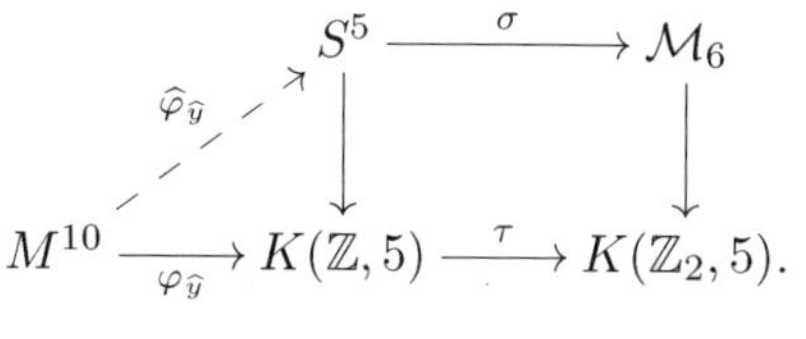

The element $\Psi(\widehat{y})[\iota, \iota] \in H^{10}(M^{10}; \pi_9(S^5))$ is the obstruction to the existence of $\widehat{\varphi}_{\widehat{y}}$. Then by obstruction theory

$$\sigma_*(\Psi(\widehat{y})[\iota, \iota]) = \Psi(\widehat{y})[\sigma_*\iota, \sigma_*\iota] = \Phi(y) = 0 \in H^{10}(M^{10}; \pi_9(\mathcal{M}_6)) \cong \pi_9(\mathcal{M}_6).$$

But Lemma 122 implies that $[\sigma_*\iota, \sigma_*\iota] \neq 0$, whence $\Psi(\widehat{y}) = 0$. Proposition 120 now implies that the Arf invariant $\overline{\Psi}$ satisfies $\overline{\Psi}(y) = 0$ if $y \in \sigma^*(H^5(W;\mathbb{Z}_2)) =: \mathscr{I}$. But this is half of the vector space $H^5(M;\mathbb{Z}_2)$ and therefore, the Arf invariant of $\overline{\Psi}$ is zero. It remains to prove the lemma.

Lemma 122. Let $\iota : S^{2m+1} \to S^{2m+1}$ be the identity map where $m \neq 1, 3, 7$. Then $[\iota, \iota] \in \pi_{4m+1}(S^{2m+1})$ is non-zero and generates a subgroup isomorphic to $\mathbb{Z}_2$. Let $\mathcal{M}_{2m+2}$ be the Moore space formed by attaching $\mathbb{D}^{2m+2}$ to S^{2m+1} by a map of degree 2 and let $\alpha : S^{2m+1} \hookrightarrow \mathcal{M}_{2m+2}$ be the inclusion map. If $m \neq 1, 3, 7$ is not of the form $2^l - 1$ then the image $\alpha_*([\iota, \iota]) = [\alpha_*\iota, \alpha_*\iota]$ is non-zero in $\pi_{4m+1}(\mathcal{M}_{2m+2})$.

Proof. The first claim is well known. For the second claim, consider the unit tangent bundle $T_1 S^{2m+2}$ of S^{2m+2}. We shall denote it by $V_{2m+3,2}$ as it is also the space of orthonormal 2-frames in $\mathbb{R}^{2m+3}$. The long exact sequence of homotopy groups imply that $V_{2m+3,2}$ is $2m$-connected. Looking at the Gysin sequence associated to the bundle $S^{2m+1} \hookrightarrow V_{2m+3,2} \to S^{2m+2}$ we get:

$$\begin{aligned} \cdots \longrightarrow H^{i+2m+1}(V_{2m+3,2};\mathbb{Z}) &\longrightarrow H^i(S^{2m+2};\mathbb{Z}) \xrightarrow{\cup\rceil} H^{i+2m+2}(S^{2m+2};\mathbb{Z}) \\ \longrightarrow H^{i+2m+2}(V_{2m+3,2};\mathbb{Z}) &\longrightarrow \cdots . \end{aligned}$$

Note that the Euler class $\rceil$ of the bundle is twice the Poincaré dual of the fundamental class $[S^{2m+2}]$. In particular, the map

$$(\cup\rceil) : H^0(S^{2m+2};\mathbb{Z}) \longrightarrow H^{2m+2}(S^{2m+2};\mathbb{Z})$$

is multiplication by 2. This implies that $H^{2m+2}(V_{2m+3,2};\mathbb{Z}) = \mathbb{Z}_2$ and by Poincaré duality, $H_{2m+1}(V_{2m+3,2};\mathbb{Z}) = \mathbb{Z}_2$. Now by Hurewicz theorem, $\pi_{2m+1}(V_{2m+3,2};\mathbb{Z}) = \mathbb{Z}_2$. In particular, this $\mathbb{Z}_2$ is generated by the image of ι under $i_* : \pi_{2m+1}(S^{2m+1}) \to \pi_{2m+1}(V_{2m+3,2})$. Comparing homotopy groups we see that $\mathcal{M}_{2m+2}$ is a subset of the $(2m+2)$-skeleton of $V_{2m+3,2}$. Therefore, the map $i : S^{2m+1} \hookrightarrow V_{2m+3,2}$ factors through the inclusion $\alpha : S^{2m+1} \hookrightarrow \mathcal{M}_{2m+2}$. Consequently, in order to prove $[\alpha_*\iota, \alpha_*\iota] \neq 0$ it suffices to prove $[i_*\iota, i_*\iota] \neq 0$.

Now look at the long exact sequence in the homotopy groups:

$$\begin{aligned} \cdots \longrightarrow \pi_{4m+2}(S^{2m+2}) &\xrightarrow{d} \pi_{4m+1}(S^{2m+1}) \xrightarrow{i_*} \pi_{4m+1}(V_{2m+3,2}) \\ \longrightarrow \pi_{4m+1}(S^{2m+2}) &\longrightarrow \cdots \end{aligned}$$

We are in the stable range and $\pi_{4m+2}(S^{2m+2}) = \pi^s_{2m}$. When $m = 2$ we know that $\pi^s_4 = 0$ and i_* is injective. But this is not true in general and it requires some work to prove that $[\iota, \iota]$ is not in the image of d which then implies, by exactness, that $i_*[\iota, \iota] \neq 0$.

If $m \neq 2^l - 1$ for any integer l then $n := 2m + 1$ satisfies $n + 1 \neq 2^k$ for any integer k. By [19] we know that $[\iota, \iota]$ is a direct summand in $\pi_{2n-1}(S^n)$ for these values of n. We rewrite the long exact sequence from above along with the suspension map:

$$\cdots \longrightarrow \pi_j(S^{n+1}) \xrightarrow{d} \pi_{j-1}(S^n) \xrightarrow{i_*} \pi_{j-1}(V_{n+2,2}) \longrightarrow \cdots$$
$$\text{with } \Sigma : \pi_{j-1}(S^n) \to \pi_j(S^{n+1})$$

It can be shown that $(d \circ \Sigma)(\alpha) = (2\iota_n) \circ \alpha$ (cf. §11, Lemma 11.1 [25]) for $\alpha \in \pi_{j-1}(S^n)$ and $\iota_n = \mathrm{id} : S^n \to S^n$.

Remark 123. For $[\alpha] \in \pi_m(S^n, s_0)$ the naturally defined map

$$\pi_n(X, x_0) \longrightarrow \pi_m(X, x_0), \quad [\beta] \mapsto [\beta \circ \alpha] \tag{8}$$

need not be a group homomorphism. One may consider $(X, x_0) = (S^2, s_0), m = 3, n = 2$ and let $h : S^3 \to S^2$ denote the Hopf map. We can verify that an element $[\varphi] \in \pi_2(S^2, s_0)$ of degree k is sent to $k^2[h] \in \pi_3(S^2, s_0)$ under the map (8).

However, if α is a suspension then (8) is a homomorphism (cf. [25], p. 333). An application of Freudenthal suspension theorem yields the surjectivity of the right arrow below while the delicate suspension theorem [58] (via the EHP sequence) provides the surjectivity of the left arrow below:

$$\pi_{2n-2}(S^{n-1}) \longrightarrow \pi_{2n-1}(S^n) \longrightarrow \pi_{2n}(S^{n+1}) \longrightarrow 0.$$

Therefore, for $\alpha = \Sigma a \in \pi_{2n-1}(S^n)$ we see that

$$(d \circ \Sigma)(\alpha) = (2\iota_n) \circ \alpha = 2(\iota_n \circ \alpha) = 2\alpha.$$

If $d(\beta) = [\iota, \iota]$ with $\beta = \Sigma b$ then $[\iota, \iota] = 2b$ by the above. But this cannot happen as $\langle [\iota, \iota] \rangle \cong \mathbb{Z}_2$ is a direct summand, whence $[\iota, \iota]$ is not in the image of d. □

We recall that all of the discussion we have had so far about removing assumption (7) is for manifolds in dimension $4m + 2$ with $m \neq 2^l - 1$. We had only specialized to $m = 2$ (cf. Proposition 120 and the discussion preceeding Lemma 122) for the ease of writing.

We are finally ready to prove Kervaire's main result.

Proof of Theorem 104. We have a well defined group homomorphism, the Kervaire-Arf map

$$c : \mathbb{Z}_6 \cong \pi_{10}^s \longrightarrow \mathbb{Z}_2.$$

The equality used above is due to Toda [55]. Since $\mathbb{Z}_6 \cong \mathbb{Z}_2 \oplus \mathbb{Z}_3$, to show that $c = 0$ it suffices to prove that $c(\alpha) = 0$, where α is the element of order 2. We may take $\pi_{10}^s = \pi_{22}(S^{12})$. Toda had also showed that

$$\alpha \sim (S^{22} \xrightarrow{\Sigma^{19} h} S^{21} \xrightarrow{\gamma} S^{12}).$$

for some choice of γ and $h : S^3 \to S^2$ is the Hopf fibration. We will assume that this map γ can be perturbed to be a smooth map and then look at the inverse image of a regular value. Then $\gamma^{-1}(*) =: N^9$ is a framed manifold. By the easy surgery theorem, N^9 is framed cobordant to a 3-connected manifold. Since 9 is odd, by the hard surgery theorem (§6, Theorem 6.6 [30]), we can actually make it framed cobordant to a 4-connected framed manifold. But by the Hurewicz theorem, this resulting manifold is homotopy equivalent (actually homeomorphic) to S^9. We may replace N^9 by $\Sigma^9 \in \Theta_9$. We have

$$\Sigma^9 \subset \mathbb{R}^{21} \subset S^{21} \xrightarrow{\gamma} S^{12}$$

and may assume that γ satisfies $\gamma^{-1}(*) = \Sigma^9$. We want to examine the element $(\Sigma^{19}h)^{-1}(\Sigma^9)$. Although $\Sigma^{19}h : S^{22} \to S^{12}$ is not a fibre bundle, there is a big open set on which it is a circle bundle. By the Whitney embedding theorem, we may find another isotopic embedding of Σ^9 which lies entirely in this big open set. Therefore,

$$S^1 \hookrightarrow (\Sigma^{19}h)^{-1}(\Sigma^9) \longrightarrow \Sigma^9$$

is a principal S^1-bundle. This also admits a cross-section as there are no obstructions and the bundle is trivial. Now we do 1-dimensional framed surgery to $\{*\} \times S^1 \subset \Sigma^9 \times S^1 =: M^{10}$. This kills $\pi_1(M^{10}) \cong \mathbb{Z}$ and creates no 2-dimensional homology. The result of this surgery is a manifold N^{10} which is easily seen to be 7-connected and hence a homotopy sphere. Therefore, the Kervaire invariant of M^{10} is zero. □

References

[1] *Problems in differential and algebraic topology. Seattle Conference, 1963*, Ann. of Math. (2), 81 (1965), pp. 565–591.

[2] J. F. ADAMS, *Vector fields on spheres*, Ann. of Math. (2), 75 (1962), pp. 603–632.

[3] ——, *On the groups J(X). IV*, Topology, 5 (1966), pp. 21–71.

[4] S. AKBULUT AND J. D. MCCARTHY, *Casson's invariant for oriented homology 3-spheres*, vol. 36 of Mathematical Notes, Princeton University Press, Princeton, NJ, 1990. An exposition.

[5] C. ARF, *Untersuchungen über quadratische Formen in Körpern der Charakteristik 2. I*, J. Reine Angew. Math., 183 (1941), pp. 148–167.

[6] M. G. BARRATT, J. D. S. JONES, AND M. E. MAHOWALD, *Relations amongst Toda brackets and the Kervaire invariant in dimension* 62, J. London Math. Soc. (2), 30 (1984), pp. 533–550.

[7] R. BOTT, *The stable homotopy of the classical groups*, Ann. of Math. (2), 70 (1959), pp. 313–337.

[8] W. BROWDER, *The Kervaire invariant of framed manifolds and its generalization*, Ann. of Math. (2), 90 (1969), pp. 157–186.

[9] ——, *Surgery on simply-connected manifolds*, Springer-Verlag, New York-Heidelberg, 1972. Ergebnisse der Mathematik und ihrer Grenzgebiete, Band 65.

[10] E. H. Brown, Jr. and F. P. Peterson, *The Kervaire invariant of* $(8k+2)$*-manifolds*, Bull. Amer. Math. Soc., 71 (1965), pp. 190–193.

[11] S. Cappell, A. Ranicki, and J. Rosenberg, eds., *Surveys on surgery theory. Vol. 1*, vol. 145 of Annals of Mathematics Studies, Princeton University Press, Princeton, NJ, 2000. Papers dedicated to C. T. C. Wall.

[12] ——, eds., *Surveys on surgery theory. Vol. 2*, vol. 149 of Annals of Mathematics Studies, Princeton University Press, Princeton, NJ, 2001. Papers dedicated to C. T. C. Wall.

[13] P. E. Conner and E. E. Floyd, *Differentiable periodic maps*, Ergebnisse der Mathematik und ihrer Grenzgebiete, N. F., Band 33, Academic Press Inc., New York; Springer-Verlag, Berlin-Göttingen-Heidelberg, 1964.

[14] F. T. Farrell, L. Göttsche, and W. Lück, eds., *Topology of high-dimensional manifolds. No. 1, 2*, vol. 9 of ICTP Lecture Notes, Abdus Salam International Centre for Theoretical Physics, Trieste, 2002. Papers from the School on High-Dimensional Manifold Topology held in Trieste, May 21–June 8, 2001, Available electronically at `http://www.ictp.trieste.it/~pub off/lectures/vol9.html`.

[15] S. C. Ferry, A. Ranicki, and J. Rosenberg, eds., *Novikov conjectures, index theorems and rigidity. Vol. 2*, vol. 227 of London Mathematical Society Lecture Note Series, Cambridge University Press, Cambridge, 1995. Including papers from the conference held at the Mathematisches Forschungsinstitut Oberwolfach, Oberwolfach, September 6–10, 1993.

[16] M. H. Freedman, *The topology of four-dimensional manifolds*, J. Differential Geom., 17 (1982), pp. 357–453.

[17] M. H. Freedman and F. Quinn, *Topology of 4-manifolds*, vol. 39 of Princeton Mathematical Series, Princeton University Press, Princeton, NJ, 1990.

[18] H. Freudenthal, *Über die Klassen der Sphärenabbildungen I. Große Dimensionen*, Compositio Math., 5 (1938), pp. 299–314.

[19] M. Gilmore, *Some Whitehead products on odd spheres*, Proc. Amer. Math. Soc., 20 (1969), pp. 375–377.

[20] A. Haefliger, *Plongements différentiables de variétés dans variétés*, Comment. Math. Helv., 36 (1961), pp. 47–82.

[21] M. A. Hill, M. J. Hopkins, and D. C. Ravenel, *A solution to the Arf-Kervaire invariant problem*, in Proceedings of the Gökova Geometry-Topology Conference 2010, Int. Press, Somerville, MA, 2011, pp. 21–63.

[22] M. W. Hirsch, *Obstruction theories for smoothing manifolds and maps*, Bull. Amer. Math. Soc., 69 (1963), pp. 352–356.

[23] F. Hirzebruch, *Topological methods in algebraic geometry*, Third enlarged edition. New appendix and translation from the second German edition by R. L. E. Schwarzenberger, with an additional section by A. Borel. Die Grundlehren der Mathematischen Wissenschaften, Band 131, Springer-Verlag New York, Inc., New York, 1966.

[24] H. Hopf, *Über die Abbildungen der dreidimensionalen Sphäre auf die Kugelfläche*, Math. Ann., 104 (1931), pp. 637–665.

[25] S.-T. Hu, *Homotopy theory*, Pure and Applied Mathematics, Vol. VIII, Academic Press, New York-London, 1959.

[26] J. JONES AND E. REES, *Kervaire's invariant for framed manifolds*, in Algebraic and geometric topology (Proc. Sympos. Pure Math., Stanford Univ., Stanford, Calif., 1976), Part 1, Proc. Sympos. Pure Math., XXXII, Amer. Math. Soc., Providence, R.I., 1978, pp. 141–147.

[27] J. D. S. JONES, *The Kervaire invariant of extended power manifolds*, Topology, 17 (1978), pp. 249–266.

[28] M. A. KERVAIRE, *An interpretation of G. Whitehead's generalization of H. Hopf's invariant*, Ann. of Math. (2), 69 (1959), pp. 345–365.

[29] ——, *A manifold which does not admit any differentiable structure*, Comment. Math. Helv., 34 (1960), pp. 257–270.

[30] M. A. KERVAIRE AND J. W. MILNOR, *Groups of homotopy spheres. I*, Ann. of Math. (2), 77 (1963), pp. 504–537.

[31] R. C. KIRBY AND L. C. SIEBENMANN, *Foundational essays on topological manifolds, smoothings, and triangulations*, Princeton University Press, Princeton, N.J.; University of Tokyo Press, Tokyo, 1977. With notes by John Milnor and Michael Atiyah, Annals of Mathematics Studies, No. 88.

[32] W. B. MASSEY AND G. W. WHITEHEAD, Bull. Amer. Math. Soc., 57 (1951), pp. 491–492.

[33] J. MILNOR, *On manifolds homeomorphic to the 7-sphere*, Ann. of Math. (2), 64 (1956), pp. 399–405.

[34] ——, *Differentiable structures on spheres*, Amer. J. Math., 81 (1959), pp. 962–972.

[35] J. W. MILNOR, *Topology from the differentiable viewpoint*, Princeton Landmarks in Mathematics, Princeton University Press, Princeton, NJ, 1997. Based on notes by David W. Weaver, Revised reprint of the 1965 original.

[36] J. W. MILNOR AND M. A. KERVAIRE, *Bernoulli numbers, homotopy groups, and a theorem of Rohlin*, in Proc. Internat. Congress Math. 1958, Cambridge Univ. Press, New York, 1960, pp. 454–458.

[37] J. W. MILNOR AND J. D. STASHEFF, *Characteristic classes*, Princeton University Press, Princeton, N. J., 1974. Annals of Mathematics Studies, No. 76.

[38] S. P. NOVIKOV, *Rational Pontrjagin classes. Homeomorphism and homotopy type of closed manifolds. I*, Izv. Akad. Nauk SSSR Ser. Mat., 29 (1965), pp. 1373–1388.

[39] ——, *Topological invariance of rational classes of Pontrjagin*, Dokl. Akad. Nauk SSSR, 163 (1965), pp. 298–300.

[40] G. PERELMAN, *The entropy formula for the Ricci flow and its geometric applications.* available at `http://arxiv.org/abs/math.DG/0211159`.

[41] ——, *Finite extinction time for the solutions to the Ricci flow on certain three-manifolds.* available at `http://arxiv.org/abs/math.DG/0307245`.

[42] ——, *Ricci flow with surgery on three-manifolds.* available at `http://arxiv.org/abs/math.DG/0303109`.

[43] H. POINCARÉ, *Œuvres. Tome VI*, Les Grands Classiques Gauthier-Villars. [Gauthier-Villars Great Classics], Éditions Jacques Gabay, Sceaux, 1996. Géométrie. Analysis situs (topologie). [Geometry. Analysis situs (topology)], Reprint of the 1953 edition.

[44] L. S. PONTRYAGIN, *Smooth manifolds and their applications in homotopy theory*, in American Mathematical Society Translations, Ser. 2, Vol. 11, American Mathematical Society, Providence, R.I., 1959, pp. 1–114.

[45] A. A. RANICKI, *Algebraic L-theory and topological manifolds*, vol. 102 of Cambridge Tracts in Mathematics, Cambridge University Press, Cambridge, 1992.

[46] V. A. ROHLIN, *Classification of mappings of an $(n+3)$-dimensional sphere into an n-dimensional one*, Doklady Akad. Nauk SSSR (N.S.), 81 (1951), pp. 19–22.

[47] ——, *On a mapping of the $(n+3)$-dimensional sphere into the n-dimensional sphere*, Doklady Akad. Nauk SSSR (N.S.), 80 (1951), pp. 541–544.

[48] ——, *New results in the theory of four-dimensional manifolds*, Doklady Akad. Nauk SSSR (N.S.), 84 (1952), pp. 221–224.

[49] J.-P. SERRE, *Homologie singulière des espaces fibrés. Applications*, Ann. of Math. (2), 54 (1951), pp. 425–505.

[50] ——, *Groupes d'homotopie et classes de groupes abéliens*, Ann. of Math. (2), 58 (1953), pp. 258–294.

[51] S. SMALE, *Generalized Poincaré's conjecture in dimensions greater than four*, Ann. of Math. (2), 74 (1961), pp. 391–406.

[52] E. H. SPANIER, *Algebraic topology*, Springer-Verlag, New York-Berlin, 1981. Corrected reprint.

[53] N. STEENROD, *The topology of fibre bundles*, Princeton Landmarks in Mathematics, Princeton University Press, Princeton, NJ, 1999. Reprint of the 1957 edition, Princeton Paperbacks.

[54] R. THOM, *Quelques propriétés globales des variétés différentiables*, Comment. Math. Helv., 28 (1954), pp. 17–86.

[55] H. TODA, *Composition methods in homotopy groups of spheres*, Annals of Mathematics Studies, No. 49, Princeton University Press, Princeton, N.J., 1962.

[56] C. T. C. WALL, *Surgery on compact manifolds*, Academic Press, London-New York, 1970. London Mathematical Society Monographs, No. 1.

[57] G. W. WHITEHEAD, *On the homotopy groups of spheres and rotation groups*, Ann. of Math. (2), 43 (1942), pp. 634–640.

[58] ——, *On the Freudenthal theorems*, Ann. of Math. (2), 57 (1953), pp. 209–228.

[59] J. H. C. WHITEHEAD, *Combinatorial homotopy. I*, Bull. Amer. Math. Soc., 55 (1949), pp. 213–245.

[60] ——, *Combinatorial homotopy. II*, Bull. Amer. Math. Soc., 55 (1949), pp. 453–496.

[61] H. WHITNEY, *The self-intersections of a smooth n-manifold in $2n$-space*, Ann. of Math. (2), 45 (1944), pp. 220–246.

[62] W.-T. WU, *Classes caractéristiques et i-carrés d'une variété*, C. R. Acad. Sci. Paris, 230 (1950), pp. 508–511.

Survey Papers

Introduction to Modern Mathematics
ALM 33, pp. 95–107

Uniqueness for a Class of Embedded Weingarten Hypersurfaces in S^{n+1}*

Ben Andrews[†], Zhijie Huang[‡], and Haizhong Li[§]

Abstract

We apply the "non-collapsing" technique, previously applied by Brendle in the proof of the Lawson conjecture and by Andrews and Li in the proof of the Pinkall-Sterling conjecture, to higher dimensional hypersurfaces satisfying a linear relation between the principal curvatures, under the additional assumption that the hypersurface has two distinct principal curvatures at each point. As special cases, the result gives simple new proofs of results of Otsuki for minimal hypersurfaces, and of Li and Wei for hypersurfaces with vanishing m-th mean curvature.

2000 Mathematics Subject Classification: Primary 53C42; Secondary 53A10.

Keywords and Phrases: Embedded CMC tori, rotational embedded CMC hypersurfaces, the non-collapsing technique, embedded, Weingarten Hypersurfaces.

1 Introduction

In classical differential geometry, the study of minimal surfaces is one of the most basic subjects. The study of minimal surfaces in space forms, such as the Euclidean space $\mathbb{R}^3$ or the sphere S^3, is of particular interest. In 1970, H.B. Lawson [L2] proved that for any positive integer g, there exists at least one compact embedded minimal surface with genus g in S^3. If $g > 1$ is not a prime, such embedded minimal surface is not unique. He also conjectured in [L3] that for $g = 1$, the Clifford torus is the only compact embedded minimal torus in S^3. This conjecture was proved by Simon Brendle [B1] in 2012 (see also the survey [B2]):

Theorem 1.1 (Brendle [B1])**.** *If $F : \Sigma \to S^3$ is an embedded minimal torus, then F is congruent to Clifford torus.*

*The research of the first author was partially supported by Discovery grant DP120102462 of the Australian Research Council. The research of the third author was supported by NSFC No. 11271214.

[†]Mathematical Sciences Institute, Australia National University; and Mathematical Sciences Center, Tsinghua University. Email: Ben.Andrews@anu.edu.au

[‡]Department of mathematical sciences, Tsinghua University, 100084, Beijing, P. R. China. Email: huangzj07@mails.tsinghua.edu.cn

[§]Department of Mathematical Sciences and Mathematical Sciences Center, Tsinghua University, 100084, Beijing, P. R. China. Email: hli@math.tsinghua.edu.cn

An important ingredient in Brendle's proof was the first author's work on "non-collapsing" in mean curvature flow [A], in which a maximum principle argument was used to compare the size of balls touching the hypersurface to the mean curvature at each point. By modifying this argument to involve the maximum principal curvature, Brendle made the argument applicable to minimal surfaces, and used it to prove the rigidity statement. Soon afterwards, the first and third authors [AL] extended the argument to embedded constant mean curvature tori and proved that these must be surfaces of rotation, confirming a 1989 conjecture of U. Pinkall and I. Sterling [PS]. The results of Andrews-Li [AL] are:

Theorem 1.2 (Andrews-Li [AL]). 1. *Every embedded CMC torus Σ in S^3 is a surface of rotation: There exists a two-dimensional subspace Π of $\mathbb{R}^4$ such that Σ is invariant under the group S^1 of rotations fixing Π.*

2. *If Σ is an embedded CMC torus which is not congruent to a Clifford torus, then there exists a maximal integer $m \geqslant 2$ such that Σ has m-fold symmetry: Precisely, Σ is invariant under the group $\mathbb{Z}_m$ generated by the rotation which fixes the orthogonal plane $\Pi^\perp$ and rotates Π through angle $2\pi/m$.*
3. *For given $m \geqslant 2$, there exists at most one such CMC torus (up to congruence).*
4. *For given $m \geqslant 2$, there exists an embedded CMC torus with mean curvature H and maximal symmetry $S^1 \times \mathbb{Z}_m$ if $|H|$ lies strictly between $\cot \frac{\pi}{m}$ and $\frac{m^2-2}{2\sqrt{m^2-1}}$.*
5. *If $H \in \{0, \frac{1}{\sqrt{3}}, -\frac{1}{\sqrt{3}}\}$, then every embedded torus with mean curvature H is congruent to the Clifford torus $S^1(\frac{\sqrt{2}}{2}) \times S^1(\frac{\sqrt{2}}{2})$ and torus $S^1(\frac{1}{2}) \times S^1(\frac{\sqrt{3}}{2})$.*

The higher dimensional cases are also interesting, and in particular the study of minimal hypersurfaces in space forms has attracted considerable interest. There seems no prospect for results in high dimensions of a generality comparable to the two-dimensional case: Not only do the methods of proof break down in higher dimensions, but there are many more examples of constant mean curvature hypersurfaces known. Our results in this paper are partly motivated by those of T. Ôtsuki [Ô1, Ô2] who proved that every embedded minimal hypersurface with two distinct principal curvatures in S^{n+1} is congruent to the Clifford torus.

Theorem 1.3 (Ôtsuki [Ô1, Ô2]). *A compact embedded minimal hypersurface with two distinct principal curvatures in S^{n+1} is congruent to product*

$$S^m\left(\sqrt{\frac{m}{n}}\right) \times S^{n-m}\left(\sqrt{\frac{n-m}{n}}\right), \quad 1 \leqslant m \leqslant n-1.$$

In 2007, H. Li and G. Wei extended Ôtsuki's result to hypersurfaces Σ with $H_m = 0$, where H_m is the m-th elementary symmetric function of the principal curvatures (also called the m-th mean curvature of Σ):

Theorem 1.4 (Li-Wei [LW]). *If $1 \leqslant m \leqslant n-1$, there are no compact embedded rotational hypersurfaces of S^{n+1} with $H_m = 0$ other than*

$$S^{n-1}\left(\sqrt{\frac{n-m}{n}}\right) \times S^1\left(\sqrt{\frac{m}{n}}\right)$$

and the totally geodesic spheres.

In this paper, we prove the following, imposing as in [Ô1, Ô2, LW] an assumption on the number of distinct principal curvatures:

Theorem 1.5. *Suppose M is a compact embedded hypersurface in S^{n+1} with two distinct principal curvatures λ and μ, whose multiplicities are m and $n-m$ respectively. If for some positive number α,*

$$\lambda + \alpha\mu = 0,$$

then λ and μ are both constant and M is congruent to tori

$$S^m\left(\sqrt{\frac{1}{\alpha+1}}\right) \times S^{n-m}\left(\sqrt{\frac{\alpha}{\alpha+1}}\right), \quad 1 \leqslant m \leqslant n-1.$$

2 Hypersurfaces with two principal curvatures in S^{n+1}

Let $M^n \subset S^{n+1}$ be a compact hypersurface with second fundamental form h and two principal curvatures λ and μ, whose multiplicities are m and $n-m$ respectively. The maximum and minimum eigenvalues of a symmetric matrix are continuous, so λ and μ are continuous functions on M. Therefore the matrix $\frac{h-\mu g}{\lambda-\mu}$ is continuous since $\lambda \neq \mu$ everywhere, and has trace equal to m. Therefore m is continuous, hence constant.

In a neighbourhood of any point on M, we can choose an orthonormal frame $\{e_i : i = 1, \cdots, n\}$ for TU, and a normal vector field e_0. Let ω_A be the dual forms of e_A on S^{n+1}. The connection forms are given by

$$d\omega_A = \omega_{AB} \wedge \omega_B$$
$$d\omega_{AB} = \omega_{AC} \wedge \omega_{CB} + \Omega_{AB}$$

where $\Omega_{AB} = -\frac{1}{2}\bar{R}_{ABCD}\omega_C \wedge \omega_D$ and $\bar{R}_{ABCD}$ are the components of Riemannian curvature tensor of S^{n+1}. Let θ_A, θ_{AB} be the restriction of ω_A, ω_{AB} to M. Since $\theta_0 = 0$, take its exterior derivative and we have

$$\theta_i \wedge \theta_{i0} = 0.$$

Therefore, $\theta_{i0} = h_{ij}\theta_j$, where h_{ij} are symmetric in i, j, and they are called the components of the second fundamental form of M. We can choose $\{e_i\}$ such that h_{ij} is diagonal, i.e., $h_{ij} = \lambda_i\delta_{ij}$. Moreover, in our case, since M has only two distinct principal curvatures λ, μ with multiplicities m and $n-m$, we can assume

$$\lambda_1 = \cdots = \lambda_m = \lambda, \quad \lambda_{m+1} = \cdots = \lambda_n = \mu.$$

Lemma 2.1. *Under the assumptions above, if $m \geqslant 2$, we have*

$$e_i(\lambda) = 0, \quad \forall\, i = 1, \cdots, m.$$

Proof. Let the $h_{ij,k}$ denote the covariant derivative of h_{ij} with respect to e_k. The Codazzi equation gives us

$$h_{ij,k} = h_{ik,j}.$$

Therefore, $h_{ij,k}$ are symmetric in i, j, k. From the definition of covariant derivative, we have

$$h_{ij,k}\theta_k = dh_{ij} + h_{kj}\theta_{ki} + h_{ik}\theta_{kj} = dh_{ij} + (\lambda_i - \lambda_j)\theta_{ij}.$$

Take $j = i$, we have

$$h_{ii,k}\theta_k = dh_{ii} = d\lambda_i = e_k(\lambda_i)\theta_k,$$

that is, $h_{ii,k} = e_k(\lambda_i)$, for any i and k. If $\lambda_i = \lambda_j$ and $i \neq j$ then $h_{ij} = 0$, we have $h_{ij,k} = 0$, for any k. If $m \geqslant 2$, for any $1 \leqslant i \leqslant m$, we can choose $1 \leqslant j \neq i \leqslant m$, thus $\lambda_i = \lambda_j = \lambda$,

$$e_i(\lambda) = e_i(\lambda_j) = h_{jj,i} = h_{ij,j} = 0.$$

This completes the proof of Lemma 2.1. □

Using Lemma 2.1 and the equation $\lambda + \alpha\mu = 0$ we deduce the following:

Corollary 2.2. *If the multiplicities of λ, μ are both greater than 1, then*

$$e_i(\lambda) = e_i(\mu) = 0, \quad \forall 1 \leqslant i \leqslant n,$$

and M is congruent to the Clifford torus

$$S^m\left(\sqrt{\frac{1}{\alpha+1}}\right) \times S^{n-m}\left(\sqrt{\frac{\alpha}{\alpha+1}}\right), \quad 2 \leqslant m \leqslant n-2.$$

The last conclusion follows since M is a compact isoparametric hypersurface with two constant distinct principal curvatures λ and μ.

From now on, we concentrate on the case when one of the principal curvature, say λ, is simple. We assume M is a compact hypersurface in S^{n+1} with second fundamental form h and two principal curvatures λ and μ of multiplicity 1 and $n-1$ respectively, related by the equation $\lambda + \alpha\mu = 0$ for some $\alpha > 0$.

Lemma 2.1 implies that $e_j(\mu) = 0$ for $j = 2, \ldots, n$. By the equation we also have $e_j(\lambda) = 0$ for $j = 2, \ldots, n$.

The following lemma provides useful information on the covariant derivatives of the second fundamental form.

Lemma 2.3. *Suppose $\{e_i\}$ is a local orthonormal frame such that the components of second fundamental form h_{ij} are diagonal, i.e., $h_{ij} = \lambda_i\delta_{ij}$, and $\lambda_1 = \lambda, \lambda_i = \mu (i \geqslant 2)$. Then $h_{ij,k}$ are symmetric in i, j, k and*

$$h_{11,1} = \lambda_{,1}, \quad h_{ii,1} = \mu_{,1}, i = 2, \ldots, n.$$

and $h_{ij,k} = 0$, for i, j, k are distinct and $h_{ii,j} = 0$, for $i \geqslant 1$ and $j \neq 1$.

Proof. The covariant derivatives of h_{ij} are given by

$$h_{ij,k}\theta_k = dh_{ij} + h_{kj}\theta_{ki} + h_{ik}\theta_{kj} = dh_{ij} + (\lambda_i - \lambda_j)\theta_{ij}.$$

The Codazzi identities imply that $h_{ij,k}$ is totally symmetric. Therefore, to compute $h_{ij,k}$, we need only consider the following three types:

$$h_{ij,k} \quad h_{ii,k} \quad h_{ii,i}$$

If i, j, k are distinct, then $h_{ij,k}$ must be zero: At least two of i, j, k are greater then 1, say i, j. In this case $\lambda_i = \lambda_j = \mu$, $i \neq j$, and we have

$$h_{ij,k}\theta_k = dh_{ij} = 0,$$

that is $h_{ij,k} = 0$ for any k.

To calculate $h_{ii,k}$, we notice that $h_{ii,k}\theta_k = dh_{ii} = d\lambda_i$, and therefore

$$h_{ii,k} = e_k(\lambda_i).$$

Thus, for any $i, j = 2, \ldots, n$, we have

$$\begin{aligned} h_{11,1} &= e_1(\lambda), \quad h_{11,j} = e_j(\lambda) = 0, \\ h_{ii,1} &= e_1(\mu), \quad h_{ii,j} = e_j(\mu) = 0. \end{aligned}$$

This completes the proof. □

Let ∇ be the Levi-Civita connection of M, and $\{e_i : i = 1, \ldots, n\}$ a local orthogonal frame such that e_1 is the principal direction of λ. Then e_1 is a smooth vector field since $m = 1$ and $\lambda \neq \mu$. Define an elliptic operator $\mathcal{L}$ by

$$\mathcal{L} = a^{ij}(\nabla_{e_i}\nabla_{e_j} - \nabla_{e_i}e_j),$$

where $a = e_1 \otimes e_1 + \frac{\alpha}{n-1}\sum_{i=2}^{n} e_i \otimes e_i$. That is, a is diagonal in the basis $\{e_1, \ldots, e_n\}$, and $a^{11} = 1$, $a^{22} = \cdots = a^{nn} = \frac{\alpha}{n-1}$. Note that $\mathcal{L}$ is smoothly defined, since e_1 is.

By the Gauss equation, we have

$$R_{ijkl} = (h_{ik}h_{jl} - h_{il}h_{jk}) + \delta_{ik}\delta_{jl} - \delta_{il}\delta_{jk}.$$

Thus $R_{ipjp} = (\lambda_i\lambda_p + 1)(\delta_{ij} - \delta_{ip}\delta_{jp})$.

Proposition 2.4. *An n-dimensional ($n \geqslant 3$) connected complete hypersurface M in S^{n+1} with principal curvatures $\lambda \neq \mu$ of multiplicity 1 and $n-1$ respectively is a rotation hypersurface: There exists a two-dimensional subspace Π of $\mathbb{R}^{n+2}$ such that M is invariant under the group $O(\Pi^{\perp})$ of rotations fixing Π.*

Proof. We begin by establishing expressions for the certain covariant derivatives of the vector fields e_i. The two identities

$$0 = g(\nabla_k e_i, e_j) + g(e_i, \nabla_k e_j)$$

and

$$e_k(h(e_i,e_j)) = \nabla_k h_{ij} + h(\nabla_k e_i, e_j) + h(e_i, \nabla_k e_j)$$

yield (since h is diagonal) the expressions for $i \neq j$

$$(\lambda_i - \lambda_j)\nabla_k e_i \cdot e_j = \nabla_k h_{ij}.$$

Choosing $k, j > 1$ and $i = 1$ we deduce that

$$\nabla_k e_j \cdot e_1 = -\nabla_k e_1 \cdot e_j = -\frac{\nabla_1 h_{jk}}{\lambda - \mu}.$$

The right hand side vanishes if $j \neq k$, and equals $\frac{e_1\mu}{\lambda-\mu}$ if $j = k$.

Next we establish an identity involving the second derivatives of μ, which can be considered an analogue of a well-known identity for minimal hypersurfaces first proved by Simons [S] (also see Lawson [L1]):

Lemma 2.5. *Under the conditions of Proposition 2.4,*

$$0 = \nabla^2\mu(e_1,e_1) + \frac{e_1\mu\,(2e_1\mu - e_1\lambda)}{\lambda - \mu} + (\lambda - \mu)(1 + \lambda\mu); \tag{1}$$

and for $i > 1$,

$$0 = \nabla^2\mu(e_i,e_i) - \frac{(e_1\mu)^2}{\lambda - \mu}. \tag{2}$$

Proof. Applying the Gauss and Codazzi equations we find the following:

$$\begin{aligned}
\nabla_i\nabla_j h_{kl} &= \nabla_i\nabla_k h_{jl} \\
&= \nabla_k\nabla_i h_{jl} + R_{ikj}{}^p h_{pl} + R_{ikl}{}^p h_{jp} \\
&= \nabla_k\nabla_l h_{ij} + h_{ij}h^2_{kl} - h_{jk}h^2_{il} + h_{il}h^2_{jk} - h_{kl}h^2_{ij} \\
&\quad + g_{ij}h_{kl} - g_{jk}h_{il} + g_{il}h_{jk} - g_{kl}h_{ij},
\end{aligned}$$

where $h^2_{kl} = h_{kp}h_{lq}g^{pq}$.

Substituting $k = l = 1$ and $i = j = 2$ gives

$$\nabla_2\nabla_2 h_{11} = \nabla_1\nabla_1 h_{22} + (\lambda\mu + 1)(\lambda - \mu). \tag{3}$$

We next compute the left hand side: We have

$$\nabla_2 h_{11} = e_2\lambda = 0,$$

and so

$$\begin{aligned}
\nabla_2\nabla_2 h_{11} &= e_2\left(\nabla_2 h_{11}\right) - \nabla_{\nabla_2 e_2} h_{11} - 2\nabla_2 h(e_1, \nabla_2 e_1) \\
&= -\left(\nabla_2 e_2 \cdot e_1\right)\nabla_1\lambda - 2(\nabla_2 e_1 \cdot e_2)\nabla_1\mu \\
&= \frac{e_1\mu e_1\lambda}{\lambda - \mu} - 2\frac{(e_1\mu)^2}{\lambda - \mu}.
\end{aligned} \tag{4}$$

Finally we compute

$$\begin{aligned}\nabla_1\nabla_1 h_{22} &= e_1\left(\nabla_1 h_{22}\right) - \nabla_{\nabla_1 e_1} h_{22} - 2\nabla_1 h(e_2, \nabla_1 e_2)\\ &= e_1 e_1 \mu\\ &= \nabla^2\mu(e_1, e_1).\end{aligned} \tag{5}$$

The identity (1) follows after substituting the identities (4) and (5) into (3). The identity (2) follows more directly: Since $e_i\mu = 0$, we have

$$\nabla^2\mu(e_i, e_i) = e_i(\nabla_i\mu) - \nabla_{\nabla_i e_i}\mu = \frac{(e_1\mu)^2}{\lambda - \mu}.$$ □

Now we identify at each point of M a plane Π: Let $p = \mu x - \nu$ and $q = (e_1\mu)x - (\lambda - \mu)e_1$, and define Π to be the span of p and q. We compute

$$e_i p = e_i q = 0, \quad \text{for } i > 1;$$

and

$$e_1 p = q,$$

while

$$\begin{aligned}e_1 q = &-\lambda(\lambda - \mu)p - \frac{2e_1\mu - e_1\lambda}{\lambda - \mu} q\\ &+ \left[\nabla^2\mu(e_1, e_1) + \frac{e_1\mu(2e_1\mu - e_1\lambda)}{\lambda - \mu} + (\lambda - \mu)(1 + \lambda\mu)\right] x.\end{aligned}$$

The term in the last bracket vanishes by Lemma 2.5, so that derivatives of vectors in Π along M are in Π. Since M is connected, Π is constant.

Finally, we observe that at each point x of M the tangent space to the $(n-1)$-sphere given by the orbit of the action of $O(\Pi^\perp)$ is orthogonal to x, p and q, hence to x, ν and e_1, and so coincides with the span of $\{e_2, \ldots, e_n\}$. Therefore the action of $O(\Pi^\perp)$ is tangent to M at each point and preserves M, so M is rotationally symmetric. □

From Lemma 2.5 we can deduce the following Simons-type identity for the class of hypersurfaces we are considering:

Proposition 2.6. *Let M be a complete hypersurface in S^{n+1} with two principal curvatures λ, μ of multiplicity 1 and $n - 1$ respectively, satisfying the equation $\lambda + \alpha\mu = 0$. Then*

$$\mathcal{L}\lambda = \frac{2}{1+\alpha}\frac{|\nabla\lambda|^2}{\lambda} + \frac{1+\alpha}{\alpha}\lambda(\alpha - \lambda^2).$$

Proof. This follows from equations (1) and (2) using the relations $\lambda = -\alpha\mu$ and $e_1\lambda = -\alpha e_1\mu$, since $\mathcal{L}\mu = \nabla^2\mu(e_1, e_1) + \sum_{i=2}^{n}\frac{\alpha}{n-1}\nabla^2\mu(e_i, e_i)$. □

3 Interior ball curvature

In this section, we use the non-collapsing argument previously employed in [A, B1, AL]. Let M^n be an embedded hypersurface in $S^{n+1} \subset \mathbb{R}^{n+2}$ given by an embedding F, with two principal curvatures λ and μ of multiplicity 1 and $n-1$ respectively, satisfying the equation $\lambda+\alpha\mu=0$ for some $\alpha>0$. Noting that $\lambda\neq\mu$ and therefore $\mu\neq 0$ everywhere, we choose the direction of the unit normal in such a way that $\lambda>0>\mu$ everywhere. The hypersurface bounds a region Ω in S^{n+1}, which we choose in such a way that the unit normal vector ν of M^n is pointing out of Ω. For any point $x\in M^n$, the arguments in [AL] and [ALM] show that the existence of a ball in Ω of boundary curvature Φ which touches at x is equivalent to the inequality

$$k(x,y) := \frac{2\langle \nu(x), x-y\rangle}{|x-y|^2} \leqslant \Phi(x), \quad \text{for all } y\in M.$$

where $\nu(x)$ is the unit normal vector of M at x as a hypersurface of S^{n+1}. In particular, the largest ball in Ω which touches M at x has boundary curvature equal to $\bar{k}(x) := \sup\{k(x,y) : y\in M\setminus\{x\}\}$. If the supremum is achieved at some $\bar{y}\in M\setminus\{x\}$, then there is a ball of boundary curvature $\bar{k}(x)$ in Ω which touches at both x and $\bar{y}$. Note that this ball is simply the intersection of the ball $B(P,\bar{k}^{-1}(x))$ in $\mathbb{R}^{n+2}$ with S^{n+1}, where $P=F(x)-\bar{k}^{-1}(x)\nu(x)$ and $B(P,r)$ denotes the ball centered at P with radius r.

Proposition 3.1. *Let M be a compact embedded hypersurface in S^{n+1} which has two principal curvatures $\lambda\neq\mu$ of multiplicity 1 and $n-1$ respectively, satisfying $\lambda+\alpha\mu=0$ for some $\alpha>0$. Then on $\tilde{M}=\{x : \bar{k}(x)>\lambda(x)\}$, the function $\bar{k}$ is a viscosity subsolution of the equation*

$$\mathcal{L}\bar{k} - 2\frac{(e_1\bar{k})^2}{\bar{k}-\lambda} + \frac{1+\alpha}{\alpha}(\lambda^2-\alpha)\bar{k} \geqslant 0.$$

Proof. We explain below the meaning of viscosity solution (see also [ALM]). By Proposition 2.4, M is a rotational hypersurface, invariant under the action of the group of rotations $O(\Pi^\perp)$ fixing a two-dimensional subspace Π. Choose a three-dimensional subspace $\tilde{\Pi}$ containing Π. Then M is given by the action of $O(\Pi^\perp)$ on a curve $\gamma=M\cap\tilde{\Pi}$ in the two-dimensional sphere $S=S^{n+1}\cap\tilde{\Pi}$. The vector e_1 is tangent to γ at any $x\in\gamma$, and $\tilde{\Pi}$ is the span of x, $\nu(x)$ and $e_1(x)$ at any such point.

We fix any $\bar{x}\in\tilde{M}$ and suppose $\bar{y}$ is chosen to maximize $k(\bar{x},\bar{y})$, so that $\bar{k}(\bar{x})=k(\bar{x},\bar{y})$ (note that the assumption $\bar{x}\in\tilde{M}$ ensures that the supremum is attained at some $\bar{y}\neq\bar{x}$). Since k and λ are invariant under the isometric action of $O(\Pi^\perp)$ on M, we can rotate so that $\bar{x}\in\tilde{\Pi}$. Therefore also $\nu(\bar{x})$ and $e_1(\bar{x})$ are in $\tilde{\Pi}$. The ball $B(P,\bar{k}(\bar{x})^{-1})$ touches M at both $\bar{x}$ and $\bar{y}$. In particular we have $P=\bar{x}-\bar{k}(\bar{x})^{-1}\nu(\bar{x})=\bar{y}-\Phi(\bar{x})^{-1}\nu(\bar{y})\in\tilde{\Pi}$. We also know that $p=\mu(\bar{y})\bar{y}-\nu(\bar{y})\in\Pi\subset\tilde{\Pi}$, and $\bar{k}(\bar{x})\geqslant\lambda(\bar{y})>\mu(\bar{y})$, so it follows that $\bar{y}\in\tilde{\Pi}$ and therefore $\bar{y}\in\gamma$. The identity $P=\bar{x}-\bar{k}(\bar{x})^{-1}\nu(\bar{x})=\bar{y}-\bar{k}(\bar{x})^{-1}\nu(\bar{y})$ also gives the useful identity

$$\nu(\bar{y})=\nu(\bar{x})+\bar{k}(\bar{x})(\bar{y}-\bar{x}). \tag{6}$$

Let $\vec{l} = \frac{\bar{y}-\bar{x}}{|\bar{y}-\bar{x}|}$, $d = |\bar{y} - \bar{x}|$, and let $R_{\vec{l}}$ be the reflection in the hyperplane perpendicular to $\vec{l}$, given by $R_{\vec{l}}(v) = v - 2\langle v, \vec{l}\rangle\vec{l}$. Then by equation (6) we have

$$R_{\vec{l}}(\nu(\bar{x})) = \nu(\bar{x}) - 2\langle\nu(\bar{x}), \vec{l}\rangle\vec{l} = \nu(\bar{x}) + \bar{k}(\bar{x})d\vec{l} = \nu(\bar{y}). \tag{7}$$

We conclude that $R_{\vec{l}}$ maps $T_{\bar{x}}M$ isometrically into $T_{\bar{y}}M$. Furthermore, since $\bar{x}$ and $\bar{y}$ are both in $\tilde{\Pi}$, and $e_1(\bar{x})$ is in $\tilde{\Pi}$, we have that $R_{\vec{l}}(e_1(\bar{x})) \in \tilde{\Pi}$, and therefore $e_1(\bar{y}) := R_{\vec{l}}(e_1(\bar{x}))$ is an eigenvector of $h(\bar{y})$ with eigenvalue $\lambda(\bar{y})$. It follows also that $e_j(\bar{y}) := R_{\vec{l}}(e_j(\bar{x})) = e_j(\bar{x})$, $j \geqslant 2$, defines an orthonormal basis for the eigenspace of $h(\bar{y})$ with eigenvector $\mu(\bar{y})$.

To prove the proposition we must show the following: If $\varphi : U \to \mathbb{R}$ is a smooth function on an open neighborhood U of $\bar{x}$ in $\tilde{M}$, and $\varphi \geqslant \bar{k}$ on U with equality holding at $\bar{x}$, then φ satisfies

$$\left(\mathcal{L}\varphi - 2\frac{(e_1\varphi)^2}{\varphi - \lambda} + \frac{1+\alpha}{\alpha}(\lambda^2 - \alpha)\varphi\right)\Big|_{\bar{x}} \geqslant 0.$$

Let x_i, y_i be geodesic normal coordinates for M around $\bar{x}$ and $\bar{y}$ respectively, chosen in such a way that $\partial_i\big|_{\bar{x}} = e_i(\bar{x})$ and $\partial_i\big|_{\bar{y}} = e_i(\bar{y})$ for $i = 1, \ldots, n$.

We observe that since $\varphi(x) \geqslant \bar{k}(x) = \sup_y k(x, y)$, we have $\varphi(x) \geqslant k(x, y)$ for all $x \in U$ and all y, with equality holding at $(\bar{x}, \bar{y})$. Since both φ and k are smooth, it follows that $\frac{\partial\varphi}{\partial x^i} = \frac{\partial k}{\partial x^i}$ and $\frac{\partial k}{\partial y^i} = 0$ for each i at $(\bar{x}, \bar{y})$, while

$$D^2\varphi\big|_{\bar{x}}(v, v) \geqslant D^2k\big|_{(\bar{x},\bar{y})}((v, w), (v, w))$$

for any $v \in T_{\bar{x}}M$ and any $w \in T_{\bar{y}}M$. It follows that

$$\mathcal{L}\varphi\big|_{\bar{x}} \geqslant a^{ij}(\partial_j^x + \partial_j^y)(\partial_i^x + \partial_i^y)k\big|_{(\bar{x},\bar{y})},$$

where we denote by ∂_i^x and ∂_i^y the coordinate tangent vectors at $\bar{x}$ and $\bar{y}$ respectively. We now proceed to compute the right hand side: The first derivatives are as follows:

$$\begin{aligned}(\partial_i^x + \partial_i^y)k\big|_{(\bar{x},\bar{y})} &= \frac{2}{d^2}((\partial_i^x - \partial_i^y)\cdot\nu_x + (x - y)\cdot(h^x)_i^p\partial_p^x - k(x - y)\cdot(\partial_i^x - \partial_i^y))\\ &= \frac{2}{d^2}((\partial_i^x - \partial_i^y)\cdot(\nu_x + kd\vec{l}) - d(h^x)_i^p\partial_p^x\cdot\vec{l}).\end{aligned}$$

Here the y derivatives vanish, while the x derivatives yield $e_1\varphi = e_1k = \frac{2(k-\lambda)\partial_1^x\cdot\vec{l}}{d}$ and $e_ik = 0$ for $i > 1$. We differentiate again, multiply by a^{ij} and take the sum, yielding

$$\begin{aligned}a^{ij}(\partial_j^x + \partial_j^y)(\partial_i^x + \partial_i^y)k\big|_{(\bar{x},\bar{y})} &= \frac{2}{d^2}a^{ij}(-\nabla^p h_{ij}^x d\vec{l}\cdot\partial_p^x - 2(h^x)_i^p\partial_p^x\cdot(\partial_j^y - \partial_j^x)\\ &\quad + (h^x)_i^p d\vec{l}\cdot(h_{pj}^x\nu_x + g_{pj}x)\\ &\quad + (-h_{ij}^x\nu_x - g_{ij}x + h_{ij}^y\nu_y + g_{ij}y)\cdot(\nu_x + kd\vec{l})\\ &\quad - k(\partial_i^x - \partial_i^y)\cdot(\partial_j^x - \partial_j^y) + 2\partial_jk(\partial_i^x - \partial_i^y)\cdot d\vec{l}).\end{aligned}$$

Noting $a^{1i} = \delta_{1i}, a^{pi} = \frac{\alpha}{n-1}\delta_{pi}, p \geqslant 2$, from

$$a^{ij}_{,k}\theta_k = da^{ij} + a^{kj}\theta_{ki} + a^{ik}\theta_{kj},$$

we easily get $a^{ii}_{,k} = 0, 1 \leqslant i, j, k \leqslant n$. Thus we have that $a^{ij}\nabla^p h_{ij} = \nabla^p(a^{ij}h_{ij}) = 0$ since ∇a^{ij} is off-diagonal in i and j while h_{ij} is diagonal, and $a^{ij}h_{ij} = 0$. We also note that $\nu_x + kd\vec{l} = \nu_y$, and that $\nu_x \cdot \vec{l} = -\frac{kd}{2}$, $\vec{l} \cdot x = -\frac{d}{2}$, $a^{ij}h^x_{ik}g^{kl}h^x_{lj} = \lambda^2 + \alpha\mu^2 = \frac{1+\alpha}{\alpha}\lambda^2$, $a^{ij}g_{ij} = 1 + \alpha$, and $a^{ij}h^x_{ij} = a^{ij}h^y_{ij} = 0$. The last identity uses the fact that both h^x and h^y are diagonal, $h^x_{11} = \lambda(x)$, $h^y_{11} = \lambda(y)$, and $h^x_{ii} = \mu(x)$ and $h^y_{ii} = \mu(y)$ for $i > 1$. Finally we observe that since $\partial^y_i = R_{\vec{l}}(\partial^x_i)$, we have for each i

$$\partial^y_i - \partial^x_i = -2(\vec{l} \cdot \partial^x_i)\vec{l}$$

Thus $\partial^y_i = \partial^x_i$ for $i = 2, \dots, n$, while $\partial^y_1 - \partial^x_1 = -\frac{e_1 k}{k-\lambda}d\vec{l}$. From this we find at $(\bar{x}, \bar{y})$

$$\begin{aligned} a^{ij}(\partial^x_j + \partial^y_j)(\partial^x_i + \partial^y_i)k &= -\frac{1+\alpha}{\alpha}k\lambda^2 + (1+\alpha)k + 2\frac{(e_1 k)^2}{k-\lambda} \\ &= \frac{1+\alpha}{\alpha}(\alpha - \lambda^2)k + 2\frac{(e_1 k)^2}{k-\lambda}. \end{aligned}$$

This completes the proof of Proposition 3.1. □

4 Proof of Theorem 1.5

We can now prove Theorem 1.5, which follows directly from Corollary 2.2 and the following theorem.

Theorem 4.1. *Let $F : M^n \to S^{n+1} (n \geqslant 3)$ be a compact embedded hypersurface with two distinct principal curvatures λ, μ, whose multiplicities are 1 and $n-1$ respectively. If for some positive number α,*

$$\lambda + \alpha\mu = 0,$$

then λ, μ are all constants, and M is congruent to torus

$$S^1\left(\sqrt{\frac{1}{\alpha+1}}\right) \times S^{n-1}\left(\sqrt{\frac{\alpha}{\alpha+1}}\right).$$

Proof. By combining Proposition 2.6 and Proposition 3.1 we arrive at the following:

Lemma 4.2. *The function $f = \bar{k}/\lambda$ is a viscosity solution of the equation*

$$\mathcal{L}f - \frac{2}{(f-1)}(e_1 f)^2 - 2\frac{(f+1)}{\lambda(f-1)}e_1\lambda\, e_1 f - \frac{2f(\alpha f + 1)}{\lambda^2(1+\alpha)(f-1)}(e_1\lambda)^2 \geqslant 0$$

on the set $U = \{f > 1\}$ in M.

Proof. If ψ is a smooth function which satisfies $\psi \geqslant f$ with equality at some point $\bar{x}$, then $\varphi = \lambda\psi$ is a smooth function satisfying $\varphi \geqslant \bar{k}$ with equality at $\bar{x}$. Proposition 3.1 therefore implies that

$$\mathcal{L}\varphi - 2\frac{(e_1\varphi)^2}{\varphi - \lambda} + \frac{1+\alpha}{\alpha}(\lambda^2 - \alpha)\varphi \geqslant 0$$

at the point $\bar{x}$. Expanding this in terms of derivatives of ψ and using the result of Proposition 2.6 we arrive at

$$\mathcal{L}\psi - \frac{2}{(\psi - 1)}(e_1\psi)^2 - 2\frac{(\psi+1)}{\lambda(\psi-1)}e_1\lambda\, e_1\psi - \frac{2\psi(\alpha\psi+1)}{\lambda^2(1+\alpha)(\psi-1)}(e_1\lambda)^2 \geqslant 0$$

at the point $\bar{x}$, which proves Lemma 4.2. □

Corollary 4.3. *$\bar{k} = \lambda$ everywhere on M.*

Proof. The inequality $\bar{k} \geqslant \lambda$ holds everywhere (compare [ALM]). If there exists a point with $\bar{k} > \lambda$ then the set U is a non-empty open set on which $f > 1$, and on the boundary of U (if non-empty) we have $f = 1$. In particular f attains an interior maximum on U. Lemma 4.2 and the strong maximum principle for viscosity solutions of uniformly elliptic equations (see for example [CC]) implies that f is a constant larger than 1. But at the point of largest curvature on M we have $\bar{k} = \lambda$, hence $f = 1$. This is a contradiction, so $\bar{k} = \lambda$ everywhere on M. □

We now complete the proof, by showing that $e_1\lambda$ is identically zero, using essentially the argument from [B1]: For any $x \in \gamma$ we have $\bar{k} = \lambda$, and hence the inequality $k(x, \gamma(t)) \leqslant \lambda(x)$ holds for all s and t (with respect to a local arc-length parametrization of γ, which we choose to have $\gamma(0) = x$). This may be written as

$$Z(t) = \lambda(x)|x - \gamma(t)|^2 - 2\nu(x)\cdot(x - \gamma(t)) \geqslant 0.$$

We have $Z(0) = 0$, and differentiation gives

$$Z'(t) = 2\left(\nu(x) - \lambda(x)(x - \gamma(t)\right)\cdot\gamma'(t),$$

so that $Z'(0) = 0$. Differentiating again gives

$$Z''(t) = -2\left(\nu(x) - \lambda(x)(x - \gamma(t)\right)\cdot\left(\lambda_{\gamma(t)}\nu_{\gamma(t)} + \gamma(t)\right) + 2\lambda(x)|\gamma'(t)|^2,$$

so that $Z''(0) = -2\lambda(x) + 2\lambda(x) = 0$. It follows that $Z'''(0) = 0$ since otherwise positivity of Z if violated. This gives

$$0 = Z'''(0) = -2e_1\lambda(x).$$

Since $x \in \gamma$ is arbitrary and M is rotationally symmetric, λ is constant. By the equation μ is also constant, and M is an isoparametric hypersurface. □

Remark 4.4. *If $m = 1$, choose $\alpha = n - 1$, then our Theorem 1.5 implies that any compact embedded rotational minimal hypersurface is either totally geodesic, i.e. S^n, or the Clifford torus*

$$S^1\left(\sqrt{\frac{1}{n}}\right) \times S^{n-1}\left(\sqrt{\frac{n-1}{n}}\right),$$

which was proved by T. Ôtsuki [Ô1],[Ô2].

Remark 4.5. *In* [LW], *H. Li and G. Wei extended Ôtsuki's theorem to the case when M is rotational and $H_m = 0$, where H_m is the normalized m-th symmetric function of the principal curvatures. In fact, in this case, they get a relation between λ and μ:*

$$\mu^{m-1}\big(m\lambda + (n-m)\mu\big) = 0.$$

Our Theorem 1.5 here gives a simple proof of their result.

References

[A] Ben Andrews, *Noncollapsing in mean-convex mean curvature flow*, Geom. Topol. **16** (2012), no. 3, 1413–1418.

[AL] Ben Andrews and Haizhong Li, *Embedded constant mean curvature tori in the three-sphere*, to appear in J. Diff. Geom., available at `arXiv:1204.5007v3`.

[ALM] Ben Andrews, Mat Langford, and James McCoy, *Non-collapsing in fully non-linear curvature flows*, Ann. Inst. H. Poincaré Anal. Non Linéaire **30** (2013), no. 1, 23–32.

[B1] Simon Brendle, *Embedded minimal tori in S^3 and the Lawson conjecture*, Acta Math. **211** (2013), 177–190.

[B2] ______, *Minimal surfaces in S^3: a survey of recent results*, Bull. Math. Sci. **3** (2013), 133-171.

[CC] Luis A. Caffarelli and Xavier Cabré, *Fully nonlinear elliptic equations*, American Mathematical Society Colloquium Publications, vol. 43, American Mathematical Society, Providence, RI, 1995.

[L1] H. Blaine Lawson Jr., *Local rigidity theorems for minimal hypersurfaces*, Ann. of Math. (2) **89** (1969), 187–197.

[L2] ______, *Complete minimal surfaces in S^3*, Ann. of Math. (2) **92** (1970), 335–374.

[L3] ______, *The unknottedness of minimal embeddings*, Invent. Math. **11** (1970), 183–187.

[LW] Haizhong Li and Guoxin Wei, *Compact embedded rotation hypersurfaces of S^{n+1}*, Bull. Braz. Math. Soc. (N.S.) **38** (2007), no. 1, 81–99.

[Ô1] Tominosuke Ôtsuki, *Minimal hypersurfaces in a Riemannian manifold of constant curvature.*, Amer. J. Math. **92** (1970), 145–173.

[Ô2] ______, *On integral inequalities related with a certain nonlinear differential equation*, Proc. Japan Acad. **48** (1972), 9–12.

[PS] U. Pinkall and I. Sterling, *On the classification of constant mean curvature tori*, Ann. of Math. (2) **130** (1989), no. 2, 407–451.

[S] James Simons, *Minimal varieties in riemannian manifolds*, Ann. of Math. (2) **88** (1968), 62–105.

Introduction to
Modern Mathematics
ALM 33, pp. 109–128

Geometric Group Theory: An Introduction*

Jason Behrstock†

Abstract

These lectures are designed to provide a brief introduction to geometric group theory for advanced undergraduates and beginning graduate students. The first lecture provides some foundational definitions and an introduction to thinking about groups geometrically. Each of the remaining lectures are dedicated to a classical theorem in the field. The choice of topics is designed to illustrate the power of the geometric viewpoint in proving algebraic properties of groups.

AMS Classification Numbers: 20F65, 57M07 , 20F67, 20F69.

Keywords and Phrases: Geometric group theory.

Overview

A central aspect of geometric group theory is that finitely generated groups can be treated as geometric objects and that this view point allows one to prove results that are otherwise difficult or inaccessible through purely algebraic means.

The topics for the five lectures will be:

(1) Groups as geometric objects
(2) Curves on surfaces and the word problem
(3) Geometry of finite presentability
(4) Groups and their ends
(5) How many groups are there?

Students should be familiar (or familiarize themselves) with basic notions in group theory and topology such as: free groups, presenting a group by generators and relators, normal subgroups, fundamental groups, and covering spaces. For instance, for group theory see: "An introduction to the theory of groups", by Rotman. For topology see: "Algebraic Topology", by Hatcher (Chapters 0 and 1), available at: `http://www.math.cornell.edu/hatcher/AT/ATpage.html`

---

*Lecture notes from 2013 Summer School on Modern Mathematics, Beijing, China.

†Lehman College and The Graduate Center, City University of New York, U.S.A. Email: jason.behrstock@lehman.cuny.edu

For further introductions to various topics studied in geometric group theory, one can browse the following surveys, each of which has some easily accessible topics and several more advanced ones:

B. Bowditch, A course on geometric group theory. Download available: `http://www.warwick.ac.uk/masgak/papers/bhb-ggtcourse.pdf`

C. Drutu and M. Kapovich, Lectures on Geometric Group Theory. Download available: `http://www.math.ucdavis.edu/kapovich/EPR/ggt.pdf`

J. Meier, Groups, Graphs and Trees. Cambridge University Press.

H. Short (edit.): MSRI Notes on Hyperbolic groups. Download available: `http://www.cmi.univ-mrs.fr/hamish/Papers/MSRInotes2004.pdf`

M. Gromov, Asymptotic invariants of infinite groups. LMS lecture note series v. 182 (1993).

1 Groups as geometric objects

Let G be a group with a finite generating set S. We will always assume that the identity element, e, is not in S and that the generating set is *symmetric*, i.e., if $x \in S$ then $x^{-1} \in S$.

There are many interesting examples of finitely generated groups, below are a few basic examples to keep in mind.

- $\mathbb{Z}$ under addition. This is isomorphic to $\langle a \rangle$ under multiplication, where each element is a^n for some $n \in \mathbb{Z}$.
- $\mathbb{Z}/5\mathbb{Z}$ under addition. This is isomorphic to $\langle a; a^5 = 1 \rangle$ under multiplication, so that each element is a^n for some $n \in \{0, 1, 2, 3, 4\}$.
- $\mathbb{Z} \times \mathbb{Z} = \mathbb{Z}^2 \cong \{(n, m); n, m \in \mathbb{Z}\}$. The *free abelian group on two generators.* Equivalently, $\langle a, b; aba^{-1}b^{-1} = 1 \rangle$ so that each element is $a^n b^m$ for some $n, m \in \mathbb{Z}$.
- $\mathbb{F}_2 \cong \langle a, b \rangle$. The *free group on two generators.* Any word is of the form $a^{n_1} b^{n_2} a^{n_3} \ldots b^{n_k}$ for $n_1, \ldots, n_k \in \mathbb{Z}$ and $k \in \mathbb{N}$. The multiplication here is given by concatenating words.
- $\pi_1(S_g)$. The *fundamental group of* S_g, where S_g denotes the closed orientable surface of genus g. Example: $\pi_1(S_2) = \langle a, b, c, d; aba^{-1}b^{-1}cdc^{-1}d^{-1} \rangle$.
- $\mathcal{MCG}(S_g) \cong \mathrm{Out}(\pi_1(S_g)) \cong \mathrm{Homeo}(S_g)/\mathrm{Homeo}_0(S_g)$. The *mapping class group of* S_g.

For a finitely generated group, G, we define the *Cayley graph of G with respect to the generating set S* as the graph whose vertex and edge sets consist of the following.

- Vertices $\Longleftrightarrow$ one for each element of G
- Edges $\Longleftrightarrow$ one connecting each pair $a, b \in G$ for which there exists $c \in S$ satisfying $a = b \cdot c$

The Cayley graph of $\mathbb{F}_2$ with its canonical generating set is the regular 4–valent tree. In the case of $\mathbb{Z}^2$, there is a natural set of generators consisting of the elements $(0, 1)$ and $(1, 0)$, and their inverses; with respect to these generators, we obtain a Cayley graph whose underlying set is the collection of horizontal and vertical lines in $\mathbb{R}^2$ through the integer x and y axes; i.e., this Cayley graph looks like graph paper. Further examples appear in Figures 1 and 2.

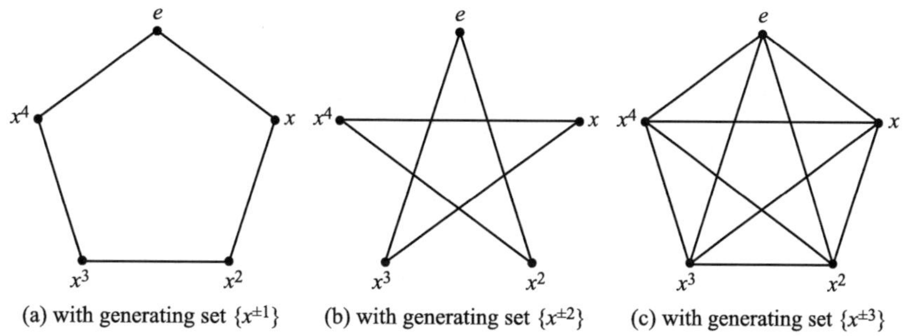

Figure 1. Cayley graphs of $\mathbb{Z}/5\mathbb{Z} = \langle x; x^5 \rangle$ with different generating sets. Note that all line segments should be considered to have equal length.

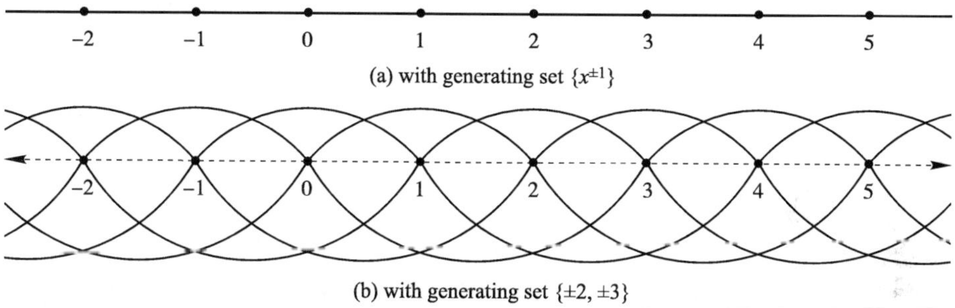

Figure 2. Cayley graphs of $\mathbb{Z}$ with different generating sets. Note that all arcs between vertices are of equal length.

Associated to any finitely generated group, we may associate the *word metric on G with respect to the finite generating set S* as follows: For each $x, y \in G$ define:

$$d_{G,S}(x, y) = \left\|x^{-1}y\right\|$$

where the norm on the right side is defined to be the smallest number of letters (with multiplicity) from the set S needed to represent the group element $x^{-1}y$.

Remark 1.1. It is easily checked that this metric is *left-invariant*, that is, multiplication on the left is an isometry of the metric. More precisely, for each $a, b, g \in G$ we have $d_{G,S}(ga, gb) = d_{G,S}(a, b)$

Exercise 1.2. *Prove the following are equivalent:*

(1) $d_{G,S}(x, y) = n$
(2) $x = y \cdot w$ *where w is a word of length n*
(3) *Consider the Cayley graph of G with respect to S and metrize this graph by making each edge isometric to $[0, 1]$. In this graph the shortest length of a path connecting x to y is n.*

As one can see by generalizing the above examples, any given group may have many distinct Cayley graphs, depending on the choice of finite generating set. In Figure 1 we see different Cayley graphs for the group $\mathbb{Z}/5\mathbb{Z}$. In Figure 2, we see Cayley graphs of $\mathbb{Z}$ with the generating sets $\{\pm 1\}$ and $\{\pm 2, \pm 3\}$. As an exercise, you can draw Cayley graphs with respect to the generating sets $\{\pm 1, \pm 3\}$, $\{\pm 1, \pm 2, \pm 3\}$, or ... Indeed, the attentive reader will notice that we can even build Cayley graphs with infinite generating set, but those are beasts of a different nature and will not be discussed further in these notes as their geometry is quite a bit different than those arising from finite generating sets (we encourage the reader to justify this statement).

Considering these examples raises the question of whether there is any relation between different word metrics on the same group. Contrary to what might be your expectation from these examples, it turns out that the answer is "Yes!" when given an appropriate formulation. The formulation which we consider uses the following definition which makes precise the notion that two Cayley graphs for the same group "look roughly the same from far away." Although our main interest will be in Cayley graphs, we formulate the following definition in the more general context of arbitrary metric spaces.

Definition 1.3. Let (X, d_X) and (Y, d_Y) be metric spaces. A map $\phi\colon X \to Y$ is called a *quasi-isometric embedding* if there exist constants $K \geqslant 1$ and $C \geqslant 0$ such that for all $a, b \in X$:

$$\frac{1}{K} d_X(a,b) - C \leqslant d_Y(\phi(a), \phi(b)) \leqslant K d_X(a,b) + C.$$

This map is called a *quasi-isometry* if, additionally, every point of Y lies in the C–neighborhood of the image of ϕ.

Definition 1.4. Let $f\colon X \to Y$ be a quasi-isometry. We say a quasi-isometry $g\colon Y \to X$ is a *quasi-inverse* to f if the maps satisfy the following:

$$d_X(g \circ f, 1_X) < \infty \text{ and } d_Y(f \circ g, 1_Y) < \infty$$

where 1_X denotes the identity map on X and distance is given by the supremum over all points in the relevant space.

The following exercise is a good way to get used to working with quasi-isometries, even though it will not be needed explicitly in these lectures.

Exercise 1.5. *Given a quasi-isometry between metric spaces, $f\colon X \to Y$, prove that f has a quasi-inverse.*

We will now prove that for any finitely generated group the word metric is unique up to quasi-isometry, i.e., any two word metrics on a fixed finitely generated group are quasi-isometric.

Notation 1.6. *When the group is clear we simplify the notation $d_{G,S}$ to d_S or even d when this will not cause ambiguity. Also, for $w \in G$ we use the notation $\|w\|_S = d_S(e, w)$.*

Proposition 1.7. *If S, T are two finite generating sets for a group, G, then (G, d_S) and (G, d_T) are quasi-isometric.*

Proof. We will show the identity map from d_S to d_T is a quasi-isometry.

Since T is a generating set, any element of G, and, in particular, any element of S can be written as a product of elements of T. Accordingly we can set $K_1 = \max_{s \in S} d_T(e, s)$ and this will be a positive integer. Now, since S generates, any $w \in G$ can be factored as a product of elements of S and the shortest way to do so is a word in elements of S of length $\|w\|_S$. Given this factorization as elements of S, we can write each of the elements of S as a word in the elements of T with length at most K_1. We have thus shown that for any $w \in G$ we have $\|w\|_T \leqslant K_1 \|w\|_S$.

Setting $K_2 = \max_{t \in T} d_S(e, t)$, the argument above with the roles of S and T reversed shows that for any $w \in G$ we have $\|w\|_S \leqslant K_2 \|w\|_T$.

Taking $K = \max\{K_1, K_2\}$ and $C = 0$, we then have:

$$\frac{1}{K} d_S(a, b) - C \leqslant d_T(a, b) \leqslant K d_S(a, b) + C$$

for all $a, b \in S$. □

Example 1.8. *Any two finite groups are quasi-isometric.*

Milnor–Švarc Lemma and its applications. After recording a few preliminary definitions we will state a result which is sometimes called the *Fundamental Observation of Geometric Group Theory.* The proof of this result will be postponed until the third lecture.

Definition 1.9. A *geodesic metric space* is a metric space X with the property that for any two points $x, y \in X$ there exists an isometric embedding, $g\colon [0, d(x, y)] \to X$, of an interval into X which satisfies $g(0) = x$ and $g(d(x, y)) = y$.

We say that a metric space X is *proper* if closed balls in X are compact. Note that any proper metric space is locally compact and complete.

A discrete group, G, acting by homeomorphisms on a locally compact topological space is said to act *properly* if for any compact subset $K \subseteq X$, the set

$$\{g \in G\colon gK \cap K \neq \varnothing\}$$

is finite. If the quotient of a space by a group action is compact, we say the group action is *cocompact*

Exercise 1.10. *Prove that a Cayley graph of a finitely generated group is a proper geodesic metric space. Further, we noted previously that a group acts on itself isometrically by left multiplication; check that in doing so G acts on its Cayley graph properly and cocompactly.*

The following is a quintessential geometric group theory result: starting with no hypothesis on a group except that it acts in a reasonable way on a space with some mild geometric/topological hypotheses, we then use geometry to deduce an algebraic result, which in this case is that the group in question is finitely generated.

Lemma 1.11 (Milnor–Švarc). *Let G be a group acting, on the left, properly and isometrically on a proper geodesic metric space, X. If $G\backslash X$ is compact, then G is finitely generated and is quasi-isometric to X.*

We say two groups are *commensurable* if they "differ by a finite subgroup"; more precisely this means that one can be obtained by the other by passage to a finite index subgroup and/or by quotienting by a finite normal subgroup. Below are two immediate corollaries of the Milnor–Švarc Lemma which show that *commensurable* groups are quasi-isometric.

Corollary 1.12. *Let G be a finitely generated group and G' a subgroup of finite index. Then G' is also finitely generated and is quasi-isometric to G.*

Corollary 1.13. *Let G be a finitely generated group and N a finite normal subgroup of G, then G is quasi-isometric to $N\backslash G$.*

The next corollary is used so frequently, that it is itself sometimes referred to as the Milnor–Švarc Lemma.

Corollary 1.14. *Let M be a compact Riemannian manifold, then $\pi_1(M)$ is quasi-isometric to $\widetilde{M}$, the universal covering space of M.*

The next two examples are immediate consequences of the above corollary; we single them out for attention because they are examples which we will consider again in later lectures.

Example 1.15. *Let T^2 be the torus. Its fundamental group $\pi_1(T^2) \cong \mathbb{Z}^2$ is quasi-isometric to $\mathbb{R}^2$.*

Example 1.16. *Let S_g be the closed orientable surface of genus $g \geqslant 2$, then $\pi_1(S_g)$ is quasi-isometric to the hyperbolic plane, $\mathbb{H}^2$.*

To verify the second example, choose a Riemannian metric of constant curvature -1 on S_g and consider the associated action of $\pi_1(S_g)$ on $\mathbb{H}^2$.

Remark 1.17. Generalizing this example, we note that in each dimension, all compact manifolds of constant curvature -1 have quasi-isometric fundamental groups, since they are each quasi-isometric to $\mathbb{H}^n$. Using the (highly non-trivial) result that there exists hyperbolic 3–dimensional manifolds whose volumes are not rationally related, from such examples one can obtain many quasi-isometric groups which are not commensurable.

Exercise 1.18. *For each $m, n \geqslant 2$, the groups $\mathbb{F}_m$ and $\mathbb{F}_n$ are commensurable. In particular, for all $m, n \geqslant 2$ the groups $\mathbb{F}_m$ and $\mathbb{F}_n$ are quasi-isometric.*

Hint. This follows from the algebraic fact that for all $k \geqslant 2$ the group $\mathbb{F}_k$ is a subgroup of $\mathbb{F}_2$. We sketch a topological proof of this algebraic fact. Consider $\mathbb{F}_2 = \pi_1(S^1 \vee S^1)$, where $S^1 \vee S^1$ denotes the wedge of two circles. Each $(k-1)$-sheeted cover of $S^1 \vee S^1$ is a graph with $(k-1)$ vertices and $2(k-1)$ edges. Thus the Euler characteristic of any such cover is $1-k$ and hence its fundamental group is $\mathbb{F}_k$. It follows that $\mathbb{F}_k$ is an index $k-1$ subgroup of $\mathbb{F}_2$ and hence that $\mathbb{F}_k$ is quasi-isometric to $\mathbb{F}_2$. □

Exercise 1.19. *Any two infinite regular trees whose valence at each vertex is at least 3 are quasi-isometric. The case where the valence is a multiple of 4 is a consequence of the previous exercise about F_n. Find a geometric argument that treats all cases and thus gives an alternate proof of the above exercise as well.*

2 Curves on surfaces and the word problem

Problem 2.1 (Contractibility problem)**.** *Let S_g denote the closed orientable surface of genus g and let γ denote a closed curve on S_g. Can γ be contracted to a point?*

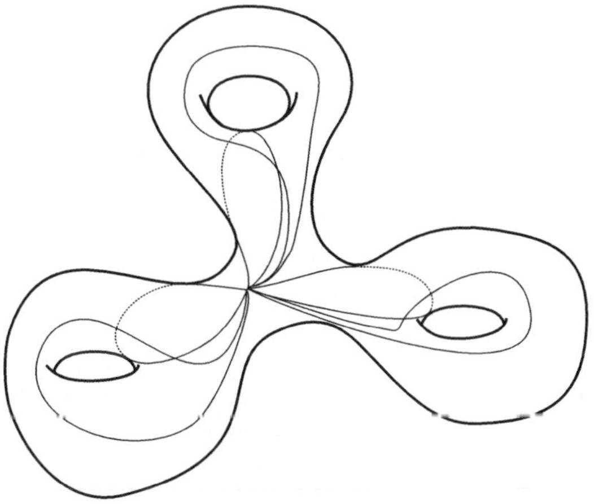

Figure 3. A curve, γ, on the closed orientable genus three surface, S_3.

We start with an observation used by Dehn in the early 1900's and perhaps known earlier. Given a surface S, we write $\widetilde{S}$ to denote the universal cover of S and $\widetilde{\gamma}$ to refer to a lift of γ to $\widetilde{S}$.

Observation. *γ is a contractible closed curve in S_g $\Leftrightarrow$ $\widetilde{\gamma}$ is a closed curve in $\widetilde{S}_g$.*

Proof of observation. ($\Rightarrow$) Since γ can be contracted to a point, we can do so fixing a basepoint. Such a contraction then lifts to a contraction of $\widetilde{\gamma}$ fixing both endpoints. Hence these two endpoints of $\widetilde{\gamma}$ must coincide and thus we have that $\widetilde{\gamma}$ is a closed curve in $\widetilde{S}_g$.

($\Leftarrow$) Since the universal cover of a closed surface is homeomorphic to either $\mathbb{S}^2$ or $\mathbb{R}^2$, we know that if $\widetilde{\gamma}$ is a closed curve in $\widetilde{S}_g$ then $\widetilde{\gamma}$ is contractible in $\widetilde{S}_g$. Viewing $\widetilde{S}_g$ as being tessellated by polygons given by translations of a fundamental domain, we see that by contracting $\widetilde{\gamma}$ through one polygon in $\widetilde{S}_g$ at a time, that this contraction in the cover projects to a contraction of γ in S_g. □

Problem 2.2 (Alternative formulation of Contractibility Problem)**.** *In $\pi_1(S_g)$, is $[\gamma] = 1$? Here $[\gamma]$ denotes the homotopy class of the curve γ.*

Recall that $\pi_1(S_g) = \langle a_1, \ldots, a_g, b_1, \ldots, b_g; a_1b_1a_1^{-1}b_1^{-1}, \ldots, a_gb_ga_g^{-1}b_g^{-1}\rangle$. Formally, we can think of a group as the quotient of the free group (*generators*) by the normal subgroup generated by the *relators*. When the set of generators is finite we say the group is *finitely generated*, when the set of relators is also finite, then we say the group is *finitely presented*; by an abuse of grammar, it is common parlance to also use these phrases in the case when there exists such a presentation, even if a particular presentation is not indicated. Given any *word*[1], w, in the generators (and their inverses), we can consider the associated element in the group, which we denote $[w]$. Note that there are many words which all have the same associated group element. A word may contain an "obvious shortenings" if it admits a subword of the form $a^{-1}a$ or aa^{-1} where a is a generator, since the word obtained by removing this subword still represents the same group element. If there are no obvious shortenings we say the word is *reduced.* A word is *cyclically reduced* if it and all of its cyclic conjugates are reduced.

Problem 2.3 (Word Problem; Dehn, 1912). *In a finitely presented group, give a method whereby it may be decided in a finite number of steps whether or not any particular given word in the generators represents the identity element.*

Dehn noticed that the labelled net of polygons which form $\widetilde{S}$ can also be seen as a Cayley graph for $\pi_1(S)$. Thus the contractibility problem for S is equivalent to the word problem for $\pi_1(S)$. This led to Dehn's 1912 proof of this problem using the geometry of $\mathbb{H}^2$; later that same year he gave a more topological/algebraic proof. We will give a version of the latter argument after first mentioning what happens in the simplest cases (which, it so happens, are also the only closed surfaces which do not admit hyperbolic structures).

Example 2.4. *The two-dimensional sphere, $\mathbb{S}^2$, which is the closed oriented surface of genus 0, satisfies: $\pi_1(\mathbb{S}^2) = \{1\}$. Thus every word in this group is already equal to 1 and the word problem admits a trivial solution.*

The two-dimensional torus $\mathbb{T}^2$, which is the closed oriented surface of genus 1, satisfies: $\pi_1(\mathbb{T}^2) = \langle a, b; aba^{-1}b^{-1}\rangle \cong \mathbb{Z} \times \mathbb{Z}$, so for any word we can take as a "normal form" the unique representative of the form a^mb^n. One can arrive at this normal form by switching the order of adjacent letters $a^{\pm 1}$ and $b^{\pm 1}$. The solvability of the word problem then follows from this and the fact that $a^mb^n = 1 \Leftrightarrow m{=}n{=}0$.

It remains to discuss the case of closed orientable surfaces with $g \geqslant 2$.

Remark 2.5. Note that there exists a Cayley graph for the fundamental group of a closed orientable surfaces with $g \geqslant 2$ which consists of a planar net of $4g$-gons with $4g$ at each vertex. We will work with this particular Cayley graph, which will be described in more detail below.

Lemma 2.6. *If γ is a closed path in the Cayley graph of $\pi_1(S_g)$ with $g \geqslant 2$, then γ has either a "spike" (see Figure 4) or a subpath consisting of more than half the edges in the boundary of a polygon, in succession.*

[1] A *word* is a finite sequence of elements from a given alphabet.

Proof. We assume γ does not contain any spikes; note that this is equivalent to assuming that γ represents a cyclically reduced word. It thus suffices to show that there is a subpath consisting of more than half the edges in the boundary of a polygon.

Figure 4. A "spike" is a subpath that consists of an edge followed by the same edge in the opposite direction.

We now describe a planar embedding of a Cayley graph for $\pi_1(S_g)$ (see Figure 5). Fix a collection $\mathcal{C}_1, \mathcal{C}_2, \ldots, \mathcal{C}_n, \ldots$ of nested circles with increasing radii. Each circle will be subdivided into circumferential edges, and there will be edges from $\mathcal{C}_i$ to $\mathcal{C}_{i+1}$, called *radial edges.*

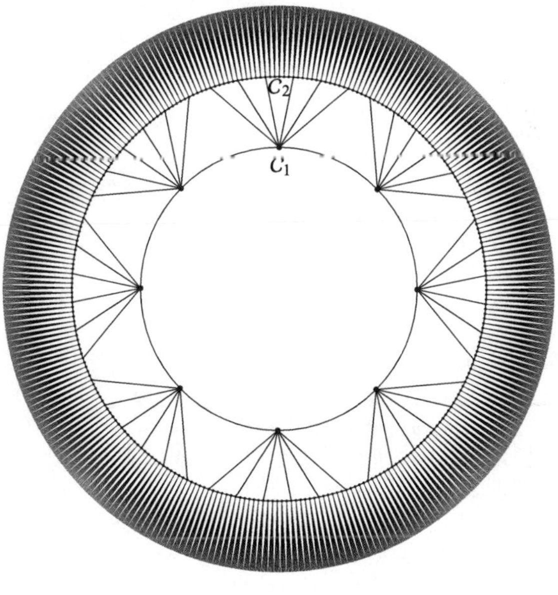

Figure 5. A portion of the Cayley graph of $\pi_1(S_2)$.

- Divide $\mathcal{C}_1$ into $4g$ arcs to form the first $4g$-gon.
- Each vertex on $\mathcal{C}_1$ will emit $4g-2$ radial edges to $\mathcal{C}_2$ thereby dividing $\mathcal{C}_2$ into $4g(4g-2)$ arcs.
- Subdivide each arc in $\mathcal{C}_2$ so that every region between $\mathcal{C}_1$ and $\mathcal{C}_2$ is a $4g$-gon (i.e., insert either $4g-3$ or $4g-4$ vertices depending on whether the polygon contains an arc of $\mathcal{C}_1$ or not).

- Repeat the second and third step to construct radial edges from $\mathcal{C}_2$ to $\mathcal{C}_3$, from $\mathcal{C}_3$ to $\mathcal{C}_4$, etc.

By Remark 2.5, this is a (topological embedding of a) Cayley graph for $\pi_1(S_g)$ into the plane. A fact which we will not use, but which we note for the reader, is that this embedding strongly distorts distances.

Consider a closed path γ in this graph starting at a vertex of $\mathcal{C}_1$. Let $\mathcal{C}_k$ be the outermost circumferential portion reached by γ; we may assume $k > 1$ since otherwise γ must travel around the full polygon $\mathcal{C}_1$. The curve γ must travel across a radial edge e from $\mathcal{C}_{k-1}$ to $\mathcal{C}_k$, at which point, since it is not a spike, it must travel along $\mathcal{C}_k$, and at last return $\mathcal{C}_{k-1}$ via a radial edge from $\mathcal{C}_k$ to $\mathcal{C}_{k-1}$. Now e and the initial segment of γ following e determine a $4g$-gon. Along this $4g$-gon, γ must travel along either $4g-1$ edges (when this $4g$-gon has no edge on $\mathcal{C}_{k-1}$) or $4g-2$ edges (when this $4g$-gon has one edge on $\mathcal{C}_{k-1}$); this follows from the assumption that no spike exists and that $\mathcal{C}_k$ is the largest circle that γ reaches. Since $g \geqslant 2$ we have that $4g-2 > \frac{4g}{2}$ and thus we have found a polygon more than half of whose boundary is traversed by a subpath of γ. $\square$

The following is known as *Dehn's algorithm*; it was originally formulated in the context of fundamental groups of surfaces, as it is used here, but it can also be applied in a significantly more general setting known as *small cancellation groups*, of which fundamental groups of surfaces are an important example. We will return to small cancellation groups in Lecture 5.

Dehn's Algorithm. *Start with a non-trivial word w in the generators of $\pi_1(S)$, and let $[w] \in \pi_1(S)$ denote the element which it represents.*

(1) *Cyclically reduce w, i.e., remove all spikes.*

(2) *If w contains a subword which corresponds to a subpath consisting of more than half the edges in the boundary of a polygon, replace this subword with the inverse of the subword corresponding to the subpath consisting of those edges in the boundary of that polygon which are not in the previous subword.*

Algebraically this corresponds to finding a subword $w_1 \subset w$ and r, a cyclic permutation of a defining relator satisfying $w_1 w_2 = r$ and $2|w_1| > |r|$. In w we then replace w_1 by $r^{-1}w_1 = w_2^{-1}$ which gives a shorter word which represents $[w] \in \pi_1(S)$.

(3) *Repeat the first two steps until there are no longer any spikes nor subpaths which consist of more than half the edges in the boundary of a polygon (i.e., there is neither an "obvious shortenings" nor a subword consisting of more than half of a relator).*

Theorem 2.7 (Dehn). *There exists a solution to the Word Problem for the fundamental group of a closed orientable surfaces.*

Proof. In genus 0 and 1 we have already provided a solution in Example 2.4. Now we consider a surface S with genus greater than 1.

Fix a non-trivial word w in the generators of $\pi_1(S)$ and let γ be the corresponding curve in a Cayley graph for $\pi_1(S)$. We can remove all spikes without changing the class in $\pi_1(S)$, so without loss of generality we may assume that our word w is reduced. We now proceed to step two in Dehn's Algorithm and search w for subwords which are more than half a relator.

Now, if γ is a closed curve (or, equivalently, if $[w] = 1$) this path contains a subpath which travels along more than half of a relator, by Lemma 2.6. Hence, if there is no such subword, then $[w] \neq 1 \in \pi_1(S)$. On the other hand, if there exists such a subpath, then we continue following Dehn's Algorithm. If we start Dehn's Algorithm with a closed curve, each iteration produces a shorter closed curve in the Cayley graph (or equivalently a shorter word w satisfying $[w] = 1$) and thus in a finite number of steps w will be reduced to the trivial word, otherwise we will have a word that can't be shortened, certifying non-triviality. □

Exercise 2.8. *Using the techniques of this Lecture, give a solution to the Word Problem for closed non-orientable surfaces.*

3 Geometry of finite presentability

In this lecture we will develop some geometry related to finite presentability and also provide a proof of the Milnor–Švarc Lemma, which was stated in Lecture 1.

Theorem 3.1. *If Γ is a finitely presented group and Γ' is a finitely generated group quasi-isometric to Γ, then Γ' must be finitely presented.*

To prove this theorem we will use the following lemma which gives a topological condition characterizing when a set of words forms a complete set of relation for a given group. We let $\mathbb{F}(A)$ denote the free group generated on the alphabet given by the elements of A and their inverses. Also, recall that given a subset $\mathcal{R} \subset \mathbb{F}(A)$ the *normal subgroup generated by* $\mathcal{R}$, denoted $\langle\langle \mathcal{R} \rangle\rangle$, is defined to be $\langle \{g^{-1}rg : r^{\pm 1} \in \mathcal{R} \text{ and } g \in \mathbb{F}(A)\} \rangle$

Lemma 3.2. *Let Γ be a group with finite generating set A and let $\mathcal{R}$ be a subset of the kernel of the natural homomorphism $\mathbb{F}(A) \to \Gamma$. Build a 2–complex by attaching 2–cells to each edge loop in the Cayley graph of (Γ, A) which is labelled by a reduced word of $\mathcal{R}$. This 2–complex is simply-connected if and only if $\langle\langle \mathcal{R} \rangle\rangle = \ker(\mathbb{F}(A) \twoheadrightarrow \Gamma)$.*

Sketch of proof. Cayley(Γ, A) is the quotient of the tree Cayley$(\mathbb{F}(A), A)$ by the free action of the subgroup given by $\ker(\mathbb{F}(A) \twoheadrightarrow \Gamma)$. Thus there exists a natural identification of $\ker(\mathbb{F}(A) \twoheadrightarrow \Gamma)$ with $\pi_1(\text{Cayley}(\Gamma, A))$. Now apply the Seifert–van Kampen theorem from algebraic topology. □

Proposition 3.3. *A group admits a finite presentation if and only if it acts properly discontinuously and cocompactly by isometries on a simply-connected geodesic space.*

Sketch of proof. ($\Leftarrow$): This direction is a bit harder and uses different tools then we need for the rest of the lectures, so we will omit it.

($\Rightarrow$): Metrize the simply-connected 2–complex from Lemma 3.2 as a piecewise Euclidean complex where all edges have length 1 and all 2–cells are regular polygons. The action on the Cayley graph then extends to the desired action on the complex. □

Now we are ready to prove the main theorem of this lecture that finite presentability is preserved by quasi-isometries:

Sketch of proof of Theorem 3.1. Fix Γ, Γ' where $\Gamma = \langle A; \mathcal{R}\rangle$ is finitely presented and Γ' is quasi-isometric to Γ. Let ℓ be the length of the longest word in $\mathcal{R}$. By Lemma 3.2, the 2–complex $\mathcal{K}$, obtained by attaching 2–cells to the Cayley graph for (Γ, A) along all edge loops of length $\leqslant \ell$, is simply-connected. Note that to produce the complex $\mathcal{K}$ the collection of 2–cells added to the Cayley graph is locally finite and all of uniformly bounded diameter and hence $\mathcal{K}$ is quasi-isometric to Γ.

To show Γ' admits a finite presentation, by the above Proposition, it suffices to find a simply-connected 2–complex $\mathcal{K}'$ on which the group acts properly discontinuously and cocompactly by isometries. To do this we start with a fixed Cayley graph for Γ' and add to this a locally finite collection of 2–cells to yield a simply-connected 2–complex $\mathcal{K}'$. In particular, we add a 2–cell to each loop in the Cayley graph for Γ' which is of length at most M, for some sufficient large M. It remains to show that this complex which we obtain is simply-connected. Note that since Γ' is quasi-isometric to Γ there exist a quasi-isometry $\phi\colon \mathcal{K}' \to \mathcal{K}$.

Now, we want to show that $\mathcal{K}'$ is simply-connected, i.e., that any loop in $\mathcal{K}'$ is the boundary of a disk in $\mathcal{K}'$. We will sketch the argument here under the additional assumption that ϕ and ϕ^{-1} are continuous; we leave it as an exercise for the careful reader to prove that the *a priori* non-continuous quasi-isometry ϕ can be assumed to be continuous.

Consider a loop $\gamma \subset \mathcal{K}'$ and note that $\phi(\gamma)$ is a loop in $\mathcal{K}'$ (this is using the continuity of ϕ which you proved could be assumed). Now, since $\mathcal{K}'$ is simply-connected, there exists a disk $D \subset \mathcal{K}'$ whose boundary is $\phi(\gamma)$. Then map this disk back to $\mathcal{K}$ by the inverse of the quasi-isometry, which we have also assumed to be continuous. This map of a disk into $\mathcal{K}'$ yields a filling for the original loop in $\mathcal{K}'$. □

By keeping count of the number of cells of the 2–complex in the above proof one can show:

Corollary 3.4 (Alonso). *Among finitely presented groups, the property of having a solvable word problem is preserved by quasi-isometries.*

We now give a proof of a result introduced in the first Lecture:

Lemma 3.5 (Milnor–Švarc). *Let G be a group acting, on the left, properly and isometrically on a proper geodesic metric space, X. If $G\backslash X$ is compact, then G is finitely generated and is quasi-isometric to X.*

Proof. Let $\pi\colon X \to G\backslash X$ be the canonical projection. Endow $G\backslash X$ with the metric $d(p,q) = \inf\{d(x,y)\colon x \in \pi^{-1}(p), y \in \pi^{-1}(q)\}$. Since $G\backslash X$ is compact its diameter, $R = \sup\{d(p,q)\colon p,q \in G\backslash X\}$, is finite.

Fix a base point $x_0 \in X$ and set $B = \{x \in X : d(x_0, x) < R\}$. Note that since X is proper B is compact. The sets $\{\gamma B : \gamma \in G\}$ form a covering of X. Define

$$S = \{\gamma \in G : \gamma \neq e \text{ and } \gamma B \cap B \neq \varnothing\}$$

noting that $S = S^{-1}$, and, since the action is proper, $|S| < \infty$.

Set $r = \inf_{\gamma \in G - (S \cup \{e\})} d(B, \gamma B)$. Without loss of generality, we assume $G - (S \cup \{e\})$ is non-empty, since otherwise the result follows immediately since G would be finite and X compact.

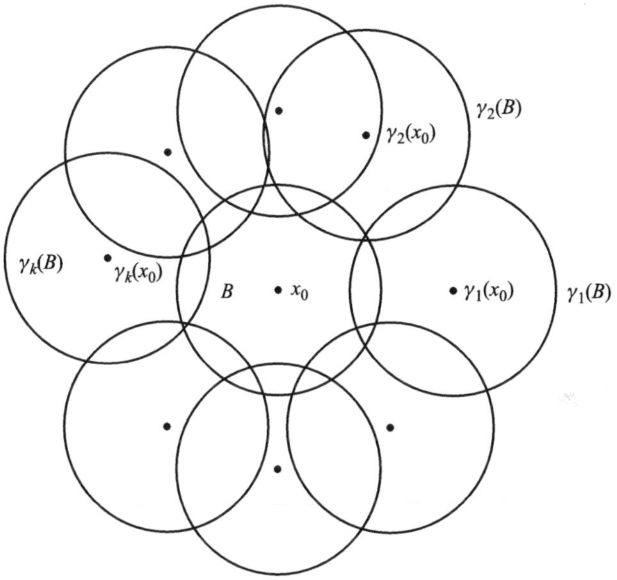

Figure 6. B and γB with $\gamma \in S$.

Claim (1). $r > 0$.

Proof. Choose $\gamma' \in G - (S \cup \{e\})$ and set $r' = d(B, \gamma' B)$. By the definition of S and the compactness of B, we have that $r' > 0$. Now consider the set $\{\gamma \in G - (S \cup \{e\}) : d(B, \gamma B) \leqslant r'\}$ which is non-empty (since it contains γ') and finite (since G acts properly). Thus r is actually the minimum over a finite set of positive numbers. This establishes the claim. ■

Claim (2). *S generates G and there exists $\lambda > 1$ such that for all $\gamma \in G$ we have*

$$\frac{1}{\lambda} d(x_0, \gamma x_0) \leqslant d_S(e, \gamma) \leqslant \frac{1}{r} d(x_0, \gamma x_0) + 1.$$

Proof. Fix $\gamma \in G$. Let k be the smallest integer such that $d(x_0, \gamma x_0) < k \cdot r + R$. Since X is geodesic, there exist $x_1, x_2, \ldots, x_k, x_{k+1} = \gamma x_0$ such that $d(x_0, x_1) \leqslant R$ and $d(x_i, x_{i+1}) \leqslant r$ for all $i \in \{1, 2, \ldots, k\}$.

Since $\{gB\}_{g\in G}$ covers X, we can choose $\gamma_1, \gamma_2, \ldots, \gamma_{k+1} \in G$ such that $x_i \in \gamma_i B$ for $1 \leqslant i \leqslant k+1$ with $\gamma_1 = e$ and $\gamma_{k+1} = \gamma$. Set $s_i = \gamma_i^{-1}\gamma_{i+1}$. Note that we then have that $\gamma = s_1 s_2 \ldots s_k$. We will now show that for each i we have $s_i \in S \cup \{e\}$. Since G acts on the left isometrically on X, we have $d(\gamma_i^{-1}x_i, \gamma_i^{-1}x_{i+1}) < r$ where $\gamma_i^{-1}x_i \in B$ and $\gamma_i^{-1}x_{i+1} = s_i\gamma_{i+1}^{-1}x_{i+1} \in s_i B$. Since $r = \inf_{\gamma\in G-(S\cup\{e\})} d(B, \gamma B)$ this then implies that $\gamma_i \in S \cup \{e\}$.

Hence $d_S(e, \gamma) \leqslant k$. Since we chose k to satisfy $(k-1)r + R \leqslant d(x_0, \gamma x_0)$, we have:

$$d_S(e,\gamma) \leqslant \frac{1}{r} d(x_0, \gamma x_0) + 1 - \frac{R}{r}.$$

By induction on $d_S(e, \gamma)$, it then follows that:

$$d(x_0, \gamma x_0) \leqslant \lambda d_S(e, \gamma)$$

where $\lambda = \sup_{s\in S}\{d(x_0, s x_0)\}$. This establishes the claim. ■

Now define $f\colon G \to X$ by $f\colon \gamma \mapsto \gamma x_0$.

By claim (2), for any $\gamma_1, \gamma_2 \in G$ we have

$$\frac{1}{K} d(f(\gamma_1), f(\gamma_2)) \leqslant d_S(\gamma_1, \gamma_2) \leqslant K \cdot d(f(\gamma_1), f(\gamma_2)) + 1$$

where $K = \max\{\frac{1}{r}, \lambda\}$.

And we have $d(f(G), X) \leqslant R$ since $\{\gamma B : \gamma \in G\}$ is a covering of X. □

4 Groups and their ends

Definition 4.1. Let X be a connected, proper and geodesic metric space (e.g., a locally finite graph). The *number of ends* of X, denoted $e(X)$, is defined to be the limit as $n \to \infty$ of the number of infinite diameter connected components of $X - B_n(x_0)$ where x_0 is a fixed point in X and $B_n(x_0)$ is the ball of radius n about x_0.

Exercise 4.2. *Prove that for any space X as above this limit either exists or is infinite and that the value of the limit is independent of the choice of $x_0 \in X$.*

Exercise 4.3. *If two connected, proper, geodesic metric spaces are quasi-isometric, then they have the same number of ends.*

Hint. Consider a path connecting two points which stays away from some large ball in the space X. Let $\phi\colon X \to Y$ be a quasi-isometry. Show that the image of the path (which need not be a path since quasi-isometries need not be continuous), can be "connected up" to form a path disjoint from a (possible smaller, but still fairly large) ball in Y. □

Definition 4.4. The *number of ends* of a finitely generated group is defined to be the number of ends of a(ny) Cayley graph with respect to some finite generating set.

Example 4.5. *Consider* $\mathbb{F}_2 = \langle a, b\rangle$ *and the Cayley graph of this group with respect to the generators* $\{a, b\}$*. For any non-negative integer* n*, removing a ball of radius* n *leaves* 4×3^n *unbounded connected components. Thus* $\mathbb{F}_2$ *has infinitely many ends.*

The poof of the next Proposition is left as an easy exercise.

Proposition 4.6. *A finitely generated group has zero ends if and only if it is finite.*

Considering Cayley graphs with respect to their standard generating set, we easily obtain the following two examples.

Example 4.7. $\mathbb{Z}$ *has 2 ends and, for each integer* $n \geqslant 2$*, the group* $\mathbb{Z}^n$ *has 1 end.*

Surprisingly, the four simple examples given above exhaust all possibilities for values of $e(G)$ where G is a finitely generated group, as shown by the following:

Theorem 4.8 (Freudenthal–Hopf; 1944)**.** *Every finitely generated group has either 0, 1, 2, or infinitely many ends.*

Proof. Suppose G is a group with $3 \leqslant e(G) < \infty$ ends, for simplicity of notation we set $e(G) = k$. We fix a finite generating set for G and a corresponding Cayley graph, which as an abuse of notation, we also denote G.

By the exercise given above, since $k > 0$ we have $|G| = \infty$.

Since G has k ends, there exists a ball $B \subset G$ large enough so that $G - B$ has k unbounded connected components. Further, since G is infinite, there exists $g \in G$ satisfying $d(e, g) > 2 \cdot \operatorname{diam}(B)$ and with g in an unbounded component of $G - B$. Note, in particular, that this yields $B \cap gB = \varnothing$.

Since g acts as an isometry, one of the k unbounded connected components of $G - B$ contains gB and removing gB from this component divides it into at least $k - 1$ unbounded connected components. Thus $B \cup gB$ is a compact set in G whose removal leaves at least $2(k - 1)$ unbounded components. Hence the space obtained by removing the ball of radius $2d(e, g)$ from G has at least $2(k - 1)$ unbounded connected components. Since $k > 3$ we then have a contradiction, since by hypothesis $e(G) = k$ and whence $k = e(G) \geqslant 2(k - 1) > k$. □

The following theorem, first proven by Wall in 1967, yields a straightforward way to recognize exactly which groups have two ends. There are a number of proofs of this result many of which are elementary. I like the following argument, the details of which appear in the book listed above by John Meier.

Theorem 4.9 (Wall)**.** *A group* G *has two ends if and only if* G *contains a finite index subgroup isomorphic to* $\mathbb{Z}$*.*

Proof. ($\Leftarrow$): This follows from the fact that $\mathbb{Z}$ has two ends and that the number of ends is preserved by commensurability since it is preserved by quasi-isometries.

($\Rightarrow$): Recall the definition of symmetric difference, $A \triangle B = (A \cup B) - (A \cap B)$, and two easy consequences, whose proofs are left as exercises:

(1) For all A, B, D we have $A \triangle B = (A \triangle D) \triangle (B \triangle D)$;

(2) For all X and $A, B \subset X$ we have $A \triangle B = (X - A) \triangle (X - B)$.

Let $F \subset G$ be a finite graph for which $G - F$ has exactly two unbounded components. We will use the notation F^{c} to denote the complement of F in G. Without loss of generality we may assume F^{c} has exactly two connected components (we do this by, if necessary, adding all bounded components to F). The set F divides G into two unbounded components; we think of these as "left" and "right" sides and call the vertices of one of these sides R.

For any $g \in G$ we can consider the translates $g \cdot R = \{g \cdot r \colon r \in R\}$ and $g \cdot R^{\mathrm{c}} = \{g \cdot l \colon l \notin R\}$. Note: $g \cdot R^{\mathrm{c}} = (g \cdot R)^{\mathrm{c}}$.

To continue the proof, we need the following two lemmas. When we speak of the cardiality of a subset of G we will mean the number of vertices in that subset.

Lemma 4.10. *Because G has two ends, either $R \triangle gR$ or $(R \triangle gR)^{\mathrm{c}}$ is finite.*

Proof. $G - (F \cup gF)$ contains 2 unbounded components. Each of R and $g \cdot R$ union at most finitely many additional points (i.e., some of F, $g \cdot F$, or the bounded components) is equal to exactly one of these components.

Hence, if both R and $g \cdot R$ contain the same unbounded component, then $R \triangle gR$ is finite; otherwise, $(R \triangle gR)^{\mathrm{c}}$ is finite. □

We leave the proof of the next Lemma as an exercise.

Lemma 4.11. *The subset of elements $H = \{g \in G \colon R \triangle gR\}$ is finite and forms a subgroup which is of index at most 2 in G.*

Let $h \in H$ satisfy $hF \cap F = \varnothing$, then either $hF \subset R$ or $hF \subset R^{\mathrm{c}}$. Hence in this case we have either

$$R \cap hR^{\mathrm{c}} = \varnothing \text{ and } R^{\mathrm{c}} \cap hR \neq \varnothing$$

or

$$R \cap hR^{\mathrm{c}} \neq \varnothing \text{ and } R^{\mathrm{c}} \cap hR = \varnothing.$$

We think of the first case above as "translating to the left," as the image of R^{c} under h is disjoint from R and the image of R under h intersects R^{c}, similarly, think of the second case as "translating to the right."

We now formulate and prove the following proposition which will provide the key step in proving the theorem:

Proposition 4.12. *Let G and H be as above, then there exists a homomorphism $\phi \colon H \to \mathbb{Z}$ whose kernel is finite.*

Proof. Let $h \in H$, so that $R \triangle hR$ is finite. Thus $|R \cap hR^{\mathrm{c}}| < \infty$ and $|R^{\mathrm{c}} \cap hR| < \infty$.

We define the map $\phi \colon H \to \mathbb{Z}$ by setting:

$$\phi(h) = |R \cap hR^{\mathrm{c}}| - |R^{\mathrm{c}} \cap hR|.$$

We begin by the following straightforward observation for any $g, h \in H$:

$$R \cap hR^{\mathrm{c}} = (R \cap hR^{\mathrm{c}} \cap hgR^{\mathrm{c}}) \sqcup \underbrace{(R \cap hR^{\mathrm{c}} \cap hgR)}_{\mathrm{A}}$$

$$\text{and } R^{\mathrm{c}} \cap hR = \underbrace{(R^{\mathrm{c}} \cap hR \cap hgR^{\mathrm{c}})}_{\mathrm{B}} \sqcup (R^{\mathrm{c}} \cap hR \cap hgR).$$

Hence

$$\phi(h) = |R \cap hR^c \cap hgR^c| + \underbrace{|R \cap hR^c \cap hgR|}_{\text{A}}$$
$$-\underbrace{|R^c \cap hR \cap hgR^c|}_{\text{B}} - |R^c \cap hR \cap hgR|.$$

The first equality here is by definition of ϕ, the second is since the h acts on G as an isometry: $\phi(g) = |R \cap gR^c| - |R^c \cap gR| = |hR \cap hgR^c| - |hR^c \cap hgR|$. Since each of these terms can be written as the sum of two terms, one measuring the intersection with R and the other with R^c, we can rewrite $\phi(g)$ as:

$$\phi(g) = |R \cap hR \cap hgR^c| + \underbrace{|R^c \cap hR \cap hgR^c|}_{\text{B}}$$
$$-\underbrace{|R \cap hR^c \cap hgR|}_{\text{A}} - |R^c \cap hR^c \cap hgR|.$$

Thus, we see that in $\phi(h) + \phi(g)$ the A and B terms are paired with opposite signs and thus we have: $\phi(h) + \phi(g) = |R \cap hgR^c| - |R^c \cap hgR| = \phi(hg)$, i.e., ϕ is a homomorphism from H to $\mathbb{Z}$.

Since $|F| < \infty$, there exists only finitely many $h \in H$ satisfying $hF \cap F \neq \varnothing$. Since Lemma 4.11 shows that $gF \cap F = \varnothing$ implies $\phi(g) \neq 0$, we know an element can only be in the kernal if it is in F. Then, since F is finite, we are done. □

To finish the proof of Wall's Theorem we now just need to observe that the previous Proposition yields a finite index subgroup $H \subset G$ and a homomorphism $\phi\colon H \to \mathbb{Z}$ with finite kernel. If $h \in H$ satisfies $\phi(h) \neq 0$, then h generates a finite index subgroup isomorphic to $\mathbb{Z}$, which is also a finite index subgroup in G. □

Groups with infinitely many ends also admit a particularly nice characterization, which we now state. This result is one of the first major theorems in modern geometric group theory and its proof is beyond the scope of these lectures.

Theorem 4.13 (Stallings Ends Theorem). *Let G be a torsion-free[2] finitely generated group. Then G has infinitely many ends if and only if G can be written as the free product of a pair of groups H and K which are both non-trivial and not both $\mathbb{Z}/2\mathbb{Z}$.*

5 How many groups are there?

Theorem 5.1 (B. Neumann). *There exists uncountably many finitely generated groups up to isomorphism.*

Neumann's original proof of this theorem in 1937, as well as most subsequent proofs rely on delicate arguments using torsion. In this Lecture we discuss an alternate approach, namely, we will describe the following stronger and purely geometric result, which has the Neumann Theorem as an immediate corollary:

[2]The torsion-free hypothesis can be removed; if this is done the phrase "free product" must be replaced by "amalgamation along a finite group," a phrase which might be unfamiliar to readers of these notes, which is why we took the special case in the main text.

Theorem 5.2 (Bowditch). *There exists uncountably many quasi-isometry classes of torsion-free small-cancellation 2–generator groups.*

We will begin by defining the small-cancellation condition. To do this we will fix a finite presentation $\langle A; \mathcal{R}\rangle$ with the property that for each relator $r \in \mathcal{R}$, the set $\mathbb{R}$ contains r^{-1} and also all cyclic conjugates of r. (Note that to any finite presentation, we can add finitely many additional relators to put it in this form.)

Definition 5.3. A cyclically-reduced, non-trivial word $w \in \mathbb{F}(A)$ is called a *piece* if there exist $r_1, r_2 \in \mathcal{R} \subset \mathbb{F}(A)$ such that $r_1 = wr_1'$ and $r_2 = wr_2'$ (i.e., if there exists two distinct relations for which w is an initial subword of both). We now define two of the most commonly used *small-cancellation conditions*:

We say $\langle A; \mathcal{R}\rangle$ satisfies the $C(p)$*–condition* if no element of $\mathcal{R}$ is a product of fewer than p pieces. We say $\langle A; \mathcal{R}\rangle$ satisfies the $C'(\frac{1}{p})$*–condition* if each piece w occurring in $r \in \mathcal{R}$ satisfies $p \cdot \ell(w) < \ell(r)$.

Remark 5.4. $C'(\frac{1}{p})$–condition implies $C(p)$–condition.

Example 5.5. *In earlier lectures we discussed $\pi_1(S_g)$ with the presentation*

$$\langle a_1, \ldots, a_g, b_1, \ldots, b_g; \prod_{i=1}^{g} a_g b_g a_g^{-1} b_g^{-1}\rangle.$$

To study pieces we need to modify this presentation by adding additional relators corresponding to each of the $4g-1$ other cyclic conjugates of this relator (for instance, $b_1 a_1^{-1} b_1^{-1} \cdots a_g b_g a_g^{-1} b_g^{-1} a_1^{-1}$, etc) as well as the inverse of this relator (namely, the word $\prod_{i=0}^{g-1} b_{g-i} a_{g-i} b_{g-i}^{-1} a_{g-i}^{-1}$) and each of its cyclic conjugates. In this presentation we see that a maximal piece is a single letter, so this presentation satisfies $C(4g)$ and $C'(\frac{1}{4g-1})$.

Exercise 5.6. $\mathbb{Z}^3 = \langle a, b, c; aba^{-1}b^{-1}, aca^{-1}c^{-1}, bcb^{-1}c^{-1}\rangle$. *Write the presentation which contains all inverses and cyclic conjugates of each relator. Then show that this presentation satisfies $C(4)$.*

A powerful tool in working with small-cancellation groups is a result called Greendlinger's Lemma which provides that in a group which satisfies a sufficiently strong small-cancellation condition, for instance $C'(\frac{1}{6})$, then any word representing the identity element contains a subword which is more than half of either a relator or a cyclic conjugate of a relator.

Hence, any $C'(\frac{1}{6})$-presented group admits a *Dehn presentation*: a presentation $\langle A; \mathcal{R}\rangle$ such that for any reduced non-trivial word $w \in \mathbb{F}(A)$ which represents the identity in $\langle A; \mathcal{R}\rangle$ there exist a relator $r = r_1 r_2 \in \mathcal{R}$ with $w = w_1 r_1 w_2$ and $\ell(r_1) > \ell(r_2)$ (and thus w can be represented by the shorter word $w = w_1 r_2^{-1} w_2$).

Upshot: the word problem is solvable in any group with a Dehn presentation by applying *Dehn's Algorithm*, as we saw in Lecture 1.

Definition 5.7. For Bowditch's Theorem 5.2 we use *small-cancellation* to mean $C'(\frac{1}{6})$, in particular, if r is a common subword of two relators r_1 and r_2 then either $r_1 = r_2$ or $6|r| < \min\{|r_1|, |r_2|\}$.

Let $\mathcal{P}(\mathbb{N})$ denote the power set of $\mathbb{N}$. Given $F, F' \in \mathcal{P}(\mathbb{N})$, we write $F \sim F'$ if their symmetric difference $F \triangle F'$ is finite. This is an equivalence relation on an uncountable set with each equivalence class countable. Thus there exist uncountably many equivalence classes, indeed the number of distict equivalence classes is the cardinality of $2^{\mathbb{N}}$.

Definition 5.8. Given $F \in \mathcal{P}(\mathbb{N})$, define $S(F) = \{2^{2^n} : n \in F\} \subset \mathbb{N}$. For any $p \in \mathbb{N}$, define

$$w_p = a(a^p b^p)^7 \in \mathbb{F}(a, b).$$

Given $S \subset \mathbb{N}$, define the group $\Gamma(S) = \langle a, b; \{w_p : p \in S\}\rangle$.

Proposition 5.9. *For any $F \in \mathcal{P}(\mathbb{N})$, the group $\Gamma(S(F))$ is $C'(\frac{1}{6})$.*

Proof. The length of w_p is $14p + 1$. For any $p' \neq p$, the longest piece which w_p and $w_{p'}$ share is $a^p b^p$ (here we assume, without loss of generality, that $p < p'$) and hence has length $2p$, and $6 \times 2p < 14p + 1$. □

Remark 5.10. A result in small-cancellation theory states that any torsion must arise in the cyclic symmetry of a relator (e.g., a group has an order two element if and only if it has a relator of the form w^2). Hence, $\Gamma(S(F))$ is torsion-free.

Theorem 5.11. *If $F, F' \in \mathcal{P}(\mathbb{N})$ and $\Gamma(S(F))$ is quasi-isometric to $\Gamma(S(F'))$, then $F \sim F'$.*

Sketch of proof. We are only going to give the idea of the argument here. The two main ingredients in the argument are the fact that the group is $C'(\frac{1}{6})$ and the super-exponential growth of length of relators.

In a group with a presentation satisfying $C'(\frac{1}{6})$, the lengths of relators determine the size of "holes" in the Cayley graph, i.e., long loops in the graph for which there is no shortcut. The sizes of these holes geometrically determine a subset of $\mathbb{N}$; a quasi-isometry of the group only changes these sizes by a bounded multiplicative and additive amount. Hence, arranging these holes to grow super-exponentially ensures that from the quasi-isometry type we can recover, up to finite error, the original set of sizes, as we shall see in the following lemma. □

Let $A \subset \mathbb{N}$ and $q \in \mathbb{N}$, define $q \cdot A = \{qa : a \in A\}$.

Lemma 5.12. *Let $F, F' \subset \mathbb{N}$ for which $q \cdot S(F)$ and $q \cdot S(F')$ satisfy the property that there exists $k \in \mathbb{N}$ such that for each $m \in q \cdot S(F)$ satisfying $m > k$ there exists $m' \in q \cdot S(F')$ satisfying $\frac{m}{k} \leqslant m' \leqslant km$. Then $F \sim F'$.*

Proof. If not, then there exist $n \in F \setminus F'$, such that $2^{2^{n-1}} > k$. Since $2^{2^{n-1}} q \in q \cdot S(F)$, there exists $m \in F'$ such that $\frac{2^{2^n} q}{k} \leqslant 2^{2^m} q \leqslant 2^{2^n} kq$. Thus $2^{2^{n-1}} < \frac{2^{2^n}}{k} \leqslant 2^{2^m} \leqslant 2^{2^n} k < 2^{2^{n+1}}$. Hence $m = n$ which contradicts our assumption that $n \notin F'$. □

Acknowledgements

The author would like to thank all the organizers and participants of the 2013 Summer School on Modern Mathematics for making such a wonderful intellectual exchange possible. In particular, I'd like to especially thank Lizhen Ji for his helpful questions and for his encouragement to write up these notes. I'm very thankful to Hai Yu, a participant in the Summer School, for typing up a draft of these lecture notes, making the diagrams, and for numerous comments which helped to improve the exposition. Also, thanks to the anonymous referee for making several helpful comments. Behrstock was supported as an Alfred P. Sloan Fellow and by NSF grant #DMS-1006219.

Introduction to
Modern Mathematics
ALM 33, pp. 129–142

The Geometry and Topology of Coxeter Groups

Michael W. Davis*†

Abstract

These notes are intended as an introduction to the theory of Coxeter groups. They closely follow my talk in the Lectures on Modern Mathematics Series at the Mathematical Sciences Center in Tsinghua University on May 10, 2013. They were prepared from the beamer presentation which I used during my talk.

AMS Classification Numbers: Primary: 20F55, 20F65.

Keywords and Phrases: Coxeter group, reflection group, Artin group.

1 Geometric reflection groups

By a *geometric reflection group* we mean a discrete group generated by reflections across the faces of a convex polytope in $\mathbb{S}^n$, $\mathbb{E}^n$ or $\mathbb{H}^n$ (the n-sphere, Euclidean n-space or hyperbolic n-space, respectively). The simplest reflection group is the cyclic group of order two acting as a reflection on the real line. The next simplest are the dihedral groups.

Dihedral groups. A *dihedral group* is any group which is generated by two involutions, call them s and t. Such a group is determined up to isomorphism by the order m of st (m is an integer $\geqslant 2$ or ∞). Let $\mathbf{D}_m$ denote the dihedral group corresponding to m.

For $m \neq \infty$, $\mathbf{D}_m$ can be represented as the subgroup of $O(2)$ which is generated by reflections across lines L and L', making an angle of π/m. (See Figure 1.)

1.1 Some history

In 1852 Möbius determined the finite subgroups of $O(3)$ that are generated by isometric reflections on the 2-sphere. The fundamental domain for such a group

*The author thanks the Mathematical Sciences Center in Tsinghua University for the opportunity to visit during the spring of 2013. These notes were written during that period. The author also was partially supported by NSF grant DMS 1007068.

†Department of Mathematics, The Ohio State University, 231 W. 18th Ave. Columbus, Ohio 43210, United States. Email: davis.12@math.osu.edu

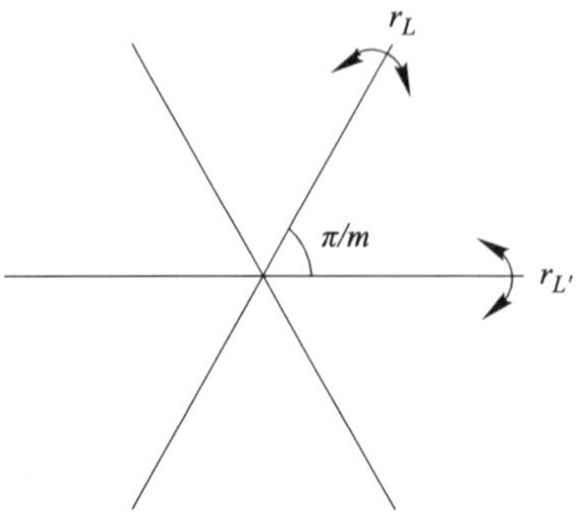

Figure 1. A finite dihedral group.

on the 2-sphere is a spherical triangle with angles $\frac{\pi}{p}$, $\frac{\pi}{q}$, $\frac{\pi}{r}$, with p, q, r integers $\geqslant 2$. Since the sum of the angles is $> \pi$, we have $\frac{1}{p} + \frac{1}{q} + \frac{1}{r} > 1$. For $p \geqslant q \geqslant r$, the only possibilities are: $(p, 2, 2)$ for any $p \geqslant 2$ and $(p, 3, 2)$ with $p = 3$, 4 or 5. The last three cases are the symmetry groups of the Platonic solids. (See Figure 2.)

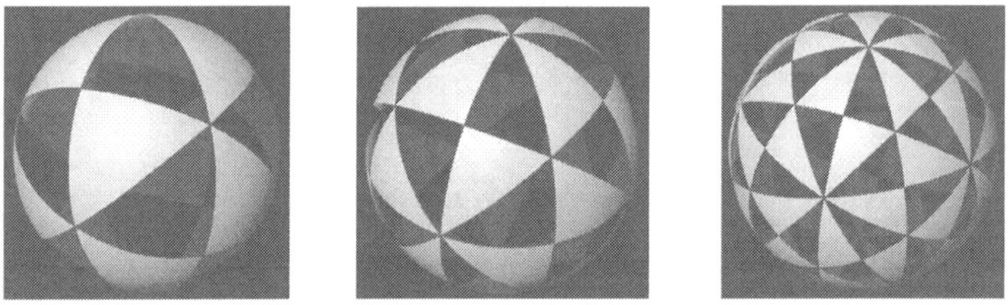

Figure 2. $(2, 3, 3)$, $(2, 3, 4)$ and $(2, 3, 5)$ triangle groups.

Later work by Riemann and Schwarz showed there were discrete groups of isometries of $\mathbb{E}^2$ or $\mathbb{H}^2$ generated by reflections across the edges of triangles with angles integral submultiples of π. (See Figure 3 for a hyperbolic triangle group.) Poincaré and Klein showed there were similar groups for polygons in $\mathbb{H}^2$. (See Figure 4 for a right-angled pentagon group.)

Figure 3. $(3, 6, 6)$ triangle group.

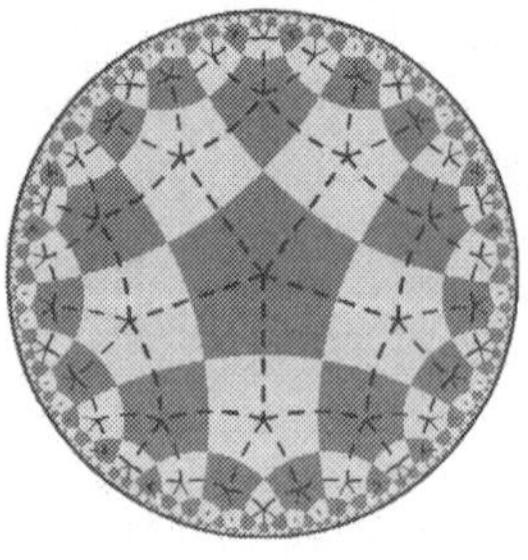

Figure 4. Right-angled pentagon group.

In the second half of the nineteenth century work began on finite reflection groups on $\mathbb{S}^n$, $n > 2$, generalizing Möbius' results for $n = 2$. It developed along two lines. First, around 1850, Schläfli classified regular polytopes in $\mathbb{R}^{n+1}$, $n > 2$. The symmetry group of such a polytope was a finite group generated by reflections and, as in Möbius' case, the projection of a fundamental domain to $\mathbb{S}^n$ was a spherical simplex with dihedral angles integral submultiples of π. Second, around 1890, Killing and E. Cartan classified complex semisimple Lie algebras in terms of their root systems. In 1925, Weyl showed the symmetry group of such a root system was a finite reflection group. These two lines were united by Coxeter [7] in the 1930's. He classified discrete reflection groups on $\mathbb{S}^n$ or $\mathbb{E}^n$.[1] (See Figure 5)

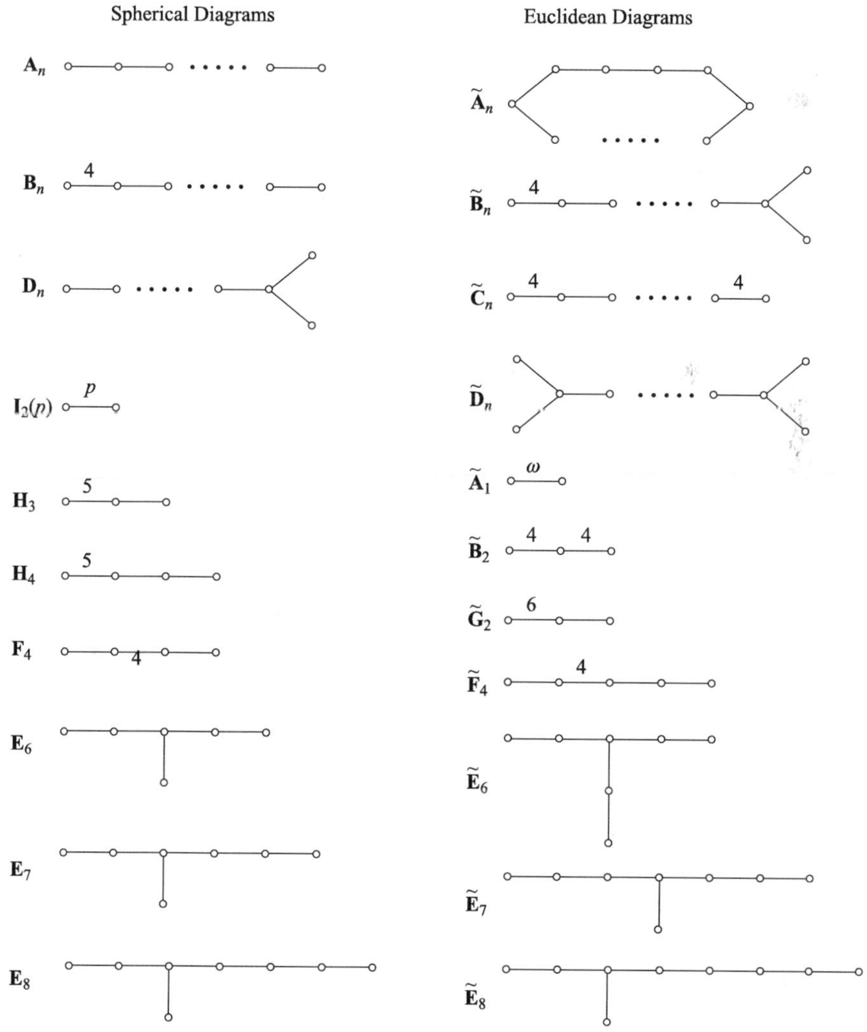

Figure 5. Coxeter diagrams.

[1]This subsection is based on the Historical Notes in [3].

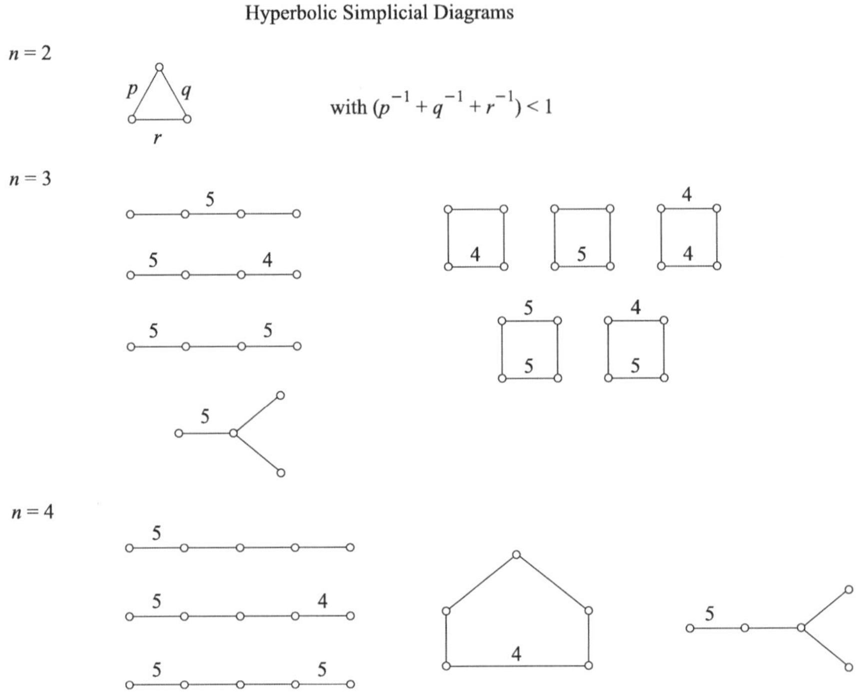

Figure 6. Hyperbolic simplicial diagrams.

1.2 Properties

Let K be a fundamental polytope for a geometric reflection group. For $\mathbb{S}^n$, K is a simplex (i.e., the generalization of a triangle to a higher dimension). For $\mathbb{E}^n$, K is a product of simplices, for example, K could be a product of intervals. For $\mathbb{H}^n$ there are other possibilities, e.g., a right-angled pentagon in $\mathbb{H}^2$ (see Figure 4) or a right-angled dodecahedron in $\mathbb{H}^3$. For a survey of the known examples of geometric reflection groups on $\mathbb{H}^n$, see Vinberg's article [18].

Conversely, given a convex polytope K in $\mathbb{S}^n$, $\mathbb{E}^n$ or $\mathbb{H}^n$ so that all dihedral angles have form $\pi/\text{integer}$, there is a discrete group W generated by isometric reflections across the codimension one faces of K (cf. [8, Thm. 6.4.3]).

Let S be the set of reflections across the codimension one faces of K. Then S generates W. For $s, t \in S$, let $m(s,t)$ be the order of st. The faces corresponding to s and t intersect in a codimension two face if and only if $m(s,t) \neq \infty$, and for $s \neq t$, the dihedral angle along that face is $\pi/m(s,t)$. (The $S \times S$ symmetric matrix $(m(s,t))$ is the *Coxeter matrix.*) Moreover, W has a presentation of the form

$$\langle S \mid (st)^{m(s,t)}, \ \text{where } (s,t) \in S \times S \rangle \tag{1.1}$$

Coxeter diagrams. Associated to (W, S), there is a labeled graph Γ called its *Coxeter diagram.* The vertex set of Γ is S. Connect distinct vertices s and t by an edge if and only if $m(s,t) \neq 2$. Label the edge by $m(s,t)$ if $m(s,t)$ is > 3 or $= \infty$ and leave it unlabeled if it is $= 3$. (W, S) is *irreducible* if Γ is connected. (The components of Γ give the irreducible factors of W.)

Figure 5 shows Coxeter's classification in [7] of irreducible spherical and irreducible cocompact Euclidean reflection groups. Figure 6 shows Lannér's classification in [13] of the hyperbolic reflection groups with fundamental polytope equal to a simplex in $\mathbb{H}^n$.

1.3 The relationship with Lie theory

The class of finite Coxeter groups coincides with the class of geometric reflection groups on spheres. These groups are intimately tied to Lie theory.

Suppose G is a simple Lie group with Lie algebra $\mathfrak{g}$. The Lie algebra $\mathfrak{h}$ of a maximal torus is a "Cartan subalgebra." The derivative of the adjoint action of G on $\mathfrak{g}$ leads to the adjoint action of $\mathfrak{g}$ on itself by Lie brackets, defined by $\mathrm{ad}(x)(y) := [x, y]$. Restricting this action to $\mathfrak{h}$, we get a decomposition of $\mathfrak{g}$ into simultaneous eigenspaces:

$$\mathfrak{g} = \bigoplus_{\alpha} \mathfrak{g}_\alpha,$$

where the α are certain linear functionals on $\mathfrak{h}$, called the *roots*. The set of such $\alpha \in \mathfrak{h}^*$ is a *root system*. The classification of simple Lie algebras over $\mathbb{C}$ reduces to the classification of finite root systems. As mentioned earlier, Weyl showed that the automorphism group of a root system is a finite reflection group. (Each root defines a hyperplane and the group generated by reflections across these hyperplanes is a reflection group on $\mathfrak{h}$. For this reason, automorphism groups of root systems are usually called "Weyl groups.") The spherical diagrams in Figure 5 are automorphism groups of root systems except in the cases H_3, H_4, and the dihedral groups $\mathrm{I}_2(5)$ and $\mathrm{I}_2(p)$, $p > 6$. The classification of root systems is almost the same as the remaining list of spherical diagrams except that there are two types of root systems for B_n. These are usually denoted by B_n and C_n.

The connection between spherical Coxeter groups and root systems has many ramifications. For one, the field need not be the field of complex numbers. For example, algebraic groups over finite fields (i.e., finite groups of "Lie type") have corresponding spherical Coxeter diagrams. "Spherical buildings" are associated to these groups. Euclidean reflection groups arise in a number of related ways. One is as the automorphism group of the weight lattice in $\mathfrak{h}^*$. (This automorphism group is an extension of the finite Coxeter group by a lattice in $\mathfrak{h}^*$.) Euclidean reflection groups also occur in the theory of algebraic groups over local fields. The corresponding building is called "Euclidean" or "affine."

Finally, there are the theories of Kac-Moody algebras, groups and buildings. These are related to the more general Coxeter groups that are considered in the next section. In particular, the Tits representations of 2.2 are related to root systems for Kac-Moody Lie algebras.

2 Abstract reflection groups

The theory of abstract reflection groups was invented by Tits [16] around 1961. His term for an abstract reflection group was a "Coxeter group." This theory of Coxeter groups is explained in the beautiful book of Bourbaki [3]; Ken Brown's book [4] is also an excellent source.

Question 2.1. *Given a group W and a set S of involutions which generate W, when does (W, S) deserve to be called an "abstract reflection group"?*

There are two answers to Question 2.1. Suppose $\mathrm{Cay}(W, S)$ is the Cayley graph of (W, S) (i.e., its vertex set is W and $\{w, v\}$ spans an edge if and only if $v = ws$ for some $s \in S$).

***First answer*:** For each $s \in S$, the fixed point set of s on $\mathrm{Cay}(W, S)$ separates $\mathrm{Cay}(W, S)$.

***Second answer*:** W has a presentation of the same form as (1.1):

$$\langle S \mid (st)^{m(s,t)}, \text{ where } (s,t) \in S \times S \rangle. \tag{2.1}$$

Here $(m(s,t))$ is a symmetric $(S \times S)$-matrix with 1s on the diagonal and off-diagonal elements integers $\geqslant 2$ or the symbol ∞. Such a matrix is called a *Coxeter matrix.*

It turns out that these two answers are equivalent! (This is implicit in [3]. It also is stated explicitly and proved in [8, Chapter 3].)

Some of the terms used above should be defined.

Definition 2.2. (*Cayley graphs*). Given a group G and a set of generators S, let $\mathrm{Cay}(G, S)$ denote the graph with vertex set G which has a (directed) edge from g to gs, for all $g \in G$ and for all $s \in S$. The group G acts on $\mathrm{Cay}(G, S)$ (written $G \curvearrowright \mathrm{Cay}(G, S)$), the action is simply transitive on the vertex set, and the edges starting at a given vertex can be labeled by the elements of S.

Definition 2.3. (*Presentations*). Suppose S is a set of letters and $\mathcal{R}$ is a set of words in S. Let F_S be the free group on S and let $N_{\mathcal{R}}$ be the smallest normal subgroup containing $\mathcal{R}$. Set $G := F_S/N_{\mathcal{R}}$ and write

$$G = \langle S \mid \mathcal{R} \rangle.$$

It is a *presentation* for G.

2.1 Coxeter systems

Definition 2.4. If either of the previous two answers holds, (W, S) is a *Coxeter system* and W is a *Coxeter group.*

The terms "Coxeter group" and "Coxeter system" were coined by Tits [16]. The second answer is usually taken to be the official definition: W has a presentation of the form given by (2.1).

Definition 2.5. Given a subset $T \subset S$, put $W_T := \langle T \rangle$ and call it a *special subgroup*. T is *spherical* if W_T is finite. Let $\mathcal{S}$ denote the poset of spherical subsets of S. (N.B. $\emptyset \in \mathcal{S}$.)

Definition 2.6. The *nerve* of (W, S) is the simplicial complex L $(= L(W, S))$ with vertex set S and with $T \subset S$ a simplex if and only if T is spherical. (So, the poset of simplices in L, including the empty simplex, is $\mathcal{S}$.)

Definition 2.7. (*Right-angled Coxeter systems or* RACS*s*). A Coxeter system (W, S) is *right-angled* (a RACS for short) if all off-diagonal $m(s,t)$ are $= 2$ or ∞. W is a *right-angled Coxeter group* (a RACG).

Suppose Γ is a graph with vertex set S. Define an $(S \times S)$ Coxeter matrix $(m(s,t))$ with off-diagonal elements given by

$$m(s,t) = \begin{cases} 2, & \text{if } \{s,t\} \in \operatorname{Edge}\Gamma \\ \infty, & \text{otherwise.} \end{cases}$$

This defines a RACS, (W, S). A subset $T \subset S$ is spherical if and only if it spans a complete subgraph of Γ. So, the nerve L is the associated flag complex of Γ (called the "clique complex" of Γ by combinatorialists).

Question 2.8. *Does every Coxeter system have a geometric realization?*

Answer 2.9. *Yes. In fact, there are two different realizations:*

(1) *the Tits representation,*
(2) *the cell complex* Σ.

Both realizations use the following basic construction.

The basic construction. A *mirror structure* on a space X is a family of closed subspaces $\{X_s\}_{s\in S}$. For $x \in X$, put $S(x) = \{s \in S \mid x \in X_s\}$. Define

$$\mathcal{U}(W, X) := (W \times X)/\sim,$$

where $\sim$ is the equivalence relation: $(w, x) \sim (w', x') \iff x = x'$ and $w^{-1}w' \in W_{S(x)}$ (the special subgroup generated by $S(x)$). In other words, $\mathcal{U}(W, X)$ is formed by gluing together copies of X (the *chambers*), one for each element of W. The group W acts on $\mathcal{U}(W, X)$ via $w \cdot [v, x] = [wv, x]$, where $[v, x]$ denotes the equivalence class of (v, x). (Think of X as the fundamental polytope and the X_s as its codimension one faces.)

Definition 2.10. Suppose (W, S) is a Coxeter system. The action of W on a space $\mathcal{U}$ is a *reflection group* if there is a mirror structure $\{X_s\}_{s\in S}$ on a space X such that $\mathcal{U}$ is equivariantly homeomorphic to $\mathcal{U}(W, X)$. (In the past I considered various possible definitions of a "reflection group" - always with an eye to proving that the action was equivalent to the action of W on $\mathcal{U}(W, X)$. Finally, I decided this should be taken as the definition.)

The properties a geometric realization should have: It should be an action of W on a space $\mathcal{U}$ so that

(i) W is a reflection group on $\mathcal{U}$.
(ii) The stabilizer of each $x \in \mathcal{U}$ is finite.
(iii) $\mathcal{U}$ is contractible.
(iv) $\mathcal{U}/W$ $(= X)$ is compact.

2.2 The first realization: the Tits representation

Definition 2.11. (*Linear reflections*). Two data determine a (not necessarily orthogonal) linear reflection on $\mathbb{R}^n$:

(1) a linear form $\alpha \in (\mathbb{R}^n)^*$ (the fixed hyperplane is $\alpha^{-1}(0)$).
(2) a (-1)-eigenvector $h \in \mathbb{R}^n$ (normalized so that $\alpha(h) = 2$

The formula for the reflection is $v \mapsto v - \alpha(v)h$.

A symmetric bilinear form. Let $(e_s)_{s \in S}$ be the standard basis for $\mathbb{R}^S$. Given a Coxeter matrix $m(s,t)$ define a symmetric bilinear form B on R^S by

$$B(e_s, e_t) = -2\cos(\pi/m(s,t)).$$

(In the case of geometric reflection group, if the e_s denote the unit normal vectors to the codimension one faces of a fundamental polytope, then the matrix on the right hand side of this formula is the matrix of inner products, $(e_s \cdot e_t)$.) For each $s \in S$, we have a linear reflection $r_s : v \mapsto v - B(e_s, v)e_s$. Tits showed this defines a linear action $W \curvearrowright \mathbb{R}^S$. We are interested in the dual representation $\rho : W \to GL(\mathbb{R}^S)$ defined by $s \mapsto \rho_s := (r_s)^*$.

Theorem 2.12. (Properties of Tits representation $W \to GL(\mathbb{R}^S)$, cf. [3]).

(i) *The ρ_s are reflections across the faces of the standard simplicial cone $C \subset (\mathbb{R}^S)^*$.*
(ii) *$\rho : W \hookrightarrow GL(\mathbb{R}^S)$ is a faithful representation.*
(iii) *WC $(:= \bigcup_{w \in W} wC)$ is a convex cone. Let $\mathcal{I}$ denote the interior of this cone.*
(iv) *$\mathcal{I} = \mathcal{U}(W, C^f)$, where C^f denotes the complement of the nonspherical faces of C (a face is* spherical *if its stabilizer is finite). So, W is a "discrete reflection group" on $\mathcal{I}$.*

Selberg's Lemma asserts that a finitely generated linear group is residually finite and virtually torsion-free. So, a consequence of Theorem 2.12 is the following.

Corollary 2.13. *W is residually finite and virtually torsion-free.*

Advantages of the Tits representation: $\mathcal{I}$ is contractible (since it is convex) and W acts properly (i.e., with finite stabilizers) on it.

Disadvantage: $\mathcal{I}/W$ is not compact (since C^f is not compact).

Remarks 2.14. (1) By dividing by scalar matrices, we get a representation $W \hookrightarrow PGL(\mathbb{R}^S)$. So, $W \curvearrowright P\mathcal{I}$, the image of $\mathcal{I}$ in projective space. When W is infinite and irreducible, this is a proper convex subset of $\mathbb{R}P^{n-1}$, where $n = \#S$.

(2) In [17] Vinberg showed one can get linear representations across the faces of more general polyhedral cones. As before, $W \curvearrowright \mathcal{I}$, where $\mathcal{I}$ is the interior of

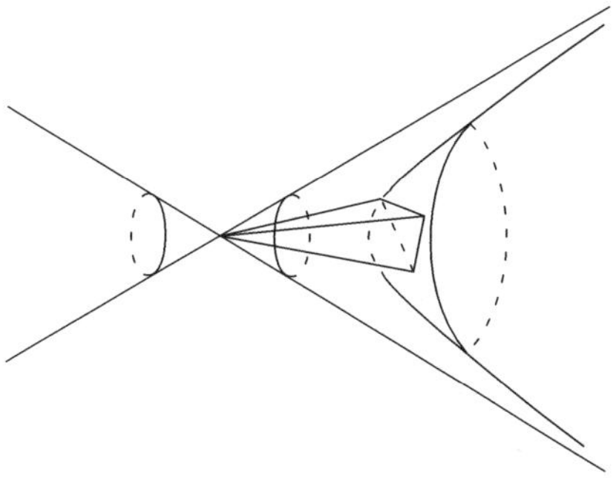

Figure 7. A hyperbolic triangle group.

a convex cone; $P\mathcal{I}$ is a open convex subset of $\mathbb{R}P^{n-1}$. The fundamental chamber is a convex polytope with some faces deleted. Sometimes it can be a compact polytope, for example, a pentagon. (See Figure 4 again).

Yves Benoist has written a series of papers (see [2]) about these projective representations $W \hookrightarrow PGL(n, \mathbb{R})$. In particular, there are interesting examples with fundamental chamber a compact polytope which are not equivalent to a Euclidean or hyperbolic reflection group. In particular, Moussong [14, 15] had given examples of word hyperbolic Coxeter groups whose nerves were boundary complexes of simplicial polytopes yet could not be realized as reflection groups on any hyperbolic space. Benoist showed that certain of Moussong's examples admit cocompact projective representations as reflection groups.

Question 2.15. *For (W, S) to be a reflection group in $PGL(n+1, \mathbb{R})$ with fundamental chamber a compact convex polytope $P^n \subset \mathbb{R}P^n$ there is an obvious necessary condition: its nerve $L(W, S)$ (defined by the spherical subsets of S) must be dual to ∂P^n. In other words, the simplicial complex $L(W, S)$ must be the boundary complex of a simplicial polytope. Is this condition sufficient?*

Almost certainly the answer is no. In the opposite direction, one could ask the following.

Question 2.16. *Are there irreducible, non-affine examples of such $W \subset PGL(n+1, \mathbb{R})$ with $P^n \subset \mathbb{R}P^n$ a compact convex polytope, for n arbitrarily large?*

2.3 The second realization: the cell complex Σ

The second answer to Question 2.8 is to construct a contractible cell complex Σ on which W acts properly and cocompactly as a group generated by reflections. Its advantage is that Σ/W will be compact.

There are two dual constructions of Σ:

(1) Build the correct fundamental chamber K with mirror structure $\{K_s\}_{s\in S}$, then apply the basic construction to get $\Sigma = \mathcal{U}(W, K)$.
(2) Fill in the Cayley graph of (W, S).

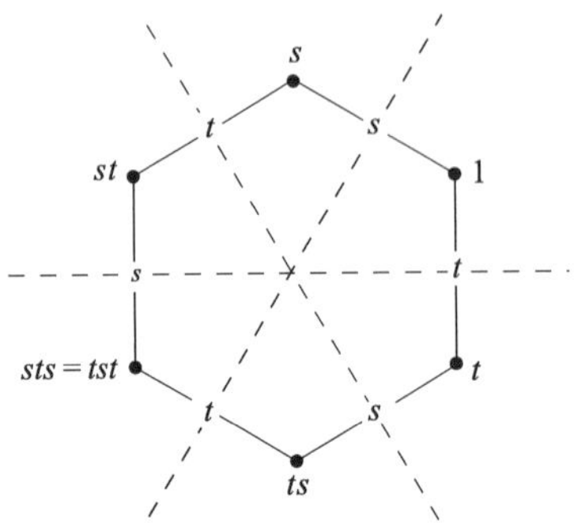

Figure 8. The Cayley 2-complex of a finite dihedral group.

Figure 9. Cayley 2-complex of $(3,3,3)$-triangle group.

Filling in the Cayley graph. Let $W_{\{s,t\}}$ be the dihedral subgroup $\langle s,t\rangle$. Whenever $m(s,t) < \infty$ each coset of $W_{\{s,t\}}$ spans a polygon in $\mathrm{Cay}(W,S)$. If we fill in these polygons, we get a simply connected 2-dimensional complex, which will be the 2-skeleton of Σ. If we want to obtain a contractible space, then we need to fill in higher dimensional polytopes ("cells"). Corresponding to a spherical subset T with $\#T = k$, there is a k-dimensional convex polytope called a *Coxeter zonotope.* (A *zonotope* means a convex polytope which is dual to the face poset of an arrangement of hyperplanes in $\mathbb{R}^n$.) When $k = 2$, a zonotope is the polygon with an even number of edges.

Coxeter zonotopes. Suppose W_T is finite reflection group on $\mathbb{R}^T$. Choose a point x in the interior of fundamental simplicial cone and let P_T be convex hull of $W_T x$. The 1-skeleton of P_T is $\mathrm{Cay}(W_T, T)$ (See Figures 8 and 10.)

Example 2.17. When $W_T = (\mathbb{Z}/2)^n$, P_T is an n-cube.

Let $W\mathcal{S}$ be the *poset of spherical cosets*, i.e., it is the disjoint union of all cosets of spherical special subgroups (partially ordered by inclusion):

$$W\mathcal{S} := \coprod_{T\in\mathcal{S}} W/W_T. \tag{2.2}$$

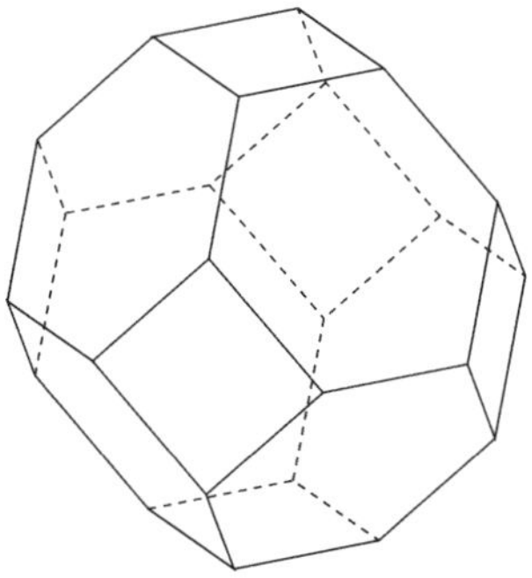

Figure 10. The permutohedron

The idea is to build a cell complex with one cell (a Coxeter zonotope) for each spherical coset.

Definition 2.18. (*Geometric realization of a poset*). Associated to any poset $\mathcal{P}$ there is a simplicial complex $|\mathcal{P}|$ called its *geometric realization*. The vertex set of $|\mathcal{P}|$ is $\mathcal{P}$. A set of $k+1$ distinct vertices $\{p_0, p_1, \dots, p_k\}$ spans a k-simplex if and only if it is totally ordered, i.e., if, possibly after renumbering, $p_0 < \cdots < p_k$.

Put $\Sigma := |W\mathcal{S}|$. There is a cell structure on Σ with

$$\{\text{cells in } \Sigma\} = W\mathcal{S}.$$

This follows from fact that poset of cells in P_T is $\cong W_T\mathcal{S}_{\leqslant T}$. The cells of Σ are defined as follows: the geometric realization of subposet of cosets $\leqslant wW_T$ is $\cong$ barycentric subdivision of P_T.

Theorem 2.19. (Properties of this cell structure on Σ, cf. [8]).

(i) *W acts cellularly on Σ.*
(ii) *Σ has one W-orbit of cells for each spherical subset $T \in \mathcal{S}$; the dimension of the cell is* $\mathrm{Card}(T)$.
(iii) *The 0-skeleton of Σ is W.*
(iv) *The 1-skeleton of Σ is* $\mathrm{Cay}(W, S)$.
(v) *The 2-skeleton of Σ is the Cayley 2-complex of the presentation.*
(vi) *If W is right-angled (i.e., if each $m(s,t)$ is $1, 2$ or ∞), then each Coxeter cell is a cube.*
(vii) (Gromov [11] and Moussong [14]): *the induced piecewise Euclidean metric on Σ is* CAT(0) *(meaning that it is simply connected and nonpositively curved).*
(viii) *Σ is contractible. (This follows from the fact it is* CAT(0)*).*
(ix) *The W-action is proper (by construction each isotropy subgroup is conjugate to some spherical W_T).*
(x) *Σ/W is compact.*
(xi) *If W is finite, then Σ is a Coxeter zonotope.*

Theorem 2.20. (A typical application of CAT(0)-ness, cf. [9, Remark(5b.2), p. 384]). *There is a closed manifold M with a nonpositively curved (polyhedral) metric so that M is not homotopy equivalent to a nonpositivley curved Riemannian manifold (the reason being that the universal cover of M is not homeomorphic to Euclidean space.)*

The dual construction of Σ. Recall $\mathcal{S}$ is the poset of spherical subsets of S. The *fundamental chamber* K is defined by $K := |\mathcal{S}|$. (K is the cone on the barycentric subdivision of L.) The mirror structure is given by $K_s := |\mathcal{S}_{\geqslant\{s\}}|$. The proof of the following proposition is an easy exercise - it is essentially just a matter of unraveling the definitions.

Proposition 2.21. (cf. [8]). $\Sigma := \mathcal{U}(W, K)$. *So, K is homeomorphic to Σ/W.*

The construction of Σ is very useful for constructing examples. The basic reason is that the chamber K is the cone over a fairly arbitrary simplicial complex (for example, any barycentric subdivision can occur). This means we can construct Σ with whatever local topology we like.

Relationship with geometric reflection groups. If W is a geometric reflection group on $\mathbb{X}^n = \mathbb{E}^n$ or $\mathbb{H}^n$, then K can be identified with the fundamental polytope, Σ with $\mathbb{X}^n$ and the cell structure of Σ is dual to the tessellation of Σ by translates of K. (See Figure 4; the dotted black lines form the 1-skeleton of Σ.)

Relationship with Tits representation. Suppose W is infinite. Then K is subcomplex of $b\Delta$, the barycentric subdivision of the simplex $\Delta \subset C$. Consider the vertices which are barycenters of spherical faces. They span a subcomplex of $b\Delta$. This subcomplex is K. It is a subspace of Δ^f. So, $\Sigma = \mathcal{U}(W, K) \subset \mathcal{U}(W, \Delta^f) \subset \mathcal{U}(W, C^f) = \mathcal{I}$. Thus, Σ is the "cocompact core" of $\mathcal{I}$.

2.4 The relationship between Coxeter groups and Artin groups

Given a Coxeter matrix $(m(s,t))$ there is an associated *Artin group* A with set of generators $\{x_s\}_{s\in S}$ and with relations defined by

$$x_s x_t \cdots = x_t x_s \cdots$$

for all s, t with $s \neq t$ and $m(s,t) \neq \infty$. Here the words on each side on the equation are alternating words of length $m(s,t)$ in x_s and x_t. There is a natural epimorphism from A to the associated Coxeter group W sending x_s to s. For example, if W is the symmetric group A_n on $n+1$ letters than A is the braid group on $n+1$ strands. (Artin was the first to study braid groups.)

The Tits representation $W \hookrightarrow GL(N; \mathbb{R})$, where $N = \#S$, has complexification $W \hookrightarrow GL(N; \mathbb{C})$. The subset $\mathbb{R}^N + i\mathcal{I}$ is W-stable. For each $r \in R$ (where R is the set of conjugates of S), we have the complexified hyperplane H_r that is fixed by r. Denote the complement of the union of these reflecting hyperplanes by

$$M := (\mathbb{R}^N + i\mathcal{I}) - \bigcup_{r\in R} H_r.$$

The group W acts freely on M and $\pi_1(M/W) = A$. The major unknown question in the area is the $K(\pi,1)$-Problem for Artin groups. This asks whether M is a model for $K(A,1)$ (see [5, 6]). The answer is affirmative for Artin groups of spherical type, for right-angled Artin groups (as defined below) and for many other cases (see [5]).

In analogy with Definition 2.7, an Artin group is *right-angled* if all the off-diagonal $m(s,t)$ are 2 or ∞. The term "right-angled Artin group" is commonly abbreviated RAAG.

There is a beautiful model for the classifying space $K(A,1)$ of a RAAG A. Let T^S denote the product of S copies of S^1. For each simplex $\sigma \in L$ $(= L(W,S))$, let T^σ denote the subproduct over the vertices of σ of copies of S^1. Then $\bigcup_{\sigma\in L} T^\sigma$ is a $K(A,1)$. Its universal cover is a CAT(0) cube complex (see [6]).

It is proved in [10] that every RAAG is commensurable with a RACG. (In fact, any RAAG has a RACG as a subgroup of finite index as well as a RACG as finite index super group.) Using this, one can show that RAAGs have very strong separability properties. For example, since RACGs are linear groups via the Tits representation, so are RAAGs. In particular, RAAGs are residually finite. Furthermore, Haglund proved that every quasiconvex subgroup of a RACG is a virtual retract of it; so, the same is true for any RAAG.

Some recent spectacular advances in geometric group theory and in 3-manifold theory involve RACGs and RAAGs. First, Haglund and Wise proved that if a group Γ acts on a sufficiently nice CAT(0) cube complex, then Γ virtually embeds in a RAAG or RACG. These nice cube complexes are the "special" cube complexes of [12]. An overview of this theory can be found in [19]. In [1] Ian Agol proved a conjecture of Wise asserting that if a word hyperbolic groups acts on a CAT(0) cube complex, then it can also act on a special CAT(0) cube complex. Using this, he proved Thurston's remaining conjectures on 3-manifolds and their fundamental groups. In particular, he proved Thurston's Virtual Fibering Conjecture and Virtual Haken Conjecture. For this work Agol and Wise were awarded the Veblen Prize in 2013.

References

[1] I. Agol, *The virtual Haken Conjecture*, arXiv:1204.2810.

[2] Y. Benoist, *A survey on divisible convex sets. Geometry, analysis and topology of discrete groups*, 118, Adv. Lect. Math. (ALM), **6**, Int. Press, Somerville, MA, 2008.

[3] N. Bourbaki, *Lie Groups and Lie Algebras, Chapters 4–6*, Springer-Verlag, New York, 2002.

[4] K.S. Brown, *Cohomology of Groups*, Springer-Verlag, Berlin and New York, 1982.

[5] R. Charney and M.W. Davis, *The $K(\pi,1)$-problem for hyperplane complements associated to infinite reflection groups*, J. Amer. Math. Soc. **8** (1995), 597–627.

[6] R. Charney and M.W. Davis, *Finite $K(\pi, 1)$s for Artin groups*, in *Prospects in Topology* (ed. F. Quinn), Annals of Math. Studies **138**, Princeton Univ. Press, Princeton, 2000, pp. 110–124.

[7] H.S.M. Coxeter, *Discrete groups generated by reflections*, Ann. of Math. **35** (1934), 588–621.

[8] M.W. Davis, *The Geometry and Topology of Coxeter Groups*, Princeton Univ. Press, London Math. Soc. Monograph Series, vol. **32**, 2008.

[9] M.W. Davis and T. Januszkiewicz, *Hyperbolization of polyhedra*, J. Differential Geometry **34** (1991), 347–388.

[10] M.W. Davis and T. Januszkiewicz, *Right-angled Artin groups are commensurable with right-angled Coxeter groups*, J. of Pure and Appl. Algebra **153** (2000), 229–235.

[11] M. Gromov, *Hyperbolic groups.* In *Essays in Group Theory* (ed. S. Gersten), MSRI Publications vol. **8**, pp. 75–264, Springer-Verlag, Berlin, 1987.

[12] F. Haglund and D. Wise, *Special cube complexes*, Geom. Funct. Anal. **17** (2008), no. 5, 1551–1620.

[13] F. Lannér, *On complexes with transitive groups of automorphisms*, Comm. Sem. Math. Univ. Lund **11** (1950), 1–71.

[14] G. Moussong, *Hyperbolic Coxeter groups*, Ph.D. thesis, Ohio State University, 1988.

[15] G. Moussong, *Some non-symmetric manifolds*, Colloquia Mathematica Societatis János Bolyai **56** Differential Geometry, Eger, Hungary, 1989.

[16] J. Tits, *Groupes et géométries de Coxeter*, unpublished manuscript, 1961.

[17] È.B. Vinberg, *Discrete linear groups generated by reflections*, Math. USSR Izvestija **5**, no. 5 (1971), English translation, 1083–1119.

[18] È.B. Vinberg, *Hyperbolic reflection groups*, Russian Math. Surveys **40** (1985), 31–75.

[19] D. Wise, *From riches to raags: 3-manifolds, right-angled Artin groups, and cubical geometry* CBMS Regional Conference Series in Mathematics, **117**. Published for the Conference Board of the Mathematical Sciences, Washington, DC; by the AMS, Providence, RI, 2012.

Introduction to
Modern Mathematics
ALM 33, pp. 143–169

Three Flavors of Twisted Invariants of Knots

Jérôme Dubois*, Stefan Friedl†, and Wolfgang Lück‡

Abstract

The Alexander polynomial of a knot has been generalized in three different ways to give twisted invariants. The resulting invariants are usually referred to as twisted Alexander polynomials, higher-order Alexander polynomials and L^2-Alexander invariants of knots. We quickly recall the definitions and we summarize and compare some of their properties. We also report on work by the authors on L^2-Alexander torsions and we conclude the paper with several conjectures on L^2-Alexander torsions.

2000 Mathematics Subject Classification: 57M25, 57M27.

Keywords and Phrases: Knot theory, knot genus, fibered knots, Alexander polynomial, Reidemeister torsion, twisted invariants, L^2-invariants.

1 Introduction

Alexander [Al28] introduced in 1928 the eponymous polynomial $\Delta_K(t) \in \mathbb{Z}[t^{\pm 1}]$ of a knot K in the three–dimensional sphere S^3. In contrast to its mysterious twin, the Jones polynomial, the formal properties and the topological content of the Alexander polynomial and its many generalizations are for the most part well-understood. For example, Seifert [Se34] showed that the Alexander polynomial can be normalized such that $\Delta_K(t^{-1}) = \Delta_K(t)$ and $\Delta_K(1) = 1$, and that any polynomial satisfying these two conditions can be realized as the Alexander polynomial of a knot. Furthermore, if K_1 and K_2 are oriented knots, then we can consider the connected sum $K_1 \# K_2$ and we obtain the following equality

$$\Delta_{K_1 \# K_2}(t) = \Delta_{K_1}(t) \cdot \Delta_{K_2}(t). \tag{1}$$

*Université Blaise Pascal-Laboratoire de Mathématiques UMR 6620-CNRS, Campus des Cézeaux-B.P. 80026, 63171 Aubiére cedex, France. Email: jerome.dubois@math.univ-bpclermont.fr

†Fakultät für Mathematik, Universität Regensburg, 93040 Regensburg, Germany. Email: sfriedl@gmail.com

‡Mathematisches Institut, Universität Bonn, Endenicher Allee 60, 53115 Bonn, Germany. Email: wolfgang.lueck@him.uni-bonn.de

In terms of topological information we have for any knot K the inequality

$$\deg(\Delta_K(t)) \leqslant 2\,\mathrm{genus}(K), \tag{2}$$

where genus(K) denotes the minimal genus of a Seifert surface for K. Also, if K is a fibered knot, then

$$\deg(\Delta_K(t)) = 2 \text{ genus } (K) \text{ and } \Delta_K(t) \text{ is monic.} \tag{3}$$

The Alexander polynomial also contains information on symmetries of knots [Mu71, Hat81] and on knot concordance [FM66, Ka78, FQ90]. The original definition of the Alexander polynomial has been extended to much more general settings and the generalizations of the Alexander polynomial have been effectively used in the study of links, 3-manifolds [Mc02], algebraic varieties and singularities [Di92].

Nonetheless, it is well-known that in each case the Alexander polynomial has only partial information. For example, the fact that there exist (infinitely many) non-trivial knots with trivial Alexander polynomial shows that Alexander polynomials are not a complete invariant and it also shows that Equation (2) is in general not an equality. Over the last years a huge effort has been put into finding invariants which generalize the Alexander polynomial and which contain more information, especially more topological information.

Arguably the most important and successful generalization is Heegaard Floer homology due to Ozsváth–Szabó [OS04a, OS04b] and its offspring knot Floer homology [Ras03, OS04c] and sutured Floer homology [Ju06]. These invariants always detect the knot genus, and more generally the Thurston norm, and they also detect fibered knots and 3-manifolds [Ni07, Gh08, Ju08, AN09, Ni09a, Ni09b, AlFJ13]. The Heegaard Floer setup has been amazingly effective in dealing with many problems in topology, it is impossible for us to list all the results. We therefore refer instead to the recent beautiful survey paper by Juhász [Ju13]. Despite, or arguably because of, the power of Heegaard Floer invariants there are at least two issues. For one, despite some progress [MaOS09, SaW10, LCSVV13] the invariants tend to be difficult to calculate in more complicated situations. Furthermore, the Heegaard Floer invariants can not be generalized to invariants of higher-dimensional manifolds or to invariants of groups.

The quest for useful generalizations of the Alexander polynomial is therefore not over yet. In recent years there has been a lot of interest in twisted versions of the Alexander polynomial. These twisted invariants come in three flavors:

(a) The *twisted Alexander polynomial* introduced by Lin [Lin01] and Wada [Wa94] associates to an oriented knot K and a linear representation of its group $\alpha\colon \pi_1(S^3 \setminus K) \to (k, \mathbb{F})$ over a commutative field $\mathbb{F}$ an invariant $\Delta_K^\alpha(t) \in \mathbb{F}[t^{\pm 1}]$.

(b) The *higher-order Alexander polynomial* of Cochran [Co04] associates to an oriented knot K and an epimorphism $\gamma\colon \pi_1(S^3 \setminus K) \to \Gamma$ onto a torsion-free elementary-amenable group Γ an invariant Δ_K^γ which is a Laurent polynomial with coefficients in a certain skew field.

(c) The L^2*-Alexander invariant* of Li–Zhang [LiZ06a, LiZ06a] associates to an oriented knot K in S^3 a function $\Delta_K^{(2)}\colon \mathbb{R}_{>0} \to \mathbb{R}_{\geqslant 0}$.

The goal of this paper is to discuss these three twisted invariants. More precisely, for each invariant we will outline the definition and we will recall some of the key properties, with a special focus on the relationship to the knot genus and fiberedness.

The first two invariants are by now fairly well-known, but the third invariant is hitherto little studied. We will introduce a slight variation on the L^2-Alexander invariant, namely the L^2-Alexander torsion $\tau^{(2)}(K)(t)$ of a knot which is by definition a function $\mathbb{R}_{>0} \to \mathbb{R}_{\geqslant 0}$. It follows from the definitions that $\tau^{(2)}(K)(1)$ equals the usual L^2-torsion of the knot exterior which by work of Lück–Schick [LüS99] implies that $\tau^{(2)}(K)(1)$ is in fact a repackaging of arguably the most important geometric invariant of a knot complement, namely the volume of a knot. More precisely, given a knot K they showed that

$$\tau^{(2)}(K)(1) = \exp\left(\frac{1}{6\pi}\text{Vol}(K)\right),$$

where we define the volume of K as

$$\text{Vol}(K) := \sum \text{volumes of the hyperbolic pieces in the JSJ decomposition of } S^3 \setminus K.$$

Put differently, the function $t \mapsto \tau^{(2)}(K)(t)$ can be viewed as a deformation of the volume of a knot. Our main goal will then be to make the following conjecture precise and to give some evidence towards it.

Conjecture 1.1. *Let K be a knot.*

(1) *The full L^2-Alexander torsion $\tau^{(2)}(K)(t)$ determines the genus of K.*
(2) *The full L^2-Alexander torsion determines whether or not K is fibered.*
(3) *If K is fibered, then $\tau^{(2)}(K)(t)$ determines the entropy of the monodromy.*

The paper is organized as follows. In Section 2 we recall the recasting of the Alexander polynomial as a torsion invariant, which we refer to as the Alexander torsion. This point of view was introduced by Milnor [Mi62, Mi66] and exploited very successfully by Turaev [Tu86, Tu01, Tu02a]. In our discussion of twisted invariants we will in fact discuss twisted versions of the Alexander torsion. The differences to the aforementioned Alexander polynomials and invariants are minimal, but it is a well-established fact that Reidemeister torsion has better formal properties than orders of modules. In Section 3 we recall the definition and main properties of the twisted Alexander torsion and in Section 4 we do the same for higher-order Alexander torsion of knots. Finally in Section 5 we will turn our attention to the L^2-Alexander torsion of a knot. We first give an outline of the key properties of the Fuglede-Kadison determinant and of L^2-torsions. We then introduce the L^2-Alexander torsions of a knot and we state some of the results and computations obtained by the first author and Wegner [DubW10, DubW13], Ben-Aribi [BA13a, BA13b] and by the authors [DubFL14a, DubFL14b]. We conclude this paper with a long list of open questions on the L^2-Alexander torsions of knots. In particular we will discuss Conjecture 1.1 in more detail.

Acknowledgments. The second author wishes to thank the organizers of the Fourth Conference of the Tsinghua Sanya International Mathematics Forum and in particular the organizers of the workshop 'speculations and wild conjectures in low dimensional differential topology'. The first two authors would like to thank IMJ-PRG for its hospitality during the stay of the second author in Paris in March 2012. The paper is financially supported by the Leibniz-Award of the third author granted by the Deutsche Forschungsgemeinschaft DFG. We are also grateful to Fathi Ben Aribi and Stefano Vidussi for helpful comments.

2 The Alexander polynomial and Alexander torsion

The Alexander polynomial has many different definitions, the equivalence of which is not always entirely obvious. We now recall the definition which for most theoretical purposes is the most useful and which also lends itself most easily to generalizations to other classes of manifold and groups.

Let $K \subset S^3$ be an oriented knot. We denote by νK an open tubular neighborhood of K and we denote by $X_K := S^3 \setminus \nu K$ the knot exterior. We refer to $\pi_K = \pi_1(S^3 \setminus \nu K)$ as the group of K and we denote by $\widetilde{X_K}$ the universal cover of X_K. Throughout this paper we always think of X_K as equipped with a CW-structure. Note that π_K acts via deck transformations on the left on $\widetilde{X_K}$ and thus induces a left $\mathbb{Z}[\pi_K]$-module structure on $C_*(\widetilde{X_K}) := C_*(\widetilde{X_K};\mathbb{Z})$. Using the canonical involution on $\mathbb{Z}[\pi_K]$ we henceforth view $C_*(\widetilde{X_K})$ as a right $\mathbb{Z}[\pi_K]$-module.

We denote by $\phi_K\colon \pi_K \to \langle t\rangle = \mathbb{Z}$ the abelianization map which sends the oriented meridian of K to t. We then consider the chain complex

$$C_*(\widetilde{X_K}) \otimes_{\mathbb{Z}[\pi_K]} \mathbb{Z}[t^{\pm 1}]$$

of $\mathbb{Z}[t^{\pm1}]$-modules. Here $g \in \pi_K$ acts on $\mathbb{Z}[t^{\pm1}]$ by multiplication by $t^{\phi_K(g)}$. We then write

$$H_k(X_K;\mathbb{Z}[t^{\pm1}]) = H_k\left(C_*(\widetilde{X_K}) \otimes_{\mathbb{Z}[\pi_K]} \mathbb{Z}[t^{\pm1}]\right).$$

The *Alexander polynomial of* K is then defined as the order of the $\mathbb{Z}[t^{\pm1}]$-module $H_1(X_K;\mathbb{Z}[t^{\pm1}])$, i.e., it is defined as

$$\Delta_K(t) := \operatorname{order}_{\mathbb{Z}[t^{\pm1}]}\left(H_1(X_K;\mathbb{Z}[t^{\pm1}])\right).$$

We refer to [Tu01, Hi12] for details on orders. It follows in particular from the theory of orders that $\Delta_K(t)$ is well-defined up to multiplication by a unit in $\mathbb{Z}[t^{\pm1}]$, i.e., up to multiplication by an element of the form $\pm t^k$ with $k \in \mathbb{Z}$.

We now turn to the Alexander torsion of a knot. We consider the chain complex

$$C_*(\widetilde{X_K}) \otimes_{\mathbb{Z}[\pi_K]} \mathbb{Q}(t)$$

of $\mathbb{Q}(t)$-modules. By picking an order of the cells, an orientation of each cell, and by picking a lift of each cell of X_K to $\widetilde{X_K}$ we can view the above chain complex as a chain complex of based free $\mathbb{Q}(t)$-modules. One can further observe that this chain complex is acyclic (see [Mi62]), so that it is possible to compute its Reidemeister torsion. The Alexander torsion of K, which is sometimes also referred to as Milnor torsion, is then defined as

$$\tau(K)(t) := \operatorname{Tor}\left(C_*(\widetilde{X_K}) \otimes_{\mathbb{Z}[\pi_K]} \mathbb{Q}(t)\right) \in \mathbb{Q}(t).$$

We refer to [Mi66, Tu86, Tu01] for more on torsions of chain complexes. This invariant depends on the choice of the CW-structure, the ordering of the cells, the orientation of the cells and the choice of lifts of the cells to the universal cover. Nonetheless, it follows from standard arguments that $\tau(K)(t)$ is well-defined up to multiplication by an element of the form $\pm t^k, k \in \mathbb{Z}$.

The Alexander torsion $\tau(K)(t)$ thus has the same indeterminacy as the Alexander polynomial. In fact Milnor [Mi62, Mi66] showed that the Alexander torsion $\tau(K)(t) \in \mathbb{Q}(t)$ satisfies the following equality

$$\tau(K)(t) = (1-t)^{-1} \cdot \Delta_K(t).$$

Even though the difference between $\tau(K)(t)$ and $\Delta_K(t)$ is marginal, Turaev [Tu86, Tu01, Tu02a] showed that this shift in point of view greatly simplifies many proofs and that it is 'the right point of view'. It is also much easier to generalize $\tau(K)(t)$ to twisted settings and to prove properties of these new invariants.

In the remainder of this section we translate several of the aforementioned properties of the Alexander polynomial into properties of the Alexander torsion. For example, the fact that $\Delta_K(t)$ can be chosen to be symmetric is equivalent to the statement that for any representative of $\tau(K)(t)$ we have

$$\tau(K)(t^{-1}) = -t^l \tau(K)(t) \tag{4}$$

for some odd l. In the following we define the degree of a non-zero polynomial $p(t) = \sum_{i=k}^{l} a_i t^i$ with $a_k \neq 0$ and $a_l \neq 0$ as $\deg(p(t)) = l - k$. For a non-zero rational function $f(t) = p(t)/q(t)$ we define its degree as $\deg(f(t)) = \deg(p(t)) - \deg(q(t))$. We extend this to $\deg(0) := -\infty$. Note that with this convention Equality (4) implies that $\deg(\tau(K)(t))$ is odd. Furthermore, Inequality (2) translates into

$$\deg(\tau(K)(t)) \leqslant 2\operatorname{genus}(K) - 1. \tag{5}$$

We furthermore say that a rational function is *monic* if it is the quotient of two monic polynomials, i.e., polynomials for which the top coefficient is ± 1. If K is a fibered knot, then the conditions stated in Equation (3) now translate into

$$\deg(\tau(K)(t)) = 2\operatorname{genus}(K) - 1 \text{ and } \tau(K)(t) \text{ is monic.} \tag{6}$$

3 Twisted Alexander torsion

3.1 Definition

Let $K \subset S^3$ be an oriented knot and let $\alpha \colon \pi_K \to (k, \mathbb{F})$ be a representation over a commutative field $\mathbb{F}$. We consider the chain complex

$$C_*(\widetilde{X_K}) \otimes_{\mathbb{Z}[\pi_K]} (\mathbb{F}(t) \otimes \mathbb{F}^k)$$

of $\mathbb{F}(t)$-modules, where $g \in \pi_K$ acts again on $\mathbb{F}(t)$ by multiplication by $t^{\phi_K(g)}$ and it acts furthermore on $\mathbb{F}^k$ via the representation α. We pick a basis for $\mathbb{F}^k$ and lifts of the cells of X_K to $\widetilde{X_K}$. The tensor products of the basis elements and the lifts then turn the above into a based $\mathbb{F}(t)$-chain complex. If the above twisted chain complex is non-acyclic, then we write $\tau(K,\alpha)(t) = 0$. Otherwise the *twisted Alexander torsion of* (K,α) is defined as

$$\tau(K,\alpha)(t) := \operatorname{Tor}\left(C_*(\widetilde{X_K}) \otimes_{\mathbb{Z}[\pi_K]} (\mathbb{F}(t) \otimes \mathbb{F}^k)\right) \in \mathbb{F}(t).$$

Note that $\tau(K,\alpha)(t) \in \mathbb{F}(t)$ is well-defined up to multiplication by an element of the form $\pm t^l$, $l \in \mathbb{Z}$. In analogy to (4) it was shown in [Ki96, HiSW10, FrKK12, Hi12] that if α is a unitary representation or if α is a representation taking values in $(2,\mathbb{C})$, then for any representative of $\tau(K,\alpha)(t)$ we have

$$\tau(K,\alpha)(t^{-1}) = (-t)^l \tau(K,\alpha)(t)$$

where $l \equiv k \bmod 2$. In particular, if $\tau(K,\alpha)(t) \neq 0$, then the parity of $\deg(\tau(K,\alpha)(t))$ is the same as the parity of k. We refer to [Mo11, MS13] for more on degrees of twisted Alexander torsions.

Kitano [Ki96] showed that the above definition of the twisted Reidemeister torsion $\tau(K,\alpha)(t) \in \mathbb{F}(t)$ of an oriented knot K is equivalent to Wada's invariant [Wa94], which in turn is closely related to the twisted Alexander torsion introduced by Lin [Lin01]. We refer to [Ki96, KiL99a, FrV10] for the precise relationship between the twisted Reidemeister torsions and various related invariants.

3.2 Applications and properties

The twisted Reidemeister torsion $\tau(K,\alpha)(t)$ and its generalizations to links, 3-manifolds and groups have been studied extensively over the last years. These invariants have in particular been applied to knot concordance [KiL99a, KiL99b, Ta02, HerKL10, ColKL13], periodicity of knots [HiLN06, Ell08], detecting various types of knots and links [SiW06, FrV07, FrV13], a certain partial ordering on knots [HoKMS11, HoKMS12] and the study of singularities in particular and algebraic geometry in general [CogF07, Coh08]. We refer to the survey paper [FrV10] for more details.

In the following we will highlight a few results which have appeared after the survey paper [FrV10] was written and we also highlight a few results which will be of interest to us when we compare the three flavors of twisted invariants.

Alexander polynomials and its generalizations are particularly suitable for the study of the knot genus and fibered knots. The following theorem says in particular that twisted Alexander torsions detect the genus of a given knot.

Theorem 3.1. *Let $K \subset S^3$ be an oriented knot. Then for any representation $\alpha\colon \pi_K \to (k,\mathbb{F})$ over a commutative field we have*

$$\deg(\tau(K,\alpha)(t)) \leqslant k(2\operatorname{genus}(K) - 1).$$

Furthermore, there exists a representation $\alpha\colon \pi_K \to (k, \mathbb{F})$ *over a commutative field such that*

$$\deg(\tau(K, \alpha)(t)) = k(2\,\text{genus}(K) - 1).$$

Here, the first statement is proved in [Tu02a, FrK06, Fr13] whereas the second statement is proved in [FrV12b]. The latter result builds on the Virtually Compact Special Theorem of Wise [Wi09, Wi12a, Wi12b] and Przytycki–Wise [PW12] and on the Virtual Fibering Theorem of Agol [Ag08] (see also [FrKt12]).

Similarly we have the following theorem which says that twisted Alexander torsions detect whether or not a given knot is fibered.

Theorem 3.2. *Let* $K \subset S^3$ *be an oriented knot. If* K *is fibered, then for any representation* $\alpha\colon \pi_K \to (k, \mathbb{F})$ *over a commutative field we have*

$$\deg(\tau(K, \alpha)(t)) = k(2\,\text{genus}(K) - 1)$$

and $\tau(K, \alpha)(t)$ *is monic. Conversely, if* K *is not fibered, then there exists a representation* $\alpha\colon \pi_K \to (k, \mathbb{F})$ *over a commutative field such that*

$$\tau(K, \alpha)(t) = 0.$$

Here, the first part of the theorem was shown in [GKiM05] (see also [GM03, Ch03, Fr13]) and the second part was shown in [FrV12a] (see also [FrV08, FrV11]), the proof of which again builds on the recent work of Wise [Wi09, Wi12a, Wi12b].

The alert reader will have noticed that neither Theorem 3.1 nor Theorem 3.2 specifies the representation which detects the genus and which detects non-fiberedness. This is not a problem for some of the theoretical applications of the theorems, e.g. in the applications to splittings of knot groups [FrSW13], symplectic 4-manifolds [FrV08, FrV11, FrV12a], rank gradients of 3-manifold groups [DFV14] and fundamental groups of non-fibered knots [SiW09a, SiW09b].

It is also straightforward to see that both theorems give rise to an algorithm which determines the genus and the fiberedness of a given knot. We refer to [FrV12b] for details.

3.3 Questions and conjectures

As we mentioned in the previous section, neither Theorem 3.1 nor Theorem 3.2 specifies the representation which has the desired property. If we want efficient algorithms for determining the knot genus and fiberedness it would be helpful to have more precise information regarding the representations. The following conjecture was formulated in [DunFJ12].

Conjecture 3.3. *Let* $K \subset S^3$ *be a hyperbolic oriented knot. Let* $\alpha\colon \pi_K \to (2, \mathbb{C})$ *be a lift of the discrete and faithful representation, then*

$$\deg(\tau(K, \alpha)(t)) = 2(2\,\text{genus}(K) - 1).$$

Furthermore, K *is fibered if and only if* $\tau(K, \alpha)(t)$ *is monic.*

A proof of this conjecture would result in an extremely fast algorithm for determining the knot and fiberedness of a hyperbolic knot. In [DunFJ12] the conjecture was verified for all hyperbolic knots up to 15 crossings. Further positive evidence towards this conjecture was also given in [Mo12, KiKM13, MoT13, Tr13]. We refer to [DunFJ12] for various other conjectures and open questions regarding the twisted Alexander torsion of hyperbolic knots.

An elementary satellite knot construction shows that for any Alexander polynomial of a knot there exist infinitely many distinct knots with the same Alexander polynomial. Put differently, Alexander polynomials are far from being a complete invariant for knots. One of the first uses of twisted Alexander torsions, see [Lin01, DunFJ12], was to show that they can be used to distinguish the Conway knot from the Kinoshita-Terasaka knot. The following question is asked in [Hi12, Section 6.6.1].

Question 3.4. *Is it possible to distinguish any two distinct prime knots using twisted Alexander torsions?*

Note that it requires some thought to make this question precise, since the twisted Alexander torsion $\tau(K,\alpha)(t)$ is an invariant of a knot together with a representation α. Some evidence for a positive answer is provided by [SiW06, FrV07, FrV13] where it is shown that twisted Alexander torsions detect the unknot, the trefoil and the Figure-8 knot. We also refer to [Ei07] for related ideas.

4 Noncommutative invariants

4.1 Definition

Given a knot $K \subset S^3$ we say that an epimorphism $\gamma\colon \pi_K \to \Gamma$ is *admissible* if the abelianization $\phi_K\colon \pi_K \to H_1(X_K;\mathbb{Z}) \cong \mathbb{Z}$ factors through γ. Note that by replacing $\gamma\colon \pi_K \to \Gamma$ with $\gamma \times \phi_K$ we can turn any homomorphism into an admissible homomorphism. Admissibility is thus not a big restriction. Throughout this section let $\gamma\colon \pi_K \to \Gamma$ be an admissible epimorphism to a torsion-free elementary-amenable group. Since Γ is torsion-free and elementary-amenable the ring $\mathbb{Z}[\Gamma]$ admits by [DLMSY03, KrLM88] an Ore localization which we denote by $\mathbb{K}(\Gamma)$. We then consider the chain complex

$$C_*(\widetilde{X_K}) \otimes_{\mathbb{Z}[\pi_K]} \mathbb{K}(\Gamma)$$

of right $\mathbb{K}(\Gamma)$-modules, where $g \in \pi_K$ acts on $\mathbb{K}(\Gamma)$ via left multiplication by $\gamma(g)$. We again pick a lift of each cell in X_K to $\widetilde{X_K}$. The corresponding cells then turn the above into a based chain complex of right $\mathbb{K}(\Gamma)$-modules. If the chain complex is not acyclic, then we define $\tau(K,\gamma) := 0$. Otherwise we consider the corresponding Whitehead-Reidemeister torsion

$$\tau(K,\gamma) := \tau\left(C_*(\widetilde{X_K}) \otimes_{\mathbb{Z}[\pi_K]} \mathbb{K}(\Gamma)\right) \in K_1(\mathbb{K}(\Gamma)).$$

We refer to [Ro94] for the definition of the K_1-group of a ring. Note that if we write $\mathbb{K}(\Gamma)^\times = \mathbb{K}(\Gamma) \setminus \{0\}$, then the Dieudonné determinant induces by [Ro94] a canonical isomorphism

$$K_1(\mathbb{K}(\Gamma)) \xrightarrow{\cong} \mathbb{K}(\Gamma)^\times_{ab} := \mathbb{K}(\Gamma)^\times/[\mathbb{K}(\Gamma)^\times, \mathbb{K}(\Gamma)^\times].$$

The *higher-order Alexander torsion of* (K, α) is then defined as the image of $\tau(K, \gamma) \in \{0\} \cup \mathbb{K}(\Gamma)^\times_{ab}$. It is well-defined up to multiplication by an element of the form $\pm g, g \in \Gamma$. The higher-order Alexander torsion was first defined in [Fr07], it is a slight variation of the higher-order Alexander polynomial introduced by Cochran [Co04] and Harvey [Har05].

It is perhaps puzzling at first glance that no 't' appears in this definition at all. It is thus a priori not clear why this should be viewed as a higher-order *Alexander* torsion. Also note that the higher-order Alexander torsion takes values in a rather unwieldy algebraic object. This has caused serious problems and in fact the only useful invariant which has ever been extracted is the degree of $\tau(K, \gamma)$. In the following we will thus recall the definition of $\deg(\tau(K, \gamma))$.

By the admissibility of γ the abelianization epimorphism $\phi_K : \pi_K \to \mathbb{Z}$ factors through γ. We denote the resulting epimorphism $\Gamma \to \mathbb{Z}$ by ϕ_K as well. Given a non-zero $p = \sum_{g \in \Gamma} a_g g \in \mathbb{Z}[\Gamma]$ we now write

$$\deg(p) := \max\{\phi_K(g) - \phi_K(h) \mid a_g \neq 0 \text{ and } a_h \neq 0\}.$$

Furthermore, for $pq^{-1} \in \mathbb{K}(\Gamma)$ with $p, q \in \mathbb{Z}[\Gamma]$ and $p, q \neq 0$ we define

$$\deg(pq^{-1}) = \deg(p) - \deg(q).$$

We again extend this to $\deg(0) := -\infty$. It follows from $\deg(fg) = \deg(f) + \deg(g)$ that

$$\deg(\tau(K, \gamma)) \in \mathbb{Z} \cup \{-\infty\}$$

is well-defined. As an example, if we take $\gamma = \phi_K$ to be the abelianization, then $\tau(K, \phi_K) = \tau(K)(t)$ 'on the nose', and $\deg(\tau(K, \phi_K)) = \deg(\tau(K)(t))$.

The fact that $\tau(K, \gamma)$ always has a degree is justification enough for us to refer to it as an Alexander invariant. In [FrK08, FrKK12] Equality (4) was generalized to any γ, more precisely, it was shown that $\deg(\tau(K, \gamma))$ is always odd.

4.2 Applications and properties

The higher-order invariants, and its generalizations to more general manifolds have been studied in the context of knot concordance [CocT08], singular plane curves [LeM06, LeM08], Morse-Novikov theory [Kiy10] and homology cobordisms of surfaces [Sa06, Sa08]. Arguably the greatest interest in higher-order invariants stems from their connection to the knot genus and the Thurston norm. It is perhaps not entirely surprising that higher-order Alexander torsions give lower bounds on the knot genus. Indeed, the following theorem was proved in [Co04], with generalizations and extensions given in [Tu02b, Har05, FrH07, Fr07].

Theorem 4.1. *Let $K \subset S^3$ be an oriented knot. Then for any admissible epimorphism $\gamma : \pi_K \to \Gamma$ onto a torsion-free elementary-amenable group we have*

$$\deg(\tau(K, \gamma)) \leqslant 2\,\mathrm{genus}(K) - 1.$$

In general higher-order invariants are very difficult to calculate, see e.g. [Ho13] for some algorithms and a discussion of the inherent difficulties. Nonetheless, the examples given in [Co04, Har05, Ho13] show that higher-order invariants give very powerful lower bounds on the knot genus.

From a purely theoretical point of view the most interesting fact about higher-order Alexander torsions is the following theorem, due to Cochran [Co04], with extensions given in [Har06, Fr07], which says loosely speaking that 'the bigger the quotient, the better the lower bound on the knot genus'. More precisely, the following theorem holds.

Theorem 4.2. *Let $K \subset S^3$ be an oriented knot. Let $\gamma\colon \pi_K \to \Gamma$ be an epimorphism onto a torsion-free solvable group group and let $\delta\colon \Gamma \to \Omega$ be another epimorphism onto a torsion-free solvable group such that $\gamma \circ \delta$ is admissible. Then the following inequality holds:*

$$\deg(\tau(K,\gamma)) \geqslant \deg(\tau(K,\gamma\circ\delta)).$$

Recall that if $\gamma\colon \pi_K \to \Gamma$ is an admissible epimorphism, then by definition the abelianization ϕ_K factors through γ. It thus follows from Theorem 4.2 that the degree of a higher-order Alexander torsion is always at least the degree of the ordinary Alexander torsion.

Finally, we turn to fibered knots. If $K \subset S^3$ is a fibered knot, then for any admissible epimorphism $\gamma\colon \pi_K \to \Gamma$ onto a torsion-free elementary-amenable we have by [Co04, Har05, Fr07] the following equality:

$$\deg(\tau(K,\gamma)) = 2\,\mathrm{genus}(K) - 1.$$

This gives only very limited information on fiberedness. For example, if K is a non-fibered knot with $\deg(\Delta_K(t)) = 2\,\mathrm{genus}(K)$, then it follows from Theorem 4.2 that degrees of higher-order torsion can not detect that K is not fibered. The problem with higher-order Alexander torsions is that there is no established notion of 'monicness'. One way around this problem is to define 'monicness' in a radically different way, namely as the vanishing of a non-commutative Novikov homology as introduced in [Si87]. This point of view will be discussed in more detail in [Fr14]. We also refer to [GS11] for another approach to using noncommutative invariants for detecting non-fibered knots.

4.3 Questions and conjectures

The conclusion of Theorem 4.2 seems to suggest that the degree of the higher-order Alexander torsion corresponding to an epimorphism $\gamma\colon \pi_K \to \Gamma$ onto a torsion-free elementary-amenable group is the optimal lower bound on the genus that one can obtain from twisted invariants where the twisting factors through γ. We therefore propose the following conjecture.

Conjecture 4.3. *Let $K \subset S^3$ be an oriented knot. Let $\gamma\colon \pi_K \to \Gamma$ be an epimorphism onto a torsion-free elementary-amenable group and let $\delta\colon \Gamma \to \mathrm{GL}(k,\mathbb{F})$ be* *a representation over a commutative field. Then the following inequality holds:*

$$\deg(\tau(K,\gamma)) \geqslant \frac{1}{k}\deg(\tau(K,\gamma\circ\delta)(t)).$$

We next turn to the question on whether higher-order Alexander torsions can detect the genus of a knot. The invariants of Cochran [Co04, Har05] were initially defined for epimorphisms onto PTFA groups, which are special classes of torsion-free solvable groups. It is straightforward to see that if K is a knot with trivial Alexander polynomial, then the abelianization is the only epimorphism onto a non-trivial torsion-free solvable group. It follows that all the higher-order invariants of K corresponding to epimorphisms onto torsion-free solvable groups are equal to $(1-t)^{-1}$, in particular they can not determine the genus of K.

There is hope though if we consider higher-order invariants corresponding to epimorphisms onto torsion-free elementary-amenable groups. In fact we propose the following conjecture.

Conjecture 4.4. *For any knot K there exists an epimorphism $\gamma\colon \pi_K \to \Gamma$ onto a torsion-free elementary-amenable group, such that*

$$\deg(\tau(K,\gamma)) = 2\,\mathrm{genus}(K) - 1.$$

In order for this conjecture to have a chance to be true we need that knot groups have 'enough' epimorphisms onto torsion-free elementary-amenable group. Before we state our next conjecture we recall that if $\mathcal{P}$ is a property of groups, then a group π is called *residually $\mathcal{P}$* if given any $g \in \pi$ there exists an epimorphism $\gamma\colon \pi \to \Gamma$ such that $\gamma(g)$ is non-trivial and such that Γ has Property $\mathcal{P}$.

We propose the following conjecture, which also appears in [AsFW13] as a question.

Conjecture 4.5. *Given any knot K the group π_K is residually torsion-free elementary-amenable.*

Some evidence for a positive answer to this conjecture is provided by the work of Przytycki-Wise [PW12] and Wise [Wi09, Wi12a, Wi12b] which implies that any knot group is virtually residually torsion-free nilpotent, i.e., any knot group admits a finite index normal subgroup which is residually torsion-free nilpotent. (We refer to [AsFW13] for precise references.)

5 The L^2-Alexander torsion

In this last section we will discuss L^2-Alexander torsions. The original definition of L^2-Alexander invariants was given by Li-Zhang [LiZ06a, LiZ06b, LiZ08] and so far it has been studied only in a few papers [Du11, DubW10, DubW13, BA13a, BA13b]. Since this invariant is a relative newcomer and still largely unknown we will discuss this invariant in more detail than the previous two twisted invariants.

5.1 Definition of the L^2-torsion of a chain complex

Before we start with the definition of the L^2-Alexander torsion we need to recall some key properties of the Fuglede-Kadison determinant and the definition of the L^2-torsion of a chain complex of $\mathbb{R}[\Gamma]$-modules. Throughout the section we refer to [Lü02] and to [DubFL14a] for details and proofs.

Throughout this section let Γ be a group. Let A be a matrix over $\mathbb{R}[\Gamma]$. Then there exists the notion of A being of 'determinant class'. (To be slightly more precise, we view the $k \times l$-matrix A as a homomorphism $\mathcal{N}(\Gamma)^l \to \mathcal{N}(\Gamma)^k$, where $\mathcal{N}(\Gamma)$ is the von Neumann algebra of Γ, and then there is the notion of being of 'determinant class'.) We treat this entirely as a black box, but we note that if Γ is residually amenable, e.g. if Γ is a 3-manifold group [Hem87] or if Γ is solvable, then by [Lü94, Sc01, Cl99, EleS05] any matrix over $\mathbb{Q}[\Gamma]$ is of determinant class. It is in fact possible that all matrices which appear in our setup are of determinant class.

Let A be a matrix over $\mathbb{R}[\Gamma]$. (Note that we do not assume that A is a square matrix.) If A is not of determinant class then for the purpose of this paper we define $\det_\Gamma(A) = 0$. On the other hand, if A is of determinant class, then we define

$$\det_\Gamma(A) := \text{Fuglede-Kadison determinant of } A.$$

We will not provide a formal definition of the Fuglede-Kadison determinant, the non specialist reader can think of it as a continuous generalization of the usual determinant of an operator with a finite spectrum to an operator with an infinite spectrum. Furthermore, in this paper we will not provide any proofs, but to give the reader a flavor of the Fuglede-Kadison determinant we now list some properties.

(a) If A is of determinant class, then $\det_\Gamma(A) > 0$.
(b) If $\Gamma = \{e\}$ is the trivial group and A is a square matrix over $\mathbb{R} = \mathbb{R}[\{e\}]$ such that the ordinary determinant $\det(A) \in \mathbb{R}$ is non-zero, then it follows from [Lü02, Example 3.12] that $\det_{\{e\}}(A) = |\det(A)|$.
(c) If A is a matrix over $\mathbb{R}[\Gamma]$ and if Γ is a subgroup of a group G, then $\det_\Gamma(A) = \det_G(A)$.
(d) The Fuglede-Kadison determinant stays unchanged under the following operations.
    (i) Adding a column of zeros or a row of zeros to a matrix.
    (ii) Swapping two columns or rows.
    (iii) Right multiplication of a column by some $\pm g$ with $g \in G$.
    (iv) Left multiplication of a row by some $\pm g$ with $g \in G$.
(e) If $\Gamma = \langle x \rangle$ is an infinite cyclic group and if A is a square matrix over $\mathbb{R}[\Gamma] = \mathbb{R}[x^{\pm 1}]$ such that the ordinary determinant $\det(A) \in \mathbb{R}[x^{\pm 1}]$ is non-zero, then it follows from [Lü02, Example 3.22] and [Rai12, Section 1.2] that

$$\det_{\langle x \rangle}(A) = m(\det(A)) \tag{7}$$

where $\det(A) \in \mathbb{R}[x^{\pm 1}]$ is the usual determinant of A over the ring $\mathbb{R}[x^{\pm 1}]$, and where given a non-zero polynomial $p(x) \in \mathbb{R}[x^{\pm 1}]$ we denote by $m(p(x))$

its Mahler measure. Recall, that if $p(x) = c_n x^n + c_{n-1}x^{n-1} + \cdots + c_1 x + c_0$ (with $c_n \neq 0$ and $c_0 \neq 0$) and if $a_1, \ldots, a_n$ are the roots of $p(x)$, then it follows from Jensen's formula that

$$m(p(x)) = |c_n| \cdot \prod_{j=1}^{n} \max(|a_j|, 1). \tag{8}$$

We refer to [SiW04] for more details and references.

(f) For any two square matrices A and B of determinant class of the same size which have 'full rank' we have $\det_\Gamma(AB) = \det_\Gamma(A) \cdot \det_\Gamma(B)$. Here, we say that a square $k \times k$-matrix over $\mathbb{R}[\Gamma]$ has 'full rank' if the L^2-Betti number of the kernel of the corresponding automorphism of $\mathbb{R}[\Gamma]^k$ is zero.

After this quick low-carb introduction to the Fuglede-Kadison determinant we now turn to L^2-torsions of chain complexes over group rings. In the following let

$$C_* = \left(0 \to \mathbb{R}[\Gamma]^{n_k} \xrightarrow{A_k} \mathbb{R}[\Gamma]^{n_k} \xrightarrow{A_{k-1}} \cdots \mathbb{R}[\Gamma]^{n_1} \xrightarrow{A_1} \mathbb{R}[\Gamma]^{n_0} \to 0\right)$$

be a chain complex of free based left-$\mathbb{R}[\Gamma]$-modules. (In particular we view the elements of $\mathbb{R}[G]^{n_i}$ as row vectors on which the matrices A_i acts by right multiplication.) We can then consider the corresponding L^2-Betti numbers $b_i^{(2)}(C_*) \in \mathbb{R}_{\geqslant 0}$. If any of these L^2-Betti number is non-zero or if any of the boundary maps is not of determinant class, then we define the L^2-torsion $\tau^{(2)}(C_*) := 0$. Otherwise we define the L^2-torsion of C_* to be

$$\tau^{(2)}(C_*) := \prod_{i=1}^{k} \det_\Gamma(A_i)^{(-1)^i} \in \mathbb{R}_{>0}.$$

5.2 The L^2-torsion of a knot

Let $K \subset S^3$ be an oriented knot and let $\gamma\colon \pi_K \to \Gamma$ be an epimorphism onto a group. We consider the chain complex

$$C_*^\gamma(X_K; \mathbb{R}[\Gamma]) := \mathbb{R}[\Gamma] \otimes_{\mathbb{Z}[\pi_K]} C_*(\widetilde{X_K})$$

where $g \in \pi_K$ acts on $C_*(\widetilde{X_K})$ using the deck transformation action and where $g \in \pi_K$ acts on $\mathbb{R}[\Gamma]$ by right-multiplication by $\gamma(g)$.

Remark. The astute reader will have noticed that in contrast to the previous twisted invariants we now view $C_*(\widetilde{X_K})$ as a left $\mathbb{Z}[\pi_K]$-module. In fact it makes basically no difference whether we view $C_*(\widetilde{X_K})$ as a left $\mathbb{Z}[\pi_K]$-module or a right $\mathbb{Z}[\pi_K]$-module. But in most papers on twisted and higher-order Alexander torsions the later convention is used, whereas in all papers on L^2-Alexander torsions the former convention is used.

We once again pick a lift of each cell in X_K to $\widetilde{X_K}$. Note that the chain complex $C_*^\gamma(X_K; \mathbb{R}[\Gamma])$ is a chain complex of free left-$\mathbb{R}[\Gamma]$-complexes where the lifts of the cells give naturally rise to a basis. We can therefore define

$$\tau^{(2)}(K, \gamma) := \tau^{(2)}\left(\mathbb{R}[\Gamma] \otimes_{\mathbb{Z}[\pi_K]} C_*(\widetilde{X_K})\right) \in \mathbb{R}_{\geqslant 0}.$$

Note that $\tau^{(2)}(K,\gamma) \in \mathbb{R}_{\geqslant 0}$ is well-defined with no indeterminacy. We refer to $\tau^{(2)}(K,\gamma)$ as the *L^2-torsion of (K,γ)*. If γ is the identity, then we write $\tau^{(2)}(K) = \tau^{(2)}(K,\gamma)$, and we refer to is as the *full L^2-torsion of K*.

As we mentioned in the introduction, Lück–Schick [LüS99] showed that the full L^2-torsion of K is in fact a repackaging of the volume of a knot, namely they showed that

$$\tau^{(2)}(K) = \exp\left(\frac{1}{6\pi}\operatorname{Vol}(K)\right). \tag{9}$$

5.3 The L^2-Alexander torsion of a knot

In the last section we stated that the volume of a knot can be recovered as an L^2-invariant of its exterior. Even though this is a very pretty fact, this result is also a little disappointing in so far as it shows that the full L^2-torsion does not give us any new information on a knot. The problem with L^2-torsions is furthermore that they are 'just' numbers, they therefore have little structure.

Following Li–Zhang [LiZ06a, LiZ06b, LiZ08] we will address these issues by twisting the representations using characters. More precisely, let $K \subset S^3$ be an oriented knot and let $\gamma\colon \pi_K \to \Gamma$ be an admissible homomorphism, i.e., γ is an homomorphism such that the abelianization ϕ_K factors through γ. Furthermore let $t \in \mathbb{R}_{>0}$. We then consider the representation

$$\begin{aligned} \gamma_t\colon \pi_K &\to \operatorname{Aut}(\mathbb{R}[\Gamma]) \\ g &\mapsto (f \mapsto t^{\phi(g)}\gamma(g)\cdot f) \end{aligned}$$

and we consider the chain complex

$$C_*^{\gamma_t}(X_K;\mathbb{R}[\Gamma]) := \mathbb{R}[\Gamma] \otimes_{\mathbb{Z}[\pi_K]} C_*(\widetilde{X_K})$$

where $g \in \pi_K$ acts on $\mathbb{R}[\Gamma]$ on the right via the representation γ_t. We pick a lift of each cell in X_K to $\widetilde{X_K}$. For each $t \in \mathbb{R}_{>0}$ the above is now a chain complex of free left $\mathbb{R}[\Gamma]$-modules where the lifts of the cells give naturally rise to a basis. We can therefore define

$$\tau^{(2)}(K,\gamma)(t) := \tau^{(2)}\left(C_*^{\gamma_t}(X_K;\mathbb{R}[\Gamma])\right) \in \mathbb{R}_{\geqslant 0}.$$

Put differently, we just defined a function

$$\begin{aligned} \tau^{(2)}(K,\gamma)\colon \mathbb{R}_{>0} &\to \mathbb{R}_{\geqslant 0} \\ t &\mapsto \tau^{(2)}(K,\gamma)(t) \end{aligned}$$

which we will refer to as the *L^2-Alexander torsion of (K,γ)*. This function is well-defined up to multiplication by a function of the form $t \mapsto t^k$, for some $k \in \mathbb{Z}$, see e.g. [LiZ06a, LiZ06b, DubW13].

We now say that two functions $f, g : \mathbb{R}_{>0} \to [0,\infty)$ are *equivalent*, written as $f \doteq g$, if there exists a $k \in \mathbb{Z}$, such that

$$f(t) = t^k g(t) \text{ for all } t \in \mathbb{R}_{>0}.$$

By the above, the equivalence class of $\tau^{(2)}(K,\gamma)(t)$ is a well–defined invariant of (K,γ).

One of the reasons why L^2-invariants are so interesting is the fact that any group admits a canonical epimorphism onto a group, namely the identity map. In the following we refer to the corresponding L^2-Alexander torsion as the *full* L^2-Alexander torsion of K. More precisely, the full L^2-Alexander torsion of K is defined as

$$\tau^{(2)}(K)(t) := \tau^{(2)}(K, \mathrm{id}_{\pi_K})(t).$$

Note that in contrast to the twisted Reidemeister torsion and the higher-order Alexander torsion which always depend on the choice of a representation, the full L^2-Alexander torsion is finally again a canonical invariant. Moreover, the invariant looks promising as it 'sees the whole group'.

The full L^2-Alexander torsion of K is a slight variation on the L^2-Alexander invariant $\Delta_K^{(2)}(t)\colon \mathbb{R}_{>0} \to \mathbb{R}_{\geqslant 0}$ which was first introduced by Li–Zhang [LiZ06a, LiZ06b, LiZ08]. In fact in [DubFL14a] we show that

$$\Delta_K^{(2)}(t) \doteq \tau^{(2)}(K)(t) \cdot \max\{1,t\}.$$

5.4 Properties

We have seen that the Alexander polynomial and its generalizations are symmetric. In [DubFL14b] we will show that if $K \subset S^3$ is an oriented knot and $\gamma\colon \pi_K \to \Gamma$ an admissible epimorphism onto a group, then $\tau^{(2)}(K,\gamma)(t)$ is also symmetric, in the sense that for any representative $\tau^{(2)}(K,\gamma)(t)$ of the L^2-Alexander torsion of (K,γ) we have

$$\tau^{(2)}(K,\gamma)(t^{-1}) = t^n \cdot \tau^{(2)}(K,\gamma)(t) \tag{10}$$

for some odd n.

It is obvious from the definitions that the full L^2-Alexander torsion of a knot K satisfies $\tau^{(2)}(K)(1) = \tau^{(2)}(K)$. Combining this with Equality (9) we obtain that

$$\tau^{(2)}(K)(1) = \exp\left(\frac{1}{6\pi}\mathrm{Vol}(K)\right). \tag{11}$$

It follows in particular that $\tau^{(2)}(K)(1)$ is non-zero. The full L^2-Alexander torsion thus can be viewed as a deformation of the volume of a knot.

The L^2-Alexander torsion has been determined only for a small number of knots. First of all, a straightforward calculation, using Equations (7) and (8) shows that

$$\tau^{(2)}(\text{unknot})(t) \doteq \max\{1,t\}^{-1}.$$

Dubois–Wegner [DubW10, DubW13] used Fox differential calculus to generalize this result and to show implicitly that if $T_{p,q}$ denotes the (p,q)-torus knot, where p,q are positive coprime integers, then for any admissible epimorphism $\gamma\colon \pi_K \to \Gamma$ we have

$$\tau^{(2)}(T_{p,q},\gamma)(t) \doteq \max\{1, t^{(p-1)(q-1)-1}\} = \max\{1, t^{2\,\mathrm{genus}(K)-1}\}. \tag{12}$$

In [BA13a, BA13b] Ben Aribi also used Fox differential calculus to prove a formula for the L^2-Alexander torsion for cables of knots. Furthermore, Ben Aribi [BA13b, Theorem 1.2] shows that the full L^2-Alexander torsion behaves well under the connected sum operation. More precisely, he showed that for any two oriented knots J and K the following equality holds

$$\tau^{(2)}(J\#K)(t) \doteq \tau^{(2)}(J)(t) \cdot \tau^{(2)}(K)(t) \tag{13}$$

which can be viewed as an analogue of Equation (1). These results will be generalized in [DubFL14a] to the study of L^2-Alexander torsions of graph manifolds and 3-manifolds with non-trivial JSJ decomposition.

The two results of Ben Aribi, together with Equation (9) and [Go83, Corollary 4.2] imply that the full L^2-torsion of a non-trivial knot is not equivalent to $\max\{1,t\}^{-1}$. Put differently, we see that the work of Lück–Schick and Ben Aribi implies that the full L^2-Alexander torsion detects the unknot:

Theorem 5.1. *A knot $K \subset S^3$ is the trivial knot if and only if $\tau^{(2)}(K)(t) \doteq \max\{1,t\}^{-1}$.*

5.5 The L^2-Alexander torsion and fibered knots

Let K be a fibered knot. We denote by $f\colon \Sigma \to \Sigma$ the corresponding monodromy of the fiber surface Σ. We can associate to f its entropy $h(f) \in \mathbb{R}_{\geqslant 0}$, as defined say in [FLP79, p. 185]. Since a fibered knot admits a unique fibration we can define $h(K) := h(f)$.

Note that if K is an iterated torus knot, then K is fibered with $h(K) = 0$. On the other hand, if K is a hyperbolic fibered knot, then the monodromy is pseudo-Anosov, and by [FLP79, p. 195] the entropy equals the logarithm of the dilatation of f.

The following theorem is proved in [DubFL14a]. It gives a partial computation of the L^2-Alexander torsion for fibered knots.

Theorem 5.2. *Suppose K is a fibered knot and $\gamma\colon \pi_K \to \Gamma$ is an admissible epimorphism onto a group that is residually amenable (e.g. Γ is solvable or a 3-manifold group). Then there exists a representative $\tau^{(2)}(K,\gamma)(t)$ of the L^2-Alexander torsion of (K,γ) such that for $T := \exp(h(K))$ we have*

$$\tau^{(2)}(K,\gamma)(t) = \begin{cases} 1, & \text{if } t < \dfrac{1}{T} \\ t^{2\,\mathrm{genus}(K)-1}, & \text{if } t > T. \end{cases}$$

In the following we say that a function $f\colon \mathbb{R}_{>0} \to [0,\infty)$ is *monomial in the limit* if there exist $d, D \in \mathbb{R}$ and non-zero real numbers c, C such that

$$\lim_{t\to 0} \frac{f(t)}{t^d} = c \text{ and } \lim_{t\to\infty} \frac{f(t)}{t^D} = C.$$

For a function that is monomial in the limit as above we define its *degree* as

$$\deg(f(t)) = D - d.$$

We furthermore refer to C as the *top coefficient of f* and we refer to c as the *bottom coefficient of f*. Finally we say that f is *monic* if its top and bottom coefficient are both equal to 1. Note that the above definitions do not depend on the equivalence class of the function.

With these definitions we can formulate the following corollary to Theorem 5.2.

Corollary 5.3. *If K is a fibered knot and if $\gamma\colon \pi_K \to \Gamma$ is an admissible epimorphism, then $\tau^{(2)}(K,\gamma)(t)$ is monomial in the limit, it is monic and it has degree $2\,\mathrm{genus}(K)-1$.*

5.6 The L^2-Alexander torsion and the knot genus

Let $f\colon \mathbb{R}^+ \to [0,\infty)$ be a function which is non-zero for sufficiently small t and for sufficiently large t. We then define

$$\deg(f) = \limsup_{t\to\infty} \frac{\ln(f(t))}{\ln(t)} - \liminf_{t\to 0} \frac{\ln(f(t))}{\ln(t)} \in \mathbb{R}.$$

Note that for functions that are monomial in the limit this definition of degree is the same as in the previous section.

In [DubFL14a] we prove the following theorem that can be viewed as a generalization of Inequality (5).

Theorem 5.4. *For any oriented knot K and any admissible epimorphism $\pi_K \to \Gamma$ onto a virtually abelian group the following inequality holds*

$$\deg\left(\tau^{(2)}(K,\gamma)(t)\right) \leqslant 2\,\mathrm{genus}(K) - 1.$$

5.7 The L^2-Alexander torsion corresponding to the abelianization

From the results of the previous sections it is not easy to guess what the general structure of the L^2-Alexander torsions might look like. We therefore change the point of view and we now consider the L^2-Alexander torsion $\tau^{(2)}(K,\phi_K)(t)$, i.e., the L^2-Alexander torsion corresponding to the abelianization map $\phi_K\colon \pi_K \to \mathbb{Z}$. In [DubFL14a] we prove the following proposition which says in particular that the Alexander polynomial $\Delta_K(t)$ determines $\tau^{(2)}(K,\phi_K)(t)$.

Proposition 5.5. *Let K be a knot and let $\Delta(z) = \Delta_K(z) \in \mathbb{Z}[z^{\pm 1}]$ be a representative of the Alexander polynomial of K. We write*

$$\Delta(z) = C \cdot z^m \cdot \prod_{i=1}^{k} (z - a_i),$$

with some $C \in \mathbb{Z}\setminus\{0\}, m \in \mathbb{Z}$ and $a_1,\dots,a_k \in \mathbb{C}\setminus\{0\}$. Then

$$\tau^{(2)}(K,\phi_K)(t) \doteq C \cdot \prod_{i=1}^{k} \max\{|a_i|,t\} \cdot \max\{1,t\}^{-1}.$$

Outline of the proof. It follows relatively easily from the definitions that there exists a matrix $A(z)$ over $\mathbb{Z}[z^{\pm 1}]$ with $\det(A(z)) = \Delta(z)$ and such that

$$\tau^{(2)}(K, \phi_K)(t) \doteq \det_{\langle z \rangle}(A(tz)) \cdot \det_{\langle z \rangle}(tz - 1)^{-1}.$$

The proposition can then be deduced from Equalities (7) and (8). □

With these definitions we can formulate the following immediate corollary to Proposition 5.5.

Corollary 5.6. *Let K be a knot, then $\tau^{(2)}(K, \phi_K)(t)$ is monomial in the limit with*

$$\deg\left(\tau^{(2)}(K, \phi_K)(t)\right) = \deg\left(\Delta_K(t)\right) - 1.$$

Furthermore, $\tau^{(2)}(K, \phi_K)(t)$ is monic if and only if $\Delta_K(t)$ is monic.

We return to the torus knots as an instructive example. Recall that if $K = T_{p,q}$ is the (p,q)-torus knot, where p, q are coprime positive integers, then its Alexander polynomial is given by

$$\Delta_{T_{p,q}}(t) = \frac{(t^{pq} - 1)(t - 1)}{(t^p - 1)(t^q - 1)}.$$

This is a polynomial of degree $(p-1)(q-1)$ and all the zeros are roots of unity. Proposition 5.5 thus states that

$$\tau^{(2)}(K, \phi_K)(t) \doteq \max\{1, t^{(p-1)(q-1)-1}\}.$$

In fact we had already seen in Equality (12) that this equality holds for $\tau^{(2)}(K, \gamma)$ and any admissible epimorphism γ. Note that if we consider the torus knots $T_{3,7}$ and $T_{4,5}$, then it is now straightforward to see that all the L^2-Alexander torsions agree, but that the ordinary Alexander polynomials are different.

5.8 Questions and conjectures

The calculations presented in (12), Theorem 5.2 and Proposition 5.5 are pretty much the only calculations of L^2-Alexander torsions of knots known to the authors. We will not be intimidated by the scarcity of examples not to state a long list of ambitious conjectures.

We start out with the discussion what the functions $\tau^{(2)}(K, \gamma)(t)$ can look like. We have the following conjecture.

Conjecture 5.7. *Let K be an oriented knot and let $\gamma \colon \pi_K \to \Gamma$ be an admissible epimorphism. Then the following hold:*

(a) $\tau^{(2)}(K, \gamma)(t)$ is monomial in the limit.
(b) $\tau^{(2)}(K, \gamma)(t)$ is continuous.
(c) $\tau^{(2)}(K, \gamma)(t) \cdot \max\{1, t\}$ is convex.

In order to simplify the discussion we will for the remainder of this section assume that Conjecture 5.7 (a) holds, so that we can throughout this section talk of degree, monicness, top coefficient and bottom coefficient. (Note that by Equation (10) the top and bottom coefficient of L^2-Alexander torsions are always the same.)

The following conjecture, which we will discuss in more detail in a future paper, can be viewed as saying that L^2-Alexander torsions are a generalization of the higher-order Alexander torsions.

Conjecture 5.8. *For any oriented knot K and any epimorphism $\gamma\colon \pi_K \to \Gamma$ onto a non-trivial torsion-free elementary-amenable group we have*

$$\deg\left(\tau^{(2)}(K,\gamma)(t)\right) = \deg(\tau(K,\gamma)).$$

In Theorem 4.2 we saw that 'the bigger the quotient-the better the lower bound on the genus'. One problem with higher-order Alexander torsions is that there is no maximal torsion-free elementary-amenable quotient of a knot group. On the other hand such a problem does not exist for the L^2-Alexander torsion, we can just use the identity of the fundamental group. The corresponding full L^2-Alexander torsion should thus give the best possible lower bound on the genus, and we expect that it in fact always determines the knot genus. More precisely, we propose the following conjecture.

Conjecture 5.9. *For any oriented knot K the full L^2-Alexander torsion satisfies*

$$\deg\left(\tau^{(2)}(K)(t)\right) = 2\,\mathrm{genus}(K) - 1.$$

We are quite optimistic about this conjecture, in fact in [DubFL14a] we use the work of Agol [Ag08], Przytycki–Wise [PW12] and Wise [Wi09, Wi12a, Wi12b] to show that given any oriented knot K there exists an epimorphism $\gamma\colon \pi_K \to \Gamma$ onto a virtually abelian group with $\deg\left(\tau^{(2)}(K,\gamma)(t)\right) = 2\,\mathrm{genus}(K) - 1$.

Theorem 5.2 also shows that L^2-Alexander torsions have information about fiberedness. We therefore ask the following question.

Question 5.10. *If K is an oriented knot such that the full L^2-Alexander torsion $\tau^{(2)}(K)(t)$ is monic, does this imply that K is fibered?*

The attentive reader might have noticed that this is phrased as a question, thus somewhat less optimistically than our conjectures. In fact there is some computational evidence, due to Ben Aribi, that the answer might in fact be no.

If K is not fibered, then it is an interesting problem to determine what (if any) geometric or dynamic information is contained in the top coefficient of $\tau^{(2)}(K)(t)$. We now turn to the case that K is fibered. We conjecture that the full L^2-Alexander torsion determines the dilatation of the monodromy. More precisely, we propose the following conjecture.

Conjecture 5.11. *Let K be a fibered oriented knot with monodromy f. We normalize $\tau^{(2)}(K)(t)$ such that $\tau^{(2)}(K)(1) = 1$. Then*

$$\sup\left\{T \in \mathbb{Q}_{\geqslant 0} \,|\, \tau^{(2)}(K)(t)|_{(0,T)} \text{ is constant}\right\} = \exp(-h(f)).$$

Note that by Theorem 5.2 we know that the inequality '$\geqslant$' holds. Proving the inequality '$\leqslant$' seems significantly harder.

We have two pieces of evidence for the inequality '$\leqslant$' of the conjecture. First of all, it follows from Equation (12) that it holds if the monodromy has finite order, i.e., if K is a torus knot. Furthermore, if we consider $\tau^{(2)}(K, \phi_K)(t)$ with the normalization $\tau^{(2)}(K)(1) = 1$, then it follows fairly easily from Corollary 5.6 that

$$\sup\left\{T \in \mathbb{Q}_{\geqslant 0} \mid \tau^{(2)}(K, \phi_K)|_{(0,T)} \text{ is constant}\right\} = \exp(-m),$$

where m is the maximal absolute value of an eigenvalue of the induced automorphism f_* of $H_1(\Sigma; \mathbb{R})$. But it is well-known e that $m = h(f_*\colon H_1(\Sigma; \mathbb{Z}) \to H_1(\Sigma; \mathbb{Z}))$.

In Section 5.7 we saw that L^2-Alexander torsions can not distinguish certain pairs of torus knots. Our final question now asks whether at least hyperbolic knots are determined by their full L^2-Alexander torsion. Here one has to be a little careful about what one means by 'determine'. We say that two knots J and K are equivalent if there exists a diffeomorphism h of S^3 with $h(J) = K$ as sets, i.e. we do not demand that the orientations match. It is straightforward to see that the full L^2-Alexander torsions of equivalent knots are equivalent functions. We propose the following question.

Question 5.12. *If J and K are two oriented hyperbolic knots with equivalent full L^2-Alexander torsions, does this imply that J and K are equivalent?*

A typical example for pairs of knots which are difficult to distinguish is given by mutants. Many invariants, e.g. the Alexander polynomial and the Jones polynomial do not distinguish mutants. It follows from work of Ruberman [Ru87] and Equation (11) that the evaluation of the full L^2-Alexander torsion at $t = 1$ stays invariant under mutation. On the other hand the genus is not invariant under mutation, for example the Conway knot has genus 3 and its mutant, the Kinoshita-Terasaka knot, has genus 2. In light of Conjecture 5.9 we thus expect that in general the full L^2-Alexander torsion is not invariant under mutation.

References

[Ag08] I. Agol, *Criteria for virtual fibering*, J. Topol. 1 (2008), No. 2, 269–284.

[AN09] Y. Ai and Y. Ni, *Two applications of twisted Floer homology*, Int. Math. Res. Not. 2009, No. 19, 3726–3746.

[Al28] J. W. Alexander, *Topological invariants of knots and links*, Trans. Amer. Math. Soc. 30 (1928), No. 2, 275–306.

[AlFJ13] I. Altman, S. Friedl and A. Juhász, *Sutured Floer homology, fibrations, and taut depth one foliations*, Preprint (2013), arXiv:1312.0823.

[AsFW13] M. Aschenbrenner, S. Friedl and H. Wilton, *3-manifold groups*, Preprint (2013), arXiv:1205.0202.

[BA13a] F. Ben Aribi, *The L^2-Alexander invariant detects the unknot*, C. R. Math. Acad. Sci. Paris 351 (2013), 215–219.

[BA13b] F. Ben Aribi, *The L^2-Alexander invariant detects the unknot*, arXiv:1311.7342, to appear Ann. Sc. Norm. Super. Pisa Cl. Sci.

[Ch03] J. Cha, *Fibred knots and twisted Alexander invariants*, Trans. Amer. Math. Soc. 355 (2003), 4187–4200.

[Cl99] B. Clair, *Residual amenability and the approximation of L^2-invariants*, Michigan Math. J. 46 (1999), Issue 2, 331–346.

[Co04] T. Cochran, *Noncommutative knot theory*, Algebr. Geom. Topol. 4 (2004), 347–398.

[CocT08] T. Cochran and T. Kim, *Higher-order Alexander invariants and filtrations of the knot concordance group*, Trans. Amer. Math. Soc. 360 (2008), No. 3, 1407–1441.

[CogF07] A. Cogolludo and V. Florens, *Twisted Alexander polynomials of plane algebraic curves*, J. Lond. Math. Soc. (2) 76 (2007), No. 1, 105–121.

[Coh08] D. Cohen and A. Suciu, *The boundary manifold of a complex line arrangement*, Geometry & Topology Monographs 13 (2008), 105–146.

[ColKL13] J. Collins, P. Kirk and C. Livingston, *The concordance classification of low crossing number knots*, Preprint (2013), arXiv:1309.7252.

[DFV14] J. DeBlois, S. Friedl and S. Vidussi, *Rank gradients of infinite cyclic covers of 3-manifolds*, Mich. J. Math. 63 (2014), 65–81.

[Di92] A. Dimca, *Singularities and topology of hypersurfaces*, Universitext, Springer-Verlag, New York (1992).

[DLMSY03] J. Dodziuk, P. Linnell, V. Mathai, T. Schick and S. Yates, *Approximating L^2-invariants and the Atiyah conjecture*, Communications on Pure and Applied Mathematics 56 (2003), no. 7, 839–873.

[Du11] J. Dubois, *Computational aspects in Reidemeister torsion and Chern-Simons theories*, Chern-Simons gauge theory: 20 years after, 43–64, AMS/IP Stud. Adv. Math., 50, Amer. Math. Soc., Providence, RI (2011).

[DubFL14a] J. Dubois, S. Friedl and W. Lück, *The L^2–Alexander torsion of 3-manifolds*, preprint (2014), arXiv:1410.6918.

[DubFL14b] J. Dubois, S. Friedl and W. Lück, *L^2-Alexander torsions are symmetric*, Preprint (2014), arXiv:1411.2292.

[DubW10] J. Dubois and C. Wegner, *L^2-Alexander invariant for torus knots*, C. R. Math. Acad. Sci. Paris 348 (2010), No. 21-22, 1185–1189.

[DubW13] J. Dubois and C. Wegner, *Weighted L^2-invariants and applications to knot theory*, to appear in Communications in Contemporary Mathematics (2014).

[DunFJ12] N. Dunfield, S. Friedl and N. Jackson, *Twisted Alexander polynomials of hyperbolic knots*, Experiment. Math. 21 (2012), 329–352.

[Ei07] M. Eisermann, *Knot colouring polynomials*, Pac. J. Math. 231 (2007), 305–336.

[EleS05] G. Elek and E. Szabó, *Hyperlinearity, essentially free actions and L^2-invariants*, Math. Ann. 332 (2005), 421–441.

[Ell08] R. Elliot, *Alexander Polynomials of Periodic Knots: A Homological Proof and Twisted Extension*, Princeton Undergraduate Thesis (2008).

[FLP79] A. Fathi, F. Laudenbach and V. Poénaru, *Travaux de Thurston sur les surfaces*, Astérisque, 66-67, Soc. Math. France, Paris (1979).

[FM66] R. H. Fox and J. W. Milnor, *Singularities of 2–spheres in 4–space and cobordism of knots*, Osaka J. Math 3 (1966), 257–267.

[FQ90] M. Freedman and F. Quinn, *Topology of 4-manifolds*, Princeton Mathematical Series 39, Princeton University Press, Princeton, NJ (1990).

[Fr07] S. Friedl, *Reidemeister torsion, the Thurston norm and Harvey's invariants*, Pac. J. Math. (2007), 271–296.

[Fr13] S. Friedl, *Twisted Reidemeister torsion, the Thurston norm and fibered manifolds*, Geometriae Dedicata 172, (2014), 135–145.

[Fr14] S. Friedl, *Novikov homology and higher order Alexander polynomials*, in preparation (2014).

[FrH07] S. Friedl and S. Harvey, *Non-commutative multivariable reidemeister torsion and the thurston norm*, Alg. Geom. Top. 7 (2007), 755–777.

[FrK06] S. Friedl and T. Kim, *The Thurston norm, fibered manifolds and twisted Alexander polynomials*, Topology 45 (2006), 929–953.

[FrK08] S. Friedl and T. Kim, *The parity of the Cochran-Harvey invariants of 3-manifolds*, Trans. Amer. Math. Soc. 360 (2008), 2909–2922.

[FrKK12] S. Friedl, T. Kim, and T. Kitayama, *Poincaré duality and degrees of twisted Alexander polynomials*, Indiana Univ. Math. J. 61 (2012), 147–192.

[FrKt12] S. Friedl and T. Kitayama, *The virtual fibering theorem for 3-manifolds*, Preprint (2012), arXiv:1210.4799. To be published by l'Enseignement Mathématique.

[FrLM09] S. Friedl, C. Leidy and L. Maxim, *L^2-Betti numbers of plane algebraic curves*, Michigan Math. Journal, 58 (2009), No. 2, 291–301.

[FrSW13] S. Friedl, D. Silver and S. Williams, *Splittings of knot groups*, Preprint (2013), arXiv:1308.6497.

[FrV07] S. Friedl and S. Vidussi, *Nontrivial Alexander polynomials of knots and links*, Bull. Lond. Math. Soc. 39 (2007), 614–622.

[FrV08] S. Friedl and S. Vidussi, *Symplectic $S^1 \times N^3$, surface subgroup separability, and vanishing Thurston norm*, J. Amer. Math. Soc. 21 (2008), 597–610.

[FrV10] S. Friedl and S. Vidussi, *A survey of twisted Alexander polynomials*, The Mathematics of Knots: Theory and Application (Contributions in Mathematical and Computational Sciences), editors: Markus Banagl and Denis Vogel (2010), 45–94.

[FrV11] S. Friedl and S. Vidussi, *Twisted Alexander polynomials detect fibered 3–manifolds*, Ann. Math. 173 (2011), 1587–1643.

[FrV12a] S. Friedl and S. Vidussi, *A Vanishing Theorem for Twisted Alexander Polynomials with Applications to Symplectic 4-manifolds*, to be published by the J. Eur. Math. Soc.

[FrV12b] S. Friedl and S. Vidussi, *The Thurston norm and twisted Alexander polynomials*, Preprint (2012), arXiv:1204.6456. To be published by J. reine angew. Math.

[FrV13] S. Friedl and S. Vidussi, *Twisted Alexander invariants detect trivial links*, Preprint (2013), arXiv:1306.3487.

[Gh08] P. Ghiggini, *Knot Floer homology detects genus-one fibered knots*, Amer. J. Math. 130, No. 5 (2008), 1151–1169.

[GKiM05] H. Goda, T. Kitano and T. Morifuji, *Reidemeister Torsion, Twisted Alexander Polynomial and Fibred Knots*, Comment. Math. Helv. 80 (2005), No. 1, 51–61.

[GM03] H. Goda and T. Morifuji, *Twisted Alexander polynomial for* SL(2, $\mathbb{C}$)*-representations and fibered knots*, C. R. Math. Acad. Sci. Soc. R. Can. 25 (2003), No. 4, 97–101.

[GS11] H. Goda and T. Sakasai, *Factorization formulas and computations of higher-order Alexander invariants for homologically fibered knots*, J. Knot Theory Ramifications 20 (2011), No. 10, 1355–1380.

[Go83] C. McA. Gordon, *Dehn surgery and satellite knots*, Trans. Amer. Math. Soc. 275 (1983), 687–708.

[Har05] S. Harvey, *Higher–order polynomial invariants of 3–manifolds giving lower bounds for the Thurston norm*, Topology 44 (2005), 895–945.

[Har06] S. Harvey, *Monotonicity of degrees of generalized Alexander polynomials of groups and 3-manifolds*, Math. Proc. Camb. Phil. Soc. 140 (2006), 431–450.

[Hat81] R. Hartley, *Knots with free period*, Canad. J. Math. 33 (1981), No. 1, 91–102.

[Hem87] J. Hempel, *Residual finiteness for 3-manifolds*, Combinatorial group theory and topology (Alta, Utah, 1984), 379 396, Ann. of Math. Stud., 111, Princeton Univ. Press, Princeton, NJ (1987).

[HerKL10] C. Herald, P. Kirk and C. Livingston, *Metabelian representations, twisted Alexander polynomials, knot slicing, and mutation*, Math. Z. 265 (2010), 925–949.

[Hi12] J. Hillman, *Algebraic invariants of links*, second edition, Series on Knots and Everything, 52. World Scientific Publishing Co., Inc., River Edge, NJ (2012).

[HiLN06] J. Hillman, C. Livingston and S. Naik, *The Twisted Alexander Polynomial and Periodicity of Knots*, Alg. Geom. Top. 6 (2006), 145–169.

[HiSW10] J. Hillman, D. Silver and S. Williams, *On Reciprocality of Twisted Alexander Invariants*, Alg. Geom. Top. 10 (2010), 2017–2026.

[HoKMS11] K. Horie, T. Kitano, M. Matsumoto and M. Suzuki, *A partial order on the set of prime knots with up to 11 crossings*, J. Knot Theory Ramifications 20 (2011), No. 2, 275–303.

[HoKMS12] K. Horie, T. Kitano, M. Matsumoto and M. Suzuki, *Errata: A partial order on the set of prime knots with up to 11 crossings*, J. Knot Theory Ram. 21 (2012), No. 4, 1292001.

[Ho13] P. Horn, *On computing higher-order Alexander modules of knots*, Preprint (2013), arXiv:1303.1545.

[Ju06] A. Juhász, *Holomorphic discs and sutured manifolds*, Alg. & Geom. Top. 6 (2006), 1429–1457.

[Ju08] A. Juhász, *Floer homology and surface decompositions*, Geom. Top. 12 (2008), 299–350.

[Ju13] A. Juhász, *A survey of Heegaard Floer homology*, Preprint (2013), arXiv:1310.3418.

[Ka78] A. Kawauchi, *On the Alexander polynomials of cobordant links*, Osaka J. Math. 15 (1978), No. 1, 151–159.

[KiKM13] T. Kim, T. Kitayama and T. Morifuji, *Twisted Alexander polynomials on curves in character varieties of knot groups*, Int. J. Math. 24 (2013) 1350022.

[KiL99a] P. Kirk and C. Livingston, *Twisted Alexander invariants, Reidemeister torsion and Casson–Gordon invariants*, Topology 38, No. 3 (1999), 635–661.

[KiL99b] P. Kirk and C. Livingston, *Twisted knot polynomials: inversion, mutation and concordan*, Topology 38 (1999), No. 3, 663–671.

[Ki96] T. Kitano, *Twisted Alexander polynomials and Reidemeister torsion*, Pacific J. Math. 174 (1996), No. 2, 431–442.

[KiM05] T. Kitano and T. Morifuji, *Divisibility of twisted Alexander polynomials and fibered knots*, Ann. Sc. Norm. Super. Pisa Cl. Sci. (5) 4 (2005), No. 1, 179–186.

[KiS05a] T. Kitano and M. Suzuki, *A partial order in the knot table*, Experiment. Math. 14 (2005), 385–390.

[KiS05b] T. Kitano and M. Suzuki, *Twisted Alexander polynomials and a partial order on the set of prime knots*, Geometry and Topology Monographs 13 (2008), Groups, Homotopy and Configuration Spaces (Tookyo 2005), 307–322.

[KiS08] T. Kitano and M. Suzuki, *A partial order in the knot table II*, Acta Mathematica Sinica 24 (2008), 1801–1816.

[KiSW05] T. Kitano, M. Suzuki and M. Wada, *Twisted Alexander polynomial and surjectivity of a group homomorphism*, Algebr. Geom. Topol. 5 (2005), 1315–1324.

[Kiy10] T. Kitayama, *Non-commutative Reidemeister torsion and Morse-Novikov theory*, Proc. Amer. Math. Soc. 138 (2010), 3345–3360.

[KrLM88] P. H. Kropholler, P. A. Linnell and J. A. Moody, *Applications of a new K-theoretic theorem to soluble group rings*, Proc. Amer. Math. Soc. 104 (1988), No. 3, 675–684.

[LCSVV13] P. Lambert-Cole, M. Stone and D. Vela-Vick, *Braids and combinatorial knot Floer homology*, Preprint (2013), arXiv:1312.5586.

[LeM06] C. Leidy and L. Maxim, *Higher-order Alexander invariants of plane algebraic curves*, Int. Math. Res. Not. 2006 (2006), Article ID 12976, 23 pages.

[LeM08] C. Leidy and L. Maxim, *Obstructions on fundamental groups of plane curve complements*, Real and Complex Singularities, Contemporary Mathematics 459 (2008), 117–130.

[LiZ06a] W. Li and W. Zhang, *An L^2-Alexander invariant for knots*, Commun. Contemp. Math. 8 (2006), No. 2, 167–187.

[LiZ06b] W. Li and W. Zhang, *An L^2-Alexander-Conway invariant for knots and the volume conjecture*, Differential geometry and physics, 303–312, Nankai Tracts Math., 10, World Sci. Publ., Hackensack, NJ, 2006.

[LiZ08] W. Li and W. Zhang, *Twisted L^2-Alexander-Conway invariants for knots*, Topology and physics, 236–259, Nankai Tracts Math., 12, World Sci. Publ., Hackensack, NJ, 2008.

[Lin01] X. S. Lin, *Representations of knot groups and twisted Alexander polynomials*, Acta Math. Sin. (Engl. Ser.) 17 (2001), No. 3, 361–380.

[Lil93] P. Linnell, *Division rings and group von Neumann algebras*, Forum Math. 5 (1993), No. 6, 561–576.

[LoL95] J. Lott and W. Lück, *L^2-topological invariants of 3-manifolds*, Invent. Math. 120 (1995), No. 1, 15–60.

[Lü94] W. Lück, *Approximating L^2-invariants by their finite-dimensional analogues*, Geom. Funct. Anal. 4 (1994), 455–481.

[Lü02] W. Lück, *L^2-invariants: theory and applications to geometry and K-theory*, Ergebnisse der Mathematik und ihrer Grenzgebiete. 3. Folge. A Series of Modern Surveys in Mathematics, 44. Springer-Verlag, Berlin, 2002.

[LüS99] W. Lück and T. Schick, *L^2-torsion of hyperbolic manifolds of finite volume*, Geom. Funct. Anal. 9 (1999), No. 3, 518–567.

[MaOS09] C. Manolescu, P. Ozsváth and S. Sarkar, *A combinatorial description of knot Floer homology*, Ann. Math. (2) 169 (2009), No. 2, 633–660.

[Mc02] C. T. McMullen, *The Alexander polynomial of a 3–manifold and the Thurston norm on cohomology*, Ann. Sci. Ecole Norm. Sup. (4) 35 (2002), No. 2, 153–171.

[Mi62] J. Milnor, *A duality theorem for Reidemeister torsion*, Ann. of Math. (2) 76 (1962) 137–147.

[Mi66] J. Milnor, *Whitehead torsion*, Bull. Amer. Math. Soc. 72 (1966), 358–426.

[Mo11] T. Morifuji, *A note on degrees of twisted Alexander polynomials*, RIMS Kokyuroku 1747 (2011), 121–125.

[Mo12] T. Morifuji, *On a conjecture of Dunfield, Friedl and Jackson*, C. R. Acad. Sci. Paris, Ser I 350 (2012), 921–924.

[MS13] T. Morifuji and M. Suzuki, *A note on degrees of twisted Alexander polynomials. II.*, Bull. Korean Math. Soc. 50 (2013), No. 3, 929–934.

[MoT13] T. Morifuji and A. Tran, *Twisted Alexander polynomials of 2-bridge knots for parabolic representations*, Preprint (2013), arXiv:1301.1101.

[Mu71] K. Murasugi, *On periodic knots*, Comment. Math. Helv. 46 (1971), 162–174.

[Ni07] Y. Ni, *Knot Floer homology detects fibred knots*, Invent. Math. 170 (2007), No. 3, 577–608.

[Ni09a] Y. Ni, *Erratum: Knot Floer homology detects fibred knots*, Invent. Math. 170 (2009), No. 1, 235–238.

[Ni09b] Y. Ni, *Heegaard Floer homology and fibred 3-manifolds*, Amer. J. Math. 131 (2009), No. 4, 1047–1063.

[OS04a] P. Ozsváth and Z. Szabó, *Holomorphic disks and topological invariants for closed 3-manifolds*, Annals of Mathematics 159 (2004), No. 3, 1027–1158.

[OS04b] P. Ozsváth and Z. Szabó, *Holomorphic disks and 3-manifold invariants: properties and applications*, Annals of Mathematics 159 (2004) No. 3, 1159–1245.

[OS04c] P. Ozsváth and Z. Szabó, *Holomorphic disks and knot invariants*, Adv. in Math. 186 (2004), 58–116.

[PW12] P. Przytycki and D. Wise, *Mixed 3-manifolds are virtually special*, Preprint (2012), arXiv:1205.6742.

[Rai12] J. Raimbault, *Exponential growth of torsion in abelian coverings*, Algebr. Geom. Topol. 12 (2012), 1331–1372.

[Ras03] J. Rasmussen, *Floer homology and knot complements*, PhD thesis, Harvard University, 2003.

[Ro94] J. Rosenberg, *Algebraic K-theory and its applications*, Graduate Texts in Mathematics, 147. Springer-Verlag, New York (1994).

[Ru87] D. Ruberman, *Mutation and volumes of knots in S^3*, Invent. Math. 90 (1987), 189–215.

[Sa06] T. Sakasai, *Higher-order Alexander invariants for homology cobordisms of a surface*, Intelligence of Low Dimensional Topology 2006, Series on Knots and Everything 40, 271–278.

[Sa08] T. Sakasai, *The Magnus representation and higher-order Alexander invariants for homology cobordisms of surfaces*, Algebraic & Geometric Topology 8 (2008), 803–848.

[SaW10] S. Sarkar and J. Wang, *An algorithm for computing some Heegaard Floer homologies*, Ann. of Math. 171 (2010), No. 2, 1213–1236.

[Sc01] T. Schick, *L^2-determinant class and approximation of L^2–Betti numbers*, Trans. Amer. Math. Soc. Volume 353, Number 8 (2001), 3247–3265.

[Se34] H. Seifert, *Über das Geschlecht von Knoten*, Math. Ann. 110 (1934), 571–592.

[Si87] J.-C. Sikorav, *Homologie de Novikov associée à une classe de cohomologie réelle de degré un*, Thèse Orsay, 1987.

[SiW04] D. Silver and S. Williams, *Mahler measure of Alexander polynomials*, J. London Math. Soc. (2) 69 (2004), 767–782.

[SiW06] D. Silver and S. Williams, *Twisted Alexander Polynomials Detect the Unknot*, Algebraic and Geometric Topology, 6 (2006), 1893–1907.

[SiW09a] D. Silver and S. Williams, *Nonfibered knots and representation shifts*, Proceedings of Postnikov Memorial Conference, Banach Center Publications, 85 (2009), 101–107.

[SiW09b] D. Silver and S. Williams, *Twisted Alexander Polynomials and Representation Shifts*, Bull. London Math. Soc., 41 (2009), 535–540.

[St74] R. Strebel, *Homological methods applied to the derived series of groups*, Comment. Math. Helv. 49 (1974), 302–332.

[Ta02] A. Tamulis, *Knots of ten or fewer crossings of algebraic order 2*, J. Knot Theory Ramifications 11 (2002), 211–222.

[Th86] W. P. Thurston, *A norm for the homology of 3–manifolds*, Mem. Amer. Math. Soc. 339 (1986), 99–130.

[Tr13] A. Tran, *On twisted Alexander polynomials associated to* $_2(\mathbb{C})$*-representations of knot groups*, Preprint (2013), arXiv:1302.1631.

[Tu86] V. Turaev, *Reidemeister torsion in knot theory*, Russian Math. Surveys 41 (1986), No. 1, 119–182.

[Tu01] V. Turaev, *Introduction to combinatorial torsions*, Birkhäuser, Basel, (2001).

[Tu02a] V. Turaev, *Torsions of 3–manifolds*, Progress in Mathematics, 208. Birkhauser Verlag, Basel, 2002.

[Tu02b] V. Turaev, *A homological estimate for the Thurston norm*, unpublished note (2002), arXiv:math. GT/0207267

[Wa94] M. Wada, *Twisted Alexander polynomial for finitely presentable groups*, Topology 33, no. 2 (1994), 241–256.

[Wi09] D. Wise, *The structure of groups with a quasi-convex hierarchy*, Electronic Res. Ann. Math. Sci 16 (2009), 44–55.

[Wi12a] D. Wise, *The structure of groups with a quasi-convex hierarchy*, 189 pages, Preprint (2012), downloaded on October 29, 2012 from `http://www.math.mcgill.ca/wise/papers.html`

[Wi12b] D. Wise, *From riches to RAAGs: 3-manifolds, right–angled Artin groups, and cubical geometry*, CBMS Regional Conference Series in Mathematics, 2012.

Introduction to
Modern Mathematics
ALM 33, pp. 171–201

Homogeneous Dynamics: A Study Guide

Manfred Einsiedler* and Thomas Ward†

Abstract

These notes give a summary of the course "Ergodic Theory and Applications in Number Theory" at the Summer School in Modern Mathematics at Tsinghua University in Beijing, June 23–27 (2013). As in the summer school, we will need to be brief at times and refer to the references for a detailed treatment. Nonetheless we wish to survey some results and ideas on a close to geodesic journey from the most basic concepts of ergodic theory to some more sophisticated and recent results in homogeneous dynamics.

2000 Mathematics Subject Classification: 37A35, 37A45, 37D40, 11J25, 11J13.

Keywords and Phrases: Ergodic theory, homogeneous dynamics, nilflow.

1 Introduction

In these notes we will try and give an overview and road map to the area of homogeneous dynamics without becoming too diverted by details. For much of the background material we refer to [3] or [5], and for the more advanced material we refer to [4] or [7]. Sections 3–7 of these notes are essentially taken from a preliminary version of [4].

1.1 What is ergodic theory?

A vague answer to this question is that ergodic theory is *the study of dynamical systems from a probabilistic point of view.* In the next few sections we will expand on the meaning of these concepts. In the next section we will also introduce–at times in vague terms–some other key concepts for dynamical systems.

1.2 What is a dynamical system?

We will use more sophisticated settings later, but for now we define a dynamical system (X, T) to be a space X, usually equipped with a topology, together with a

*ETH Zürich, Departement Mathematik, Rämistrasse 101, 8092 Zürich, Switzerland. Email: manfred.einsiedler@math.ethz.ch

†Executive Office, Palatine Centre, Durham University, DH1 3LE, England. Email: t.b.ward@durham.ac.uk

map $T : X \to X$, usually assumed to be continuous. We will also refer to such a map as a transformation.

In a dynamical system (X,T) we will often be interested in the orbit

$$\mathsf{O}_T(x) = \{x, Tx, T^2x, \dots\}$$

of a point $x \in X$. If X is a topological space we can try to describe the closure $\overline{\mathsf{O}_T(x)}$ of the orbit of a point $x \in X$. It is also interesting to know whether the orbit spends more time in certain parts of the space, or whether it spreads out throughout X so as to spend a proportion of time in any set proportional to a natural notion of 'size' of the set. We will return to these questions later, and will make the latter concept–*equidistribution*–precise.

Example 1. Let $\mathbb{T} = \mathbb{R}/\mathbb{Z}$, which we may also identify with $[0,1)$ with the quotient topology inherited from $\mathbb{R}$. Thus, for example, $1 - \frac{1}{n} \to 0 \in [0,1)$ as $n \to \infty$. For any $p \in \mathbb{Z}$ we define the map

$$\begin{aligned} T_p : \mathbb{T} &\longrightarrow \mathbb{T} \\ x &\longmapsto px \pmod{\mathbb{Z}}. \end{aligned}$$

The case $p = 10$ is particularly easy to describe. Given a number $x \in [0,1)$ we may write its decimal expansion as

$$x = 0.a_1a_2 \cdots \in [0,1),$$

with decimal digits $a_n \in \{0,1,\dots,9\}$ for all $n \geqslant 1$. Then we have

$$T_{10}(x) = 0.a_2a_3 \cdots \in [0,1).$$

Thus if the orbit of $x \in [0,1)$ under T_{10} is dense, then every digit from 0 to 9 will appear in the decimal expansion of x. In fact every finite block of digits, 2013 for example, has to appear infinitely often in the decimal expansion of such a point.

For this dynamical system it is clear that very different types of orbit are possible. It is easy to construct many decimal expansions that do not contain a given block like 2013. On the other hand it is also easy to find points with dense orbits. There are many ways to see this, here are two.

1. Enumerate all finite blocks of decimal digits in increasing order of length,

$$0, 1, 2, \dots, 9, 00, 01, 02, \dots, 99, 000, \dots$$

and then concatenate them to produce a number

$$x = 0.012 \cdots 9000102 \cdots 99000001002 \cdots .$$

2. Enumerate the natural numbers in their natural order,

$$1, 2, 3, 4, 5, 6, 7, 8, 9, 10, 11, 12, 13, 14, \dots$$

and then concatenate them to produce a number

$$x = 0.1234567891011121314 \cdots .$$

Exercise 2. Show that $x \in [0, 1)$ has a finite orbit under T_p for $p \geqslant 2$ if and only if $x \in \mathbb{Q}$. Describe which points are *periodic* (that is, have the property that there is some $n \geqslant 1$ for which $T_p^n(x) = x$).

Example 3. A different, and much better behaved, dynamical system on the circle $\mathbb{T}$ is given by the *circle rotation*

$$R_\alpha : \mathbb{T} \longrightarrow \mathbb{T}$$
$$x \longmapsto x + \alpha.$$

If $\alpha = \frac{p}{q} \in \mathbb{Q}$ (written in lowest terms) then $R_\alpha^q = I$ and all orbits are periodic (with q elements). As this behavior is too simple, we will always assume that α is not rational, in which case R_α is called an *irrational rotation.*

Exercise 4. Show that the orbit of every point under an irrational rotation is dense in the circle.

The sense in which Example 3 is *better behaved* than Example 1 is an instance of the important notion of *measure rigidity*, which will emerge as a central theme of these notes. For the moment, notice that the orbits of the map R_α for a fixed α are all very similar–for rational α they are finite sets of fixed cardinality, and for irrational α they are dense. Moreover, for fixed α any orbit may be obtained from any other orbit by a simple rotation.

Example 5. Let

$$M_z : \mathbb{C} \longrightarrow \mathbb{C}$$
$$w \longmapsto zw$$

be the multiplication by z map on the complex plane. If $|z| < 1$ then every orbit converges to $0 \in \mathbb{C}$. If $|z| > 1$ then the orbit of every point apart from 0 diverges to infinity. If $|z| = 1$ then every orbit stays on a circle. Thus for most choices of the parameter z this map does not have interesting orbits, and for $|z| = 1$ the orbits remain on a circle of fixed radius, and the map M_z restricted to each such circle is isomorphic to a rotation on $\mathbb{T}$.

One way to view the system $(\mathbb{C}, M_z)$ for $|z| = 1$ is as a combination of constituent systems, each of which is a copy of a circle rotation, and it is simpler to study its constituent systems. However, in general we cannot expect such a geometrically straightforward decomposition of a dynamical system into constituent parts which cannot readily be decomposed further.

1.3 What is the probabilistic point of view?

A deeper examination of maps like T_p from Example 1 suggests that the possible behaviors of orbits means that the study of orbits and their closures is in general impossibly difficult. As a result one is led to the question of whether it is more straightforward to describe the orbits of *typical* points. One of the most satisfactory answers to this question is given by the pointwise ergodic theorem of Birkhoff from 1931. In order to state this, we need the following compatibility notion that picks out those probability measures on the underlying space of a dynamical system that respect the dynamics.

Definition 6. Let (X, T) be a dynamical system, and assume that X is equipped with a σ-algebra $\mathcal{B}$ of measurable subsets[1]. A probability measure μ defined on $\mathcal{B}$ is called *T-invariant* (or T is called *μ-preserving*) if

$$\mu(T^{-1}B) = \mu(B)$$

for all $B \in \mathcal{B}$.

Theorem 7 (Birkhoff, Pointwise Ergodic Theorem)**.** *Let (X, T, μ) be a measure-preserving dynamical system, and let $f \in L^1(X, \mu)$. Then*

$$\frac{1}{N}\sum_{n=0}^{N-1} f(^n x) \longrightarrow f^*(x)$$

in $L^1(X, \mu)$ and μ-almost everywhere, where $f^* \in L^1(X, \mu)$ is a T-invariant function.

The difficult part of this theorem is the convergence almost everywhere (hence the name pointwise). The convergence in $L^1(X, \mu)$ with respect to the L^1 norm is easier, and can be derived from von Neumann's mean ergodic theorem established in the next exercise. While the behavior of T on individual points is described by the map T itself, it is convenient to study the behavior of T on functions via the associated operator U_T defined by $U_T(f)(x) = f(Tx)$.

Exercise 8. Let (X, T, μ) be a measure-preserving dynamical system, and suppose for simplicity that T is invertible.

(a) Show that $U_T : L^2(X, \mu) \to L^2(X, \mu)$ is *unitary*, meaning that

$$\langle U_T f, U_T g\rangle = \langle f, g\rangle$$

for all $f, g \in L^2(X, \mu)$.

(b) Show that[2]

$$\frac{1}{N}\sum_{n=0}^{N-1} f \circ T^n = \frac{1}{N}\sum_{n=0}^{N-1} U_T^n f \longrightarrow P(f)$$

for every $f \in L^2(X, \mu)$, where $P(f)$ denotes the orthogonal projection of f onto the subspace

$$\mathcal{I}_T = \{f \in L^2(X, \mu) \mid U_T f = f\}.$$

[1] In most situations we will encounter, X will be a topological space and $\mathcal{B}$ will be the Borel σ-algebra, which is defined to be the smallest σ-algebra that contains all the open sets.

[2] Consider the cases $f \in \mathcal{I}_T$ and $f \in \mathcal{I}_T^{\perp}$ separately, and then use the orthogonal decomposition. For the latter subspace consider the functions $f = g - U_T(g)$ and show that these are dense in $\mathcal{I}_T^{\perp}$.

As in Exercise 8, the function f^* in Theorem 7 should be thought of as the 'orthogonal projection' of $f \in L^1(X,\mu)$ to the space

$$\{g \in L^1(X,\mu) \mid U_T g = g\}.$$

This projection (which *a priori* does not make sense because $L^1(X,\mu)$ does not have the geometry of a Hilbert space) is called the *conditional expectation* and is denoted by $E(f\big|\mathcal{E})$ where

$$\mathcal{E} = \{B \in \mathcal{B} \mid T^{-1}B = B\}$$

is the σ-algebra of T-invariant sets. We refer to [3, Chap. 5] for the details.

Definition 9. Let μ be a T-invariant probability measure on X. We say that μ is *ergodic* if any set $B \in \mathcal{B}$ with $T^{-1}B = B$ has $\mu(B) \in \{0,1\}$. Equivalently, μ is ergodic if $U_T f^* = f^*$ for some $f^* \in L^2(X,\mu)$ implies that $f^*(x) = \int_X f^*\,\mathrm{d}\mu$ for μ-almost every x (that is, f^* is equal to a constant μ-almost everywhere).

The equivalence of these two definitions of ergodicity is relatively easy and may be found in [3, Chap. 2]. Notice that the first formulation of ergodicity expresses the idea that (X,T,μ) cannot be decomposed into invariant subsets that are non-trivial with respect to μ. For this reason ergodicity has also been called *indecomposability*. With this notion of indecomposability we can give a strengthening of the conclusion of Theorem 7 as follows.

Corollary 10. *Let $T : X \to X$ be a continuous transformation of a σ-compact metric space. If μ is a T-invariant and ergodic probability measure on X, then μ-almost every point $x \in X$ is* generic *for T.*

Here $x \in X$ is said to be generic if

$$\frac{1}{N}\sum_{n=0}^{N-1} f(T^n x) \longrightarrow \int_X f\,\mathrm{d}\mu$$

for all $f \in C_c(X)$.

We note that Corollary 10 is not a completely trivial consequence of Theorem 7 because there are uncountably many functions in $C_c(X)$, each of which potentially requires a null set of badly-behaved points to be avoided. The single null set arising in Corollary 10 does not depend on the function f. The main idea in the proof is to get around this by taking advantage of the fact that $C_c(X)$ is *separable*, meaning that it has a countable subset that is dense with respect to the supremum norm. This observation together with a relatively easy approximation argument gives the corollary (see [3, Chap. 4] for the details).

Example 11. The Lebesgue measure λ on $[0,1) \cong \mathbb{T}$ is invariant under the map T_p for all $\geqslant 1$. One way to see this is to note that for $0 \leqslant a < b < 1$ we have

$$T_p^{-1}\left([a,b)\right) = [\tfrac{a}{p},\tfrac{b}{p}) \sqcup [\tfrac{a+1}{p},\tfrac{b+1}{p}) \sqcup \cdots \sqcup [\tfrac{a+p-1}{p},\tfrac{b+p-1}{p})$$

is a disjoint union, so

$$\lambda\left(T_p^{-1}([a,b))\right) = \lambda([a,b)) = b - a.$$

Using this, one can show by an approximation argument that we also have

$$\lambda(T^{-1}B) = \lambda(B)$$

for all Borel sets $B \subset [0,1)$ since the intervals of the form $[a,b)$ generate the Borel σ-algebra.

The measure λ is also ergodic with respect to T_p for $p \geqslant 2$. One way to see this is to use Fourier analysis for functions on the circle as follows. Assume that $f \in L^2(\mathbb{T},\lambda)$ is invariant under T_p, and write

$$f(x) = \sum_{n\in\mathbb{Z}} c_n \mathrm{e}^{2\pi i n x}$$

for the Fourier expansion of f at $x \in \mathbb{T}$. Thus, working in L^2, we have

$$f(x) = \sum_{n\in\mathbb{Z}} c_n \mathrm{e}^{2\pi i n x} = f(T_p x) = \sum_{n\in\mathbb{Z}} c_n \mathrm{e}^{2\pi i n p x}.$$

This equality in L^2 means that the Fourier coefficients of both functions must coincide, so

$$c_n = c_{np} = c_{np^2} = \cdots$$

for all $n \in \mathbb{Z}$. On the other hand

$$\sum_{n\in\mathbb{Z}} |c_n|^2 = \|f\|_2 < \infty$$

by the Plancherel theorem. It follows that $|c_n| \to 0$ as $n \to \infty$ so, in particular, $c_n = 0$ for $n \neq 0$. Thus $f = c_0$ is a constant as required.

Applying Theorem 7 to the map T_{10} we can now deduce that λ-almost every $x \in \mathbb{T}$ has the property that the decimal block of digits 2013 appears in the decimal expansion of x not only infinitely often, but does so with the asymptotic frequency $\frac{1}{10^4}$.

Exercise 12. Show that the Lebesgue measure λ on $[0,1) \cong \mathbb{T}$ is invariant and ergodic for any irrational rotation $R_\alpha : \mathbb{T} \to \mathbb{T}$.

Using the exercise above we prove an even stronger statement about irrational translations.

Corollary 13 (Unique ergodicity for irrational rotations). *For $\alpha \notin \mathbb{Q}$, every point $x \in \mathbb{T}$ is generic for R_α.*

Proof. Fix a function $f \in C(\mathbb{T})$ and some $\varepsilon > 0$. Since f is uniformly continuous, there exists some $\delta > 0$ such that[3]

$$\mathsf{d}(x,y) < \delta \implies |f(x) - f(y)| < \varepsilon$$

[3] We have not formally defined the translation-invariant metric d, but the reader should find it easy to construct this as the metric inherited from the usual one on $\mathbb{R}$.

for all $x, y \in \mathbb{T}$. Given $x \in \mathbb{T}$, we can use Exercise 12 and Corollary 10, almost every $y \in \mathbb{T}$ is generic for R_α. In particular, there exists some generic point y with $\mathsf{d}(x, y) < \delta$. Then

$$\mathsf{d}(R_\alpha^n x, R_\alpha^n y) < \delta$$

for all $n \geqslant 0$, and so

$$\left| \frac{1}{N} \sum_{n=0}^{N-1} f(R_\alpha^n x) - \frac{1}{N} \sum_{n=0}^{N-1} f(R_\alpha^n y) \right| < \varepsilon,$$

which for large enough N gives

$$\left| \frac{1}{N} \sum_{n=0}^{N-1} f(R_\alpha^n x) - \int_0^1 f \, \mathrm{d}\lambda \right| < 2\varepsilon$$

as required. □

Exercise 14. Prove Corollary 13 directly, using the characters $x \mapsto \mathrm{e}^{2\pi \mathrm{i} x n}$ for $x \in \mathbb{T}$ and $n \in \mathbb{Z}$.

Corollary 15. *There exists some $n \geqslant 1$ (in fact, there are infinitely many) such that the decimal expansion of 2^n starts with the digits*[4] 2013.

Exercise 16. Prove Corollary 15 using Corollary 13 (use the fact that if the rotation parameter $\alpha = \log_{10} 2$ then $\log_{10} 2^n = n\alpha$ and consider what the latter identity means modulo 1).

Exercise 17. Generalize[5] the discussion above to show that there exists some n for which the decimal expansion of both 2^n and of 3^n start with the digits 2013.

We end this introduction to the basics of ergodic theory with a short proof of the pointwise ergodic theorem, due to Hasselblatt and Katok [5].

Proof of Theorem 7. Let $f^* = E(f \middle| \mathcal{E}_T)$, which as mentioned above should be thought of as the projection of f onto the space of T-invariant functions. If μ is ergodic, then $f^* = \int_X f \, \mathrm{d}\mu$.

We fix $\varepsilon > 0$ and define

$$g = f - f^* - \varepsilon,$$

and

$$S_n(g) = \sum_{k=0}^{n-1} g \circ T^k.$$

We will show that

$$\limsup_{n \to \infty} \frac{1}{n} S_n(g) \leqslant 0 \tag{1}$$

[4]As the reader will have noticed, 2013 was the year in which the summer school took place and these notes were written, and is being used simply to represent an arbitrary finite sequence of digits that does not start with 0.

[5]You will need to consider a rotation on $\mathbb{T}^2$.

for μ-almost every $x \in X$. Using the fact that

$$S_n(g) = S_n(f) - nf^* - n\varepsilon$$

and the same statement for $-f$ then gives the theorem. In turn (1) will follow from the following lemma.

Lemma 18. *Let*

$$A = \{x \in X \mid \sup_{k\geqslant 1} S_k(g) = \infty\}.$$

Then $\mu(A) = 0$.

Proof. We need to assemble some facts about S_k, A, and the function

$$M_n = \max\{S_k g \mid k = 0, \dots, n\}.$$

- First note that $S_k(g)(Tx) = S_{k+1}(g)(x) - g(x)$ so that $T^{-1}A = A$, i.e. A is a T-invariant set.
- As is customary, we set $S_0(g) = 0$ so that $M_n \geqslant 0$ for all $n \geqslant 0$.
- Using the definition of M_n we see that

$$M_{n+1}(x) = \max\{0, g(x) + M_n(Tx)\}.$$

- It follows that

$$M_{n+1}(x) - M_n(Tx) = \max\{-M_n(Tx), g(x)\}. \tag{2}$$

- Using the definition of A and of M_n, this now shows that

$$\lim_{n\to\infty} (M_{n+1}(x) - M_n(Tx)) = g(x)$$

  for all $x \in A$.
- We also have

$$g(x) \leqslant M_{n+1}(x) - M_n(Tx) \leqslant \max\{0, g(x)\}$$

  by (2) and the inequality $M_n \geqslant 0$.
- The last fact shows that

$$M_{n+1}(x) - M_n(Tx)$$

  is dominated by an $L^1(X, \mu)$ function independently of n. Hence we may apply the dominated convergence theorem to obtain

$$\lim_{n\to\infty} \int_A (M_{n+1} - M_n \circ T)\, \mathrm{d}\mu = \int_A g\, \mathrm{d}\mu.$$

- We calculate

$$\int_A g\, \mathrm{d}\mu = \int_A f\, \mathrm{d}\mu - \int_A f^*\, \mathrm{d}\mu - \varepsilon\mu(A) = -\varepsilon\mu(A) \tag{3}$$

  by definition of f^*.

- Moreover,

$$\begin{aligned}\int_A (M_{n+1}-M_n\circ T)\,\mathrm{d}\mu &= \int_A M_{n+1}\,\mathrm{d}\mu - \int_A M_n\circ T\,\mathrm{d}\mu \\ &= \int_A M_{n+1}\,\mathrm{d}\mu - \int (\mathbb{1}_A M_n)\circ T\,\mathrm{d}\mu \quad (\text{since } T^{-1}A=A) \\ &= \int_A M_{n+1}\,\mathrm{d}\mu - \int_A M_n\,\mathrm{d}\mu \quad (\text{since } \mu \text{ is } T-\text{invariant}) \\ &= \int_A \underbrace{(M_{n+1}-M_n}_{\geqslant 0}\,\mathrm{d}\mu. \quad \text{by definition of } M_n\end{aligned}$$

Combining the last three facts we obtain

$$-\varepsilon\mu(A) \geqslant 0,$$

and hence $\mu(A) = 0$ as claimed. □

This completes the proof of Theorem 7. □

Example 19. We now present another measure-preserving transformation on (almost all of) $[0, 1]$, which will turn out to be ergodic. At first sight this example may appear less natural than the multiplication by p map and the rotation by α map on the circle above, but it also is of algebraic origin–though the algebra in question is rather hidden (see [3, Ch. 9]). The *Gauss map* is defined by

$$\begin{aligned}T : (0,1) &\longrightarrow (0,1) \\ x &\longmapsto \left\{\frac{1}{x}\right\},\end{aligned}$$

where $\{\cdot\}$ denotes the fractional part of a real number. Notice that strictly speaking this map does not define a dynamical system, as the orbit of any rational point eventually reaches 0, where the map is not defined. Restricting T to the set $[0, 1]\smallsetminus\mathbb{Q}$ does however give a dynamical system, and this is the map we will think about. The map T preserves the *Gauss measure* μ defined by

$$\mu([a,b]) = \int_a^b \frac{1}{\log 2(1+x)}\,\mathrm{d}x,$$

and therefore defines a measure-preserving dynamical system.

We cannot resist mentioning one striking consequence of the study of the Gauss map, and in particular of the ergodicity of the Gauss measure with respect to T. For every irrational $x \in (0, 1)$ there exists a sequence of best rational approximations $(\frac{p_n(x)}{q_n(x)})$ (defined using the continued fraction expansion of x), and μ-almost every $x \in (0, 1)$ satisfies

$$\lim_{n\to\infty} \frac{1}{n} \log\left| x - \frac{p_n(x)}{q_n(x)}\right| = -\frac{\pi^2}{6\log 2}.$$

We refer to [3, Chap. 3] for the details and the history of this kind of result.

2 What is a nilflow?

Example 3 may be described in the following way. We start with a locally compact group $\mathbb{R}$, with a discrete subgroup $\mathbb{Z}$ of finite covolume (this means that the Lebesgue measure on $\mathbb{R}$ gives finite measure to any measurable fundamental domain for $\mathbb{Z} < \mathbb{R}$), and an element $\alpha \in \mathbb{R}$. This defines a measure-preserving map on the quotient space $\mathbb{R}/\mathbb{Z} = \mathbb{T}$ by rotation under α, and this is the map R_α. Much of our purpose here is concerned with understanding what phenomena arise if $\mathbb{R}$ is replaced by a non-commutative locally compact group, $\mathbb{Z}$ by a lattice in it, and Lebesgue measure by the Haar measure.

As a first step towards the study of algebraic actions on homogeneous spaces defined by non-commutative groups, we now discuss briefly a special case of *flows on nilmanifolds*.

Let[6]

$$H = \left\{ \begin{pmatrix} 1 & a & c \\ & 1 & b \\ & & 1 \end{pmatrix} \mid a, b, c \in \mathbb{R} \right\}$$

be the Heisenberg group (which is an example of a *two-step nilpotent Lie group* or a *unipotent algebraic group* depending on the language one is used to speaking). Similarly, let

$$\Gamma = \left\{ \begin{pmatrix} 1 & \ell & n \\ & 1 & m \\ & & 1 \end{pmatrix} \mid \ell, m, n \in \mathbb{Z} \right\}$$

be the group of integer points in H. Then the space

$$X = \Gamma \backslash H = \{\Gamma h \mid h \in H\}$$

is (by definition) a *nilmanifold*. This space is a compact manifold which can, for example, be obtained by gluing opposite faces of the cube $[0,1)^3$ together in the appropriate way. However, in some of the gluing of faces a twist is applied (without the twist, gluing the faces of the cube together would produce the 3-torus $\mathbb{T}^3$). In fact we use elements $\gamma \in \Gamma$ to identify elements $g \in H$ using the equivalence relation

$$\gamma g = \begin{pmatrix} 1 & \ell & n \\ & 1 & m \\ & & 1 \end{pmatrix} \begin{pmatrix} 1 & a & c \\ & 1 & b \\ & & 1 \end{pmatrix} = \begin{pmatrix} 1 & a+\ell & c+n+\ell b \\ & 1 & b+m \\ & & 1 \end{pmatrix} \sim g. \tag{4}$$

Given

$$\alpha = \begin{pmatrix} 0 & \alpha_1 & \beta \\ & 0 & \alpha_2 \\ & & 0 \end{pmatrix}$$

in the Lie algebra of H, we define a flow

$$\begin{aligned} R_t : X &\longrightarrow X \\ x = \Gamma g &\longmapsto x \exp(t\alpha) = \Gamma g \left(I + t + \frac{1}{2} t^2 \alpha^2 + \cdots \right) \end{aligned}$$

[6] We will sometimes forget to write entries in a matrix if these are zero.

for all $t \in \mathbb{R}$. (As H is two-step nilpotent we could actually simplify the expression for the exponential map and simply write $\exp(t\alpha) = I + t + \frac{1}{2}t^2$.) Here we call the dynamical system a *flow* because we have an action of $\mathbb{R}$ instead of the $\mathbb{Z}$-action induced by a single transformation.

Theorem 20 (A uniquely ergodic nilflow). *If* $\alpha_2 \neq 0$ *and* $\frac{\alpha_1}{\alpha_2} \notin \mathbb{Q}$ *then*

$$\{R_t \mid t \in \mathbb{R}\}$$

is a uniquely ergodic flow on X. *This means that*

$$\frac{1}{T}\int_0^T f\left(R_t(x)\right) \mathrm{d}t \longrightarrow \int_X f \,\mathrm{d}\lambda_X$$

as $T \to \infty$ *for any* $f \in C_c(X)$, *where* λ_X *is the image of the Lebesgue measure*[7] *on* $[0,1)^3$ *in* X.

Proof. Define a factor map

$$\pi : X \longrightarrow \mathbb{T}^2$$

by

$$\pi : \Gamma \begin{pmatrix} 1 & a & c \\ & 1 & b \\ & & 1 \end{pmatrix} \longmapsto \begin{pmatrix} a \\ b \end{pmatrix} + \mathbb{Z}^2 \in \mathbb{T}^2.$$

The calculation in (4) shows that π is well-defined. This map is often called the *maximal abelian factor*. By essentially the same argument as that used in Corollary 13, the maximal abelian factor is ergodic with respect to the induced flow

$$\widehat{R_t} : \mathbb{T}^2 \ni \begin{pmatrix} a \\ b \end{pmatrix} \longmapsto \begin{pmatrix} a + \alpha_1 t \\ b + \alpha_2 \end{pmatrix} \in \mathbb{T}^2,$$

and in fact is uniquely ergodic (that is, there is no other invariant measure for the flow). Our main task therefore is to extend this unique ergodicity result from $\mathbb{T}^2$ to X. This cannot be done using the standard theory of Fourier series as harmonic analysis works quite differently on the space X.

As a first step, we show that the flow R_t is ergodic. Suppose therefore that $f \in L^2(X, \lambda_X)$ is an invariant function for R_t. Let

$$g_s = \begin{pmatrix} 1 & s & \\ & 1 & \\ & & 1 \end{pmatrix},$$

and notice that[8]

$$\|f(x) - f(xg_s)\|_2 \longrightarrow 0$$

[7] The reader should check that λ_X is preserved by R_t for all $t \in \mathbb{R}$.

[8] In these expressions we are (inaccurately but efficiently) using $f(x)$ as shorthand for the function $x \mapsto f(x)$ rather than the value of the function f at x.

as $s \to 0$. This requires a proof, but we will leave this step to the reader and only note that elements of $L^2(X, \lambda_X)$ can be approximated by continuous functions (see also [3, Sect. 11.3.2]). Using the invariance property

$$f(x) = f(x \exp(t\alpha))$$

of f gives

$$\begin{aligned}\|f(x) - f(xg_s)\|_2 &= \|f(x \exp(t\alpha)) - f(xg_s \exp(t\alpha))\|_2 \\ &= \|f(y) - f(y \exp(-t\alpha) g_s \exp(t\alpha))\|_2\end{aligned}$$

where we first used the invariance and then the substitution

$$y = x \exp(t\alpha),$$

which amounts to a unitary transformation on $L^2(X, \lambda_X)$. We now calculate

$$\exp(-t\alpha) g_s \exp(t\alpha) = \begin{pmatrix} 1 & s & \alpha_2 ts \\ & 1 & 0 \\ & & 1 \end{pmatrix}. \tag{5}$$

We set $s = \frac{1}{n}$ and notice that $t = t_n$ can be chosen arbitrarily. We choose t_n so that

$$\frac{\alpha_2 t_n}{n} = r$$

for some arbitrary $r \in \mathbb{R}$. Therefore the matrix in (5) approaches

$$\begin{pmatrix} 1 & 0 & r \\ & 1 & 0 \\ & & 1 \end{pmatrix}$$

as $n \to \infty$, and we get

$$\left\| f(x) - f\left(x \begin{pmatrix} 1 & 0 & r \\ & 1 & 0 \\ & & 1 \end{pmatrix} \right) \right\|_2 = 0,$$

or equivalently that f is invariant under

$$\begin{pmatrix} 1 & 0 & r \\ & 1 & 0 \\ & & 1 \end{pmatrix}$$

for all $r \in \mathbb{R}$. This step is an instance of the *Mautner phenomenon* (which we will discuss again in Section 4).

This shows that

$$f(x) = f\left(\Gamma \begin{pmatrix} 1 & a & c \\ & 1 & b \\ & & 1 \end{pmatrix} \right)$$

almost surely does not depend on c, so f is a function on the maximal abelian factor $\mathbb{T}^2$. However, as we already know that the induced flow is ergodic on $\mathbb{T}^2$ this shows that f must be constant almost everywhere, and the flow R_t is thus seen to be ergodic.

Once more we know ergodicity, and so almost every point is generic. We wish to upgrade this ergodicity to unique ergodicity. This is not always possible (see, for example, the map from Exercise 2 which clearly has many different invariant measures supported on periodic orbits). Moreover, the simple commutative or isometric argument from Corollary 13 also does not work in our current situation. Despite this, it is possible in this situation to upgrade the argument to show unique ergodicity using an observation of Furstenberg. We give a slightly modified version of this argument.

Notice first that for each $y \in \mathbb{T}^2$ the fibre $\pi^{-1}y$ is isomorphic to $\mathbb{T}$, and write $\lambda_{\pi^{-1}y}$ to denote the Lebesgue measure on $\pi^{-1}y$. We note that

$$\frac{1}{T}\int_0^T (R_t)_* \lambda_{\pi^{-1}y}\, \mathrm{d}t \longrightarrow \lambda_X \tag{6}$$

in the weak*-topology on the space of probability measures $\mathcal{M}(X)$ on X, which via Riesz representation is identified with a subset of the dual of the Banach space $C(X)$. In fact we already know (6), but not quite in the language used above. The statement can be translated as follows. Given a continuous function $f \in C(X)$, the measure on the left-hand side of (6) is defined to be the measure that integrates f to

$$\frac{1}{T}\int_0^T \int_{\pi^{-1}y} f\, \mathrm{d}\lambda_{\pi^{-1}(\widehat{R_t}y)}\, \mathrm{d}t.$$

Here we can identify the inner integral with the function F defined by

$$F(y) = \int_{\pi^{-1}y} f\, \mathrm{d}\lambda_{\pi^{-1}y} \in C(\mathbb{T}^2)$$

evaluated at $\widehat{R_t}(y)$. However, on $\mathbb{T}^2$ we already have unique ergodicity and so

$$\frac{1}{T}\int_0^T F\left(\widehat{R_t}y\right)\, \mathrm{d}t \longrightarrow \int_{\mathbb{T}^2} F = \int_X f\, \mathrm{d}\lambda_X$$

by Fubini's theorem. Combining these we see the weak*-convergence in (6).

Let $x \in X$ be arbitrary, let $y = \pi(x)$ and fix $\delta > 0$. Then we can split $\lambda_{\pi^{-1}y}$ into a convex combination

$$\lambda_{\pi^{-1}y} = (2\delta)\lambda_{I_\delta(x)} + (1-2\delta)\lambda_{\pi^{-1}y\smallsetminus I_\delta(x)}, \tag{7}$$

where $I_\delta(x)$ denotes the δ-neighbourhood of x in $\pi^{-1}y$, $\lambda_{I_\delta(x)}$ is the Lebesgue measure on $I_\delta(x)$ normalized to be a probability measure, and $\lambda_{\pi^{-1}y\smallsetminus I_\delta(x)}$ is defined similarly.

Now consider the analogue of the left-hand side of (6) for $\lambda_{I_\delta(x)}$, giving

$$\frac{1}{T}\int_0^T (R_t)_* \lambda_{I_\delta(x)}\, \mathrm{d}t \in \mathcal{M}(X). \tag{8}$$

By the Tychonoff–Alaoglu theorem on $C(X)^*$ we can choose a sequence (T_n) with $T_n \to \infty$ as $n \to \infty$ such that (8) converges to come probability measure $\nu_{x,\delta}$. From the construction it is easy to see that $\nu_{x,\delta}$ is R_t-invariant (because the interval $[0,T]$ is almost translation-invariant). Furthermore, we can choose another subsequence so that the same expression for $\nu_{\pi^{-1}y \setminus I_\delta}$ also converges to an invariant probability measure $\widetilde{\nu_{x,\delta}}$. Comparing this with (6) and (7) we get

$$\lambda_X = 2\delta\nu_{x,\delta} + (1-2\delta)\widetilde{\nu_{x,\delta}}.$$

We did not mention this before, but an ergodic measure is an extreme point of the convex set of invariant probability measures (see [3, Th. 4.4] for the details). It follows that

$$\lambda_X = \nu_{x,\delta},$$

and what is also curious is that the limit measure obtained is independent of the chosen subsequence (T_n). Now if a sequence in a compact topological space has the property that the limit of convergent subsequence is independent of the subsequence, then the original sequence (or, in our case, map from $(0,\infty)$ to $\mathcal{M}(X)$) must itself converge.

Finally, fix some $f \in C(X)$ and $\varepsilon > 0$. Choose $\delta > 0$ as in the condition for uniform continuity of f, and notice that $z \in I_\delta$ has the property that

$$\mathsf{d}\left(R_t(x), R_t(z)\right) < \delta$$

for all $t \in \mathbb{R}$. Now we can conclude the argument as in the proof of Corollary 13. □

We note that nilflows (or rather their discrete analogues, nilrotations) have recently become extremely important in the area of interaction between ergodic theory, Ramsey theory, and number theory (see the surveys of Bergelson [1], [2] for an overview of Ramsey theory in this context).

Exercise 21. Let

$$N = \left\{ \begin{pmatrix} 1 & a_{12} & a_{13} & a_{14} \\ & 1 & a_{23} & a_{24} \\ & & 1 & a_{34} \\ & & & 1 \end{pmatrix} \mid a_{ij} \in \mathbb{R} \right\},$$

$$\Gamma = \left\{ \begin{pmatrix} 1 & a_{12} & a_{13} & a_{14} \\ & 1 & a_{23} & a_{24} \\ & & 1 & a_{34} \\ & & & 1 \end{pmatrix} \mid a_{ij} \in \mathbb{Z} \right\},$$

and define $X = \Gamma\backslash N$. Characterize those flows $R_t(x) = x\exp(t\alpha)$ for $t \in \mathbb{R}$ which are ergodic by studying the maximal abelian quotient (and the appropriate Mautner phenomenon). Can you generalize Theorem 20 to these nilflows?

3 A brief review of dynamics on the modular surface

Starting in this section we will consider more general group actions. Let G be a group and let X be a set (called the space). Then a *G-action* is simply a homomorphism Φ from G to the group of bijections of X. We will also write $g{\boldsymbol\cdot}x = \Phi(g)(x)$ for the action of $g \in G$ on $x \in X$. Usually the space has some structure and we require the action to preserve this structure. If G is a *topological group* (where the group operations are continuous) and X is a topological space, then we call the G-action *continuous* if

$$(g, x) \in G \times X \to g{\boldsymbol\cdot}x \in X$$

is continuous. If X has a measure, then we call the G-action *measure-preserving* if $\Phi(g)$ is measure-preserving for all $g \in G$.

3.1 The space

We recall (see, for example, [3, Ch. 9]) that the upper half-plane

$$\mathbb{H} = \{z = x + \mathrm{i}y \in \mathbb{C} \mid y = \Im(z) > 0\}$$

equipped with the Riemannian metric

$$\langle (z, u), (z, v)\rangle_z = \frac{(u \cdot v)}{y^2}$$

for $(z, u), (z, v) \in \mathrm{T}_z\mathbb{H} = \{z\} \times \mathbb{C}$ is the *upper half-plane model* of the hyperbolic plane (where $u \cdot v$ denotes the inner product after identifying u and v with elements of $\mathbb{R}^2$). Moreover, the group $\mathrm{SL}_2(\mathbb{R})$ acts on $\mathbb{H}$ transitively and isometrically via the Möbius transformation

$$g = \begin{pmatrix} a & b \\ c & d \end{pmatrix} : z \longmapsto g{\boldsymbol\cdot}z = \frac{az + b}{cz + d}. \tag{9}$$

The stabilizer of $\mathrm{i} \in \mathbb{H}$ is $\mathrm{SO}(2)$, so that

$$\mathrm{SL}_2(\mathbb{R}) / \mathrm{SO}(2) \cong \mathbb{H}$$

under the map sending $g\,\mathrm{SO}(2)$ to $g{\boldsymbol\cdot}\mathrm{i}$.

The action of $\mathrm{SL}_2(\mathbb{R})$ is differentiable, and so gives rise to a derived action on the tangent bundle $\mathrm{T}\mathbb{H} = \mathbb{H} \times \mathbb{C}$ by

$$\mathrm{D}\,g : (z, u) \longmapsto \left(g{\boldsymbol\cdot}z, \frac{1}{(cz + d)^2} u \right)$$

where

$$g = \begin{pmatrix} a & b \\ c & d \end{pmatrix}.$$

This action gives rise to the simply transitive action of

$$\mathrm{PSL}_2(\mathbb{R}) = \mathrm{SL}_2(\mathbb{R})/\{\pm 1\}$$

on the unit tangent bundle

$$\mathrm{T}^1\mathbb{H} = \{(z,v) \in \mathrm{T}\mathbb{H} \mid \|(z,v)\|_z^2 = \langle (z,v),(z,v)\rangle_z = 1\},$$

so that

$$\mathrm{PSL}_2(\mathbb{R}) \cong \mathrm{T}^1\mathbb{H}$$

by sending g to $\mathrm{D}\,g(\mathrm{i},\mathrm{i})$.

The region E illustrated by shading in Figure 1. is a fundamental region for the action of the discrete subgroup $\mathrm{PSL}_2(\mathbb{Z})$ on $\mathbb{H}$ (strictly speaking we should describe carefully which parts of the boundary of the hyperbolic triangle shaded belong to the domain but as the boundary is a nullset one usually ignores that issue — we will comply with this tradition), see Exercise 22.

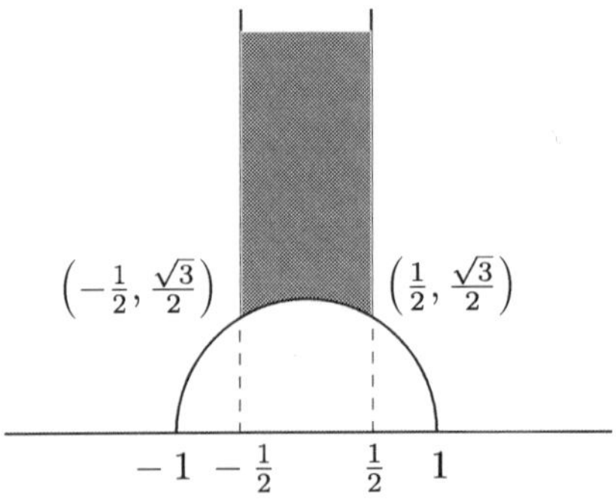

Figure 1. A fundamental domain $E \subset \mathbb{H}$ for the action of $\mathrm{SL}_2(\mathbb{Z})$.

This shows that we can define a fundamental domain for $\mathrm{PSL}_2(\mathbb{Z})$ in

$$\mathrm{PSL}_2(\mathbb{R}) \cong \mathrm{T}^1\mathbb{H}$$

by taking all vectors (z,u) whose base point z lies in E, giving the set

$$F = \{g \in \mathrm{PSL}_2(\mathbb{R}) \mid \mathrm{D}\,g(\mathrm{i},\mathrm{i}) = (z,u) \text{ with } z \in E\}.$$

(Strictly speaking we should describe more carefully which vectors attached to points $z \in \partial E$ are allowed in F.) Furthermore, we can lift the set $F \subset \mathrm{PSL}_2(\mathbb{R})$ to a surjective[9] set $F \subset \mathrm{SL}_2(\mathbb{R})$ for $\mathrm{SL}_2(\mathbb{Z})$. We claim that this argument shows that

$$\mathrm{PSL}_2(\mathbb{Z})\backslash\mathrm{PSL}_2(\mathbb{R}) \cong \mathrm{SL}_2(\mathbb{Z})\backslash\mathrm{SL}_2(\mathbb{R})$$

has finite volume. In order to see this, we recall some basic facts from [3, Ch. 9]:

[9] That is, F has the property that the natural quotient map from $\mathrm{SL}_2(\mathbb{R})$ to $\mathrm{SL}_2(\mathbb{Z})\backslash\,\mathrm{SL}_2(\mathbb{R})$ is surjective when restricted to F.

- $SL_2(\mathbb{R})$ is unimodular (see Exercise 23).
- $SL_2(\mathbb{R}) = NAK$ with[10]

$$N = \left\{ \begin{pmatrix} 1 & * \\ & 1 \end{pmatrix} \right\}, A = \left\{ \begin{pmatrix} a & \\ & a^{-1} \end{pmatrix} \mid a > 0 \right\}$$

and $K = SO(2)$, in the sense that every $g \in SL_2(\mathbb{R})$ can be written uniquely[11] as a product $g = nak$ with $n \in N$, $a \in A$ and $k \in K$.
- Let $B = NA = AN$ be the subgroup $B = \left\{ \begin{pmatrix} a & t \\ & a^{-1} \end{pmatrix} \mid a > 0, t \in \mathbb{R} \right\}$. The Haar measure $m_{SL_2(\mathbb{R})}$ decomposes in the coordinates $g = bk$, meaning that

$$m_{SL_2(\mathbb{R})} \propto m_B \times m_K$$

where $\propto$ denotes proportionality (with the constant of proportionality dependent only on the choices of Haar measures). Moreover, the left Haar measure m_B decomposes in the coordinate system

$$b(x, y) = \begin{pmatrix} 1 & x \\ & 1 \end{pmatrix} \begin{pmatrix} y^{1/2} & \\ & y^{-1/2} \end{pmatrix}$$

with $x \in \mathbb{R}$, $y > 0$, as

$$\mathrm{d}m_B = \frac{1}{y^2} \,\mathrm{d}x \,\mathrm{d}y.$$

- We also note that $b(x, y) \cdot \mathrm{i} = \begin{pmatrix} 1 & x \\ & 1 \end{pmatrix} \cdot (\mathrm{i}y) = x + \mathrm{i}y$, and that the Haar measure m_B on B is identical to the hyperbolic area measure on $\mathbb{H}$ under the map $b(x, y) \mapsto b(x, y) \cdot \mathrm{i} = x + \mathrm{i}y$.

Combining these facts we get

$$m_{SL_2(\mathbb{R})}(F) < \int_{-1/2}^{1/2} \int_{\sqrt{3}/2}^{\infty} \int_0^{2\pi} \frac{1}{y^2} \,\mathrm{d}\theta \,\mathrm{d}y \,\mathrm{d}x < \infty.$$

The argument above also helps us to understand the space

$$\mathsf{X}_2 = SL_2(\mathbb{Z}) \backslash SL_2(\mathbb{R})$$

globally: it is, apart from some difficulties arising from the distinguished points

$$\mathrm{i}, \tfrac{1}{2} + \tfrac{\sqrt{3}}{2}\mathrm{i} \in E,$$

the unit tangent bundle of the surface[12] $SL_2(\mathbb{Z}) \backslash \mathbb{H}$. This surface may be thought of as being obtained by gluing the two vertical sides in Figure 1. together using the action of

$$\begin{pmatrix} 1 & \pm 1 \\ & 1 \end{pmatrix} \in SL_2(\mathbb{Z})$$

[10] We sometimes indicate by $*$ any entry of a matrix which is only restricted to be a real number, and omit entries that are zero.

[11] This is a simple instance of the more general Iwasawa decomposition of a connected real semi-simple Lie group.

[12] Because of the distinguished points this surface is a good example of an *orbifold*, but not an example of a manifold.

and the third side to itself using the action of

$$\begin{pmatrix} & -1 \\ 1 & \end{pmatrix} \in \mathrm{SL}_2(\mathbb{Z}).$$

In particular, X_2 is non-compact.

3.2 The geodesic flow—the subgroup A

As we have seen, the unit tangent bundle of a 'hyperbolic surface' $\Gamma\backslash\mathbb{H}$ can be identified with a quotient of $\mathrm{SL}_2(\mathbb{R})$ by the discrete subgroup $\Gamma < \mathrm{SL}_2(\mathbb{R})$. More generally if G is a linear Lie group and $\Gamma < G$ is a discrete subgroup then one can define the quotient space $X = \Gamma\backslash G$. On such a space there is still a natural G-action defined by $g{\scriptstyle\bullet}x = xg^{-1}$ for $x \in X$ and $g \in G$.

We recall that

$$g_t : x \longmapsto x \begin{pmatrix} e^{t/2} & \\ & e^{-t/2} \end{pmatrix} = \begin{pmatrix} e^{-t/2} & \\ & e^{t/2} \end{pmatrix} {\scriptstyle\bullet} x$$

defines the geodesic flow on X_2, whose orbits may also be described in the fundamental region as in Figure 2..

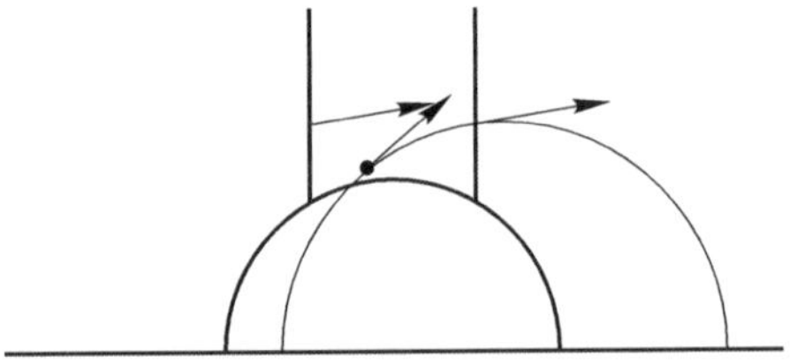

Figure 2. The geodesic flow follows the circle determined by the arrow which intersects $\mathbb{R} \cup \{\infty\} = \partial\mathbb{H}$ normally, and is moved back to F via a Möbius transformation in $\mathrm{SL}_2(\mathbb{Z})$ once the orbit leaves F.

The diagonal subgroup

$$A = \left\{ \begin{pmatrix} e^{-t/2} & \\ & e^{t/2} \end{pmatrix} \mid t \in \mathbb{R} \right\}$$

is also called the *torus* or *Cartan subgroup*. We claim that A acts ergodically on X_2 with respect to the Haar measure m_{X_2} (see Section 4 and [3, Sec. 9.5]). Even so, there are many different types of A-orbits, which include the following:

- Divergent trajectories, for example the orbit $\mathrm{SL}_2(\mathbb{Z})A$ which corresponds to the vertical geodesic through (i, i) in $\mathrm{SL}_2(\mathbb{Z})\backslash\mathrm{T}^1\mathbb{H}$.
- Compact trajectories, for example $\mathrm{SL}_2(\mathbb{Z})g_{\mathrm{golden}}A$ is compact, where the

matrix $g_{\mathrm{golden}} \in K$ has the property[13] that

$$g_{\mathrm{golden}}^{-1} \begin{pmatrix} 1 & 1 \\ 1 & 2 \end{pmatrix} g_{\mathrm{golden}} = \begin{pmatrix} \frac{3+\sqrt{5}}{2} & \\ & \frac{3-\sqrt{5}}{2} \end{pmatrix} \in A.$$

Now notice that

$$\mathrm{SL}_2(\mathbb{Z}) g_{\mathrm{golden}} \begin{pmatrix} \frac{3+\sqrt{5}}{2} & \\ & \frac{3-\sqrt{5}}{2} \end{pmatrix} = \mathrm{SL}_2(\mathbb{Z}) \begin{pmatrix} 1 & 1 \\ 1 & 2 \end{pmatrix} g_{\mathrm{golden}} = \mathrm{SL}_2(\mathbb{Z}) g_{\mathrm{golden}}.$$

This identity shows that the orbit $\mathrm{SL}_2(\mathbb{Z}) g_{\mathrm{golden}} A$ is compact (see also Figure 3. in which $\lambda = \frac{1+\sqrt{5}}{2}$).

- The set of dense trajectories, which includes (but is much larger than) the set of equidistributed trajectories of generic points in $\mathrm{SL}_2(\mathbb{Z}) \backslash \mathrm{SL}_2(\mathbb{R})$.
- Orbits that are neither dense nor closed.

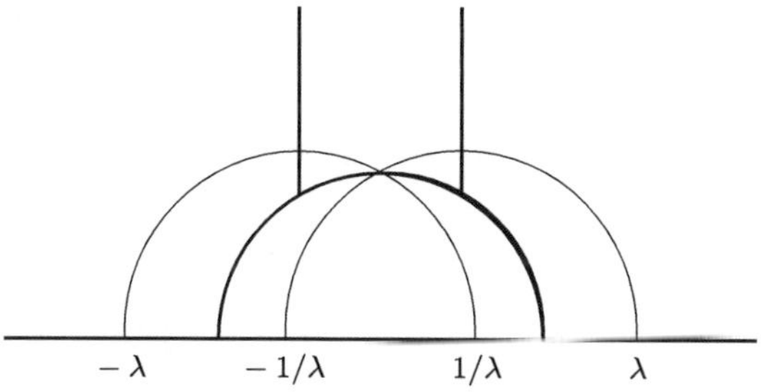

Figure 3. The union of the two geodesics considered in X_2 with both directions allowed is a periodic A-orbit, and comprises the orbit $\mathrm{SL}_2(\mathbb{Z}) g_{\mathrm{golden}} A$.

Finally we would like to point out that there is a correspondence between rational (or arithmetic) objects and closed A-orbits as in the first two types of A-orbit considered above (see Exercise 25 and 26).

3.3 The horocycle flow–the subgroup $U^- = N$

We recall that the (stable) horocycle flow on X_2 is defined by the action

$$h_s : x \longmapsto x \begin{pmatrix} 1 & -s \\ & 1 \end{pmatrix} = u(s) \cdot x$$

for $s \in \mathbb{R}$. Here the matrices

$$\begin{pmatrix} 1 & s \\ & 1 \end{pmatrix} = u(s)$$

[13]The eigenvalues of $\begin{pmatrix} 1 & 1 \\ 1 & 2 \end{pmatrix}$ are $\frac{3\pm\sqrt{5}}{2}$, and there is such a matrix $g_{\mathrm{golden}} \in K$ because $\begin{pmatrix} 1 & 1 \\ 1 & 2 \end{pmatrix}$ is symmetric.

are unipotent (that is, only have 1 as an eigenvalue) and the corresponding subgroup

$$U^- = \left\{ \begin{pmatrix} 1 & s \\ & 1 \end{pmatrix} \mid s \in \mathbb{R} \right\}$$

is precisely the stable horospherical subgroup of the geodesic flow, in the sense that

$$U^- = \left\{ g \in \mathrm{SL}_2(\mathbb{R}) \mid \begin{pmatrix} \mathrm{e}^{-t/2} & \\ & \mathrm{e}^{t/2} \end{pmatrix} g \begin{pmatrix} \mathrm{e}^{t/2} & \\ & \mathrm{e}^{-t/2} \end{pmatrix} \to I_2 \text{ as } t \to \infty \right\}.$$

This implies that

$$\mathsf{d}\left(g_t(x), g_t(u(s)\boldsymbol{\cdot}x)\right) \to 0$$

as $t \to \infty$ for any $x \in \mathsf{X}_2$ and $s \in \mathbb{R}$.

Geometrically, the horocycle orbits $U^-\boldsymbol{\cdot}x = xU^-$ can be described as circles touching the real axis with the arrows (that is, the tangent space component) normal to the circle pointing inwards or as horizontal lines with the arrows pointing upwards, as in Figure 4..

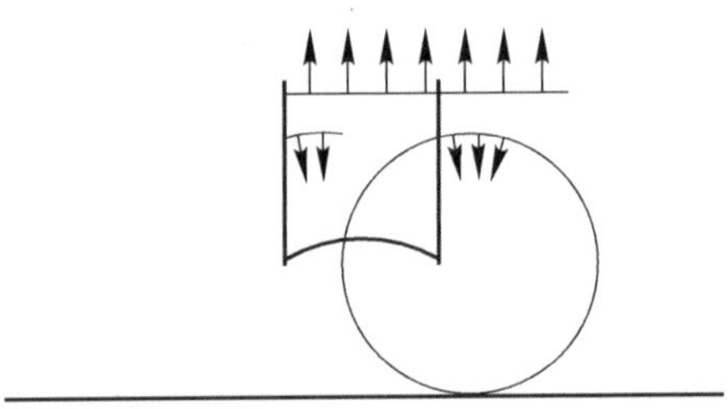

Figure 4. The picture shows the two types of horocycle orbits; the orbits in X_2 can again be understood by using the appropriate Möbius transformation whenever the orbit leaves the fundamental domain.

We recall that U^- also acts ergodically on X_2 with respect to the Haar measure m_{X_2} (see Section 4 and [3, Sec. 11.3]). However, unlike the case of A-orbits, the classification of U^--orbits on X_2 is shorter. The possibilities are as follows:

- Compact trajectories, for example $\mathrm{SL}_2(\mathbb{Z})U^-$ is compact and corresponds to the horizontal orbit through $(\mathrm{i}, \mathrm{i}) \in \mathrm{T}^1\mathbb{H}$.
- Dense trajectories, which are automatically also equidistributed with respect to m_{X_2}.

This gives the complete list of types of U^--orbits, and once more gives substance to the claim that there is a correspondence between rational objects and closed orbits (see Exercise 27).

3.4 Exercises for Section 3

Exercise 22. Let E be as in Figure 1.

1. Use $\begin{pmatrix} 1 & 1 \\ 0 & 1 \end{pmatrix}$ and $\begin{pmatrix} 0 & -1 \\ 1 & 0 \end{pmatrix}$ to show that $\mathrm{SL}_2(\mathbb{Z})\cdot E$ is 'uniformly open', meaning that there exists some $\delta > 0$ such that $z \in \mathrm{SL}_2(\mathbb{Z})\cdot E$ implies that
$$B_\delta(z) \subset \mathrm{SL}_2(\mathbb{Z})\cdot E.$$
Conclude that $\mathrm{SL}_2(\mathbb{Z})\cdot E = \mathbb{H}$.
2. Suppose that both z and $\gamma\cdot z$ lie in E for some $\gamma \in \mathrm{SL}_2(\mathbb{Z})$. Show that either $\gamma = \pm I$ or $z \in \partial E$.
3. Conclude that E can be modified (by defining which parts of the boundary of E should be included) to become a fundamental domain.

Exercise 23. Let $d \geqslant 2$. Show that

$$m_{\mathrm{SL}_d(\mathbb{R})}(B) = m_{\mathbb{R}^{d^2}}(\{tb : t \in [0,1], b \in B\})$$

for any measurable $B \subset \mathrm{SL}_d(\mathbb{R})$ defines a (bi-invariant) Haar measure on $\mathrm{SL}_d(\mathbb{R})$, where $m_{\mathbb{R}^{d^2}}$ is the Lebesgue measure on the matrix algebra $\mathrm{Mat}_{dd}(\mathbb{R})$ viewed as the vector space $\mathbb{R}^{d^2}$.

Exercise 24. Show that $\mathrm{SL}_2(\mathbb{R})$ is generated by the unipotent subgroups

$$\begin{pmatrix} 1 & * \\ & 1 \end{pmatrix} \text{ and } \begin{pmatrix} 1 & \\ * & 1 \end{pmatrix}.$$

Exercise 25. Show that $\mathrm{SL}_2(\mathbb{Z})gA$ is a divergent trajectory (that is, the map sending $a \in A$ to $\mathrm{SL}_2(\mathbb{Z})ga$ is a proper map) if and only if $ga \in \mathrm{SL}_2(\mathbb{Q})$ for some $a \in A$.

Exercise 26. Show that to any compact A-orbit in $\mathrm{SL}_2(\mathbb{Z})\backslash \mathrm{SL}_2(\mathbb{R})$ one can attach a real quadratic number field K such that the length of the orbit is $\log|\xi|$, where ξ in $\mathcal{O}_K^*$ is a unit in the order $\mathcal{O}_K$ of K. Prove that there are only countably many such orbits.

Exercise 27. Show that $\mathrm{SL}_2(\mathbb{Z})gU^-$ is compact if and only if

$$g(\infty) \in \mathbb{Q} \cup \{\infty\}.$$

Show that if $\mathrm{SL}_2(\mathbb{Z})gU^-$ is compact, then any other compact orbit is of the form $\mathrm{SL}_2(\mathbb{Z})gaU^-$ for some $a \in A$.

Exercise 28. Show that $\mathrm{SL}_2(\mathbb{Z})\backslash \mathrm{SL}_2(\mathbb{R}) \cong \{\mathbb{Z}^2 g \mid g \in \mathrm{SL}_2(\mathbb{R})\}$ can be identified with lattices $\mathbb{Z}^2 g \subset \mathbb{R}^2$ of co-volume $\det g = 1$. Use the isomorphism with $\mathrm{SL}_2(\mathbb{Z})\backslash \mathrm{T}^1\mathbb{H}$ discussed in this section to characterize compact subsets K of $\mathrm{SL}_2(\mathbb{Z})\backslash \mathrm{SL}_2(\mathbb{R})$ in terms of elements of the lattices $\mathbb{Z}^2 g$ for $\mathrm{SL}_2(\mathbb{Z})g \in K$. More precisely, calculate the relationship between the shortest vector $ng \in \mathbb{Z}^2 g$ and the imaginary part of $gi \in \mathbb{H}$ under the assumption that the representative $g \in \mathrm{SL}_2(\mathbb{R})$ has been chosen with $gi \in E$.

4 Mautner phenomenon

We continue our study of the special (but important) group $G = \mathrm{SL}_2(\mathbb{R})$. Any element $g \in \mathrm{SL}_2(\mathbb{R})$ is conjugate to one of the following three type of elements:

- an $\mathbb{R}$-diagonal matrix, that is one of the form $a = \begin{pmatrix} \lambda & \\ & \lambda^{-1} \end{pmatrix}$ with $\lambda \in \mathbb{R}$;
- a unipotent matrix $u = \begin{pmatrix} 1 & \pm 1 \\ & 1 \end{pmatrix}$; or
- a matrix in the compact subgroup $\mathrm{SO}(2)$, that is one of the form

$$k = \begin{pmatrix} \cos\phi & -\sin\phi \\ \sin\phi & \cos\phi \end{pmatrix}$$

  for some $\phi \in \mathbb{R}$.

For the last case we can make no claim concerning ergodicity of the action of g (because a compact group only acts ergodically when it acts transitively). However, for the first two types we will prove the Mautner phenomenon (Proposition 29), which gives ergodicity of these elements in many cases.

Proposition 29 (Mautner for $\mathrm{SL}_2(\mathbb{R})$). *Let $G = \mathrm{SL}_2(\mathbb{R})$ act unitarily on a Hilbert space $\mathcal{H}$, and suppose that $g \neq \pm I$ is unipotent or $\mathbb{R}$-diagonalizable and fixes a vector $v_0 \in \mathcal{H}$. Then all of G fixes v_0 also.*

Suppose $g \in G$ satisfies the hypotheses of Proposition 29, and $h \in G$ has the property that hgh^{-1} fixes $v_0 \in \mathcal{H}$. Then g fixes $\pi^{-1}(h)v_0$ and so $v_0 = \pi^{-1}(h)v_0$ is fixed by G as needed. Thus it is sufficient to consider one representative of each conjugacy class for the proof of Proposition 29.

The proof relies on the following key lemma.

Lemma 30 (The key lemma). *Let $\mathcal{H}$ be a Hilbert space carrying a unitary representation of a topological group G. Suppose that $v_0 \in \mathcal{H}$ is fixed by some subgroup $L \leqslant G$. Then v_0 is also fixed under every other element $h \in G$ with the property that*

$$B_\delta^G(h) \cap L B_\delta^G(I) L \neq \emptyset \tag{10}$$

for every $\delta > 0$.

Proof. By assumption, there exist three sequences (g_n) in G, (ℓ_n) in L, and (ℓ'_n) in L with $g_n \to e$ and $\ell_n g_n \ell'_n \to h$ as $n \to \infty$. This implies that

$$\|\pi(\ell_n g_n \ell'_n)v_0 - v_0\| = \|\pi(\ell_n)(\pi(g_n \ell'_n)v_0 - \pi(\ell_n^{-1})v_0)\| = \|\pi(g_n)v_0 - v_0\|$$

by invariance of v_0 under all elements of L and unitarity of $\pi(\ell_n)$. However, the left hand side converges to $\|\pi(h)v_0 - v_0\|$ by continuity[14] of the representation and the right hand side converges to 0. □

[14]This continuity is a very general property. For example, it always holds for the unitary representation derived from a continuous measure-preserving actions of G, see [3, Sect. 11.3.2].

Proof of Proposition 29. For $a = \begin{pmatrix} \lambda & \\ & \lambda^{-1} \end{pmatrix}$ with $\lambda \neq \pm 1$ a direct calculation shows that we can apply Lemma 30 with $L = a^{\mathbb{Z}}$ and any element of the unipotent subgroups $\begin{pmatrix} 1 & * \\ & 1 \end{pmatrix}$ or $\begin{pmatrix} 1 & \\ * & 1 \end{pmatrix}$ in $\mathrm{SL}_2(\mathbb{R})$. For example,

$$a^n \begin{pmatrix} 1 & s \\ & 1 \end{pmatrix} a^{-n} = \begin{pmatrix} 1 & \lambda^{2n} s \\ & 1 \end{pmatrix} \longrightarrow \begin{pmatrix} 1 & \\ & 1 \end{pmatrix}$$

if $\lambda^{2n} \to 0$ as $n \to \infty$. It follows that if a fixes some $v_0 \in \mathcal{H}$, then so do these two unipotent subgroups, and as they together generate $\mathrm{SL}_2(\mathbb{R})$ (see Exercise 24), we obtain Proposition 29 for this case.

If $u = \begin{pmatrix} 1 & 1 \\ & 1 \end{pmatrix}$ then

$$\begin{aligned} u^n \begin{pmatrix} 1+\delta & \\ & 1/(1+\delta) \end{pmatrix} u^{-n} &= \begin{pmatrix} 1 & n \\ & 1 \end{pmatrix} \begin{pmatrix} 1+\delta & \\ & 1/(1+\delta) \end{pmatrix} \begin{pmatrix} 1 & -n \\ & 1 \end{pmatrix} \\ &= \begin{pmatrix} 1+\delta & -2\delta n \\ & 1/(1+\delta) \end{pmatrix} \end{aligned}$$

can be made (since n can be chosen arbitrary) to converge to $\begin{pmatrix} 1 & s \\ & 1 \end{pmatrix}$ for $\delta \to 0$. It follows that if v_0 is fixed by $\begin{pmatrix} 1 & 1 \\ & 1 \end{pmatrix}$ then it is also fixed by $\begin{pmatrix} 1 & s \\ & 1 \end{pmatrix}$ for any $s \in \mathbb{R}$ by Lemma 30 applied with

$$L = \left\{ \begin{pmatrix} 1 & n \\ & 1 \end{pmatrix} \mid n \in \mathbb{Z} \right\}.$$

Applying Lemma 30 once more with

$$L = \left\{ \begin{pmatrix} 1 & s \\ & 1 \end{pmatrix} \mid s \in \mathbb{R} \right\}$$

to the matrix

$$\begin{pmatrix} 1 & s_1 \\ & 1 \end{pmatrix} \begin{pmatrix} 1 & \\ \delta & 1 \end{pmatrix} \begin{pmatrix} 1 & s_2 \\ & 1 \end{pmatrix} = \begin{pmatrix} 1+\delta s_1 & s_2(1+\delta s_1)+s_1 \\ \delta & 1+\delta s_2 \end{pmatrix} = g_\delta \tag{11}$$

with s_1 chosen to have

$$1 + \delta s_1 = \mathrm{e}^{\alpha},$$

and with s_2 chosen to have

$$s_2(1+\delta s_1) + s_1 = 0$$

shows that v_0 is also fixed by

$$\begin{pmatrix} \mathrm{e}^{\alpha} & \\ & \mathrm{e}^{-\alpha} \end{pmatrix} = \lim_{\delta \to 0} g_\delta.$$

Applying the previous (diagonal) case, we see once again that v_0 is fixed by all of $\mathrm{SL}_2(\mathbb{R})$. □

Suppose now $X = \Gamma\backslash \mathrm{SL}_2(\mathbb{R})$ is a compact (or, more generally, a finite volume) quotient of $\mathrm{SL}_2(\mathbb{R})$ by a discrete subgroup. Then clearly every G-invariant set on X is either empty or everything due to transitivity, and so G acts ergodically on X with respect to the Haar measure m_X (which is induced by the Haar measure on $\mathrm{SL}_2(\mathbb{R})$ normalized to be a probability measure). Applying Theorem 29 it follows[15] that any nontrivial diagonal matrix (that is, any nontrivial element of the geodesic flow) and every nontrivial unipotent matrix (that is, any nontrivial element of the horocycle flow) also acts ergodically on X with respect to the Haar measure.

5 Mixing of the geodesic flow

The following theorem gives a significant strengthening of ergodicity.

Theorem 31 (Howe–Moore). *Suppose that $G = \mathrm{SL}_2(\mathbb{R})$ acts on a probability space (X,μ), and that the action is measure-preserving and ergodic with respect to μ. Then the action of G is also* mixing, *meaning that*

$$\langle f_1 \circ g, f_2\rangle \longrightarrow \int f_1 \,\mathrm{d}\mu \int \overline{f_2}\,\mathrm{d}\mu$$

whenever $g \to \infty$ in G, for any $f_1, f_2 \in L^2(X,\mu)$.

Proof for the geodesic flow. Let $f_1 \in L^2(X,\mu)$ and let

$$a_t = \begin{pmatrix} \mathrm{e}^{-t/2} & \\ & \mathrm{e}^{t/2} \end{pmatrix}$$

be an element of the diagonal subgroup. We will consider the case $t \to \infty$; the case $t \to -\infty$ is similar. Then

$$\|f_1(xa_t)\|_2 = \|f\|_2,$$

and so we may choose (by the Tychonoff–Alaoglu theorem) a sequence (t_n) with

$$\lim_{n\to\infty} f_1(xa_{t_n}) = f_1^*(x)$$

in the weak* topology. In other words, we have

$$\lim_{n\to\infty} \langle f_1(xa_{t_n}), f_2\rangle = \langle f_1^*, f_2\rangle$$

[15] Hopefully without destroying the beauty of the argument in the eye of the reader, we wish to point out a small technical detail. Clearly every G-invariant set (measurable or not) is empty or everything but for Proposition 29 another notion of invariance is important. A set is called G-invariant modulo the Haar measure m_X if it is measurable and $m_X(B \triangle g{\bullet}B) = 0$ for all $g \in G$, or equivalently if the characteristic function is an invariant function in $L^2(X, m_X)$. What is needed is that any G-invariant set modulo m_X has measure zero or one. This requires a small argument on the same order of complexity as the uniqueness theorem for the Haar measure on locally compact groups.

for all $f_2 \in L^2(X,\mu)$. We claim that f_1^* is invariant under the action of the unipotent subgroup

$$U = \left\{ \begin{pmatrix} 1 & \\ s & 1 \end{pmatrix} \mid s \in \mathbb{R} \right\}.$$

In fact

$$\begin{aligned} \langle f_1^*(xu_s), f_2 \rangle &= \langle f_1^*, f_2(xu_{-s}) \rangle \\ &= \lim_{n\to\infty} \langle f_1(xa_{t_n}), f_2(xu_{-s}) \rangle \\ &= \lim_{n\to\infty} \langle f_1(xu_s a_{t_n}), f_2 \rangle \\ &= \lim_{n\to\infty} \langle f_1(xa_{t_n} u_{\mathrm{e}^{-t_n} s}), f_2 \rangle . \end{aligned}$$

However,

$$\| f_1(yu_{\mathrm{e}^{-t_n} s}) - f_1(y) \|_2 \longrightarrow 0$$

as $\mathrm{e}^{-t_n} s \to 0$, and since the substitution $y = xa_{t_n}$ is unitary we obtain

$$\langle f_1^*(xu_s), f_2 \rangle = \lim_{n\to\infty} \langle f_1(xa_{t_n}), f_2 \rangle = \langle f_1^*, f_2 \rangle .$$

This shows that

$$f_1^*(xu_s) = f_1^*(x)$$

for almost every x and all $s \in \mathbb{R}$. Now the assumption that G acts ergodically implies together with Proposition 29 that f_1^* coincides with a constant almost everywhere, which quickly shows

$$f_1^* = \int f_1 \, \mathrm{d}\mu.$$

This is the desired conclusion. □

The general case of $g \in \mathrm{SL}_2(\mathbb{R})$ going to infinity (not necessarily in the diagonal subgroup) surprisingly can be derived from the above case. Here one uses the Cartan decomposition of matrices which states that any g can be written as $k_1 a k_2$ where a is diagonal and $k_1, k_2 \in \mathrm{SO}(2)$ belong to the compact group of rotations $\mathrm{SO}(2)$.

Exercise 32. Show the Cartan decomposition of elements of $\mathrm{SL}_2(\mathbb{R})$ and finish the proof of Theorem 31.

6 Unique ergodicity on compact quotients

We will now show the unique ergodicity of the horocycle flow on compact quotients of $\mathrm{SL}_2(\mathbb{R})$, but we will derive this in a much more general framework for Lie groups and their horospherical subgroups.

Suppose that G is a connected linear Lie group and let $a \in G$ be an $\mathbb{R}$-diagonalizable element that acts as a mixing transformation[16] on all the quotients of G appearing below. Let

$$G_a^- = \{g \in G \mid a^n g a^{-n} \to I \text{ as } n \to \infty\}$$

be the stable horospherical subgroup of a.

Theorem 33 (Unique ergodicity of horospherical actions[17]). *Let G be a linear Lie group, $\Gamma < G$ a uniform lattice, and let $a \in G$ be $\mathbb{R}$-diagonalizable. Suppose a acts mixingly on $X = \Gamma\backslash G$. Then the action of G_a^- is uniquely ergodic.*

Proof. Let us assume compatibility of the Haar measures in the sense that

$$m_X(\pi(B)) = m_G(B)$$

for any injective[18] Borel subset $B \subset G$, and that $m_X(X) = 1$.

Since a is diagonalizable (we will even assume that a is diagonal) and G is linear, the subgroups G_a^- and

$$P_a = \{g \in G \mid a^n g a^{-n} \text{ stays bounded as } n \to -\infty\}$$

can easily be defined in terms of the vanishing of certain matrix entries, and so are closed subgroups. Together they define a local coordinate system, in the sense that $P_a G_a^-$ contains an open neighborhood of the identity[19], and the implied representation of elements of G in that neighborhood is unique. In fact, if $u_1 p_1 = u_2 p_2$ with $u_1, u_2 \in G_a^-$ and $p_1, p_2 \in P_a$ then

$$g = u_2^{-1} u_1 = p_2 p_1^{-1}$$

has $a^n g a^{-n} \to I$ as $n \to \infty$ and stays bounded as $n \to -\infty$, which together show[20] that $g = I$. Moreover, the Haar measure of G restricts to the product of a Haar measure on G_a^- and a left Haar measure on P_a.

We let $B_0 \subset G_a^-$ be a neighborhood of the identity with compact closure such that $m_{G_a^-}(\partial B_0) = 0$ and define $B_n = a^{-n} B_0 a^n$. We claim that

$$\frac{1}{m_{G_a^-}(B_n)} \int_{B_n} f(u \cdot x) \, \mathrm{d}m_{G_a^-}(u) \longrightarrow \int_X f \, \mathrm{d}m_X \qquad (12)$$

for any $f \in C(X)$ and any $x \in X$.

---

[16] Unless a specific other probability measure is identified, a property of a transformation on a homogeneous space like ergodicity, mixing, and so on, is meant with respect to the measure induced by the Haar measure on G.

[17] This is an example of a circle of results developed among others by Dani and Veech.

[18] The set B is called injective if the quotient map $\pi : G \to X$ is injective when restricted to B.

[19] This can be quickly checked using the Lie algebras of G_a^- (and of P_a), which are simply the sum of the eigenspaces of Ad_a for all eigenvalues of absolute value less than one (respectively greater than or equal to one).

[20] This is a consequence of considering the eigenvalue decomposition of the matrix $g - I$ in $\mathrm{Mat}_d(\mathbb{R})$ for the linear map $\mathrm{Mat}_d(\mathbb{R}) \ni v \longmapsto ava^{-1}$.

Assuming this for now, it follows that $\mu = m_X$ is the only G_a^--invariant probability measure. Indeed if μ is another such measure then

$$\int_X f \,\mathrm{d}\mu = \int_X \frac{1}{m_{G_a^-}(B_n)} \int_{B_n} f(u \centerdot x)\,\mathrm{d}m_{G_a^-}(u)\,\mathrm{d}\mu(x) \longrightarrow \int_X f\,\mathrm{d}m_X$$

by dominated convergence. As this would hold for any $f \in C(X)$ we deduce that $\mu = m_X$, as claimed.

Now fix a point $x \in X = \Gamma\backslash G$ and a function $f \in C(X)$. By compactness f is uniformly continuous, so given $\varepsilon > 0$ there is a $\delta > 0$ for which

$$\mathsf{d}(h, e) < \delta \implies |f(h \centerdot y) - f(y)| < \varepsilon \tag{13}$$

where d is a left-invariant metric on G (giving rise to the metric on X). Now we can choose a compact neighborhood $V \subset P_a$ of the identity whose boundary has measure zero with

$$\mathsf{d}(a^{-n}ha^n, e) < \delta$$

for $h \in V$ and $n \geqslant 0$. Then

$$\frac{1}{m_{G_a^-}(B_n)} \int_{B_n} f(u \centerdot x)\,\mathrm{d}m_{G_a^-}(u)$$

is within ε of

$$\frac{1}{m_{G_a^-}(B_n) m_{P_a}(a^{-n}Va^n)} \int_{B_n} \int_{a^{-n}Va^n} f(hu \centerdot x)\,\mathrm{d}m_{P_a}(h)\,\mathrm{d}m_{G_a^-}(u)$$

because of (13). Using $B_n = a^{-n}B_0a^n$, the latter may in turn be written as

$$\frac{1}{m_G(VB_0)} \int_{VB_0} f(a^{-n}ga^n \centerdot x)\,\mathrm{d}m_G(g), \tag{14}$$

since m_G is locally the product of m_{P_a} and $m_{G_a^-}$. Now notice that the map

$$G_a^- \ni u \mapsto u \centerdot x$$

is injective for all $x \in X$, for otherwise the injectivity radius at $a^n \centerdot x$ would shrink to zero, contradicting the compactness of X. By a simple compactness argument, we may assume that the above δ is small enough to ensure that the map

$$VB_0 \ni g \mapsto g \centerdot x$$

is injective for all $x \in X$. Thus (14) can also be written as

$$\frac{1}{m_G(VB_0)} \int_X f(a^{-n}y) \mathbb{1}_{VB_0a^n \centerdot x}(y)\,\mathrm{d}m_X. \tag{15}$$

In the sequence (or in any of its subsequences) $(a^n \centerdot x)_{n \geqslant 1}$ we can find (by compactness) a subsequence $(a^{n_k} \centerdot x)_{k \geqslant 1}$ converging to some $z \in X$. Since[21]

$$\|\mathbb{1}_{VB_0a^{n_k} \centerdot x} - \mathbb{1}_{VB_0 \centerdot z}\|_2 \longrightarrow 0$$

[21] Here we are making use of the fact that $m_G(\partial(VB_0)) = 0$, which follows since m_G is the product measure in the local coordinate system $G_a^- P_a$ of G that we use and we already know that $m_{G_a^-}(\partial B_0) = 0$ and $m_{P_a}(\partial V) = 0$.

by dominated convergence as $k \to \infty$, we see that the expression (15) converges to

$$\frac{1}{m_G(VB_0)} \int_X f \, dm_X \int \mathbb{1}_{VB_0 \cdot z} \, dm_X$$

as $n \to \infty$ because a defines a mixing transformation. This proves (12) for the given function f up to an error of ε. Since f and $\varepsilon > 0$ were both arbitrary, the theorem follows. □

Notice that once unique ergodicity is proved by using the Følner sequence

$$\left(a^{-n} B_0 a^n\right)_{n \geqslant 1},$$

then the pointwise everywhere convergence of the ergodic averages also follows for other Følner sets (see Exercises 34–35).

Exercise 34. We let $B_n = a^{-n} B_0 a^n$ be as in the proof of Theorem 33, with

$$m_{G_a^-}(\partial B_0) = 0.$$

Show that this sequence is a Følner sequence in G_a^-, that is a sequence satisfying

$$\frac{m_{G_a^-}(B_n \triangle (K B_n))}{m_{G_a^-}(B_n)} \longrightarrow 0$$

as $n \to \infty$ for every compact subset $K \subset G_a^-$.

Exercise 35. Let a and X be as in Theorem 33. Let $F_n \subset G_a^-$ be any Følner sequence and show that

$$\frac{1}{m_{G_a^-}(F_n)} \int_{F_n} f(u \cdot x) \, dm_{G_a^-}(u) \to \int_X f \, dm_X$$

as $n \to \infty$, for any $f \in C(X)$ and any $x \in X$.

7 Ratner's theorems in unipotent dynamics

We let $X = \Gamma \backslash G$, where G is a connected Lie group and $\Gamma < G$ a lattice. Let

$$U = \{u_s \mid s \in \mathbb{R}\} < G$$

be a one-parameter unipotent subgroup of G. Then the U-invariant probability measures on X can be completely classified. This was conjectured by Dani (as an analog of Raghunathan's conjecture, which will be described below) and proved by Ratner. We only state these important results and refer to the original papers [9, 10, 11, 12, 8], the monograph [7] and the survey [6] for proofs and more details.

Theorem 36 (Dani's conjecture; Ratner's measure classification). *If $X = \Gamma \backslash G$ and $U = \{u_s \mid s \in \mathbb{R}\} < G$ is a one-parameter unipotent subgroup, then every U-invariant ergodic probability measure μ on X is*[22] algebraic. *That is, there exists*

[22] An alternative term that is used is *homogeneous*.

a closed connected unimodular subgroup L with $U \leqslant L \leqslant G$ such that μ is the L-invariant normalized probability measure (that is, the normalized Haar measure) on a closed orbit $L \bullet x_0$ (for any $x_0 \in \operatorname{supp} \mu$).

In this result it is sufficient to assume that Γ is discrete or even just closed. Theorem 36 and Theorem 33 suggest other results which we now start to describe.

Theorem 37 (Ratner's equidistribution theorem). *Let $X = \Gamma \backslash G$ where Γ is a lattice, and let $U = \{u_s \mid s \in \mathbb{R}\} < G$ be a one-parameter unipotent subgroup. Then for any $x_0 \in X$ there exists some closed connected unimodular subgroup $L \leqslant G$ such that $U \leqslant L$,*

- *$L \bullet x_0$ is closed with finite L-invariant volume, and*
- $\displaystyle \frac{1}{T}\int_0^T f(u_s \bullet x_0)\,\mathrm{d}s \longrightarrow \frac{1}{\operatorname{vol}(L \bullet x_0)}\int_{L \bullet x_0} f\,\mathrm{d}m_{L \bullet x_0}$ *as $T \to \infty$.*

It is interesting to note that Theorem 37 in particular implies that any point $x \in X$ returns close to itself under a unipotent flow. That is, for any one-parameter unipotent subgroup $\{u_s \mid s \in \mathbb{R}\}$ and any $x \in X$ there is a sequence $(t_k)_{k \geqslant 1}$ for which $t_k \to \infty$ and $\mathsf{d}(x, u_{t_k} \bullet x) \to 0$ as $k \to \infty$. This close return statement is of course incomparably weaker than Ratner's equidistribution theorem, but even this weak statement does not seem to have an independent proof to our knowledge.

Theorem 37 also suggests that the closures of orbits under the action of a unipotent one-parameter subgroup should be algebraic. A more general version of that statement is the famous conjecture of Raghunathan that motivated all of the theorems above and was proved by Ratner.

Theorem 38 (Raghunathan's conjecture; Ratner's orbit closure theorem). *Suppose that $X = \Gamma \backslash G$, with G a connected Lie group and Γ a lattice. Let $H < G$ be a closed subgroup generated by one-parameter unipotent subgroups. Then for any $x_0 \in X$ the orbit closure is*[23] algebraic, *meaning that there exists some closed connected unimodular subgroup L with $H \leqslant L \leqslant G$ such that*

$$\overline{H \bullet x_0} = L \bullet x_0$$

and $L \bullet x_0$ supports a finite L-invariant measure.

It is also interesting to ask what the structure of the set of all probability measures that are invariant and ergodic under some unipotent flow really is. This generalizes a theorem of Sarnak concerning periodic horocycle orbits. At first sight, one might only ask this out of curiosity or to satisfy the urge to complete our understanding of this aspect of these dynamical systems. However, this line of enquiry turns out to be useful for applications to number-theoretic problems. A satisfying answer to this question is given by Mozes and Shah.

Theorem 39 (Mozes–Shah equidistribution theorem). *Let $X = \Gamma \backslash G$ with G a connected Lie group and Γ a lattice, and let $H_n < G$ be a sequence of subgroups*

[23] Again this is also called *homogeneous*.

generated by unipotent one-parameter subgroups. Let μ_n be an invariant ergodic probability measure for the action of H_n for all $n \geqslant 1$. Assume that[24] $\mu_n \to \mu$ *in the weak*-topology as $n \to \infty$. Then one of the following two possibilities holds.*

(1) $\mu = 0$, *and* $\operatorname{supp}\mu_n \to \infty$ *as* $n \to \infty$ *in the sense that for every compact set* $K \subset X$ *there is an* N *with* $\operatorname{supp}\mu_n \cap K = \emptyset$ *for* $n \geqslant N$.

(2) $\mu = m_{L\cdot y}$ *is the* L*-invariant probability measure on a closed finite volume orbit* $L\cdot y$ *for the closed connected group* $L = \operatorname{Stab}_G(\mu)^\circ \leqslant G$. *Moreover,* μ *is invariant and ergodic for the action of a one-parameter unipotent subgroup. Furthermore, suppose that* $x_n = \varepsilon_n\cdot x \in \operatorname{supp}\mu_n$ *for* $n \geqslant 1$ *and some* $x \in X$ *with* $\varepsilon_n \to 1$ *as* $n \to \infty$, *and suppose the connected subgroups* (L_n) *satisfy* $\mu_n = m_{L_n\cdot x_n}$ *for* $n \geqslant 1$. *Then* $xL = yL = \operatorname{supp}\mu$ *and there exists some* N *with* $\varepsilon_n^{-1}L_n\varepsilon_n \subset L$.

The additional information in each case is useful in applying this theorem. According to (1), once we know that for every measure μ_n there exists some point $x_n \in \operatorname{supp}\mu_n$ within a fixed compact set, the limit measure is a probability measure.

In (2), if we know that $H_n = H$ for all $n \geqslant 1$, then L has to contain H and the conjugates $\varepsilon_n^{-1}H\varepsilon_n$ as in (2). Together this often puts severe limitations on the possibilities that $L \leqslant G$ can take, and sometimes forces L to be G. This situation arises, for example, if we study long periodic horocycle orbits, or orbits of a maximal subgroup $H < G$. In any case, the final claim of (2) says that the convergence to the limit measure $m_{L\cdot x}$ is almost from within the orbit $L\cdot x$. In fact, after modifying the measures in the sequence only slightly by the elements ε_n we get

$$\operatorname{supp}\left((\varepsilon_n)_*^{-1}\mu_n\right) = \varepsilon_n^{-1}L_n\cdot x_n = \varepsilon_n^{-1}L_n\varepsilon_n\cdot x \subset L\cdot x = L\cdot y = \operatorname{supp}\mu$$

for $n \geqslant N$.

References

[1] V. Bergelson, Ergodic Ramsey theory—an update, in *Ergodic theory of $\mathbb{Z}^d$ actions (Warwick, 1993–1994)*, in *London Math. Soc. Lecture Note Ser.* **228**, pp. 1–61 (Cambridge Univ. Press, Cambridge, 1996).

[2] V. Bergelson, Ergodic Ramsey theory: a dynamical approach to static theorems, in *International Congress of Mathematicians. Vol. II*, pp. 1655–1678 (Eur. Math. Soc., Zürich, 2006).

[3] M. Einsiedler and T. Ward, *Ergodic theory with a view towards number theory*, in *Graduate Texts in Mathematics* **259** (Springer-Verlag London Ltd., London, 2011).

[4] M. Einsiedler and T. Ward, *Homogeneous dynamics and applications*, `http://maths.dur.ac.uk/~tpcc68/homogeneous/welcome.html`

[5] A. Katok and B. Hasselblatt, *Introduction to the modern theory of dynamical systems*, in *Encyclopedia of Mathematics and its Applications* **54** (Cambridge

[24] By Tychonoff-Alaoglu there always exists a subsequence that converges.

University Press, Cambridge, 1995). With a supplementary chapter by Anatole Katok and Leonardo Mendoza.

[6] D. Kleinbock, N. Shah, and A. Starkov, 'Dynamics of subgroup actions on homogeneous spaces of Lie groups and applications to number theory', in *Handbook of dynamical systems, Vol. 1A*, pp. 813–930 (North-Holland, Amsterdam, 2002).

[7] D. W. Morris, *Ratner's theorems on unipotent flows*, in *Chicago Lectures in Mathematics* (University of Chicago Press, Chicago, IL, 2005).

[8] S. Mozes and N. Shah, On the space of ergodic invariant measures of unipotent flows, *Ergodic Theory Dynam. Systems* **15** (1995), no. 1, 149–159.

[9] M. Ratner, On measure rigidity of unipotent subgroups of semisimple groups, *Acta Math.* **165** (1990), no. 3-4, 229–309.

[10] M. Ratner, Strict measure rigidity for unipotent subgroups of solvable groups, *Invent. Math.* **101** (1990), no. 2, 449–482.

[11] M. Ratner, On Raghunathan's measure conjecture, *Ann. of Math. (2)* **134** (1991), no. 3, 545–607.

[12] M. Ratner, Raghunathan's topological conjecture and distributions of unipotent flows, *Duke Math. J.* **63** (1991), no. 1, 235–280.

Introduction to
Modern Mathematics
ALM 33, pp. 203–226

Aspects of Nonnegatively Curved Kähler Manifolds

Xiao-Yong Fu* and Xi-Ping Zhu†

Abstract

We survey recent progress on the structure of complete Kähler manifolds with nonnegative or positive curvature and the related dimension estimates of the space of polynomial growth holomorphic functions.

2000 Mathematics Subject Classification: 53C20, 53C44, 53C55.

Keywords and Phrases: Kähler manifold, Ricci flow, volume growth, curvature decay, uniformization, polynomial growth holomorphic function, rigidity.

Introduction

The classical uniformization theorem of Riemann surfaces states that a simply connected Riemann surface is conformally equivalent to the Riemann sphere $\mathbb{CP}^1$, the complex line $\mathbb{C}$, or the open unit 1-disc D. A fundamental problem in Kähler geometry is to generalize this beautiful fact to higher dimensions. There are several ways of generalization, one is based on the observation that the sign of curvature has important implication for the complex structure of Riemann surface. More precisely, we have

1. Every compact surface with positive curvature is conformally equivalent to the Riemann sphere $\mathbb{CP}^1$.
2. Every noncompact complete surface with positive curvature is conformally equivalent to the complex line $\mathbb{C}$.
3. Every simply connected complete surface with curvature bounded from above by a negative constant is conformally equivalent to the open unit 1-disc D.

In this article, we will discuss the uniformization problem for higher dimensional nonnegatively or positively curved Kähler manifolds from this point of view. We will show how the curvature determine the holomorphic structure and metric

*Department of Mathematics, Sun Yat-sen University, Guangzhou, P. R. China 510275, Email: mcsfxy@mail.sysu.edu.cn

†Department of Mathematics, Sun Yat-sen University, Guangzhou, P. R. China 510275, Email: stszxp@mail.sysu.edu.cn

structure of the underlying manifold with emphasis on the results about noncompact Kähler manifold.

First we discuss a general scheme and the scope of problems in the study of noncompact Kähler manifolds. We are interested in understanding both the complex structure and the behavior of the Kähler metric. For the second question we want to show under suitable curvature conditions the Kähler metric is flat . For the first question, there is a long standing conjecture due to Yau.

Yau's Uniformization Conjecture ([88]). *Every complete noncompact Kähler manifold with nonnegative holomorphic bisectional curvature and positive Ricci curvature is biholomorphic to the total space of a holomorphic fibre bundle whose base is a compact locally Hermitian symmetric space and whose fibre is biholomorphic to the complex Euclidean space. If the manifold has positive bisectional curvature, then it is biholomorphic to the complex Euclidean space* $\mathbb{C}^n$.

A weaker version of the above conjecture was made by Greene-Wu and Siu.

Conjecture (Greene-Wu [35], Siu [78]). *A complete noncompact Kähler manifold of positive holomorphic bisectional curvature is a Stein manifold. If the manifold has positive Riemannian sectional curvature, then it is biholomorphic to* $\mathbb{C}^n$.

The motivation of the above conjectures comes from the following celebrated theorem on the structure of complete noncompact Riemannian manifolds with nonnegative sectional curvature.

Theorem (Cheeger-Gromoll [15], Gromoll-Meyer[30]). *Let M be a noncompact real n-dimensional smooth manifold equipped with a complete Riemannian metric of nonnegative sectional curvature. Then there exists a compact geodesically convex, totally geodesic submanifold S (called soul of M) such that M is diffeomorphic to the normal bundle of S in M. In case of positive sectional curvature, S is reduced to a single point and M is diffeomorphic to* $\mathbb{R}^n$.

There are two approaches to Yau's uniformization conjecture. One is by using the L^2-method of Andreotti-Vessentini and Hörmander. This elliptic method, which was initiated by Siu-Yau [80] and later Mok [57] is limited in the sense that the geometry at infinity of the manifold has to be very close to the Euclidean space (the reason being that an abundance of holomorphic functions with slow growth are needed and these functions can exist only when the geometry is very restrictive). Another approach, which was proposed by Shi [75] under the direction of Yau, is to use the Kähler-Ricci flow equation

$$\frac{dg_{i\bar{j}}}{dt} = -R_{i\bar{j}}$$

to deform an initial Kähler metric in the direction of its Ricci curvature. The idea is to show that the geometry improves to the point that it determines the holomorphic structure of the manifold.

When studying the structure of complete noncompact Kähler manifold with nonnegative bisectional curvature, one often imposes conditions on the volume growth and curvature decay of the manifold. Yau [91] conjectured that if the volume growth is of Euclidean, then the scalar curvature must decay quadratically in an average sense. Later, Chen-Zhu [20] asked whether there is a correlation between the volume growth rate and curvature decaying rate of the manifold.

The organization of this article is as follows. In Section 1 we collect some results on the structure of compact nonnegatively or positively curved Kähler manifolds. We will discuss the Frankel conjecture and its generalization under various curvature conditions. Here we are interested in only three different types of curvature: holomorphic bisectional curvature, orthogonal holomorphic bisectional curvature and quadratic orthogonal bisectional curvature. Section 2 concerns the volume growth and curvature decay of complete noncompact Kähler manifold with nonnegative or positive curvature. This turns out to be useful in the study of uniformization problem. In Section 3 we discuss structure theorems for complete noncompact Kähler manifold with nonnegative or positive bisectional curvature. The final section is devoted to the dimension estimate of the space of holomorphic functions with polynomial growth and the related rigidity on complete noncompact nonnegatively curved Kähler manifolds.

1 Compact nonnegatively curved Kähler manifold

We start with the following definitions.

Definition. *Let M be a Kähler manifold of complex dimension n and let $p \in M$,*

1. *M is said to have nonnegative bisectional curvature (abbreviated as $B \geqslant 0$) at p if for any $X, Y \in T_p^{1,0}(M)$, $R(X,\overline{X},Y,\overline{Y}) \geqslant 0$.*
2. *M is said to have nonnegative orthogonal bisectional curvature (abbreviated as $B^{\perp} \geqslant 0$) at p if $R(X,\overline{X},Y,\overline{Y}) \geqslant 0$ for all unitary pairs $X, Y \in T_p^{1,0}(M)$.*
3. *M is said to have nonnegative quadratic orthogonal bisectional curvature at p (abbreviated as $QB \geqslant 0$) if for any unitary frame $\{e_1, \ldots, e_n\}$ of $T_p^{1,0}(M)$ and any real numbers $\xi_1, \ldots, \xi_n$, the following inequality holds*

$$\sum_{i,j} R_{i\bar{i}j\bar{j}}(\xi_i - \xi_j)^2 \geqslant 0$$

where $R_{i\bar{i}j\bar{j}} = R(e_i, \overline{e_i}, e_j, \overline{e_j})$.

Remark. It is obvious that $B \geqslant 0 \Longrightarrow B^{\perp} \geqslant 0 \Longrightarrow QB \geqslant 0$ and that in dimension $n = 2$, the condition $B^{\perp} \geqslant 0$ and $QB \geqslant 0$ are same. In general, the condition $QB \geqslant 0$ is weaker than requiring $B^{\perp} \geqslant 0$. See Li-Wu-Zheng [54], Chau-Tam [26] for examples of Kähler manifolds satisfying $QB \geqslant 0$ but not $B^{\perp} \geqslant 0$.

The famous Frankel conjecture says that a compact n-dimensional Kähler manifold with positive bisectional curvature is biholomorphic to $\mathbb{CP}^n$. There were works by Andreotti-Frankel [29], Kobayashi-Ochiai [45], Iitaka[41], Mabuchi [56] and Hartshorne [38] for Kähler manifolds with dimension not greater than three. The final solution was achieved by Mori [62] and Siu-Yau [81]. While Mori proved the stronger Hartshorne Conjecture (assuming only the ampleness of tangent bundle) using the method of algebraic geometry of characteristic $p > 0$, the proof of Siu and Yau was differential-geometric in nature and applied also to the case of nonnegative bisectional curvature [79]. The method of Siu and Yau depends on the study of stable harmonic maps from Riemann sphere $\mathbb{CP}^1$ to the Kähler manifold.

In [89], Yau proposed the generalized Frankel conjecture that compact Hermitian symmetric spaces can be characterized by Kähler metrics with nonnegative bisectional curvature.

Generalized Frankel Conjecture. *Let (M^n, h) be an n-dimensional compact Kähler manifold of nonnegative holomorphic bisectional curvature and let $(\widetilde{M^n}, \tilde{h})$ be its universal covering space. Then there exist nonnegative integers $k, N_1, \ldots, N_l$ and irreducible compact Hermitian symmetric spaces $M_1, \ldots, M_p$ of rank $\geqslant 2$ such that $(\tilde{M}, \tilde{h})$ is biholomorphically isometric to*

$$(\mathbb{C}^k, g_0) \times (\mathbb{CP}^{N_1}, \theta_1) \times \cdots \times (\mathbb{CP}^{N_l}, \theta_l) \times (M_1, g_1) \times \cdots \times (M_p, g_p)$$

where g_0 denotes the Euclidean metric on $\mathbb{C}^k$, $g_1, \ldots, g_p$ are canonical metrics on $M_1, \ldots, M_p$, and $\theta_i, 1 \leqslant i \leqslant l$, is a Kähler metric on $\mathbb{C}P^{N_i}$ carrying nonnegative holomorphic bisectional curvature.

In the case dimension $n = 3$, the generalized Frankel conjecture was proved by Bando using Siu's result [79] and Kähler-Ricci flow. The higher dimensional version was settled by Cao and Chow [9] under the stronger assumption that the curvature operator is semipositive in the dual Nakano sense. On the other hand, Mok-Zhong [61] studied the generalized Frankel conjecture with the additional assumption that the manifold is Kähler-Einstein.

By using the splitting theorem of Howard-Smyth-Wu [39], one can reduce the proof of Generalized Frankel Conjecture to the following special case which was resolved by Mok:

Theorem 1.1 (Mok [58]). *Let (M^n, h) be an n-dimensional compact simply connected Kähler manifold of nonnegative holomorphic bisectional curvature such that the Ricci curvature is positive at one point. Suppose the second Betti number $b_2(M^n) = 1$. Then either M^n is biholomorphic to the complex projective space or (M^n, h) is biholomorphically isometric to an irreducible compact Hermitian symmetric manifold of rank $\geqslant 2$.*

Mok's proof uses Mori's results on rational curves and is not transcendental in nature. In 2009, inspired by the work of Brendle-Schoen [5], Hui-Ling Gu [32] gave a short analytic proof to the Generalized Frankel Conjecture.

We now turn to the discussion of compact Kähler manifolds with nonnegative or positive orthogonal bisectional curvature. In the late 1980's, Cao and Hamilton introduced the concept of orthogonal holomorphic bisectional curvature and observed that the positivity of orthogonal holomorphic bisectional curvature is preserved under the Ricci flow. In 2006, under the additional assumption that the first Chern class $c_1 > 0$, X.-X Chen [22] extended the Frankel conjecture to compact Kähler manifold with positive orthogonal holomorphic bisectional curvature. He proved that a compact irreducible Kähler manifold with positive orthogonal holomorphic bisectional curvature and positive first Chern class is biholomorphic to the complex projective space.

Later in 2010, Hui-Ling Gu and Zhu-Hong Zhang showed that a compact Kähler manifold with positive orthogonal bisectional curvature must satisfy $c_1 > 0$ and hence is biholomorphic to $\mathbb{CP}^n$. In the same paper, they also extended the Generalized Frankel Conjecture to compact Kähler manifold with nonnegative orthogonal bisectional curvature.

Theorem 1.2 (Gu-Zhang [33]). *Let (M^n, h) be an n-dimensional compact Kähler manifold of nonnegative orthogonal holomorphic bisectional curvature and let $(\widetilde{M^n}, \tilde{h})$ be its universal covering space. Then $(\tilde{M}^n, \tilde{h})$ is isometrically biholomorphic to*

(1) $(\mathbb{C}^k, g_0) \times (\mathbb{CP}^{n_1}, \theta_1) \times \cdots \times (\mathbb{CP}^{n_l}, \theta_l) \times (M_1, g_1) \times \cdots \times (M_p, g_p)$,

or

(2) $(Y, h_0) \times (\mathbb{CP}^{n_1}, \theta_1) \times \cdots \times (\mathbb{CP}^{n_l}, \theta_l) \times (M_1, g_1) \times \cdots \times (M_p, g_p)$,

where g_0 denotes the Euclidean metric on $\mathbb{C}^k$; $g_i (1 \leqslant i \leqslant p)$ are canonical metrics on the irreducible compact Hermitian symmetric spaces M_i of rank $\geqslant 2$, and $\theta_j (1 \leqslant j \leqslant l)$ is a Kähler metric on $\mathbb{CP}^{n_j}$ carrying nonnegative orthogonal holomorphic bisectional curvature; Y is a simply connected Riemann surface with Gauss curvature negative somewhere or a simply connected noncompact Kähler manifold with $\dim Y \geqslant 2$ *and nonnegative orthogonal holomorphic bisectional curvature satisfying that the minimum of the holomorphic sectional curvature < 0 somewhere.*

We conclude this section with a brief discussion on compact Kähler manifolds with nonnegative quadratic orthogonal bisectional curvature. The idea behind studying this condition is that we should like to understand Fano manifolds with nef tangent bundles. In [25], Chau-Tam proved that a locally irreducible compact Kähler manifold with $QB \geqslant 0$ must have positive first Chern class. They also studied the case when (M, g) is reducible and obtained a partial classification of the possible de-Rham factorizations of the universal cover of M. See Theorem 4.2 in [25] for more details.

2 Volume growth and curvature decay

Let (M^n, g) be a complete noncompact complex n-dimensional Kähler manifold with nonnegative bisectional curvature, and let $\mathrm{vol}(B(x, r))$ denote the volume of geodesic ball centered at $x \in M$ with radius r. Then the well-known volume comparison results of Bishop, Calabi and Yau imply that, there exist positive constants $c, C > 0$ such that

$$cr \leqslant \mathrm{vol}(B(x, r)) \leqslant Cr^{2n}$$

for all $r \geqslant 1$, where c depends on x and the dimension n and C depends only on n. Moreover, the above estimates are sharp.

If, in addition, the holomorphic bisectional curvature is positive at some point $x_0 \in M$, then the Busemann function is plurisubharmonic and strictly plurisubharmonic near $x_0 \in M$. Using this fact and integration by parts, Bing-Long Chen and the second author observed that the volume growth rate of the geodesic ball is at least n. More precisely, we obtained the following theorem.

Theorem 2.1 (Chen-Zhu [20]). *Let M be a complete noncompact complex n-dimensional Kähler manifold with nonnegative holomorphic bisectional curvature. If in addition the holomorphic bisectional curvature is positive at some point. Then the volume growth of M satisfies*

$$\mathrm{vol}(B(x_0,r)) \geqslant cr^n$$

for all $r \geqslant 1$, where c is a positive constant depending on x_0 and the dimension n.

Klembeck [44] and Cao [7] constructed some $U(n)$-invariant complete Kähler metrics with positive holomorphic bisectional curvature on $\mathbb{C}^n$ and it can be shown that the volume of the geodesic balls $B(O,r)$ centered at the origin O with respect to these metrics grows like r^n. Thus the above volume growth estimate of Chen-Zhu is also sharp.

Now we turn to discuss the curvature decay estimate on complete noncompact Kähler manifold with nonnegative bisectional curvature. In the following, we will denote the scalar curvature by R and define the average scalar curvature $k(x,r)$ as

$$k(x,r) := \frac{1}{\mathrm{vol}B(x,r)} \int_{B(x,r)} RdV$$

In order to apply Kähler-Ricci flow to Yau's uniformization conjecture, Shi [77] proved the following long time existence theorem for the Kähler-Ricci flow.

Theorem 2.2 (Shi [77]). *Let (M,g) be a complete noncompact Kähler manifold with bounded and nonnegative holomorphic bisectional curvature. Suppose there exist constants $C > 0$ and $0 < \varepsilon < 2$ such that*

$$k(x,r) \leqslant \frac{C}{(1+r)^{\varepsilon}}$$

for all x and r. Then the Kähler-Ricci flow equation

$$\frac{dg_{i\bar{j}}}{dt} = -R_{i\bar{j}}$$

has a longtime solution $g(t)$ on $M \times [0,\infty)$.

The above long time existence result suggests us to investigate the following question: What can we expect of the curvature decay rate on complete noncompact Kähler manifolds with positive bisectional curvature. In [20], Chen and the second author found that the curvature of a complete noncompact Kähler manifold actually satisfies the following decay estimate.

Theorem 2.3 (Chen-Zhu [20]). *Let M be a complete noncompact Kähler manifold with positive holomorphic bisectional curvature. Then for any $x \in M$, there exists a positive constant C, which may depend on x, such that*

$$k(x,r) := \frac{1}{\mathrm{vol}(B(x,r))} \int_{B(x,r)} RdV \leqslant \frac{C}{1+r} \tag{2.1}$$

for all $0 \leqslant r \leqslant \infty$.

The idea of the proof is to construct a strictly plurisubharmonic function u of linear growth. If such a u exists, then one can use L^2 theory to construct nontrivial holomorphic section S of the canonical line bundle using u as a weight function. The growth rate of u will give an estimate of the growth rate of the L^2 norm of the length $||S||$ of S. Using the Bochner type differential inequality

$$\triangle \log ||S||^2 \geqslant R \tag{2.2}$$

at the point where $S \neq 0$, we can show that $\log ||S||$ is at most linear growth by the mean value inequality of Li and Schoen. Then the growth rate of $\log ||S||$ will give an estimate of the average of R over geodesic balls because of (2.2).

Later, following this line of argument, Ni and Tam generalized the above curvature decay estimate, they obtained the following

Theorem 2.4 (Ni-Tam [72]). *Let (M, g) be a complete noncompact Kähler manifold with nonnegative holomorphic bisectional curvature.*

1. *Suppose M is simply connected, then $M = N \times M'$ holomorphically and isometrically, where N is a compact simply connected Kähler manifold. M' is a complete noncompact Kähler manifold and both N and M' have nonnegative holomorphic bisectional curvature. Moreover the scalar curvature R' of M' satisfies the linear decay estimate* (2.1).
2. *If the holomorphic bisectional curvature of M is positive at some point, then M itself satisfies the linear decay estimate* (2.1).

Hence in some sense, linear decay is the slowest decay rate for the curvature of complete noncompact Kähler manifolds with nonnegative holomorphic bisectional curvature.

In case the manifold has maximum volume growth, Yau [91] conjectured that the curvature must decay quadratically in certain average sense. Assuming the curvature is bounded, this was confirmed by Chen-Tang-Zhu [18] for complex surface and by Chen-Zhu [20] for higher dimensional Kähler manifold with nonnegative curvature operator. Finally, Ni [67] proved this conjecture for manifolds with bounded nonnegative holomorphic bisectional curvature. More precisely, Ni obtained the following

Theorem 2.5 (Ni [67]). *Let (M, g) be a complete noncompact Kähler manifold with bounded nonnegative holomorphic bisectional curvature. Suppose M has maximum volume growth. Then there exists $C > 0$ such that*

$$\frac{1}{\mathrm{vol}(B(x,r))}\int_{B(x,r)} R dV \leqslant \frac{C}{(1+r)^2}$$

for all $x \in M$ and all $r \geqslant 0$.

Note that in [70], Ni-Shi-Tam proved the following theorem which, in some sense, can be viewed as the converse of the above theorem.

Theorem 2.6 (Ni-Shi-Tam [70]). *Let (M, g) be a complete noncompact Kähler manifold with nonnegative holomorphic bisectional curvature and let $x_0 \in M$ be a* *fixed point in* M*. Suppose the Ricci curvature is positive at some point and the scalar curvature* R *satisfies* $R(x) \leqslant c\rho^{-2}(x)$ *and*

$$\frac{1}{\mathrm{vol}(B(x_0,r))}\int_{B(x_0,r)} R dV \leqslant cr^{-2},$$

where $\rho(x) = \mathrm{dist}(x_0,x)$ *is the distance from* x_0 *to* x*. Then* M *has maximum volume growth and*

$$\liminf_{r\to\infty} \frac{r^2}{\mathrm{vol}(B(x_0,r))}\int_{B(x_0,r)} R dV > 0.$$

The above theorems lead one to study the relation between curvature decay and volume growth. The following question was asked by Chen and the second author.

Question. *Is there a correlation between the volume growth rate and the average scalar curvature decaying rate?*

Although this question remains unsolved, the known examples suggest that the answer is yes. Recently, Wu and Zheng [85] classified complete $U(n)$ invariant Kähler metrics with positive holomorphic bisectional curvature on $\mathbb{C}^n$ and developed a systematic way to construct examples having positive holomorphic bisectional curvature such that the scalar curvature R satisfies various decay rates.

3 Structure of noncompact nonnegatively curved Kähler manifolds

The attempt to prove the uniformization conjecture has produced many nice results and useful techniques and methods. The purpose of this section is to present, with some indication of proof, a number of theorems concerning the metric/complex structure of noncompact Kähler manifold with nonnegative curvature. More specifically, we will discuss

1. Gap Theorems concerning the characterization of the complex Euclidean space with flat metric as nonnegatively curved Kähler manifold with sufficiently fast decay of curvature;
2. Levi Problem concerning geometric conditions under which a Kähler manifold is necessarily a Stein manifold;
3. Embedding and Uniformization Theorems concerning geometric conditions under which a Kähler manifold is biholomorphic to quasi-projective or affine variety or even the complex Euclidean space.

We begin with a structure theorem which reduces the classification of complete noncompact Kähler manifolds with bounded and nonnegative bisectional curvature to the study of Kähler manifolds with strictly positive bisectional curvature.

Theorem 3.1 (Cao-Chen-Zhu [10]). *Let* M *be a complete noncompact Kähler manifold with bounded and nonnegative holomorphic bisectional curvature. Then* *one of the following holds:*

(i) *M admits a Kähler metric with bounded and positive bisectional curvature;*

(ii) *The universal cover $\tilde{M}$ of M splits holomorphically, isometrically and nontrivially as*

$$\tilde{M} = \mathbb{C}^k \times M_1 \times \cdots \times M_{l_1} \times N_1 \times \cdots \times N_{l_2},$$

where k, l_1, l_2 are nonnegative integers, $\mathbb{C}^k$ is the complex Euclidean space with flat metric; $M_i, 1 \leqslant i \leqslant l_1$, are complete (compact or noncompact) Kähler manifolds with bounded and nonnegative bisectional curvature admitting a Kähler metric with bounded and positive bisectional curvature; $N_j, 1 \leqslant j \leqslant l_2$, are irreducible compact Hermitian symmetric spaces of rank $\geqslant 2$ with the canonical metrics.

Note that when M is compact, the above theorem is covered by Generalized Frankel Conjecture. The proof of the above theorem is a modification of the argument used by Hui-Ling Gu in her proof of the Generalized Frankel Conjecture.

In case M has nonnegative sectional curvature, Fangyang Zheng (see [72]) obtained the following structure theorem.

Theorem 3.2. *Let (M, g) be a complete noncompact Kähler manifold with nonnegative sectional curvature. Then its universal cover is of the form $\tilde{M} = \mathbb{C}^k \times \tilde{N} \times \tilde{L}$, where $\tilde{N}$ is a compact Hermitian manifold, $\tilde{L}$ is Stein and $\tilde{L}$ contains no Euclidean factor. Moreover, M is a holomorphic and Riemannian fibre bundle with fibre $\tilde{N} \times \tilde{L}$ over a flat Kähler manifold. If in addition, the Ricci curvature is positive at some point, then M is simply connected and $M = \tilde{M} = \tilde{N} \times \tilde{L}$ where $\tilde{L}$ is diffeomorphic to the Euclidean space*

3.1 Gap theorems

The first result supporting Yau's uniformization conjecture is the following theorem of Mok-Siu-Yau.

Theorem 3.3 (Mok-Siu-Yau [60]). *Let M be a complete noncompact complex n-dimensional $(n \geqslant 2)$ Kähler manifold with nonnegative holomorphic bisectional curvature. Suppose for a fixed base point x_0, M satisfies the following assumptions:*

(i) $\mathrm{Vol}(B(x_0, r)) \geqslant C_1 r^{2n}, \; 0 \leqslant r < +\infty,$

(ii) $R(x) \leqslant \dfrac{C_2}{1 + d(x_0, x)^{2+\varepsilon}}, \; x \in M,$

for some $C_1, C_2 > 0$ and for any arbitrarily small positive constant ε. Then M is isometrically biholomorphic to $\mathbb{C}^n$ with the standard flat metric.

The proof of the above theorem can be described as follows. First one solves the Poisson equation $\frac{1}{2}\triangle u = R$ with good estimates. This can be done because the geometry of M gives good estimates of Green's function and its gradient. Secondly, one can show that $\triangle||\sqrt{-1}\partial\overline{\partial}u - \mathrm{Ric}||^2 \geqslant 0$ in case of nonnegative bisectional curvature by applying a Bochner-Kodaira formula. Thirdly, using the estimate of u, one obtains integral estimate of $||\partial\overline{\partial}u||^2$ on geodesic balls. From this and the mean value inequality, we conclude that $\sqrt{-1}\partial\bar{\partial}u = \mathrm{Ric}$. In particular, u is a bounded plurisubharmonic function. Finally, one proves u is constant and hence M is flat by establishing a Liouville type theorem for the bounded plurisubharmonic function. In fact, Ni-Tam [72] proved the following Liouville property for plurisubharmonic functions.

Theorem 3.4 (Ni-Tam [72]). *Let (M, g) be a complete noncompact Kähler manifold with nonnegative holomorphic bisectional curvature. Suppose u is a continuous plurisubharmonic function on M satisfying*

$$\limsup_{x\to\infty} \frac{u(x)}{\log d(x_0, x)} = 0,$$

then u is constant.

In [64] , the result of Mok-Siu-Yau was improved to the cases covering manifolds of more general volume growth. In 2001, Chen and the second author [19] introduced the Kähler-Ricci flow method to study the gap phenomena on manifolds without maximal volume growth condition and proved the following

Theorem 3.5 (Chen-Zhu [19]). *Let M be a complete noncompact complex n-dimensional Kähler manifold with bounded and nonnegative holomorphic bisectional curvature. Suppose there exists a positive function $\xi : \mathbb{R} \mapsto \mathbb{R}$ with $\lim_{r\to\infty} \xi(r) = 0$ such that for any $x_0 \in M$,*

$$\frac{1}{\mathrm{vol}(B(x_0, r))} \int_{B(x_0,r)} R(x)dV \leqslant \frac{\xi(r)}{r^2}.$$

Then M^n is biholomorphically isometric to a complete flat manifold.

Later in 2002, using a more direct method which avoids the use of Kähler-Ricci flow, Ni-Tam [71] improved the above result of Chen-Zhu to the case which does not require the uniform decay and the boundedness of the curvature tensor. Recently, by solving the Cauchy problem for the Hodge-Laplacian heat equation for d-closed, positive (1,1) form, Ni [69] proved the following optimal gap theorem.

Theorem 3.6 (Ni [69]). *Let M be a complete noncompact complex n-dimensional ($n \geqslant 2$) Kähler manifold with nonnegative holomorphic bisectional curvature. Suppose for some $x_0 \in M$,*

$$\frac{1}{\mathrm{vol}(B(x_0, r))} \int_{B(x_0,r)} R(x)dV = o(r^{-2}).$$

Then M^n is flat.

An outline of the proof of Theorem 3.6 is as follows: First one solves the Cauchy problem

$$\begin{cases} \dfrac{\partial}{\partial t}\eta(x,t) + \Delta\eta(x,t) = 0, \\ \eta(x,0) = \mathrm{Ric}(x) \geqslant 0, \end{cases}$$

to get a long time solution $\eta(x,t)$ with $\eta(x,t) \geqslant 0$ on $M \times [0,\infty)$ such that $\eta(x,t)$ is closed for any t. Secondly, let $u(x,t) = g^{i\bar{j}}\eta_{i\bar{j}}$, since $u(x,t)$ is nonnegative and satisfies the heat equation with $u(x,0) = R(x)$, it can be expressed in terms of the heat kernel by the uniqueness theorem for nonnegative solutions (Theorem 5.1 of [55]). Finally, the monotonicity of [66] can be applied to $\eta(x,t)$, which particularly implies $tu(x,t)$ is monotonically nondecreasing in t and x. On the other hand, the curvature decay condition and the 'moment' type estimate implies that $\lim_{t\to\infty} tu(x,t) = 0$. Hence $u(x_0,t) = 0$ for all t. The strong maximum principle implies that $u(x,t) \equiv 0$ for any x and $t > 0$. The flatness then follows from $u(x,0) \equiv 0$ which is clear by continuity.

3.2 Steinness of Kähler manifold

To understand the complex structure of complete noncompact nonnegatively or positively curved Kähler manifold M, the first question is the following Levi Problem: *If M has positive bisectional curvature, is M a Stein manifold?*

The first result on the complex structure of complete noncompact Kähler manifold with positive curvature is the following theorem due to Greene-Wu .

Theorem 3.7 (Greene-Wu [34]). *Let M be a complete noncompact Kähler manifold with positive sectional curvature, then M is Stein.*

For the proof, by Grauert's solution to Levi problem, it suffices to produce a smooth strictly plurisubharmonic exhaustion function. Using positivity of sectional curvature, it can be shown that the Busemann function $B(x)$ is a continuous strictly plurisubharmonic exhaustion function. To get a smooth strictly plurisubharmonic exhaustion function, Greene-Wu developed a smoothing technique to use in approximating geodesically convex functions.

In case M has positive bisectional curvature, the Busemann function $B(x)$ is still strictly plurisubharmonic [83]. However, it is unknown that $B(x)$ is an exhaustion function because one does not have a geometric comparison theorem for geodesic distance as in the case of positive sectional curvature.

In [72], by solving heat equation with initial data $B(x)$, Ni-Tam developed a more efficient method to approximate $B(x)$ and obtained the following

Theorem 3.8 (Ni-Tam [72]). *Let (M,g) be a complete noncompact Kähler manifold with nonnegative holomorphic bisectional curvature. Suppose that M is of maximal volume growth or M has a pole, then M is Stein.*

Another approach to produce smooth strictly plurisubharmonic exhaustion function is by solving the Poincaré-Lelong equation

$$\sqrt{-1}\partial\bar{\partial}u = \mathrm{Ric}$$

as in [60, 57], where the following is proved:

Theorem 3.9 (Mok [57]). *Let (M,g) be a complete noncompact Kähler manifold of complex dimension n with nonnegative holomorphic bisectional curvature and positive Ricci curvature. Suppose for a fixed base point x_0, there exit positive constants C_1, C_2 such that*

(i) $\mathrm{Vol}(B(x_0, r)) \geqslant C_1 r^{2n}$, $\forall r > 0$,

(ii) $R(x) \leqslant \dfrac{C_2}{1 + d(x_0, x)^2}$, $\forall x \in M$,

then M is Stein.

When the bisectional curvature is bounded, the more powerful Kähler-Ricci flow method is introduced in [75] to study Yau's uniformization conjecture. In 2001, based on the work of Shi, Chen-Zhu [19] improved the above Mok's theorem and obtained the following:

Theorem 3.10 (Chen-Zhu [19]). *Let M be a complete noncompact Kähler manifold of complex dimension n with nonnegative holomorphic bisectional curvature. Suppose for a fixed base point x_0, there exit positive constants C_1, C_2 and $0 < \epsilon < 1$ such that*

(i) $\mathrm{vol}(B(x_0, r)) \geqslant C_1 r^{2n}$, $\forall r > 0$,

(ii) $R(x) \leqslant \dfrac{C_2}{1 + d(x_0, x)^{1+\epsilon}}$, $\forall x \in M$,

then M is biholomorphic to a pseudoconvex domain in $\mathbb{C}^n$ which is diffeomorphic to $\mathbb{R}^{2n}$ if $n > 2$ and homeomorphic to $\mathbb{R}^4$ if $n = 2$. In particular, M is Stein.

To the authors' knowledge, the best result so far was obtained by Chau-Tam [23], they proved the following:

Theorem 3.11 (Chau-Tam [23]). *Let (M, g) be a complete noncompact Kähler manifold of complex dimension n with bounded nonnegative holomorphic bisectional curvature. Suppose that the scalar curvature of g satisfies*

$$k(x, r) := \frac{1}{\mathrm{vol}(B(x, r))} \int_{B(x,r)} R dv \leqslant k(r)$$

for some function $k(r)$ satisfying

$$k(r) \leqslant \frac{C}{r}$$

for some $C > 0$. Then M is holomorphically covered by a pseudoconvex domain in $\mathbb{C}^n$ which is homeomorphic to $\mathbb{R}^{2n}$. Moreover, if M has positive bisectional curvature and is simply connected at infinity, then M is biholomorphic to a pseudoconvex domain in $\mathbb{C}^n$ which is homeomorphic to $\mathbb{R}^{2n}$, and in particular, M is Stein.

3.3 Uniformization theorems

By the gap theorems discussed in Section 3.1, we know that for complete noncompact Kähler manifold with nonnegative holomorphic bisectional curvature, the average of the scalar curvature cannot have faster than quadratic decay. In the case of quadratic curvature decay, Mok [57] proved the following theorem:

Theorem 3.12 (Mok [57]). *Let M be a complete noncompact Kähler manifold with positive holomorphic bisectional curvature of complex dimensional $n \geqslant 2$. Suppose that for a fixed point $x_0 \in M$, the following conditions hold:*

(i) $\mathrm{vol}(B(x_0, r)) \geqslant C_1 r^{2n}, \; 0 \leqslant r < +\infty$,

(ii) $R(x) \leqslant \dfrac{C_2}{1 + d(x_0, x)^2}, \; x \in M$,

where C_1, C_2 are positive constants. Then M is biholomorphic to an affine algebraic variety. Moreover, in case of $n = 2$, if the Riemannian sectional curvature of M is positive, then M is biholomorphic to $\mathbb{C}^2$.

The method of Mok is to construct enough polynomial growth holomorphic functions to embed M into some $\mathbb{C}^N$ so that the image will be an affine algebraic variety. The main steps of the proof can be described as follows:

Step 1: One obtains a strictly plurisubharmonic function u of logarithmic growth on M by solving the Poincaré-Lelong equation

$$\sqrt{-1}\partial\bar{\partial}u = \mathrm{Ric}.$$

Step 2: Use the standard L^2-estimate of $\bar{\partial}$ with u as weight function to construct algebraically independent holomorphic functions $f_1, \ldots, f_n$ of polynomial growth on M.

Let $P(M)$ be the algebra of holomorphic functions of polynomial growth on M and let $R(M)$ be the quotient field of $P(M)$.

Step 3: Show that $R(M)$ is a finite extension field of $\mathbb{C}(f_1, \ldots, f_n)$ by using a Poincaré-Siegel type argument. Write $R(M) = \mathbb{C}(f_1, \ldots, f_n, g/h)$ for some $g/h = k \in R(M)$ with $g, h \in P(M)$. Consider the map $F : M \to \mathbb{C}^{n+2}$ defined by $F = (f_1, \ldots, f_n, g, h)$.

Step 4: Show that F defines a birational equivalence between M and some irreducible affine algebraic subvariety of $\mathbb{C}^{n+2}$.

Step 5: Desingularize the map F.

The second part of Mok's theorem follows from a result of Ramanujam which states that an algebraic surface which is homeomorphic to $\mathbb{R}^4$ must be biholomorphic to $\mathbb{C}^2$.

Mok's proof used some algebraic geometric techniques to control the holomorphic functions of polynomial growth on M. Later, Shi [75] initiated a program to study Yau's uniformization conjecture by using the Kähler-Ricci flow and obtained the following result:

Theorem 3.13 (Shi [75]). *Let M be a complete noncompact Kähler manifold of complex dimension n with bounded positive holomorphic bisectional curvature. Suppose for a fixed base point $x_0 \in M$, the following conditions are satisfied:*

(i) $\mathrm{vol}(B(x_0, r)) \geqslant c_1 r^{2n}, \; 0 \leqslant r < +\infty$,

(ii) $\displaystyle\int_{B(x_0,r)} R(x)dx \leqslant c_2 r^{2n-2}, \; 0 \leqslant r < +\infty$,

where c_1, c_2 are positive constants. Then M is biholomorphic to a pseudoconvex domain in $\mathbb{C}^n$.

Note that by Theorem 2.5 , condition (ii) in the above theorem is automatically satisfied.

In 2004, by using the Ricci flow, Chen, Tang and the second author obtained the following theorem, which partially confirms Yau's uniformization conjecture in case of complex surface.

Theorem 3.14 (Chen-Tang-Zhu [18], Chen-Zhu [21]). *Let M be a complex two-dimensional complete noncompact Kähler manifold with bounded and positive holomorphic bisectional curvature. Suppose the volume growth of M is maximal, i.e.,*

$$\mathrm{vol}(B(x_0,r)) \geqslant cr^4, \quad 0 \leqslant r < +\infty,$$

for some point $x_0 \in M$ and some positive constant c. Then M is biholomorphic to the complex Euclidean space $\mathbb{C}^2$.

In the following, we sketch the proof given in [18]. The arguments are divided into three parts. In the first part, we show that M is a Stein manifold homeomorphic to $\mathbb{R}^4$. For this, we evolve the Kähler metric on M by the Ricci flow. Since the underlying complex structure, Kählerianess and positivity of holomorphic bisectional curvature are preserved under the Ricci flow, the Kähler metric in the theorem can be replaced by any one of the evolving metric. The advantage is that the properties of the evolving metric are improving during the flow. The maximal volume growth condition is unchanged and the curvature of evolving metric decays linearly in time under the Ricci flow. These imply that the injectivity radius of the evolving metric is getting bigger and bigger and any geodesic ball with radius less than half of the injectivity radius is almost pseudoconvex. By a perturbation argument we are then able to modify these geodesic balls to a sequence of exhausting pseudoconvex domains of M such that any two of them form a Runge pair. From this, it follows readily that M is a Stein manifold homeomorphic to $\mathbb{R}^4$.

In the second part of the proof, we consider the algebra $P(M)$ of holomorphic functions of polynomial growth on M and prove that its quotient field $R(M)$ has transcendental degree two over $\mathbb{C}$. For this, we first need to construct two algebraically independent holomorphic functions in the algebra $P(M)$. Using the L^2 estimate of $\bar{\partial}$ operator, it suffices to construct a strictly plurisubharmonic function of logarithmic growth on M. As is already known in [60, 57], such a strictly plurisubharmonic function of logarithmic growth can be obtain by solving the Poincaré-Lelong equation when the curvature decays at least quadratically in space. Fortunately we can proved that the curvature of the initial metric decays quadratically in space in the average sense and this turns out to be enough to insure the existence of a strictly plurisubharmonic function of logarithmic growth. Next, by using the time decay estimate and the injectivity radius estimate of the evolving metric, we proved that the dimension of the space of holomorphic functions in $P(M)$ of degree at most p is bounded by a constant times p^2. Combining this with the existence of two algebraically independent holomorphic functions in $P(M)$ which has been shown above, we could prove that the quotient field $R(M)$ of $P(M)$ has transcendental degree two over $\mathbb{C}$ by a classical argument of Poincaré-Siegal. In other words, $R(M)$ is a finite extension field of some $\mathbb{C}(f_1, f_2)$, where $f_1, f_2 \in P(M)$ are algebraically independent over $\mathbb{C}$. Then from the primitive element theorem, we deduce $R(M) = \mathbb{C}(f_1, f_2, g/h)$ for some $g, h \in P(M)$. Hence the mapping $F : M \to \mathbb{C}^4$ given by $F = (f_1, f_2, g, h)$ defines a birational map from M into some irreducible affine algebraic subvariety Z of $\mathbb{C}^4$.

In the last part of the proof, we basically follow the approach of Mok in [57] to establish a holomorphic map from M onto a quasi-projective variety by desingularizing the map F. Our essential contribution in this part was to establish uniform estimates on the multiplicity and the number of irreducible components of the zero divisor of a holomorphic function in $P(M)$. Again the curvature decay estimate under the Ricci flow plays a crucial role in the arguments. Based on these estimates, we could show that the mapping $F : M \to Z$ is almost surjective in the sense that it can miss only a finite number of holomorphic functions of polynomial growth. This shows that M is a quasi-projective variety. Finally, by combining with the fact that M is homeomorphic to $\mathbb{R}^4$, we conclude that M is indeed biholomorphic to $\mathbb{C}^2$ by a theorem of Ramanujam [73] on algebraic surfaces.

Recently, Chau-Tam [24] generalized this result to higher dimensions.

Theorem 3.15 (Chau-Tam [24]). *Let M^n be a complex n-dimensional complete noncompact Kähler manifold with bounded and positive holomorphic bisectional curvature. Suppose the volume growth of M^n is maximal, i.e.,*

$$\mathrm{vol}(B(x_0, r)) \geqslant cr^{2n}, \quad 0 \leqslant r < +\infty,$$

for some point $x_0 \in M^n$ and some positive constant c. Then M^n is biholomorphic to the complex Euclidean space $\mathbb{C}^n$.

Chau-Tam's proof also uses the Ricci flow. Firstly, they could understand partial long time behavior of the solution to the Ricci flow. More precisely, they observed that at any point p, $(M^n, g(t))$ behaves like a gradient Kähler Ricci soliton of expanding type, $\mathrm{Ric}(p,t)$ can be 'simultaneously digonalized' near infinity in some sense and $g(t)$ is Lyapunov regular in the sense of dynamical systems, see [2]. Next, they constructed biholomorphisms from a sequence of open sets which exhaust M^n onto a fixed ball in $\mathbb{C}^n$. By using the techniques developed by Rosay-Rudin [74] and Jonsson-Varolin [42], they could proceed to assemble these biholomorphisms into a global biholomorphism from M^n to $\mathbb{C}^n$.

So Yau's uniformization conjecture is basically solved in the case of maximal volume growth. For the non-maximal volume growth case, W. To [82] proved the following result.

Theorem 3.16 (To [82]). *Let M be a complete noncompact complex 2-dimensional Kähler manifold of positive sectional curvature. If for some base point $x_0 \in M$, there exist positive constants C_1, C_2 and p such that*

(i) $\displaystyle\int_{B(x_0,r)} \frac{1}{(1 + d(x_0, x))^{2p}} dx \leqslant C_2 \log(r + 2), \ 1 \leqslant r < +\infty$

(ii) $\displaystyle 0 < R(x) \leqslant \frac{C_1}{d^p(x_0, x)},$

(iii) $\displaystyle\int_M \mathrm{Ric}^2 < +\infty,$

then M is biholomorphic to $\mathbb{C}^2$. Moreover in case of $p \geqslant 2$, the result is valid without assuming condition (i).

Several remarks on the assumptions of the above theorem are in order. First, it is likely that the assumption (iii) is automatically satisfied for complete Kähler surfaces with positive sectional curvature. The reason is that on a complete noncompact real four-manifold with positive sectional curvature, we have the generalized Cohn-Vossen inequality

$$\int_M \Theta \leqslant \chi(\mathbb{R}^4) < +\infty$$

where the Gauss-Bonnet-Chern integrand Θ is positive. On the other hand, it seems that the exterior product Ric^2 is more or less comparable with the Gauss-Bonnet-Chern integrand Θ. Secondly, in view of Demailly's holomorphic Morse inequality [27] and the L^2-Riemann-Roch inequality in Nadel-Tsuji [63], the assumption (iii) is a natural condition for a complete Kähler manifold to be a quasi-projective manifold. Finally, the assumptions on the curvature decay and the volume growth are problematic since they require the geometry at infinity of the Kähler manifold to be somewhat uniform.

In [21], Bing-Long Chen and the second author showed that the assumption (iii) alone is sufficient to guarantee that the Kähler surface is biholomorphic to $\mathbb{C}^2$.

Theorem 3.17 (Chen-Zhu [21]). *Let M be a complete noncompact complex n-dimensional Kähler manifold with positive and bounded sectional curvature. Suppose that*

$$\int_M \mathrm{Ric}^n < +\infty.$$

Then M is biholomorphic to a quasi-projective variety, and in case of dimension $n = 2$, M is biholomorphic to $\mathbb{C}^2$.

The basic idea in [21] is to compactify the manifold M as a quasi–projective variety. Siu-Yau [80] and Mok [57] initiated the program of compactification by first using the L^2–method of Andreotti–Vessentini and Hörmander to establish a Siegel's theorem for a subfield of meromorphic functions and then desingularizing the "birational" map associated to the Siegel's theorem. The crux is to choose a subfield of meromorphic functions with suitable estimates for the desingularization. In [57] Mok solved the Poincaré–Lelong equation to show that there is plenty of holomorphic functions of polynomial growth and that the subfield of meromorphic functions arising from quotients of holomorphic functions of polynomial growth has the desirable estimates. In [59] and [82], the quotient fields were defined from holomorphic plurianticanonical sections with polynomial growth or satisfying certain integrability conditions associated to the assumptions on the curvature decay and the volume growth respectively.

Note that one can construct L^2 holomorphic sections of the plurianticanonical system $\{K^{-q}, q > 0\}$ without any assumption on the volume growth and the curvature decay. We observed in [21] that the space of L^2 sections forms a graded algebra. Moreover by using some techniques and estimates developed from the

Ricci flow, we were able to derive the desirable Bézout estimates for the intersections of the zero divisors of holomorphic L^2 sections. Based on these estimates we could bound the Gauss–Bonnet integrals on curves which are the intersection of zero divisors of L^2 holomorphic plurianticanonical sections. The theorem was then proved in the following way: first we used the Bézout estimates to obtain a Siegel type theorem for the subfield of meromorphic functions of M arising from the quotients of the L^2 sections; then we construct a "meromorphic" map from M into a projective algebraic variety; next we used the bounds of the Gauss–Bonnet integrals on the "algebraic" curves to show that the meromorphic map is almost surjective; finally we desingularized the map to get a biholomorphism between M and a quasi–projective variety.

4 Dimension estimates of holomorphic functions and related rigidity

As indicated above, in the study of the uniformization conjecture for complete noncompact complex n-dimensional Kähler manifolds with nonnegative bisectional curvature, one is led to investigate the algebra $P(M)$ of holomorphic functions of polynomial growth.

Recall that the classical Liouville theorem states that any bounded holomorphic functions on $\mathbb{C}$ is constant, and any bounded (or even positive) harmonic functions on the Euclidean space is constant. In 1975, Yau [86] extended the classical Liouville theorem to complete noncompact Riemannian manifolds with nonnegative Ricci curvature and proved that there are no nonconstant positive harmonic functions on such manifolds. A few years later, Cheng [11] pointed out that using a local version of Yau's gradient estimate developed in his joint work with Yau [12], one can show that there are no nonconstant harmonic functions of sublinear growth. These results motivated Yau [90] to make the following conjecture.

Conjecture. *Let M^m be a complete noncompact real m-dimensional Riemannian manifold with nonnegative Ricci curvature. For each positive integer d, the space $H_d(M^m)$ of harmonic functions on M^m satisfying $|u(x)| \leqslant C(1+r(x,x_0)^d)$ is finite dimensional, where C is a positive constant and $r(x,x_0)$ is the distance function to a fixed point $x_0 \in M$.*

Yau also later asked the following question

Question. *If the conjecture is true, is the dimension of $H_d(M^m)$ at most equal to the dimension of the space of harmonic polynomials of degree $\leqslant d$ in $\mathbb{R}^m$?*

Both the conjecture and the question were answered affirmatively in case $d = 1$ or $m = 2$ by Li-Tam [49, 50] and Kasue [43]. The conjecture was solved by Colding-Minicozzi [16] and Li gave a short elegant proof in [48].

The equality case in the question was first considered by Li [47]. Under the additional assumption that the manifold M^{2m} is Kähler, Li proved that if

$$\dim H_1(M^{2m}) = \dim H_1(\mathbb{R}^{2m}),$$

then M^{2m} is isometric to $\mathbb{C}^m$ with standard flat metric. Later, Cheeger-Colding-Minicozzi [13] proved that for a complete noncompact real n-dimension Riemannian manifold M with nonnegative Ricci curvature, if

$$\dim H_1(M^m) = \dim H_1(\mathbb{R}^m),$$

then M^m is isometric to $\mathbb{R}^m$.

Now let M^n be a complete noncompact complex n-dimensional Kähler manifold with nonnegative curvature. For any $d > 0$, consider the space $O_d(M^n)$ of polynomial growth holomorphic functions on M^n satisfying $|f(x)| \leqslant C(1 + r(x, x_0)^d)$, where $r(x, x_0)$ is the distance function to a fixed point $x_0 \in M^n$. The complex version of the above Yau's question can be stated as follows: 1) For each positive integer d, is the dimension of the space $O_d(M^n)$ less than or equal to the dimension of the space $O_d(\mathbb{C}^n)$ of holomorphic polynomials of degree $\leqslant d$ in $\mathbb{C}^n$? 2) If the equality holds for some d, is it true that M is biholomorphically isometric to $\mathbb{C}^n$?

To bound the dimension of holomorphic functions with polynomial growth, it suffices to bound the vanishing order by the growth rate. The following multiplicity estimate was proved by Mok (Theorem 3.1 in [57]).

Theorem 4.1 (Mok [57]). *Let M^n $(n \geqslant 2)$ be a complete noncompact complex n-dimensional Kähler manifold with positive Ricci curvature such that for some fixed point $x_0 \in M$ and positive constants C_1, C_2*

(i) $\mathrm{vol}(B(x_0, r)) \geqslant C_1 r^{2n}, \ 0 \leqslant r < +\infty,$

(ii) $R(x) \leqslant \dfrac{C_2}{1 + d(x_0, x)^2}, \ x \in M.$

Let $f \in O_d(M)$, then there exists a positive constant C independent of f such that

$$\mathrm{Ord}_{x_0}(f) \leqslant Cd.$$

Here $\mathrm{Ord}_{x_0}(f) = \max\{k \in \mathbb{N} | \mathrm{D}^\alpha f(x_0) = 0, |\alpha| < k\}$.

In 2003, Ni obtained the sharp dimension estimate of $O_d(M)$ for complete noncompact Kähler manifold with nonnegative bisectional curvature and maximal volume growth. More precisely, Ni proved the following

Theorem 4.2 (Ni [66]). *Let M^n be a complete noncompact complex n-dimensional Kähler manifold with nonnegative holomorphic bisectional curvature. Suppose that for some point $x_0 \in M$,*

$$\mathrm{vol}(B(x_0, r)) \geqslant Cr^{2n}$$

for all $r \geqslant 1$, where C is a positive constant. Then for each $d > 0$,

$$\dim_{\mathbb{C}} O_d(M^n) \leqslant \dim_{\mathbb{C}} O_{[d]}(\mathbb{C}^n),$$

where $[d]$ is the greatest integer less than or equal to d.

Ni's method is parabolic. The key ingredient is the following monotonicity formula for plurisubharmonic functions.

Theorem 4.3 (Ni [66]). *Let M^n be a complete noncompact Kähler manifold with nonnegative holomorphic bisectional curvature. Let $v(x, t)$ be a family of* *plurisubharmonic functions deformed by the heat equation* $(\frac{\partial}{\partial t} - \triangle)v(x,t) = 0$, *such that* $w(x,t) = \frac{\partial}{\partial t}v(x,t)$ *is continuous for each* $t > 0$. *Then*

$$\frac{\partial}{\partial t}(tw(x,t)) \geqslant 0.$$

This monotonicity formula follows from a gradient estimate of Li-Yau type, which resembles the trace form of Hamilton's Li-Yau-Hamilton differential inequality for the Ricci flow. Under the maximal volume growth condition, the above monotonicity formula was used in [66] to show that the vanishing order at a fixed point of a polynomial growth holomorphic function is bounded from above by its degree, from which the sharp dimension estimate follows.

We note that Ni only obtained sharp dimension estimates in the maximal volume growth case. Shortly afterwards, the general volume growth case and the rigidity of the manifold were discussed in [17]. In particular, our result in [17] shows that the dimension bound in the above complex version of Yau's conjecture is not sharp if the volume growth is not maximal. More precisely, we have

Theorem 4.4 (Chen-Fu-Yin-Zhu [17]). *Let* M^n *be a complete noncompact Kähler manifold of complex dimension* n *with nonnegative holomorphic bisectional curvature. Suppose there exists* $x_0 \in M$, $k > 0, C > 0$, *and such that*

$$\mathrm{vol}(B(x_0, r)) \leqslant C(1+r)^{2k}$$

for all $r > 0$. *Then there holds*

$$\dim_{\mathbb{C}} O_d(M^n) \leqslant \dim_{\mathbb{C}} O_{[d]}(\mathbb{C}^k)$$

for each $d > 0$. *Moreover, the inequality holds for some positive integer* d *if and only if* M^n *is biholomorphically isometric to* $\mathbb{C}^{[k]} \times M_1^{n-[k]}$ *for some complete Kähler manifold* $M_1^{n-[k]}$ *of complex dimension* $n-[k]$ *with nonnegative holomorphic bisectional curvature which carries no nontrivial holomorphic functions of polynomial growth.*

As a consequence, we have the following affirmative answer to the above complex version of Yau's conjecture.

Theorem 4.5 (Ni [66], Chen-Fu-Yin-Zhu [17]). *Let* M^n *be a complete noncompact Kähler manifold of complex dimension* n *with nonnegative holomorphic bisectional curvature. Then there holds*

$$\dim_{\mathbb{C}} O_d(M^n) \leqslant \dim_{\mathbb{C}} O_{[d]}(\mathbb{C}^n)$$

for each $d > 0$. *Moreover, the equality holds for some positive integer* d *if and only if* M^n *is biholomorphically isometric to the complex Euclidean space* $\mathbb{C}^n$ *with standard flat metric.*

Note that the examples constructed by Donnelley [28] show that similar estimate does not hold for harmonic functions. That is, one can not expect

$$\dim_{\mathbb{R}} H_d(M^n) \leqslant \dim H_{[d]}(M^n)$$

for all positive $d > 0$.

Recently, by generalizing the classical Hadamard three circle theorem to Kähler manifolds with nonnegative holomorphic sectional curvature, Liu [51] obtained two sharp monotonicity formula. As a consequence, he proves the following

Theorem 4.6 (Liu [51]). *Let M^n be a complete noncompact Kähler manifold with nonnegative holomorphic sectional curvature. Then for any $d > 0$, $\dim O_d(M) \leqslant \dim O_{[d]}(\mathbb{C}^n)$. Moreover, if the equality holds for some positive integer d, then M is biholomorphically isometric to $\mathbb{C}^n$.*

There are also some interesting results in [51]. We refer the reader to the original paper.

References

[1] S. Bando, *On three-dimensional compact Kähler manifolds of nonnegative bisectional curvature*, J. Diff. Geom., **19**, (1984), 283–297.

[2] L. Barreira and Y. B. Pesin, *Lyapounov exponents and smooth ergodic theory*, University Lecture Series v. 23, American Mathematical Society, 2001.

[3] M. Berger, *Sur les groups d'holonomie homogènes des variétés a connecxion affine et des variétés riemanniennes*, Bull. Soc. Math. France, **83**, (1955), 279–330.

[4] R. L. Bishop and S. I. Goldberg, *On the other second cohomology group of a Kähelr manifold of positive curvature*, Proc. Amer. Math. Soc., **16**, (1965), 119–122.

[5] S. Brendle and R. Schoen, *Classification of manifolds with weakly 1/4-pinched curvatures*, Acta Mathematica, **200**, (2008), 1–13.

[6] S. Brendle and R. Schoen, *Manifolds with 1/4-pinched curvature are space forms*, J. Amer. Math. Soc., **22**, (2009), 287–307.

[7] H. D. Cao, *Existence of gradient Kähler-Ricci solitons*, Elliptic and parabolic methods in geometry (Minneapolis, MN, 1994), 1–16, A K Peters, Wellesley, MA, 1996.

[8] H. D. Cao, *Limits of solutions to the Kähler-Ricci flow*, J. Diff. Geom., **45**, (1997), no. 2, 257–272.

[9] H. D. Cao and B. Chow, *Compact Kähler manifolds with nonnegative curvature operator*, Invent. Math., **83** (1986), 553–556.

[10] H. D. Cao, B. L. Chen, and X. P. Zhu, *Recent developments on Hamilton's Ricci flow*, Survey in Differential Geometry, **12**, (2008), 47–112.

[11] S. Y. Cheng, *Liouville theorem for harmonic map*, Proc. Symp. Pure Math., **36**, (1980), 147–151.

[12] S. Y. Cheng and S. T. Yau, *Differential equations on Riemannian manifolds and their geometric applications*, Comm. Pure Appl. Math., **28**, (1975), 333–354.

[13] J. Cheeger, T. Colding, and W. Minicozzi, *Linear growth harmonic functions on complete manifolds with nonnegative Ricci curvature*, Geom. Funct. Anal., **5**, (1995), no. 6, 948–954.

[14] J. Cheeger and D. Gromoll, *The splitting theorem for manifolds of nonnegative Ricci curvature*, J. Diff. Geom, **6**, (1971), 119–128.

[15] J. Cheeger and D. Gromoll, *On the structure of complete manifolds of nonnegative curvature*, Ann. of Math. (2), **96**, (1972), 413–443.

[16] T. Colding and W. Minicozzi, *Harmonic functions on manifolds,* Ann. of Math. (2), **146**, (1997), no. 3, 725–747.

[17] B. L. Chen, X. Y. Fu, L. Yin, and X. P. Zhu, *Sharp dimension estimates of holomorphic functions and rigidity*, Trans. Amer. Math. Soc., **358**, (2006), no. 4, 1435–1454.

[18] B. L. Chen, S. H. Tang, and X. P. Zhu, *A uniformization theorem of complete noncompact Kähler surfaces with positive bisectional curvature*, J. Diff. Geom., **67**, (2004), 519–570.

[19] B. L. Chen and X. P. Zhu, *On complete noncompact Kähler manifolds with positive bisectional curvature*, Math. Ann., **327**, (2003), no. 1, 1–23.

[20] B. L. Chen and X. P. Zhu, *Volume growth and curvature decay of positively curved Kähler manifolds*, Quarterly Journal of Pure and Applied Mathematics, **1** (2005), 68–108.

[21] B. L. Chen and X. P. Zhu, *Positively curved complete noncompact Kähler manifolds*, Acta Mathematica Scientia, **29B** (4), (2009), 829–845.

[22] X. X. Chen, *On Kähler manifolds with positive orthogonal bisectional curvature*, Adv. Math., **215**, (2007), no. 2, 427–445.

[23] A. Chau and L. F. Tam, *On the Steinness of a class of Kähler manifolds*, J. Diff. Geom., **79**, (2008), 167–183.

[24] A. Chau and L. F. Tam, *On the complex structure of Kähler manifolds with nonnegative curvature*, J. Diff. Geom., **73**, (2006), 491–530.

[25] A. Chau and L. F. Tam, *On quadratic orthogonal bisectional curvature*, J. Diff. Geom., **92**, (2012), no. 2, 187–200.

[26] A. Chau and L. F. Tam, *Kähler C-spaces and quadratic bisectional curvature*, J. Diff. Geom., **94**, (2013), no. 3, 409(C468.

[27] J. P. Demailly, *Holomorphic Morse inequalities*, Proc. Symp. Pure Math, **52**, (1991), Part 2.

[28] H. Donnelly, *Harmonic functions on manifolds of nonnegative Ricci curvature*, Internat. Math. Res. Notices, **8**, (2001), 429–434.

[29] T. Frankel, *Manifolds with positive curvature*, Pacific J. Math., **11**, (1961), 165–174.

[30] D. Gromoll and W. Meyer, *On complete open manifolds of positive curvature*, Ann. of Math., **90**, (1969), 75–90.

[31] A. Grothendieck, *Sur la classification des fibrés holomorphes sur la sphère de Roemann*, Amer. J. Math., **79**, (1957), 121–138.

[32] H. L. Gu, *A new proof of Mok's generalized Frankel conjecture theorem*, Proc. Amer. Math. Soc., **137**, (2009), no. 3, 1063–1068.

[33] H. L. Gu and Z. H. Zhang *An extension of Mok's theorem on the generalized Frankel conjecture*, Science in China Mathematics, **53**, (2010) no. 5, 1253–1264.

[34] R. E. Greene and H. Wu, *Some function-theoretic properties of noncompact Kähler manifolds*, Proc. Symp. Pure Math., **27**, (1975).

[35] R. E. Greene and H. Wu, *Analysis on non-compact Kähler manifolds*, Proc. Symp. Pure Math., **30**, Part 2, (1977).

[36] R. S. Hamilton, *Three manifolds with positive Ricci curvature* , J. Diff. Geom., **17**, (1982), 255–306.

[37] R. S. Hamilton, *Eternal solutions to the Ricci flow*, J. Diff. Geom., **38**, (1993), no. 1, 1–11.

[38] R. Hartshorne, *Ample subvarieties of algebraic varieties*, Lecture Notes in Math., **156**, (1970), Springer-Verlag.

[39] A. Howard, B. Smyth, and H. Wu, *On compact Kähler manifolds of nonnegative bisectional curvature*, I, Acta Math., **147**, (1981), 51–56.

[40] A. Huber, *On subharmonic functions and differential geometry in the large*, Comm. Math. Helv., **32**, (1957), 13–52.

[41] S. Iitaka, *On some new birational invariants of algebraic varieties and their application to rationality of certain algebraic varieties of dimension* 3, J. Math. Soc. Japan, **24**, (1972), 384–396.

[42] M. Jonsson and D. Varolin, *Stable manifolds of holomorphic diffeomorphisms*, Invent. Math., **149**, (2002), 409–430.

[43] A. Kasue, *Harmonic functions with growth conditions on a manifold of asymptotically nonnegative curvature, II.* Recent topics in differential and analytic geometry, 283–301, Adv. Stud. Pure Math., 18-I, Academic Press, Boston, MA, 1990.

[44] P. F. Klembeck, *A complete Kähler metric of positive curvature on C^n*, Proc. Amer. Math. Soc., **64**, (1977), 313–316.

[45] S. Kobayashi and T. Ochiai, *Three-dimensional compact Kähler manifolds with positive holomorphic bisectional curvature*, J. Math. Soc. Japan, **24**, (1972), 465–480.

[46] S. Kobayashi and T. Ochiai, *Characterizations of complex projective spaces and hyperquadrics*, J. Math. Kyoto Univ., **13**, (1973), 31–47.

[47] P. Li *Harmonic functions of linear growth on Kähler manifolds with nonnegative Ricci curvature*, Math. Res. Lett., **2**, (1995), 79–94.

[48] P. Li *Harmonic sections of polynomial growth*, Math. Res. Lett., **4**, (1997), 35–44.

[49] P. Li and L. F. Tam *Linear growth harmonic functions on a complete manifold*, J. Diff. Geom., **29**, (1989), 421–425.

[50] P. Li and L. F. Tam *Complete surfaces with finite total curvature*, J. Diff. Geom., **33**, (1991), 139–168.

[51] G. Liu, *Sharp estimates for holomorphic functions on Kähler manifolds and dimension estimates*, arXiv eprint 2013 arXiv:math. DG/1308.0710

[52] G. Liu, *On finite generation of holomorphic functions on Riemann surfaces*, arXiv eprint 2013 arXiv:math. DG/1308.0711

[53] G. Liu, *On the volume growth of Kähler manifolds with nonnegative bisectional curvature*, arXiv eprint 2014 arXiv:math. DG/1403.3834

[54] Q. Li, D. Wu, and F. Zheng *An example of compact Kähler manifold with nonnegative quadratic bisectional curvature*, Proc. Amer. Math. Soc., **141**, (2013), 2117–2126.

[55] P. Li and S.-T.Yau, *On the parabolic kernel of the Schödinger operator*, Acta Math., **156**, (1986), 153–201.

[56] T. Mabuchi, *C-actions and algebraic threefolds with ample tangent bundle*, Nagoya Math. J., **69**, (1978), 33–64.

[57] N. Mok, *An embedding theorem of complete Kähler manifolds of positive holomorphic bisectional curvature onto affine algebraic varieties*, Bull. Soc. Maht. France, **112**, (1984), 197–258.

[58] N. Mok, *The uniformization theorem for compact Kähler manifolds of nonnegative bisectional curvature*, J. Diff. Geom., **27**, (1988), 179–214.

[59] N. Mok, *An embedding theorem of complete Kähler manifolds of positive Ricci curvature onto quasi-projective varieties*, Math. Ann. **286**, (1990), 377–408.

[60] N. Mok, Y. T. Siu, and S. T. Yau, *The Poincaré-Lelong equation on complete Kähler manifolds*, Comp. Math., **44**, (1981), 183–218.

[61] N. Mok and J. Q. Zhong, *Curvature characterization of compact Hermitian symmetric spaces*, J. Diff. Geom., **23**, (1986), 15–67.

[62] S. Mori, *Projective manifolds with ample tangent bundles*, Ann. of Math. (2), **110**, (1979), 593–606.

[63] A. Nadel and H. Tsuji, *Compactification of complete Kähler manifolds of negative Ricci curvature*, J. Diff. Geom., **28**, (1988), 503–512.

[64] L. Ni, *Vanishing theorems on complete Kähler manifolds and their applications*, J. Diff. Geom., **50**, (1998), 89–122.

[65] L. Ni, *The Poisson equation and Hermitian-Einstein metrics on holomorphic vector bundles over complete noncompact Kähler manifolds*, Indiana Univ. Math. J., **51**, (2002), 679–704.

[66] L. Ni, *A monotonicity formula on complete Kähler manifolds with nonnegative bisectional curvature*, J. Amer. Math. Soc. (2), **17**, (2004), no. 4, 909 946.

[67] L. Ni, *Ancient solutions to Kähler-Ricci flow*, Math. Res. Lett., **12**, (2005), no. 5–6, 633–653.

[68] L. Ni, *Monotonicity and Kähler-Ricci flow*, Geometric evolution equations, 149–165, Contemp. Math., **36**, Amer. Math. Soc., Providence, RI, 2005.

[69] L. Ni, *An optimal gap theorem*, Invent. Math., **189**, (2012), no. 3, 737–761.

[70] L. Ni, Y. G. Shi and L. Tam *Poisson equation, Poincar-Lelong equation and curvature decay on complete Kähler manifolds*, J. Diff. Geom., **57**, (2001), no. 2, 339–388.

[71] L. Ni and L. Tam *Liouville properties of plurisubharmonic functions*, arXiv eprint 2002. arXiv:math.DG/0212364

[72] L. Ni and L. Tam *Plurisubharmonic functions and the structure of complete Kähler manifolds with nonnegative curvature*, J. Diff. Geom., **64**, (2003), no. 3, 457–524.

[73] C. D. Ramanujam, *A topological characterization of the affine plane as an algebraic variety*, Ann. of Math., **94**, (1971), 69–88.

[74] J. P. Rosay and W. Rudin, *Holomorphic maps from $\mathbb{C}^n$ to $\mathbb{C}^n$*, Trans. AMS, **310**, (1988), 47–86.

[75] W. X. Shi, *Ricci deformation of the metric on complete noncompact Kähler manifolds*, Ph. D. Thesis, Harvard University, (1990).

[76] W. X. Shi, *Complete noncompact Kähler manifolds with positive holomorphic bisectional curvature*, Bull. Amer. Math. Soc. (N.S.), **23**, (1990), 437–440.

[77] W. X. Shi, *Ricci flow and the uniformization on complete noncompact Kä manifolds*, J. Diff. Geom., **45**, (1997), 94–220.

[78] Y. T. Siu, *Pseudoconvexity and the problem of Levi,* Bull. Amer. Math. Soc., **84**, (1978), 481–512.
[79] Y. T. Siu, *Curvature characterization of hyperquadrics*, Duke Math. J., **47**, (1980), 641–654.
[80] Y. T. Siu and S. T. Yau, *Complete Kähler manifolds with non–positive curvature of faster than quadratic decay,* Ann. of Math., **105**, (1977), 225–264.
[81] Y. T. Siu and S. T. Yau, *Compact Kähler manifolds of positive bisectional curvature*, Invent. Math., **59**, (1980), 189–204.
[82] W. K. To, *Quasi–projective embeddings of noncompact complete Kähler manifolds of positive Ricci curvature and satisfying certain topological conditions,* Duke Math. J., **63**, (1991), no. **3**, 745–789.
[83] H. Wu, *An elementary method in the study of nonnegative curvature*, Acta Math., **142**, (1979), no. 1–2, 57–78.
[84] H. Wu, *On compact Kähler manifolds of nonnegative bisectional curvature II,* Acta Math., **147**, (1981), 57–70.
[85] H. Wu and F. Y. Zheng, *Examples of positively curved complete Kähler manifolds* Geometry and analysis, no. 1, 517–542, Adv. Lect. Math. (ALM), **17**, Int. Press, Somerville, MA, 2011
[86] S. T. Yau, *Harmonic functions on complete Riemannian manifolds*, Comm. Pure Appl. Math., **28**, (1975), 201–228.
[87] S. T. Yau, *some function-theoretic properties of complete Riemannian manifold and their applications to geometry,* Indiana Univ. Math. J., **25**, (1976), 659–670.
[88] S. T. Yau, *Survey on partial differential equations in differential geometry, Seminar on Diff. Geom.*, edit by S. T. Yau, Princeton Univ. Press, 1982.
[89] S. T. Yau, *Problem section, Seminar on Diff. Geom.*, edit by S. T. Yau, Princeton Univ. Press, 1982.
[90] S. T. Yau, *Nonlinear analysis in geometry,* Enseign. Math. (2), **33**, (1987), 109–158.
[91] S. T. Yau, *A review of complex differential geometry,* Proc. Sympos. Pure Math., **52**, (1991), 619–625.
[92] S. T. Yau, *Open problems in geometry,* Chern—A Great Geometer of the Twentieth Century, (Edit S. T. Yau), Internat. Press, Hong Kong, 1992.

Introduction to
Modern Mathematics
ALM 33, pp. 227–245

The Mahler Measure and K_2 of Elliptic Curves

Xuejun Guo* and Hourong Qin†

Abstract

There are deep relations between the Mahler measures of certain two variable polynomials and L-values of the curves defined by the polynomials. The first progress was made by Deninger in 1997. Then Boyd mad hundreds of conjectureson this topic. In recent years, many conjectures are proved. This note will give a survey mainly on the work of Bertolini & Darmon, Boyd, Brunalt, Deninger, Lalin, Rogers & Zudlin etc in this area. We also introduce our work (with Ji) in progress.

2000 Mathematics Subject Classification: 11G05, 11G30, 19E08.

Keywords and Phrases: The Mahler measure, L-function, K_2 group, Beilinson's regulator.

1 Introduction

The Mahler measure has applications in many areas, such as number theory, representation theory, K-theory etc. In this article, we review the basic properties of the Mahler measure and its relations with algebraic K-theory, especially on the connections between the Mahler measure and K_2 of elliptic curves.

Let E be an elliptic curve, and $L(E, s)$ the Hasse-Weil L-function of E. Then the Birch and Swinnerton-Dyer (BSD) conjecture links the behavior of $L(E, s)$ at $s = 1$ to the height pairing on the rational points of E. The BSD conjecture was one of the seven Millennium Prize Problems listed by the Clay Mathematics Institute which has offered a $1,000,000 prize for the first correct proof. The BSD conjecture has been studied by Coates-Wiles, Gross-Zagier, Kato, Kolyvagin, Rubin, Zhang and others. This conjecture has been proven to be true in some special cases. Recently, Y. Tian's work in [32] and [33] on the BSD conjecture leads him to make a great progress on the congruent number problem. However the special values of $L(E, s)$ at integers $s > 1$ remain mysterious.

*Department of Mathematics, Nanjing University, Nanjing 210093, China. Email: guoxj@nju.edu.cn

†Department of Mathematics, Nanjing University, Nanjing 210093, China. Email: hrqin@nju.edu.cn

Let E be an elliptic curve defined over $\mathbb{Q}$. Then the BSD conjecture predicts that the algebraic and analytic ranks of E are the same, i.e., the order of vanishing of $L(E, s)$ at $s = 1$ is equal to the rank of the Mordell-Weil group. In the 1980's, Bloch conjectured that $L(E, 2)$ should have connection with the elliptic dilogarithm by his work on the K-theory and algebraic cycles. He proved his conjecture for elliptic curves over $\mathbb{Q}$ with complex multiplication. Later, Beilinson made a general conjecture on the connection between $L(C, 2)$ and $K_2(C)$ for a smooth curve C. In particular, Beilinson proved that the rank of $K_2(E)$ is greater than or equal to 1 for modular elliptic curves defined over $\mathbb{Q}$. Wiles et al proved that all elliptic curves defined over $\mathbb{Q}$ are modular. Hence Beilinson's result holds for all elliptic curves defined over $\mathbb{Q}$. The BSD conjecture is the elliptic analogue of analytic class number formulas of number fields. And the Bloch-Beilinson conjecture is the elliptic analogue of the Lichtenbaum conjecture for K_2 of number fields.

By Kontsevich and Zagier, a period is an absolutely convergent integral of a rational function over a subset of $\mathbb{R}^n$ defined by polynomial inequalities and equalities. Let $\mathcal{P}$ be the set of periods and $\widehat{\mathcal{P}} = \mathcal{P}[1/\pi]$ be the ring of extended periods. Let $L(s)$ be a motivic L-function, m an arbitrary integer, and r the order of vanishing of $L(s)$ at $s = m$. The Deligne-Beilinson-Scholl Conjecture predicts that $L^r(m)$ is in $\widehat{\mathcal{P}}$. One can see [13] for details. One way to express the value $L^r(m)$ is to compute the regulator map of algebraic K-groups.

Beilinson's result on K_2 of modular elliptic curves E defined over $\mathbb{Q}$ implies that $L(E, 2) \in \widehat{\mathcal{P}}$. Combining Wiles et al's Theorem on the modularity of elliptic curves, one can see that $L(E, 2) \in \widehat{\mathcal{P}}$ for all elliptic curves defined over $\mathbb{Q}$.

The first progress on the link between algebraic K-theory and the Mahler measure was made in [10] by Deninger in 1997. In that paper, Deninger proved that for an elliptic curve E, the Beilinson regulator of certain element $\{u, v\} \in K_2E$ can be expressed as the Mahler measure of corresponding polynomials. One should note that Deninger's paper deals with the periods of mixed motives. His theorems hold for general higher algebraic K-groups, not only for K_2-group.

If Beilinson's conjecture is true, then the value of regulator is connected to the value of L-function of E at 2. In particular, Deninger made a precise conjecture on the Mahler measure of $1+x+1/x+y+1/y$. Inspired by Deninger's work, Boyd raised hundreds of conjectures in this direction by numerical computation in [5]. Until now many of Boyd's conjectures have been proven, especially the conjectures concerning elliptic curves with complex multiplications, by various authors. In this paper, we will review their results on this subject.

2 Jensen's integral formula

At first, let's recall the classical Jensen's integral formula which is important to the Mahler measure of polynomials of single variable.

Theorem 2.1 (Jensen's integral formula). *Let F be a function of the complex variable z, which is regular inside and on the circle $|z| = r$. Assume that $F(0) \neq 0$.*

Let $\xi_1, \ldots, \xi_N$ be all the zeros of $F(z)$ satisfying $|\xi_v| \leqslant r$. Then

$$\frac{1}{2\pi}\int_0^{2\pi} \log|F(re^{r\theta})|d\theta = \log|F(0)| + \sum_{v=1}^{N} \log\frac{r}{|\xi_v|}.$$

Let $f(z)$ be a polynomial given by

$$f(z) = a_0x^n + a_1x^{n-1} + \cdots + a_{n-1}x + a_n,$$

where $a_0 \neq 0, a_n \neq 0$. We assume that the roots of $f(z) = 0$ are

$$|\xi_1| \leqslant |\xi_2| \leqslant \cdots \leqslant |\xi_N| \leqslant 1 < |\xi_{N+1}| \leqslant |\xi_{N+2}| \leqslant \cdots \leqslant |\xi_n|.$$

Take $r = 1$ in Jensen's integral formula. Then we have

$$\begin{aligned}\frac{1}{2\pi}\int_0^{2\pi} \log|f(re^{r\theta})|d\theta &= \log|a_n| + \sum_{v=1}^{N} \log\frac{r}{|\xi_v|}\\ &= \log\left|\frac{a_n}{\xi_1\cdots\xi_N}\right|.\end{aligned}$$

By Vieta's formulas, we have

$$a_0\xi_1\cdots\xi_n = \pm a_n,$$

which implies that

$$\frac{1}{2\pi}\int_0^{2\pi} \log|f(re^{r\theta})|d\theta = \log|a_0\xi_{N+1}\cdots\xi_n|.$$

3 The Mahler measure

Definition 3.1 (The Mahler measure)**.** Assume that $f \in \mathbb{C}[x_1^{\pm 1}, \ldots, x_n^{\pm 1}]$ is a Laurent polynomial with complex coefficients. Then the logarithmic Mahler measure of f is defined as

$$m(f) = \int_0^1 \cdots \int_0^1 \log|f(\mathrm{e}^{2\pi i\theta_1}, \ldots, \mathrm{e}^{2\pi i\theta_n})|d\theta_1\cdots d\theta_n.$$

The Mahler measure of f is defined as

$$M(f) = \mathrm{e}^{m(f)}.$$

Note that $m(f)$ can be seen as the mean value of $\log|f|$ on the standard n-torus. The logarithmic Mahler measure is a linear map in the sense of

$$m(fg) = m(f) + m(g).$$

By Jensen's integral formula, if

$$f(x) = (x - x_1)\cdots(x - x_k),$$

then

$$m(f) = \log \prod_{1 \leqslant i \leqslant k} \max(|x_i|, 1).$$

Hence

$$M(f) = \prod_{1 \leqslant i \leqslant k} \max(|x_i|, 1).$$

We use the same notations as in [28].

Definition 3.2. If f is a rational function of the form

$$x_1^{\alpha_1} x_2^{\alpha_2} \cdots x_n^{\alpha_n} F(x_1, x_2, \ldots, x_n),$$

where $F(x_1, x_2, \ldots, x_n)$ is a polynomial of x_i's which is prime to $x_1 \cdots x_n$, then we define

$$Jf(x_1, x_2, \ldots, x_n) = F(x_1, x_2, \ldots, x_n).$$

Recall that a cyclotomic polynomial is a minimal polynomial of an n-th primitive root of unity.

Definition 3.3 (Extended cyclotomic polynomial). An extended cyclotomic polynomial over $\mathbb{Q}$ is an irreducible polynomial of the form

$$JF_0\left(\prod_{j=1}^{s} x_j^{\delta_j}\right),$$

where $\delta_1, \ldots, \delta_s$ are integers not all equal to zero, and F_0 is a cyclotomic polynomial.

Theorem 3.4 (Boyd, Lawton, Smyth independently, [3, 17, 29]). *Let $f \in \mathbb{Z}[x_1, x_2, \cdots, x_n]$ be an irreducible polynomial with integral coefficients. Then $m(f) = 0$ if and only if f is an extended cyclotomic polynomial.*

Conjecture 3.5 (Lehmer). There is a positive constant C such that for any $f \in \mathbb{Z}[x]$, if $m(f) > 0$, then $m(f) > C$.

In 1933, Lehmer found that

$$m(x^{10} + x^9 - x^7 - x^6 - x^5 - x^4 - x^3 + x + 1) = 0.16235712\cdots.$$

Until now, no one can find a polynomial $f \in \mathbb{Z}[x]$ such that $m(f)$ is positive and less than the above bound.

Theorem 3.6 (Boyd, 1981). *For $f \in \mathbb{C}[x_1, \ldots, x_n]$, we have*

$$m(f(x_1, \ldots, x_n)) = \lim_{k_2 \longrightarrow \infty} \cdots \lim_{k_n \longrightarrow \infty} m(f(x, x^{k_2}, \ldots, x^{k_n}))$$

So if Lehmer's conjecture holds for one variable, then it also holds for several variables.

4 The Mahler measure of polynomials and height

Definition 4.1 (Height, [16]). Let $f(z) = a_n z^n + a_{n-1}z^{n-1} + \cdots + a_1 z + a_0$ be a polynomial with complex coefficients. Then the height of f is defined as

$$H(f) = |a_n| \prod_{i=1}^{n} \max(1, |\alpha_i|),$$

where $\alpha_1, \ldots, \alpha_n$ are the complex roots of f.

If α is an algebraic number and $f(z)$ is its minimal polynomial with integral coefficients, then the height of α is defined to be $H(f)$.

By the definition of height and Jensen's integral formula, one can see that the Mahler measure is connected to the height of algebraic numbers.

Proposition 4.2. *Suppose that $f(x)$ is an irreducible integral polynomial with leading coefficient 1, and α is a root of $f(x) = 0$. Then*

$$M(f) = H(\alpha).$$

5 The Mahler measure of polynomials with several variables and L-function

We have seen that if $f(x)$ is an irreducible integral polynomial with leading coefficient 1, then the Mahler measure of f is connected with the height of roots of $f(x) = 0$. In this section, we will discuss the Mahler measure of polynomials with several variables.

Let

$$L(2, \chi_{-3}) = \sum_{n=1}^{\infty} \frac{\chi_{-3}(n)}{n^2},$$

where χ_{-3} is the Dirichlet character of conductor 3,

$$\chi_{-3}(n) = \begin{cases} 1, & \text{if } n \equiv 1 \pmod 3) \\ -1, & \text{if } n \equiv -1 \pmod 3) \\ 0, & \text{if } n \equiv 0 \pmod 3). \end{cases}$$

Example 5.1 (Boyd, [4]). In 1981, Boyd proved the following identity:

$$m(1 + z_1 + z_2) = \frac{3\sqrt{3}}{4\pi} L(2, \chi_{-3}).$$

Example 5.2 (Smyth, [30]). Let ζ be the classical Riemann Zeta function. Then

$$m(1 + z_1 + z_2 + z_3) = \frac{7}{2\pi^2}\zeta(3).$$

However, it seems that the Mahler measures of polynomials with several variables are not always connected to the values of L-functions.

Example 5.3 (Boyd, [5]). If $k \geqslant 2$, then

$$m(y^2 - x^3 - k) = \log |k|.$$

This identity can be proved by the elementary integral by variables substitution. Let $f(x, y) = y^2 - x^3 - k$. If $k > 2$, then

$$\begin{aligned} m(y^2 - x^3 - k) &= \int_0^1 \int_0^1 \log |f(\mathrm{e}^{2\pi i\theta_1},\ \mathrm{e}^{2\pi i\theta_2})| d\theta_1 d\theta_2 \\ &= \int_0^1 \int_0^1 \log |\mathrm{e}^{2\pi i\theta_2} - \mathrm{e}^{2\pi i\theta_1} - k| d\theta_1 d\theta_2 \\ &= m(x + y + k) \\ &= \log |k|. \end{aligned}$$

If $k = 2$, the above identity still holds by the continuity of the Mahler measure.

Note that

$$m(y^2 - x^3 \pm 1) = m(1 + z_1 + z_2) = \frac{3\sqrt{3}}{4\pi} L(2, \chi_{-3}).$$

Maillot computed the Mahler measure for the polynomials $a + bx + cy$. He proved the following theorem.

Theorem 5.4 (Maillot, [20]). *For $a, b, c \in \mathbb{C}$,*

$$\pi m(a + bx + cy) = \begin{cases} \mathcal{L}_2 \left(\left| \frac{a}{b} \right| \mathrm{e}^{i\gamma} \right) + \alpha \log |a| + \beta \log |b| + \gamma \log |c|, & \text{if } \triangle, \\ \pi \log\ \max\{|a|, |b|, |c|\}, & \text{otherwise}, \end{cases}$$

where $\triangle$ means $|a|, |b|, |c|$ can be the lengths of sides of a triangle, and $\mathcal{L}_2$ is the Bloch-Wigner function

$$\mathcal{L}_2(z) = -\Im \left(\int_0^z \log(1 - t) \frac{dt}{t} \right) + \arg(1 - z) \log(z).$$

Later, Maillot's result is generalized by Vandervelde to the Mahler measure of

$$P(x, y) = axy + bx + cy + d.$$

One can see [34] for details. The discussion of the Mahler measures of parametrizable polynomials can be found in [35].

Now one may wonder when the Mahler measure is connected with L-values, and when it is expressed as the value such as $\log |k|$. Note that the curves defined by the parametrizable polynomials are rational curves. The Mahler measures of the polynomials which are the defining equations of elliptic curves or hyper elliptic curves may be very complicated.

Boyd discussed in [5] some general conjectures on the necessary conditions for the Mahler measure of a polynomial with integral coefficients to be connected to L-values. Especially, based on the computer computation, he listed hundreds of precise conjectural relations between the Mahler measure of some families of polynomials and the corresponding L-values. We will discuss some of these conjectures in details in the following sections.

6 K-theory

In this section, we will briefly recall the necessary concepts in algebraic K-theory. One can see Quillen's original paper [21] or Sirinivas's introductory book [31] for details.

Let $\mathcal{C}$ be an exact category and $Q\mathcal{C}$ the new category which has the same objects with $\mathcal{C}$ but a morphism $X \longrightarrow Y$ is given by an isomorphism classes of diagrams

$$X \twoheadleftarrow Z \rightarrowtail Y.$$

Let $BQ\mathcal{C}$ be the classifying space of $Q\mathcal{C}$. Then the K-groups of $\mathcal{C}$ is defined as

$$K_i(\mathcal{C}) = \pi_{i+1}(BQ\mathcal{C}).$$

Let X be a Noetherian scheme. Let $\mathcal{P}(X)$ be the exact category of locally free $\mathcal{O}_X$-modules of finite ranks. Let $\mathcal{M}(X)$ be the exact category of coherent $\mathcal{O}_X$-modules. Then the K-theory and K'-theory of X are defined as

$$\begin{aligned} K_i(X) =& K_i(\mathcal{P}(X)), \\ K_i'(X) =& K_i(\mathcal{M}(X)). \end{aligned}$$

Let R be a ring, $\mathcal{P}(X)$ the category of finitely generated projective R-modules. Then the K-theory of the ring R is defined as

$$K_i(R) := K_i(\mathcal{P}(X)).$$

The K-groups of rings are usually difficult to compute. Until now, only the K-groups of finite fields are totally clear by Quillen's work.

Example 6.1.

$$\begin{aligned} K_{2i-1}(\mathbb{F}_q) &= \mathbb{Z}/(q^i - 1)\mathbb{Z}, \\ K_{2i}(\mathbb{F}_q) &= 0. \end{aligned}$$

Example 6.2.

$$\begin{aligned} &K_0(\mathbb{Z}) = \mathbb{Z}, K_1(\mathbb{Z}) = \mathbb{Z}/2\mathbb{Z}, K_2(\mathbb{Z}) = \mathbb{Z}/2\mathbb{Z}, K_3(\mathbb{Z}) = \mathbb{Z}/48\mathbb{Z}, K_4(\mathbb{Z}) = 0, \\ &K_5(\mathbb{Z}) = \mathbb{Z}, K_6(\mathbb{Z}) = 0, K_7(\mathbb{Z}) = \mathbb{Z}/240\mathbb{Z}, \ K_8(\mathbb{Z}) = 0(?), \ldots, . \end{aligned}$$

The second K-groups K_2R of rings were originally defined by Milnor in 1967.

Theorem 6.3 (Matsumoto, 1969)**.** *Let F be a field. Then K_2F is the quotient of $F^\times \otimes F^\times$ by the subgroup generated by the elements $u \otimes (1-u)$, where $u \neq 0, 1$.*

One can see [18] for details of the proof of Matsumoto's Theorem. We use the Steinberg symbol $\{u, v\}$ to denote the image of $u \otimes v$ in K_2F. The Steinberg symbols satisfy the following relations

$$\begin{aligned} \{u_1u_2, v\} &= \{u_1, v\}\{u_2, v\}, \\ \{u, v\} &= \{v, u\}^{-1}, \\ \{u, 1-u\} &= \{u, -u\} = 1. \end{aligned}$$

Example 6.4. In [18], Milnor gave the explicit computation of $K_2(\mathbb{Q})$. He proved that

$$K_2(\mathbb{Q}) = \mathbb{Z}/2\mathbb{Z} \bigoplus (\bigoplus_{p \geqslant 3} \mathbb{Z}/(p-1)\mathbb{Z}).$$

Let F be a number field and $\mathcal{O}_F$ the ring of integers of F. It is well known that $K_2\mathcal{O}_F$ is a finite group. However it is very difficult to determine the precise structure of $K_2\mathcal{O}_F$. By Quillen's long exact sequence, $K_2\mathcal{O}_F$ is isomorphic to the kernel of the tame map $(\partial_\mathfrak{p})$, i.e., we have the following exact sequence

$$0 \longrightarrow K_2\mathcal{O}_F \longrightarrow K_2F \xrightarrow{(\partial_\mathfrak{p})} \bigoplus_\mathfrak{p} k_\mathfrak{p}^\times \longrightarrow 1,$$

where $\mathfrak{p}$ runs through the prime ideals of $\mathcal{O}_F$ and $k_\mathfrak{p}^\times$ is the multiplicative group of the residue field $k_\mathfrak{p}$. Let v be the discrete valuation at $\mathfrak{p}$. Recall that $\partial_\mathfrak{p}$ is defined as

$$\partial_\mathfrak{p} : K_2F \longrightarrow k_\mathfrak{p}^\times$$
$$\{a, b\} \longmapsto (-1)^{v(a)v(b)} \frac{a^{v(b)}}{b^{v(a)}} \bmod \mathfrak{p}.$$

Let F be a number field. Recall that the Dedekind zeta function of F is defined as

$$\zeta_F(s) = \sum_\mathfrak{a} \frac{1}{N(\mathfrak{a})^s}, \ \Re(s) > 1,$$

where $\mathfrak{a}$ runs through all integral ideals of $\mathcal{O}_F$. The special value of the zeta function of F at point $s = 0$ can be interpreted by the class number h_F and the Dirichlet regulator R_F,

$$\zeta_F^*(0) = -\frac{h_F R_F}{w_F},$$

where w_F is the number of roots of unity of F. For example, we have

$$\zeta_\mathbb{Q}^*(0) = -\frac{1}{2},$$
$$h_\mathbb{Q} = 1,$$
$$R_\mathbb{Q} = 1,$$
$$w_\mathbb{Q} = 2.$$

Example 6.5 (Analytic class number formula for $\mathbb{Q}$).

$$\zeta_\mathbb{Q}^*(0) = -\frac{h_\mathbb{Q} R_\mathbb{Q}}{w_\mathbb{Q}}.$$

Note that

$$|K_0(\mathbb{Z})_{\text{tors}}| = 1,$$
$$|K_1(\mathbb{Z})_{\text{tors}}| = 2.$$

Hence the analytic class number formula for $F = \mathbb{Q}$ in the language of K-theory reads as

$$\zeta^*_{\mathbb{Q}}(0) = -\frac{|K_0(\mathbb{Z})_{\text{tors}}|R_{\mathbb{Q}}}{|K_1(\mathbb{Z})_{\text{tors}}|}.$$

The above general analytic class number formula can be rewritten in the language of K-theory as

$$\zeta^*_F(0) = -\frac{|K_0(\mathcal{O}_F)_{\text{tors}}|R^B_1(F)}{|K_1(\mathcal{O}_F)_{\text{tors}}|},$$

where $R^B_1(F)$ is the Borel regulator. The special value of the Dedekind zeta function at other points by higher K-theory is predicted by the Lichtenbaum conjecture.

Conjecture 6.6 (Lichtenbaum)**.** Let F be a number field. For all $n \geqslant 2$,

$$\zeta^*_F(1-n) = \pm\frac{|K_{2n-2}(\mathcal{O}_F)|R^B_n(F)}{|K_{2n-1}(\mathcal{O}_F)_{\text{tors}}|} \text{ up to powers of 2,}$$

where $R^B_n(F)$ is the higher Borel regulator.

If F is totally real and $n = 2$, the Lichtenbaum conjecture is reduced to the Birch-Tate conjecture.

Note that the Lichtenbaum conjecture for totally real abelian number fields is proven to be true.

Theorem 6.7 ([26])**.** *Let F be a totally real abelian number field. Then*

$$\zeta_F(1-n) = \pm 2^{r_1}\frac{|K_{2n-2}(\mathcal{O}_F)|}{|K_{2n-1}(\mathcal{O}_F)|}.$$

Note that if n is odd, then $K_{2n-1}(\mathcal{O}_F) \simeq \mathbb{Z}^{r_1} \oplus$torsion and $\zeta - F(1-n) = 0$. Let $\zeta(s)$ be the Riemann zeta function, n a positive integer. Then

$$\zeta(1-n) = -\frac{B_n}{n} = \pm 2 \cdot \frac{|K_{2n-2}(\mathbb{Z})|}{|K_{2n-1}(\mathbb{Z})|}, \tag{$*$}$$

where B_n is the n-th Bernoulli number. Recall that the Bernoulli number B_n is defined by $\frac{t}{e^t-1} = \sum\limits_{n=0}^{\infty} B_n\frac{t^n}{n!}$, so we have $B_0 = 1$, $B_1 = -1/2$, $B_2 = 1/6$, $B_3 = 0$, $B_4 = -1/30, \ldots$.

Let us turn to consider the K-theory of elliptic curves. Assume that E is an elliptic curve defined over $\mathbb{Q}$, and $\mathcal{E}$ is its Neron model. By the Bass conjecture, $K_i(\mathcal{E})$ should be finitely generated for any natural number i. Let

$$K^T_2(\mathcal{E}) = \ker(K_2(F) \longrightarrow \bigoplus_{\mathcal{D}} \mathbb{F}(\mathcal{D})^*),$$

where F is the function field of E, $\mathcal{D}$ runs through all irreducible curves on $\mathcal{E}$ and $\mathbb{F}(\mathcal{D})$ denotes the residue field at $\mathcal{D}$. The component map corresponding to $\mathcal{D}$ is

$$T_{\mathcal{D}}:\ \{a,b\} \longmapsto (-1)^{v_{\mathcal{D}}(a)v_{\mathcal{D}}(b)}\frac{a^{v_{\mathcal{D}}(b)}}{b^{v_{\mathcal{D}}(a)}} \text{ (modulo } \mathcal{D}).$$

The group $K_2^T(\mathcal{E})$/torsion is called the "integral part" of $K_2(E)$. Note that $K_2^T(\mathcal{E})$ is the image of

$$K_2(\mathcal{E}) \longrightarrow K_2(F).$$

The kernel of the above map is consisting of torsion elements. Hence we have that

$$K_2^T(\mathcal{E}) \bigotimes \mathbb{Q} \simeq K_2(\mathcal{E}) \bigotimes \mathbb{Q}.$$

In 1980's, A. A. Beilinson conjectured that for a general non-singular projective curve of genus g defined over a number field k, the dimension of $K_2(E) \bigotimes \mathbb{Q}$ should be $g[k : \mathbb{Q}]$.

In 1986, Bloch and Grayson suggested in [2] that Beilinson's conjecture should be modified to $K_2^T(\mathcal{E}) \bigotimes \mathbb{Q}$, i.e., for a general non-singular projective curve of genus g defined over the number field k, the dimension of $K_2^T(\mathcal{E}) \bigotimes \mathbb{Q}$ should be $g[k : \mathbb{Q}]$. They suggest that the correct analogue of $K_2(\mathcal{O}_F)$ is $K_2^T(\mathcal{E})$, not $K_2(E)$.

In 1986, Beilinson proved that for modular elliptic curves defined over $\mathbb{Q}$, the rank of $K_2(E) \geqslant 1$. Later it is also proved that the rank of $K_2^T(\mathcal{E}) \geqslant 1$. Beilinson's proof did not give the explicit expression of the non-torsion element of $K_2(E)$. Since Wiles et al proved all elliptic curves defined over $\mathbb{Q}$ are modular, one can see that the rank of $K_2^T(\mathcal{E}) \geqslant 1$. Until now, it is still unknown that if the rank of $K_2(C)$ is finite.

7 Beilinson's regulator and the Mahler measure

Let X be a curve over $\mathbb{C}$, $\mathbb{C}(X)$ the function field of X. Let $\Omega^1(X)$ be the set of differential 1-forms on X. Recall that the Beilinson regulator map for C is defined in the following way:

$$r : K_2(X) \longrightarrow \mathrm{Hom}_{\mathbb{Q}}(\Omega^1(X), \mathbb{R})$$

$$\{u, v\} \longrightarrow \left(\omega \longmapsto \int_{X(\mathbb{C})} \eta(u, v) \wedge \omega \right),$$

where $\eta(u, v) = \log |u| d \arg(v) - \log |v| d \arg(u)$.

Let E be an elliptic curve. Then there is a unique τ with $\Im(\tau) > 0$ such that $E(\mathbb{C}) \simeq \mathbb{C}/\Lambda$, where $\Lambda = \mathbb{Z} + \mathbb{Z}\tau$. Since any $P \in E(\mathbb{C})$ can be represented by a complex number z, we will write $P = [z]$ in the following text. Let $q = \exp(2\pi i \tau)$. Then for any $P = [z] \in E(\mathbb{C})$ the elliptic dilogarithm $\mathcal{L}_{2,q}(P)$ is defined as

$$\mathcal{L}_{2,q}(P) := \sum_{n \in \mathbb{Z}} \mathcal{L}_2(q^n \exp(2\pi i z)), z \in \mathbb{C}.$$

8 Periods

By Kontsevich and Zagier, a period is a complex number whose real and imaginary parts are values of absolutely convergent integral of rational functions with rational coefficients, over domains in $\mathbb{R}^n$ given by polynomial inequalities with rational coefficients.

Sums and products of periods are also periods, so the periods form a ring.

Until now, it is not known if $1/\pi$ is a period. Let $\mathcal{P}$ be the set of periods and $\widehat{\mathcal{P}} = \mathcal{P}[1/\pi]$ be the ring of extended periods.

Conjecture 8.1 (Deligne-Beilinson-Scholl). Let $L(s)$ be a motivic L-function, m an arbitrary integer, and r the order of vanishing of $L(s)$ at $s = m$. The Deligne-Beilinson-Scholl Conjecture predicts that $L^{(r)}(m)$ is in $\widehat{\mathcal{P}}$.

Example 8.2 (Periods). (1) All algebraic numbers, $\pi, \log 2$, $\zeta(s)$ for $s \geqslant 2$.

(2) The (logarithmic) Mahler measure of Laurent polynomials with rational coefficients are periods.

(3) $L(E, 2)$, where E is defined over $\mathbb{Q}$, and $L(E, s)$ is the Hasse-Weil L-function of the elliptic curve E.

The fact that $L(E, 2)$ is a period is proved by Beilinson for modular elliptic curves defined over $\mathbb{Q}$.

Example 8.3 (Examples of non-periods). Periods are countable. So there are real numbers which are not periods. However there is no any natural example of a computable number that has been proved not to be a period.

9 The Mahler measure of $k + x + 1/x + y + 1/y$

In 1997, Deninger made the following conjecture.

Conjecture 9.1 (Deninger, [10]). Let E be the projective completion of

$$1 + X + \frac{1}{X} + Y + \frac{1}{Y} = 0.$$

Then

$$L'(E, 0) = m\left(1 + X + \frac{1}{X} + Y + \frac{1}{Y}\right).$$

Note that E is an elliptic curve with conductor 15 and

$$E \simeq X_1(15).$$

Recall that for an elliptic curve E with conductor N_E,

$$|L'(E, 0)| = (2\pi)^{-2} N_E L(E, 2).$$

Inspired by Deninger, Boyd raised in 1998 hundreds of conjectures on the relations between the Mahler measure and L-values.

Let

$$m(k) := m\left(k + X + \frac{1}{X} + Y + \frac{1}{Y}\right).$$

Theorem 9.2 (Functional equation, Kurokawa and Ochiai in[14]). *If $k \in \mathbb{R}-\{0\}$, then*

$$m(4k^2)+m\left(\frac{4}{k^2}\right)=2m\left(2\left(k+\frac{1}{k}\right)\right).$$

Theorem 9.3 (Functional equation, Lalin and Rogers in [15]). *If k is a nonzero complex number and $|k|<1$, then*

$$m\left(2\left(k+\frac{1}{k}\right)\right)+m\left(2\left(ik+\frac{1}{ik}\right)\right)=m\left(\frac{4}{k^2}\right).$$

Corollary 9.4. *If k is a nonzero real number and $|k|>1$, then*

$$m\left(2\left(k+\frac{1}{k}\right)\right)-m\left(2\left(ik+\frac{1}{ik}\right)\right)=m\left(\frac{4}{k^2}\right).$$

Proof. By Kurokawa and Ochiai's functional equation, we have

$$m(4k^2)+m\left(\frac{4}{k^2}\right)=2m\left(2\left(k+\frac{1}{k}\right)\right).$$

Let $t=1/k$. Since $|t|<1$, by Lalin and Rogers' functional equation,

$$m\left(2\left(t+\frac{1}{t}\right)\right)+m\left(2\left(it+\frac{1}{it}\right)\right)=m\left(\frac{4}{t^2}\right).$$

Since $m(2(t+\frac{1}{t}))=m(2(k+\frac{1}{k}))$ and

$$m\left(2\left(it+\frac{1}{it}\right)\right)=m\left(-2\left(it+\frac{1}{it}\right)\right)=m\left(2\left(ik+\frac{1}{ik}\right)\right),$$

we have

$$m\left(2\left(k+\frac{1}{k}\right)\right)+m\left(2\left(ik+\frac{1}{ik}\right)\right)=m\left(\frac{4}{t^2}\right)=m(4k^2).$$

Hence

$$m\left(2\left(k+\frac{1}{k}\right)\right)-m\left(2\left(ik+\frac{1}{ik}\right)\right)=m\left(\frac{4}{k^2}\right).$$ □

Theorem 9.5. (1) *In 2011, Rogers and Zudlin proved in [25] that*

$$m(1)=\frac{15}{4\pi^2}L(f_{15},2),$$

where $f_{15}=\eta(\tau)\eta(3\tau)\eta(5\tau)\eta(15\tau)$. By the functional equation, the following identities also hold

$$m(5)=6m(1), m(16)=11m(1), m(3i)=5m(1).$$

(2) *In 2012, Rogers and Zudlin proved in* [24] *that*

$$m(2) = \frac{24}{4\pi^2} L(f_{24}, 2),$$

where $f_{24} = \eta(2\tau)\eta(4\tau)\eta(6\tau)\eta(12\tau)$. *By the functional equation, the following identity also holds*

$$m(8) = 4m(2).$$

(3) *In 2011, Mellit proved in* [19] *that*

$$m(2i) = \frac{40}{4\pi^2} L(f_{40}, 2),$$

where

$$f_{40} = q + q^5 - 4q^7 - 3q^9 + 4q^{11} - 2q^{13} + 2q^{17} + 4q^{19} + \cdots$$

is the modular form corresponding to $E: \ x - 1/x + y - 1/y = 2$.

(4) *In 2013, Zudlin proved in* [37] *that*

$$m(i) = \frac{34}{4\pi^2} L(f_{17}, 2),$$
$$m(\sqrt{2}) = \frac{14}{4\pi^2} L(f_{56}, 2),$$

where

$$f_{17} = q - q^2 - q^4 - 2q^5 + 4q^7 + 3q^8 - 3q^9 + 2q^{10} + \cdots,$$
$$f_{56} = q + 2q^5 - q^7 - 3q^9 - 4q^{11} + 2q^{13} - 6q^{17} + 8q^{19} + \cdots.$$

One can find the precise definition of f_{17} *and* f_{56} *in* [37].

If elliptic curves involved in Boyd's conjectures have complex multiplication, the proof is known due to Bloch.

Deninger's conjecture is proven in 2011 by Rogers and Zudilin by elementary methods. They used the method of telescoping modular equations. By virtue of the explicit version of Beilinson's theorem for the modular curves, we can give an alternative proof of Deninger's conjecture. Our strategy is to write $L(E, 2)$ as an Eisenstein series. Since Deninger had already wrote the Mahler measure of $1 + X + 1/X + Y + 1/Y$ as an Eisenstein series, one can compare these two series and prove that they are equal.

Theorem 9.6 (Beilinson's formula for $L(E, 2)$). *Let* f *be the weight* 2 *newform with conductor* N *corresponding to* E, χ *a primitive Dirichlet character modulo* N. *Let* $u_\chi, u_{\overline{\chi}}$ *be some properly chosen modular units on the modular curve* $X_1(N)$. *Then*

$$L(f, 2)L(f, \chi, 1) = \frac{N\pi}{2\varphi(N)} \tau(\chi) \langle r(\{u_\chi, u_{\overline{\chi}}\}), \omega_f \rangle,$$

where $\omega_f = 2\pi i f dz$, $\tau(\chi)$ *is the Gauss sum of* χ, φ *is Euler's totient function,* r *is Beilinson's regulator, and*

$$\langle r(\{u_\chi, u_{\overline{\chi}}\}), \omega_f \rangle = \int_{E(\mathbb{C})} \eta(u_\chi, u_{\overline{\chi}}) \wedge \omega_f.$$

Let $f = 1 + X + \frac{1}{X} + Y + \frac{1}{Y}$. Then

$$m(f) = \frac{4}{\pi}\mathcal{L}_{2,q}([1/4]).$$

This part is essentially contained in Deninger's paper.

Let E be the projective completion of

$$1 + X + \frac{1}{X} + Y + \frac{1}{Y} = 0.$$

Then the Mordell-Weil group

$$E(\mathbb{Q}) = \{O, Q, 2Q, 3Q\} \simeq \mathbb{Z}/4\mathbb{Z}.$$

Let χ be the primitive even character modulo 15 such that $\chi(2) = i$. Let $u_\chi, u_{\overline{\chi}}$ be the modular units in $\mathcal{O}^*(Y_1(15)) \otimes \mathbb{C}$ such that

$$\begin{aligned} \text{div } u_\chi &= \frac{L(\chi,2)}{\pi^2}(O + i(Q) - (2Q) - i(3Q)) \\ \text{div } u_{\overline{\chi}} &= \frac{L(\overline{\chi},2)}{\pi^2}(O - i(Q) - (2Q) + i(3Q)). \end{aligned}$$

Then by Beilinson's formula, we have

$$L(f,2)L(f,\chi,1) = \frac{15i}{8}\pi\tau(\chi)\langle r(\{u_\chi, u_{\overline{\chi}}\}), \omega_f\rangle,$$

which implies that $L(E,2) = \frac{16\pi}{15}\mathcal{L}_{2,q}([1/4])$.

Hence $L'(E,0) = \frac{4}{\pi}\mathcal{L}_{2,q}([1/4])$, which implies that Deninger's conjecture is true. Our proof is very different from that of Rogers and Zudilin which uses several identities of Ramanujan.

10 Construction of the integral elements

Let E be an elliptic curve defined over $\mathbb{Q}$, and O the identity element of the Mordell-Weil group $E(\mathbb{Q})$.

Let's recall Bloch's construction of elements in $K_2(E)$. For any element $u \in \mathbb{Q}(E)^*$, let $\text{div}(u)$ be the divisor of u. We assume that $O, P, Q \in E(\mathbb{Q})$. Suppose that the order of P is d, and the order of Q is e. Then by Abel's Theorem, there are rational functions f and g such that

$$\begin{aligned} \text{div}(f) &= d(P) - d(O), \\ \text{div}(g) &= e(Q) - e(O). \end{aligned}$$

We can normalize f and g such that

$$\begin{aligned} f(Q) &= 1, \\ g(P) &= 1. \end{aligned}$$

By the definition of tame mapping, one can see that the images of $\{f, g\}$ at P and Q are all trivial. So by the product formula, the tame mapping at O must be trivial also. Hence we have

$$\{f, g\} \in \ker(K_2(F) \longrightarrow \bigoplus_{x\in|E|} {k_x}^*).$$

In 1992, Ross proved that the K_2 of all but finitely many elliptic curves defined over $\mathbb{Q}$ and possessing a rational point of order at least 3 has strictly positive rank. At that time, it was not known if all elliptic curves defined over $\mathbb{Q}$ are modular.

In 2004, Ji and Qin gave the explicit expression of the non-torsion elements of $K_2(E)$ when E has a rational point of order at least 4. See [12] for details. When E has a rational point of exact order 3, the explicit expression of the non-torsion elements of $K_2(E)$ is also given, except at most one $\mathbb{R}$-isomorphism class of E.

In 2006, T. Dokchitser, R. De Jeu and D. Zagier studied the same problem for certain hyper elliptic curves,

$$y^2 = cx^{2g+1} + (b_g x^g + \cdots + b_1 x + b_0)^2$$
$$\text{(genus } g \text{ with } 2g+1 \text{ torsion)}$$

and

$$y^2 = -c^2 x^{2g+2} + (cx^{g+1} + b_g x^g + \cdots + b_1 x + b_0)^2.$$
$$\text{(genus } g \text{ with } 2g+2 \text{ torsion)}$$

All these works depend essentially on Bloch's construction. Hence the elliptic curves and hyper elliptic curves need to have enough rational torsions.

In 1994, D. Cooper, M. Culler, H. Gillet, D. D. Long and P. B. Shalen gave a new construction. Let E be an elliptic curve defined by

$$f(x, y) = \sum c_{(i,j)} x^i y^j = 0.$$

Let Δ be the Newton polygon of the polynomial $f(x, y) = 0$. Let τ be a side of Δ. We will always parameterize a side clockwise around Δ and in such a way that $\tau(0), \tau(1), \ldots$ are the consecutive lattice points in τ. To every side τ, we associate a one-variable polynomial

$$P_\tau = \sum_{k=0}^{\infty} c_{\tau(k)} t^k.$$

We call polynomial P "tempered" if for every side τ, the zeros of P_τ consist of roots of unity only.

Theorem 10.1 (CCGLS, [9]). *Let E be an elliptic curve defined by a tempered polynomial. Then there is an N such that the steinberg symbol $\{x, y\}^N \in K_2(E)$.*

Note that this construction does not depend on the torsion points. We will see that there are tempered elliptic curves without non-trivial rational torsion elements. This construction is essentially different from Bloch's method.

Theorem 10.2 (Rodriguez Villegas, [23]). *Let E be an elliptic curve defined by a tempered polynomial P. We assume that $E\bigcap T^2=\emptyset$, where T^2 is the torus $\{x||x|=1\}\times\{y||y|=1\}$. Let $H_1(E,\mathbb{Z})^-$ be the subgroup of $H_1(E,\mathbb{Z})$ where complex conjugation acts by -1, and γ a closed path not going through poles or zeros of x or y which generates $H_1(E,\mathbb{Z})^-$. Then*

$$m(P)=\frac{1}{2\pi}\int_\gamma \eta(\{x,y\}),$$

where

$$\eta(\{x,y\})=\log|x|\mathrm{darg}\ y-\log|y|\mathrm{darg}\ x.$$

Corollary 10.3. *Let E be an elliptic curve defined by a tempered polynomial P. If $m(P)\neq 0$, then there is an N such that the steinberg symbol $\{x,y\}^N\in K_2(E)$ is not torsion.*

Theorem 10.4 (Boyd, Lawton, Smyth). *Let $f(x,y)\in\mathbb{Z}[x,y]$. Then $m(f)=0$ if and only if there is a cyclotomic polynomial g such that*

$$f(x,y)=\pm x^a y^b g(x^c y^d)$$

for some natural numbers a,b,c,d.

Let E be an elliptic curve defined by a tempered polynomial P. Then there is always an N such that the steinberg symbol $\{x,y\}^N\in K_2(E)$ is not torsion.

Furthermore we have the following theorem.

Theorem 10.5 (Guo, Ji and Qin). *For the following three families*

$$\begin{aligned} x^3+y^2+1-kxy&=0, k\in\mathbb{Z};\\ x^2y+y^2x+x+y-kxy&=0, k\in\mathbb{Z};\\ (x-y)(x-1)(y-1)-kxy&=0,\ \ k\in\mathbb{Z}, \end{aligned}$$

there is a positive integer N such that the steinberg symbol

$$\{x,y\}^N\in K_2^T(\mathcal{E}).$$

Furthermore we can prove that $\{x,y\}^N$ is non-trivial in $K_2^T(\mathcal{E})\otimes\mathbb{Q}$. Note that for the family of elliptic curves $x^3+y^2+1-kxy=0$, the Mordell-Weil group is $\mathbb{Z}$. Hence this construction is essentially different from Bloch's construction.

11 The Mahler measure of polynomials with coefficients in a number field

We will briefly recall Bertolini and Darmon's work on $L(f,\chi,2)$. The non-holomorphic Eisenstein series of weight k, level N and character $\chi:(\mathbb{Z}/N\mathbb{Z})^*\longrightarrow\mathbb{C}$ (not necessarily primitive) is the function on $\mathcal{H}\times\mathbb{C}$ defined by

$$\widetilde{E}_{k,\chi}(z,s)=\sum_{m,n\in N\mathbb{Z}\times\mathbb{Z}}{}' \frac{\chi^{-1}(n)}{(mz+n)^k}\cdot\frac{y^s}{|mz+n|^{2s}}.$$

If $k > 2$, or $k = 1$ and χ is non-trivial, we define

$$\widetilde{E}_{k,\chi}(z) := \widetilde{E}_{k,\chi}(z, 0).$$

The normalized Eisenstein series $E_{k,\chi}$ is defined by the following identity

$$\widetilde{E}_{k,\chi}(z) = 2\iota_\chi N^{-k}\tau(\chi^{-1})\frac{(-2\pi i)^k}{(k-1)!}E_{k,\chi}(z),$$

where ι_χ is a non-zero scalar in $\mathbb{Q}(\chi)^*$ and

$$\tau(\chi) = \sum_{a=1}^{N_\chi} \chi(a)e^{2\pi i a/N_\chi},$$

is the Gauss sum and N_χ is the conductor of χ.

Let u_χ and u_{χ_1,χ_2} be modular units such that

$$d\log(u_\chi) = E_{2,\chi},$$
$$d\log(u_{\chi_1,\chi_2}) = E_2(\chi_1, \chi_2),$$

where $E_k(\chi_1, \chi_2)$ is the general Eisenstein series whose definition can be found in [1].

Theorem 11.1 (Bertolini and Darmon, [1]). *Let notations be the same as above. Then*

$$L^*(f, \chi_1, 2)L^*(f, \chi_2, 1) = C_{f,\chi_1,\chi_2} \cdot \int_{Y_0(N)} \bar{f} \log|u_\chi| d\log(u(\chi_1, \chi_2))dxdy.$$

We will use the same notation as in [27]. For any complex number t, let

$$m_2(t) := 2m(x + x^{-1} + y + y^{-1} + t^{1/2}).$$

By Bertolini and Darmon's result, we can prove the following theorem which is conjectured by Samart in [27].

Theorem 11.2 (Guo, Ji and Qin). *Let notations be the same as above, we have*

$$m_2(128) = \frac{1}{2}L'(f_{448,0})$$
$$m_2\left(\frac{49 + 9\sqrt{17}}{2}\right) = \frac{1}{2}(L'(f_{289}, 0) + 8L'(f_{17}, 0))$$
$$m_2\left(\frac{49 - 9\sqrt{17}}{2}\right) = \frac{1}{2}(L'(f_{289}, 0) - 8L'(f_{17}, 0)).$$

Acknowledgements

The authors are supported by NSFC (Nos. 11271177 and 11171141), Natural Science Foundation of the Jiangsu Province (BK2010007), Program for New Century Excellent Talents in University (NCET-10-0471).

References

[1] M. Bertolini and H. Darmon, Kato's Euler system and rational points on elliptic curves I: A p-adic Beilinson formula, preprint.

[2] S. Bloch and D. Grayson, K_2 and L-function of elliptic curves: Computer Calculations, Applications of algebraic K-theory to algebraic geometry and number theory, Contemp. Math. 55, 79–88. Amer. Math. Soc., Providence, RI, 1986.

[3] D. Boyd, Kroneckers theorem and Lehmers problem for polynomials in several variables, J. Number Theory 13 (1980), 116–121.

[4] D. Boyd, Speculations concerning the range of Mahler's measure, Canad. Math. Bull. 24 (1981), no.4, 453–469.

[5] D. Boyd, Mahler's measure and special values of L-functions, Experiment. Math. 7 (1998), 37–82.

[6] F. Brunault, Version explicite du théorème de Beilinson pour la courbe modulaire $X_1(N)$, C. R. Math. Acad. Sci. Paris 343 (2006), no. 8, 505–510.

[7] F. Brunault, Valeur en 2 de fonctions L de formes modulaires de poids 2: théorème de Beilinson explicite, Bull. Soc. Math. France 135 (2007), no. 2, 215–246.

[8] F. Brunault, Étude de la valeur en $s = 2$ de la fonction L dune courbe elliptique, Thèse de doctorat, Université Paris 7, décembre 2005.

[9] D. Cooper, M. Culler, H. Gillet, D. Long and P. Shalen, Plane curves associated to character varieties of 3-manifolds, Invent. Math. 118 (1994), no. 1, 47–84.

[10] C. Deninger, Deligne periods of mixed motives, K-theory and the entropy of certain $\mathbb{Z}_n$-actions, J. Amer. Math. Soc. 10 (1997), no. 2, 259–281.

[11] T. Dokchitser, R. De Jeu and D. Zagier, Numerical verification of Beilinson's conjecture for K_2 of hyperelliptic curves, Compos. Math. 142 (2006), no. 2, 339–373.

[12] Q. Ji and H. Qin, Rank of K_2 of elliptic curves, Sci. China Ser. A 52 (2009), no. 10, 2107–2120.

[13] M. Kontsevich and D. Zagier, Periods, Mathematics Unlimited-2001 and Beyond, Springer, Berlin, 2001, 771–808.

[14] N. Kurokawa and H. Ochiai, Mahler measures via the crystalization, Comment. Math. Univ. St. Pauli 54 (2005), 121–137.

[15] M. N. Lalin and D. Rogers, Functional equations for Mahler measures of genus-one curves, Algebra Number Theory 1 (2007), no. 1, 87–117.

[16] W. Lawton, Heights of algebraic numbers and Szego's theorem, Proceedings of the American Mathematical Society 49 (1975), No. 1, 47–50.

[17] W. Lawton, A generalization of a theorem of Kronecker, J. Sci. Fac. Chiangmai Univ. (Thailand) 4 (1977) 15–23.

[18] J. Milnor, Introduction to algebraic K-theory, Annals of Math. Studies **72**, Princeton University Press, Princeton 1971.

[19] A. Mellit, Mahler measures and q-series, in: Explicit methods in number theory (MFO, Oberwolfach, Germany, 17–23 July 2011), Oberwolfach Reports 8 (2011), no. 3, 1990–1991.

[20] V. Maillot, Géométrie d'Arakelov des variétés toriques et fibrés en droites intégrables, Mém. Soc. Math. Fr. (N.S.) 80 (2000).

[21] D. Quillen, Higher algebraic K-theory I., Algebraic K-theory I: Higher K-theories (Proc. Conf., Battelle Memorial Inst., Seattle, Wash., 1972), 85–147, Lecture Notes in Math., Vol. 341, Springer, Berlin 1973.

[22] M. A. Reichert, Explicit determination of nontrivial torsion structures of elliptic curves over quadratic number fields, Math. Comp. 46 (174) (1986), 637–658.

[23] F. Rodriguez-Villegas, Modular Mahler measures I, in: Topics in number theory (Univer-sity Park, PA, 1997), Math. Appl. 467 (Kluwer Acad. Publ., Dordrecht, 1999), 17–48.

[24] M. Rogers and W. Zudilin, From L-series of elliptic curves to Mahler measures, Compositio Math. 148 (2012), 385–414.

[25] M. Rogers and W. Zudilin, on the Mahler measure of 1+X+1/X+Y+1/Y, arXiv:1102.1153.

[26] J. Rognes and C. Weibel, Two-primary algebraic K-theory of rings of integers in number fields, Journal of the American Mathematical Society 13 (2000), Number 1, 1–54.

[27] D. Samart, Mahler measures as linear combinations of L-values of multiple modular forms, preprint.

[28] A. Schinzel, Polynomials with special regard to reducibility, With an appendix by Umberto Zannier, Encyclopedia of Mathematics and its Applications 77, Cambridge University Press, Cambridge, 2000.

[29] C. J. Smyth, A Kronecker-type theorem for complex polynomials in several variables, Canad. Math. Bull. 24 (1981) 447–452; Errata, Canad. Math. Bull. 25 (1982), 504.

[30] C. J. Smyth, An explicit formula for the Mahler measure of a family of 3-variable polynomials, J. Th. Nombres Bordeaux 14 (2002), 683–700.

[31] V. Srinivas, Algebraic K-theory, Reprint of the 1996 second edition, Modern Birkhäuser Classics, Birkhäuser Boston, Inc., Boston, MA, 2008.

[32] Y. Tian, Congruent Numbers and Heegner Points, Cambridge Journal of Mathematics, Volume 2 (2014), 117–161.

[33] Y. Tian, Congruent numbers with many prime factors, Proc. Natl. Acad. Sci. USA 109 (2012), no. 52, 21256–21258.

[34] S. Vandervelde, A formula for the Mahler measure of $axy + bx + cy + d$, J. Number Theory 100 (2003), no. 1, 184–202.

[35] S. Vandervelde, The Mahler measure of parametrizable polynomials, J. Number Theory 128 (2008), no. 8, 2231–2250.

[36] D. Zagier and H. Gangl, Classical and elliptic polylogarithms and special values of L-series, in The arithmetic and geometry of algebraic cycles (Banff, AB, 1998), NATO Sci. Ser. C Math. Phys. Sci., vol. 548, Kluwer Acad. Publ., Dordrecht, 2000, 561–615.

[37] W. Zudlin, Regulator of modular units and Mahler measures, arXiv:1304.3869.

Introduction to
Modern Mathematics
ALM 33, pp. 247–284

Which Group is the Most Beautiful?*

Lizhen Ji†

Abstract

In this article, we will propose a group as a candidate for the most beautiful group, describe some criteria for beauty in mathematics, and then look at its connections with various subjects in mathematics. We will also discuss some other important or special examples of groups. Hopefully, these discussions will convey the importance of groups in mathematics.

2000 Mathematics Subject Classification: 22E40, 22E20.

Keywords and Phrases: Group, symmetry, transformation group, Lie group, discrete group, beauty.

Contents

*This paper is an expanded version of the lecture notes of an undergraduate colloquium at Department of Mathematics, Tsinghua University in December 2013. I would like to thank the following people for their suggestions, comments and explanations for the most beautiful groups: N. A'Campo, J. Arthur, D. Barrett, R. Borcherds, I. Dolgachev, T. Farrell, B. Farb, W. Fulton, R. Griess, B. Gross, I. Hambleton, J. Humphreys, L. Keen, A. Koranyi, R. Langlands, W. Lück, G. Lusztig, Y. Manin, G. Margulis, W. Schmid, Y. Sinai, D. Sullivan, B. Sury, J.P. Serre, R. Taylor, D. Vogan, S. Weinberger, and S.T. Yau. D. Vogan has read carefully a preliminary version of this paper and made many constructive suggestions.

†Partially Supported by NSF grant DMS-1104696. Department of Mathematics University of Michigan, USA. Email: lji@umich.edu

1 Introduction

Groups are central objects in mathematics. Even in subjects which might seem to be unrelated to groups, they often play a fundamental role implicitly. Groups are (almost) everywhere in mathematics and science, especially in physics. They also play an important role in arts and humanities. One reason is that groups are the mathematics behind symmetry (or symmetry is described via group actions).

There are many applications of groups in mathematics. See [39] for a list of topics in group actions. The following list might give a sense of the scope of applications of groups:

1. Groups appear in geometry through the Erlangen program, in particular, the motion group in plane geometry. (For an elementary introduction to geometry in the spirit of the Erlangen program, see the books [7] [63]. See also [43] for related discussion and some references.)
2. Groups are an essential part of algebra. Before the arrival of modern mathematics, algebra really meant solutions of algebraic equations, which motivated the notion of group through Galois groups of equations and ushered in modern algebra.
3. Groups are essential in analysis, for example, in Fourier analysis (via periodicity and Fourier decomposition), translation and scaling. They also appear in complex analysis (for example, Mobius transformations), and functional analysis (operator algebras).
4. Groups are essential in number theory. One of the origins of groups came from number theory (for example Gauss' first book *Disquisitiones Arithmeticae* which contains some discussion of finite abelian groups arising from quadratic forms and ideal class groups of quadratic fields), and groups are both foundational and essential to number theory, for example from theory of modular forms, automorphic representations and Wiles' work on Fermat's last Theorem to the ongoing work on Riemann zeta function.
5. Group actions, structures of orbits and stabilizers are basic tools in combinatorics. They are also useful in finite geometries and designs.

6. Groups appear in topology both as a way to produce basic spaces and to obtain topological invariants such as characteristic classes, fundamental group, homology and cohomology groups, homotopy groups, and K-theory.
7. Groups are also essential in algebraic geometry in providing basic spaces (homogeneous varieties such as projective spaces and flag varieties) and constructing new spaces (such as moduli spaces) via the geometric invariant theory.
8. Groups are also useful in probability, dynamics, logic, computer science, numerical analysis, etc.
9. Groups are essential in physics from classical mechanics to the general relativity and quantum mechanics (both in describing and discovering elementary particles through representations of orthogonal groups). They are also useful in chemistry (crystals, etc.) and other subjects in science.

We will discuss some concrete applications of groups later.

2 What is a group?

First, we recall that a group is a set G with two operations:

1. a multiplication $G \times G \to G, \quad (g,h) \mapsto gh$, satisfying the associativity $(gh)l = g(hl)$ for $g, h, l \in G$, and there exists an identity element e, i.e., $ge = ge = g$ for all $g \in G$,
2. and an inverse map $g \mapsto g^{-1}$, $gg^{-1} = g^{-1}g = e$.

There are many kinds of groups and they arise from different contexts:

1. Finite groups, which were the first groups introduced, and they contain many different subclasses. One important example is the symmetric group S_n on n elements. Another basic group is the cyclic group $\mathbb{Z}_n = \mathbb{Z}/n\mathbb{Z}$ of order n.
2. Infinite groups, which are becoming more and more important outside algebra such as in number theory, geometry, topology, analysis and dynamics. Examples include $\mathbb{Z}$, $\mathbb{Q}$, $\mathbb{R}$, $\mathrm{SL}(2,\mathbb{Z})$, $\mathrm{SL}(2,\mathbb{R})$ etc.
3. Discrete groups, which are especially important in number theory, geometry and topology. Examples include subgroups of Lie groups $\mathbb{Z} \subset \mathbb{R}$, $\mathrm{SL}(2,\mathbb{Z}) \subset \mathrm{SL}(2,\mathbb{R})$, and component groups of topological groups such as the group of homeomorphisms (or diffeomorphisms) of manifolds, which give rise to mapping class groups.
4. Nondiscrete topological groups, which can either arise from discrete groups via completion or from Lie groups or algebraic groups. Examples such as the p-adic completion $\mathbb{Q}_p$ and the adelic group $\mathbb{R} \times \prod_p \mathbb{Q}_p$.
5. Lie groups provide a natural class of infinite groups, but the combination of a group structure with a smooth manifold structure enhances both structures and makes Lie groups useful in many contexts. The most basic Lie groups are $\mathbb{R}^2$ and $\mathrm{SL}(2,\mathbb{R})$. General semisimple Lie groups can be built up from them.
6. Algebraic groups, which are both specializations and important sources of Lie groups, finite groups and infinite discrete groups. One important point

is that they allow us to produce many other kinds of groups. Matrix groups such as $\mathrm{GL}(n, \mathbb{C})$ (or $\mathrm{GL}(n)$) are algebraic groups, and the group $\mathrm{GL}(n)$ produces other groups: a real Lie group $\mathrm{GL}(n, \mathbb{R})$, a locally compact topological group $\mathrm{GL}(n, \mathbb{Q}_p)$ which is not a Lie group, a discrete subgroup $\mathrm{GL}(n, \mathbb{Z})$, a discrete nonlocally compact group $\mathrm{GL}(n, \mathbb{Q})$, a finite group $\mathrm{GL}(n, F_p)$, etc.

7. Finite dimensional groups, especially finite dimensional Lie groups. It is not easy to construct finite dimensional topological groups which are not Lie groups. $\mathrm{SL}(n, \mathbb{Q}_p)$ is a totally disconnected locally compact topological group. It should be considered of dimension 0. Products of p-adic Lie groups and (real) Lie groups will provide other examples. The positive solution of Hilbert's fifth problem says that any topological groups which are topological manifolds are Lie groups.
8. Infinite dimensional groups, such as diffeomorphism groups of manifolds. The group of diffeomorphisms of the unit circle S^1 is a well-studied infinite dimensional Lie group.
9. Compact groups, which are special among locally compact topological groups. Examples include compact Lie groups: $\mathrm{SO}(2)$ (the rotation group in $\mathbb{R}^2$), $\mathrm{SO}(3)$ (the rotation group in $\mathbb{R}^3$), etc. Compact groups such as the subgroup $\mathrm{SL}(2, \mathbb{Z}_p) \subset \mathrm{SL}(2, \mathbb{Q}_p)$ are not Lie groups.
10. Noncompact Lie groups, such as nilpotent, solvable, reductive, and semisimple Lie groups, each of which has some special properties. The most basic noncompact Lie group is probably $\mathbb{R}$. Other examples include $\mathrm{GL}(n, \mathbb{R})$, which is a reductive Lie group, $\mathrm{SL}(n, \mathbb{R})$, which is a semisimple Lie group, the subgroup of upper triangular matrices of $\mathrm{GL}(n, \mathbb{R})$ which is not a semisimple or reductive Lie group but is a solvable Lie group, and its subgroup of matrices with 1s on the diagonal is a nilpotent Lie group.

Both group operations in the definition of a group can be motivated easily by group actions: the product comes from the composition of transformations and the inverse corresponds to the inverse transformation. A group acting on a space can be regarded as a symmetry group of the space. As we will see below, the importance of groups and rich supply of groups come from actions on suitable spaces or symmetries of the spaces.

Recall that an *action* of a group G on a space X is a map

$$G \times X \to X$$

such that

$$e \cdot x = x, \quad g \cdot (h \cdot x) = gh \cdot x.$$

In particular, g^{-1} undoes the action $g \cdot x$: $g^{-1} \cdot (g \cdot x) = x$.

If X admits some structures, then the group action of G is usually required to preserve it. In this case, G is mapped to the automorphism group of X. For example, if X is a topological space, then G acts by homeomorphisms. If X is a variety and G is an algebraic group, then the action is often required to be given by morphisms.

We can say roughly that a space X admits *symmetry* if a nontrivial group G acts on it nontrivially. It is not necessarily the case that when a bigger group acts on X, then X has more symmetry. One important condition is that group actions should preserve some structures in order to be counted as symmetries. For example, if X is a topological (or smooth) manifold, then its group of homeomorphisms (or diffeomorphisms) acts transitively on X. But it is no necessarily true that X admits a nontrivial action of a finite group. Therefore, diffeomorphisms and homeomorphisms are not "real symmetries" or "rigid" transformations. Consequently, manifolds which do not admit such finite diffeomorphism (or homeomorphism) group actions are sometimes called *asymmetric manifolds*. See [39] for references. Of course, if we endow the space X with additional structures such as a Riemannian metric or a complex structure, and group actions preserve them, then a bigger group will imply larger symmetry of the space. See [30] for a comprehensive discussion of rigid transformations.

One way to understand better or motivate the above list of groups is to think of their corresponding symmetries, or their actions on appropriate spaces.

3 Which group is the most beautiful?

Groups are beautiful since symmetric things are beautiful, or equivalently, spaces admitting symmetries are beautiful. Then a natural, or naive question is: *Which group is the most beautiful?*

Before we try to answer this question, probably we need to address some more basic questions.

1. What is beauty? How to judge or decide which (thing or person etc) is beautiful?
2. To compare different things, we need some quantities and invariants. How to measure beauty? How to find invariants and degrees of beauty?

One common answer for these questions is that probably there have been no precise, or objective, answers to these questions up to now, nor there will be in the future.

Indeed, people have been thinking and arguing about them for thousands of years, right from the dawn of civilization, and every major philosopher has thought about it and has his own opinion. Therefore, these questions are important (or interesting) and complicated.

On the other hand, maybe the best, or simple, answer to the question of what is beauty is that *I will know it when I see it!* This is especially true about beauty in life.

This is also probably true in mathematics. When we see some theorems, proofs or theories, we just *feel* that they are beautiful. It is often difficult to nail down precisely what makes one result beautiful. Beautiful things in mathematics are often related to simple, clean structure out of chaos and complicated things, elegance of statements, and neat and unexpected connections which bring pure joy or pleasure to the observer (or mathematician).

If we have to say more concretely, beauty in mathematics is often associated with the following aspects: *truth, simplicity, unity, power, specialness or exceptionality, etc.*

3.1 Truth

Even if a theorem or a proof is elegant and original, but if it is not true, it may not be of much use or does not count. Some people have some great ideas or approaches on how to solve big open problems. Often nothing happens.

There are also many published false proofs of some famous problems. For example, people have published or circulated many proofs of the Jacobian conjecture on polynomial maps. There are even papers written about many different wrong published proofs and approaches to prevent the newcomers from repeating the same mistakes.

There were also published proofs of the Poincare conjecture before Perelman finally proved it in 2002-2003, whose proof was confirmed by experts in 2006. See the papers [50] [9] for more detail.

Of course, there are exceptions too. Some wrong papers have opened up new subjects. It is also perhaps wise to remind ourself that there is no guarantee of correctness of long and complicated proofs in mathematics.

Since groups are the main objects under discussion in this paper, we say a few words on finite simple groups. Recall that a group is simple if it has no nontrivial normal subgroup, or equivalently it admits no nontrivial quotient group. In principle, all groups can be built up from simple groups via group extensions. One of the main objectives of mathematics is to classify objects and understand their underlying structures. To classify finite groups, the first important step is to classify finite simple groups.

People often say that the most difficult theorem (or rather proof) of mathematics in the twentieth century is the classification of finite simple groups. Though its solution was announced in the early 1980s, most people finally accepted its validity slightly before 2010 after some small "details" were filled in.

It might be interesting to look at comments by some mathematicians on this theorem. According to [21, p.389],

> When a proof is 5000 pages, and has been pieced together out of contributions by several different mathematicians, then it is clear that the team's claim to have proved their theorem is based in major part on their mutual confidence in each other's competence and integrity; and the acceptance of the work by the mathematical community as a whole is based in large part on its confidence in the members of the team.... This mutual confidence is perfectly reasonable and appropriate. But it certainly violates the notion of mathematical truth as indubitable."

These comments were written in the early 1980s. Making small errors is human nature and it will occur as sure as the Sun will rise everyday. Maybe another reason for our confidence and belief in mathematical proofs lies in some "structural stability" of mathematical arguments: when the arguments look so reasonable and have been used successfully many times before, small errors will not collapse the validity of proofs.

On the other hand, there is no doubt that people believe in the truth of the classification of finite simple groups. According to Vogan, "It's true that most mathematicians have no way to judge that personally, but I think that some of the experts understand extremely well the shape of the proof, and understand most of the details. For them there might be a question about whether a group is missing from the list, but not a question that the known methods are enough to solve the problem. (And they may also know enough to trust the mathematicians who certify those parts of the proof that they have not gone through in detail; so they can be completely sure in a reasonable sense.)"

3.2 Simplicity

The basic, or deep reason for the importance of simplicity is that people can only handle simple things. In mathematics, only simple results or theories can survive in the long run. This is probably also true in life.

For example, in the history of mathematics, a great theorem might have a very complicated proof to start with. But if it is really great and worthwhile, some simple proofs will come up eventually.

If a statement of a theorem runs over one or two pages with complicated assumptions or conditions, most people may not feel that it is beautiful, though it might be really important at a particular point in time, in a particular subject, or to some people. Most people prefer clean, simple and concise statements. Especially when a simple equation contains a lot of valuable information, and a simple idea solves a complicated problem, people often call it beautiful. Similarly, when a simple principle, theory or structure explains or unifies many complicated things, people call that beautiful too.

Of course, a simple concept or definition should be one which can be understood easily by most people.

Simplicity is also related to other aspects which give people pleasure or joy. For example, messy things are not beautiful to most people, and elegant things are usually clean, simply presented and appreciated. (Of course, it is also difficult to define elegance, and simplicity is definitely one aspect of it.)

3.3 Unexpected unity or connections with different subjects

Like a trip in a mountain, unexpected turns and openings are always exciting and make the trip more interesting. Similarly, novels with many unexpected twists and turns will attract and amaze the reader. In mathematics, results connecting different subjects or topics are usually considered beautiful and useful. They show unity of mathematics, and they also surprise, motivate and inspire people.

There are several different kinds of unexpected connections. It could be a solution of a difficult problem in one subject by methods from an unexpected subject. For example, many people did not expect that the Poincare conjecture in topology would eventually be solved by analysis. It could be an equivalence of two different things.

Unexpected connections also appear in some clever solutions or shortcuts. They are clever since they are not obvious or routine. It is helpful to recall that the Galois theory relating solvability of an algebraic equation with the structure of its Galois group must be surprising to the experts in the beginning. Maybe that was the reason why it took many years for it to be recognized and appreciated.

Another example of unexpectedly fruitful connection is the concept of *Lie group*: a combination of a manifold structure (differential geometry) and a group structure (algebra).

This might seem like a harmless combination in the beginning, but it has turned out to have many unexpected applications. The founder of the subject, Sophus Lie, always craved for recognition for his work in his lifetime. But he was recognized and honored beyond his expectation after death.

One more example of a powerful combination is the *geometry of numbers* initiated by Minkowski, who was very original, and was a friend of the more famous Hilbert and a teacher of Einstein. (Think of the Minkowski space in Einstein's theory of special relativity). Unfortunately he died relatively young at the peak of his career. In the subject of geometry of numbers, the geometry of the Euclidean space $\mathbb{R}^n$ (in particular the convex geometry) can be used to solve difficult problems in algebraic number theory in a relatively easy way.

3.4 Power

When a method, a result or theory can be used to solve many problems, its power will make people appreciate it, especially when the solution or approach is simple or unexpected.

For example, the method of integration by parts is such an example. For functions of one variable, it is a simple technique which every calculus student knows. But for functions of multiple variables, integration by parts is a different story. It is a fundamental tool in theories of partial differential equations and differential geometry. The difficulty is to come up with suitable functions to which we can apply the method of integration by parts. One example is the Bochner technique (formula or identity). See the survey papers [5] [68] [69] and the book [70] for detailed discussion of this fundamental technique.

The importance and power of the method of integration by parts can be judged from the assertion that many great mathematicians make a living out of it. Such a strong statement was made about the Courant Institute of Mathematics by a well-known mathematician at a conference in New York in the early 1990s. Basically, he said that the whole Courant Institute made a living out of the method of integration by parts.

During a graduate student seminar, my former advisor Shing-Tung Yau said that a great mathematician *truly* knows one lemma or technique, and a *really* great mathematician may know two. If you are interested in testing this out, you can try it on your favorite mathematicians.[1]

As you will see, the notion of groups we are discussing now is such a powerful example, which can be defined rather simply but has many unexpected applica-

[1]See the following autobiographic article for an example: Lennart Carleson, *Reflections on my life as a mathematician*, pp. 235-241, in *The Abel Prize, 2003–2007. The first five years.* Edited by Helge Holden and Ragni Piene. Springer-Verlag, Berlin, 2010. xii+326 pp. Carleson explained how one technique underlined all his major works.

tions. We will discuss some concrete applications of groups later. We can even make a stronger statement: *Without groups, there is no modern mathematics!* In other words, there is probably not much left if we remove groups from mathematics.

Another aspect of power comes from the fact that something is foundational so that it plays a basic or foundational role in many situations.

3.5 Specialness or exceptionality

There are many concepts, objects and results in mathematics. Some are special and stand out, and they often have some unexpected connections with various topics. Together they make these special objects interesting.

One of the basic objectives in mathematics is to understand and classify whatever we are studying. In geometry, there are many geometric figures. Circles and regular polygons stand out among all plane figures. Similarly, regular solids are special among all three dimensional polyhedrons. Among other things, they are special because of their symmetries.

Another aspect which can make something special is when they do not fit into some usual patterns. For example, in theories of groups (both finite groups and Lie groups), there are various special and exceptional groups. Besides their special properties, the process to understand them often leads to new theories and results. One example is the theory of Tits buildings motivated by the desire to understand geometrically exceptional simple Lie groups. Tits buildings turn out to have many applications in major areas of mathematics. See [42] for references and some discussion on buildings and their applications.

It is perhaps helpful to emphasize that to understand the exact meaning of "specialness" and "exceptionality" is a nontrivial and rewarding process.

3.6 The most beautiful group

We will propose a group, and examine it from the above criteria or perspectives on beauty in mathematics. In other words, we will *see it in action*. Then we can probably feel and agree that it might indeed be *the most beautiful group*.

My candidate of *the most beautiful group* is

$$\mathrm{SL}(2, \mathbb{Z}).$$

It is often called the *modular group*. The meaning of "modular" will be explained later. Since beauty lies in the eyes of the beholder, there are many other choices for the most beautiful group. Before explaining reasons for this particular choice, we make a list of other possible candidates.

3.7 Other candidates

Based on correspondences with several people and my reading of various sources, the following other candidates seem feasible, and each of them has some attractive reasons. Besides single groups, I have also added several classes of groups, since they share some common properties and can be considered as single groups in some sense. Of course, different people will definitely come up with different lists.

1. $\mathbb{Z}_2$, the group of order 2, which is related to the mirror symmetry. There are some obvious generalizations such as the Klein four-group, and cyclic groups of prime order p or of order p^m.
2. The symmetric group S_n, $n \geqslant 2$.
3. The symmetric group S_3 and triality, which is related to the simple Lie group D_4 with the largest outer automorphism group (i.e., the quotient group of outer/inner automorphisms).
4. The alternating group A_5 and the dodecahedral group, and more general A_n, $n \geqslant 5$.
5. $\mathrm{PSL}(2,7)$, the symmetry group of Klein's quartic curve.
6. The Mathieu group M_{24}.
7. The Monster group, the largest sporadic finite simple group.
8. The binary icosahedral group: the preimage in $\mathrm{SU}(2)$ of the icosahedral group in $\mathrm{SO}(3)$.
9. All finite subgroups of $\mathrm{SO}(3)$ and $\mathrm{SU}(2)$, and ADE classification.
10. The Hessian group inside $\mathrm{PGL}(3,\mathbb{C})$.
11. $\mathbb{Z}$, the group of integers.
12. The Leech Lattice in $\mathbb{R}^{24}$.
13. The free group F_2 on two generators.
14. $G_2(\mathbb{Z})$.
15. $\mathrm{SL}(2,\mathbb{R})$, $\mathrm{SL}(2,\mathbb{C})$, basic Lie groups.
16. The circle group S^1.
17. Classical Lie groups.
18. E_8, an exceptional simple Lie group.
19. G_2, an exceptional simple Lie group.
20. The Heisenberg group H_3, a nilpotent Lie group.
21. Coxeter groups, well described by presentations.
22. Simple, simply connected, complex groups.
23. Braid groups, well-described by presentations, which have also motivated other algebras.
24. Mapping class group of surfaces, providing natural, important but fairly complicated groups.

4 Reasons for the beauty of $\mathrm{SL}(2,\mathbb{Z})$

In this section, we will try to explain why the group $\mathrm{SL}(2,\mathbb{Z})$ is beautiful from the criteria discussed in the previous section.

By definition, it is

$$\mathrm{SL}(2,\mathbb{Z}) = \left\{ \begin{pmatrix} a & b \\ c & d \end{pmatrix} \middle| a,b,c,d \in \mathbb{Z}, ad - bc = 1 \right\},$$

i.e., it consists of 2×2-matrices with integer coefficients and determinant 1.

The group product is given by the matrix product:

$$\begin{pmatrix} a & b \\ c & d \end{pmatrix} \cdot \begin{pmatrix} \alpha & \beta \\ \gamma & \delta \end{pmatrix} = \begin{pmatrix} a\alpha + b\gamma & a\beta + b\delta \\ c\alpha + d\gamma & c\beta + d\delta \end{pmatrix},$$

which is still an integral matrix, and its inverse

$$\begin{pmatrix} a & b \\ c & d \end{pmatrix}^{-1} = \begin{pmatrix} d & -b \\ -c & a \end{pmatrix}$$

is also an integral matrix.

We claim that it is a *nonabelian, infinite group*. The first statement is clear. For example, two elements $\begin{pmatrix} 1 & 0 \\ 1 & 1 \end{pmatrix}$ and $\begin{pmatrix} 1 & 1 \\ 0 & 1 \end{pmatrix}$ do not commute.

For the second statement, we can write down many such integral matrices. For example, for every pair of coprime integers r, p, there exist integers s, t such that $sr - tp = 1$, which implies that $\begin{pmatrix} r & p \\ t & s \end{pmatrix} \in \mathrm{SL}(2, \mathbb{Z})$.

We also claim that $\mathrm{SL}(2, \mathbb{Z})$ is a discrete group with respect to a natural topology. Of course, we can view $\mathrm{SL}(2, \mathbb{Z})$ as a discrete group, i.e., as a group and a compatible topological space with the discrete topology. (In fact, every group can be given the discrete topology and becomes a discrete group). But a better way is to realize it as a discrete subgroup of the Lie group

$$\mathrm{SL}(2, \mathbb{R}) = \left\{ \begin{pmatrix} a & b \\ c & d \end{pmatrix} \middle| a, b, c, d \in \mathbb{R}, ad - bc = 1 \right\},$$

which has a nondiscrete subspace topology induced from the space of 2×2-matrices $\mathbb{R}^{2\times 2}$. The reason why the induced subspace topology on $\mathrm{SL}(2, \mathbb{Z})$ is the discrete topology is that $\mathbb{Z}$ is a discrete subgroup of $\mathbb{R}$.

It should be emphasized that $\mathrm{SL}(2, \mathbb{Z})$ can also be embedded into other topological groups as nondiscrete subgroups, for example, $\mathrm{SL}(2, \mathbb{Z}) \subset \mathrm{SL}(2, \mathbb{Q}_p)$ for every p-adic field.

As we will see later, the fact that $\mathrm{SL}(2, \mathbb{Z})$ is contained in the Lie group $\mathrm{SL}(2, \mathbb{R})$ as a discrete subgroup is important for both groups $\mathrm{SL}(2, \mathbb{Z})$ and $\mathrm{SL}(2, \mathbb{R})$. Many results about them are motivated by the similar, but simpler, pair of groups $\mathbb{Z}$ and $\mathbb{R}$.

4.1 Simplicity

Why did I make the choice of $\mathrm{SL}(2, \mathbb{Z})$? Let us consider from the above perspectives one by one.

Since it involves a concept: a particular group, there is no problem with truth (or correctness) yet, though there are issues on how to correctly define or understand it (for example, the right formulation, the right setup or perspective). We will comment on such things later.

Therefore, we first examine it from the perspective of *simplicity*. One basic question concerning groups is how to describe or understand them.

One obvious approach to a group is to give a description of all elements of the group and its group operations. For $\mathrm{SL}(2,\mathbb{Z})$, its elements were explicitly given, and group operations are also simple, as we did in the previous subsection.

At a deeper level, the answer depends on types of groups or applications we are discussing. For example, one can look at all elements of a group without seeing the underlying structures or essence. For a discrete group, one approach is to find generators and relations, i.e., understanding it in the framework of geometric group theory (or rather combinatorial group theory).

It can be checked easily that $\mathrm{SL}(2,\mathbb{Z})$ is generated by two elements $S = \begin{pmatrix} 0 & -1 \\ 1 & 0 \end{pmatrix}$ and $T = \begin{pmatrix} 1 & 1 \\ 0 & 1 \end{pmatrix}$, i.e., every element of $\mathrm{SL}(2,\mathbb{Z})$ is product of T, S and their inverses.

Note that it is more complicated to obtain relations between these generators, i.e., a presentation of $\mathrm{SL}(2,\mathbb{Z})$. If we let $U = ST$, it can be seen easily that S, U also generate $\mathrm{SL}(2,\mathbb{Z})$. Using a fundamental domain of $\mathrm{SL}(2,\mathbb{Z})$ in $\mathbb{H}^2$, it can be shown that S, U satisfy the only two relations:

$$S^2 = -\mathrm{Id}, \quad U^3 = \mathrm{Id}.$$

The group $\mathrm{SL}(2,\mathbb{Z})$ does not act faithfully on $\mathbb{H}^2$, and the kernel consists of $\{\pm\mathrm{Id}\}$. The group $\mathrm{PSL}(2,\mathbb{Z})$ can be written as a free product $\mathbb{Z}_2 \star \mathbb{Z}_3$, where $\mathbb{Z}_2$ is generated by the image of S and $\mathbb{Z}_3$ is generated by the image of U. See [38, p. 152] and [52, p. 139] for proofs. It should be pointed out that understanding a group from its presentation is a highly difficult problem. It is related to the so called word problems in group theory.

In spite of the explicit and simple presentation of $\mathrm{SL}(2,\mathbb{Z})$ given above, it is perhaps important to note that there are much simpler groups in terms of generators and relations. For example, the additive group $\mathbb{Z}_2 \cong \{0,1\}$ is generated by 1 with the relation $1+1=0$; or equivalently, the multiplicative group $\mathbb{Z}_2 \cong \{-1,+1\}$ is generated by -1 with the relation $(-1)\times(-1)=1$. Another example is $\mathbb{Z}$, an infinite cyclic group, which is generated by 1 without any relation.

Some people will say that the trivial group $\{e\}$ is even simpler, or really exceptional. Maybe it is the most (or rather *too*) interesting group because it is too simple.

As explained above, it is the combination of several characteristics which makes a concept (result, or theorem) beautiful and interesting. In some sense, $\mathbb{Z}_2$ and $\mathbb{Z}$ are too simple in comparison to $\mathrm{SL}(2,\mathbb{Z})$ and suffer also in *unity, power,* though not in *specialness.*

One aspect of rich structure of a group is reflected by abundance of subgroups and quotient groups, and more generally *subquotient groups* (which are quotients of subgroups).

The group $\mathbb{Z}_2$ has only few subgroups and quotient groups. Every subgroup of $\mathbb{Z}$ is cyclic and hence all subgroups can be described easily. Every cyclic group (i.e., a group generated by one element) is a quotient subgroup of $\mathbb{Z}$. Subquotient groups of $\mathbb{Z}$ can also be described easily. In some sense, they have rather boring group structures.

On the other hand, $\mathrm{SL}(2,\mathbb{Z})$ has a rich and complicated collection of subgroups, quotient groups and subquotient groups. For example, $\mathrm{SL}(2,\mathbb{Z})$ contains a family of principal congruence subgroups: for every positive integer N,

$$\Gamma(N) = \{g \in \mathrm{SL}(2,\mathbb{Z}) \mid g \cong \mathrm{Id} \mod N\},$$

and more general congruence subgroups which are subgroups of $\mathrm{SL}(2,\mathbb{Z})$ containing principal congruence subgroups.

Congruence subgroups of $\mathrm{SL}(2,\mathbb{Z})$ are of finite index. One natural question is whether every subgroup of finite index of $\mathrm{SL}(2,\mathbb{Z})$ contains a principal congruence subgroup. This is called the *congruence subgroup problem*, and it has a negative answer for $\mathrm{SL}(2,\mathbb{Z})$, i.e., there there subgroups of finite index of $\mathrm{SL}(2,\mathbb{Z})$ which are not congruence subgroups. The original proof by F. Klein explored the rich structure of subquotient groups of $\mathrm{SL}(2,\mathbb{Z})$: every finite group can be realized as a subquotient group of $\mathrm{SL}(2,\mathbb{Z})$. It follows from an explicit free subgroup on two generators in $\mathrm{SL}(2,\mathbb{Z})$. This proof also started the known ping-pong lemma about free groups or free products in geometric group theory.

On the other hand, the congruence subgroup problem is positive for $\mathrm{SL}(n,\mathbb{Z})$ when $n \geqslant 3$. It can also be formulated for any linear algebraic group $\mathbf{G}$ defined over $\mathbb{Q}$, i.e., $\mathbf{G} \subset \mathrm{GL}(n,\mathbb{C})$ is an algebraic variety defined with polynomials with coefficients in $\mathbb{Q}$ and is closed under the group operations. This congruence subgroup problem is still not completely solved. See [64, §3.4] for more detail and references.

In summary, $\mathrm{SL}(2,\mathbb{Z})$ is internally rich. It also has many applications and connections with different topics which we will discuss later.

It is perhaps also helpful to comment on another comparison of groups discussed above: $\mathbb{Z}_2$ and $\mathbb{Z}$ are abelian groups while $\mathrm{SL}(2,\mathbb{Z})$ is non-abelian. The uncertainty principle in physics says that one cannot determine both the position and moment of an elementary particle at the same time. It results from non-abelian property of matrix products in $\mathrm{SL}(2,\mathbb{Z})$ via the theory of matrix mechanics by Heisenberg.

4.2 How to study and understand $\mathrm{SL}(2,\mathbb{Z})$?

First we note that groups are useful because they act on interesting spaces. It is also true that they are interesting and can be understood because they act on important or appropriate spaces. In many cases, finding the right space is the crucial step. In geometric group theory, when there is no obvious or natural space for a group to act, the group can be made into a metric space, and it can act on itself by isometries.

The group $\mathrm{SL}(2,\mathbb{Z})$ acts on many spaces. In fact, we can spend easily one semester to discuss them, and some people I know are writing a book on actions of $\mathrm{SL}(2,\mathbb{Z})$. In this paper, we will mainly consider one action.

Let $\mathbb{H}^2 = \{z = x + iy \mid x \in \mathbb{R}, y > 0\}$ be the upper half-plane, which is a model of the hyperbolic plane, called the Poincare upper half-plane.

Recall that the Euclidean plane is simply connected and has curvature 0, that the sphere S^2 is simply connected and has curvature 1, and that the hyperbolic plane is simply connected and has curvature -1. They are complete in the sense that every geodesic in them can be extended infinitely in both directions. They give three basic geometries. The Erganlen program of Klein mentioned above gives a uniform treatment of them in terms of groups acting on them.

The group $\mathrm{SL}(2,\mathbb{R})$ acts on the extended complex plane $\mathbb{C}\cup\{\infty\}$ by linear fractional (or Mobius) transformations

$$\begin{pmatrix} a & b \\ c & d \end{pmatrix} \cdot z = \frac{az+b}{cz+d}.$$

Note that this is one basic example of group actions in complex analysis. The action of $\mathrm{SL}(2,\mathbb{R})$ leaves $\mathbb{H}^2$ invariant. If a group Γ acts on a space X, then the pair (Γ, X) is often called a transformation group. In some sense, the pair $(\mathrm{SL}(2,\mathbb{R}),\ \mathbb{H}^2)$, or rather the action of $\mathrm{SL}(2,\mathbb{R})$ on $\mathbb{H}^2$, is one of the most basic transformation groups.

It can be checked that $\mathrm{SL}(2,\mathbb{R})$ acts transitively on $\mathbb{H}^2$: for every two points z_1, z_2, there exists $g \in \mathrm{SL}(2,\mathbb{R})$ such that $g\cdot z_1 = z_2$. Hence $\mathbb{H}^2$ can be identified with the quotient space $\mathrm{SL}(2,\mathbb{R})/\mathrm{SO}(2)$, where $\mathrm{SO}(2) = \left\{ \begin{pmatrix} \cos\theta & -\sin\theta \\ \sin\theta & \cos\theta \end{pmatrix} \,\middle|\, \theta\in\mathbb{R}\right\} \cong S^1$ is the stabilizer of the point i in $\mathrm{SL}(2,\mathbb{R})$, i.e., the subgroup which leaves the point $i \in \mathbb{H}^2$ invariant.

Since $\mathrm{SL}(2,\mathbb{Z})$ is a discrete subgroup of $\mathrm{SL}(2,\mathbb{R})$ and $\mathrm{SO}(2)$ is a compact subgroup of $\mathrm{SL}(2,\mathbb{R})$, it follows that $\mathrm{SL}(2,\mathbb{Z})$ acts properly on $\mathbb{H}^2$: for every compact subset $C \subset \mathbb{H}^2$, there are only finitely many $\gamma \in \mathrm{SL}(2,\mathbb{Z})$ such that $\gamma C \cap C \neq \emptyset$.

This implies that the quotient space $\mathrm{SL}(2,\mathbb{Z})\backslash\mathbb{H}^2$ a Hausdorff space. (Note that each point in this quotient space is one orbit $\mathrm{SL}(2,\mathbb{Z})\cdot z$ in $\mathbb{H}^2$.)

4.3 Applications of the action of $\mathrm{SL}(2,\mathbb{Z})$ on $\mathbb{H}^2$

Now we consider connections of this group action of $\mathrm{SL}(2,\mathbb{Z})$ with several different topics in mathematics. This will show the unity, unexpected connections, and power of $\mathrm{SL}(2,\mathbb{Z})$. For more references and topics, see the book [40].

4.3.1 Moduli space of Riemann surfaces of genus 1 or elliptic curves over $\mathbb{C}$

First, we discuss connections with algebraic geometry and number theory.

It is known that every infinite discrete subgroup L of $\mathbb{R}^2$ is either of the form $\mathbb{Z}v_1$, where v_1 is a nonzero vector, or of the form $L = \mathbb{Z}v_1 + \mathbb{Z}v_2$, where v_1, v_2 are are two linearly independent vectors.

In the latter case, L is called a *lattice* of $\mathbb{R}^2$, and the quotient $L\backslash\mathbb{R}^2$ is compact, which is equivalent to the condition that the area of the quotient $L\backslash\mathbb{R}^2$ is finite. (For any Lie group G with finitely many connected components, a discrete subgroup $\Gamma \subset G$ is called a lattice if the volume of the quotient $\Gamma\backslash G$ with respect to a left invariant Haar measure of G is finite. In general, for example, when G

is a noncompact semisimple Lie group such as $\mathrm{SL}(n, \mathbb{R})$, there are lattices Γ of G which are not cocompact, i.e., the quotients $\Gamma\backslash G$ are not compact.)

Now $\mathbb{C} = \mathbb{R}^2$, and a lattice L of $\mathbb{R}^2$ gives a lattice of $\mathbb{C}$. The quotient $L\backslash\mathbb{C}$ is a *Riemann surface* of genus 1 (i.e., with one handle). (Recall that a Riemann surface is a complex manifold of complex dimension 1). Clearly, the additive group structure of $\mathbb{C}$ descends to $L\backslash\mathbb{C}$, and the origin is projected to the identity element (or a distinguished point). In number theory, a smooth, projective algebraic curve of genus one with a distinguished point O is called an elliptic curve, and admits a group structure such that O is the identity element. Therefore, every lattice L of $\mathbb{R}^2$ gives an *elliptic curve* $L\backslash\mathbb{C}$ over $\mathbb{C}$.

The theory of elliptic curves was motivated by the problem of computing the arc length of an ellipse, or how to compute integrals such as

$$\int \frac{1}{\sqrt{1-k^2\sin^2(t)}}dt, \quad \int \sqrt{1-k^2\sin^2(t)}dt,$$

which in turn was motivated by Kepler's first two laws of planetary motions: (1) The orbit of every planet is an ellipse with the Sun at one of the two foci. (2) A line joining a planet and the Sun sweeps out equal areas during equal intervals of time.

Pioneers of this theory included Abel and Jacobi. Their names now appear in many related concepts such as Abelian varieties and Jacobian varieties, which are projective varieties with compatible group structures.

We next construct and classify all such Riemann surfaces of genus 1. Two Riemann surfaces Σ_1, Σ_2 are called isomorphic if there exists a biholomorphic map $f : \Sigma_1 \to \Sigma_2$ between them.

For every point $z \in \mathbb{H}^2$, we form a lattice $L_z = \mathbb{Z} + z\mathbb{Z}$ in $\mathbb{C}$, and hence obtain a Riemann surface $L_z\backslash\mathbb{C}$ of genus 1. (Note that if z is a real number, then $\mathbb{Z} + z\mathbb{Z}$ is not a lattice of $\mathbb{C}$.) These Riemann surfaces can be classified as follows.

Proposition 4.1. 1. *For every two points z_1, z_2, two such Riemann surfaces $L_{z_1}\backslash\mathbb{C}$ and $L_{z_2}\backslash\mathbb{C}$ are isomorphic if and only if z_1, z_2 belong to one common orbit of* $\mathrm{SL}(2,\mathbb{Z})$ *in $\mathbb{H}^2$, or equivalently, they project to the same point in* $\mathrm{SL}(2,\mathbb{Z})\backslash\mathbb{H}^2$.
2. *Every Riemann surface of genus* 1 *is of such a form $L_z\backslash\mathbb{C}$ for some $z \in \mathbb{H}^2$.*

Therefore, we have the following corollary:

Corollary 4.2. *The space of isomorphism classes of compact Riemann surfaces of genus* 1 *(or elliptic curves) can be identified with the quotient space* $\mathrm{SL}(2,\mathbb{Z})\backslash\mathbb{H}^2$.

This space is usually denoted by $\mathcal{M}_1$, called the *moduli space* of compact Riemann surfaces of genus 1. This explains the name of modular group for $\mathrm{SL}(2,\mathbb{Z})$.

We can also define compact Riemann surfaces of genus $g \geqslant 1$, i.e., 1-dimensional complex manifold with g handles. Their moduli space $\mathcal{M}_g$, the space of isomorphism classes of such Riemann surfaces, is one of the most important objects in algebraic geometry and was first introduced by Riemann in 1857. Since then, many mathematicians have tried to study and understand it. But many things about

$\mathcal{M}_g$ are still not fully understood. For a comprehensive introduction to moduli spaces of Riemann surfaces, see [2].

The moduli space $\mathcal{M}_1$, i.e., the modular curve $\mathrm{SL}(2,\mathbb{Z})\backslash\mathbb{H}^2$, also admits another important class of generalization: Shimura curves, which are very important in arithmetic algebraic geometry (such as the Langlands program) and are related to congruence subgroups of $\mathrm{SL}(2,\mathbb{Z})$ and other related discrete subgroups of $\mathrm{SL}(2,\mathbb{R})$. See [62] for detail.

4.3.2 Equivalence classes of binary quadratic forms

In number theory, binary quadratic forms $Q(x,y) = Ax^2 + 2Bxy + Cy^2$ are basic. Assume that $A, B, C \in \mathbb{Z}$. Then a basic question asks which integer n can be *represented* as $Q(x,y)$ for some $x, y \in \mathbb{Z}$. If we make a change of variables

$$\begin{pmatrix} u \\ v \end{pmatrix} = \begin{pmatrix} a & b \\ c & d \end{pmatrix} \begin{pmatrix} x \\ y \end{pmatrix},$$

for some $\begin{pmatrix} a & b \\ c & d \end{pmatrix} \in \mathrm{SL}(2,\mathbb{Z})$, then we obtain a new binary quadratic form $Q'(u,v)$. It is clear that the two binary quadratic forms Q, Q' represent the same set of integers.

If two quadratic forms Q, Q' are related by a change of variables as above, they are called *equivalent*. As we just observed, equivalent quadratic forms represent the same set of integers.

We can also consider quadratic forms with real coefficients and the above equivalence on them. If $Q(x,y) = Ax^2+2Bxy+Cy^2$ is a positive definite quadratic form with real coefficient, then the quadratic equation $Q(x,1) = 0$ has a pair of conjugate non-real roots $\tau, \overline{\tau}$, and hence

$$Q(x,y) = A(x - \tau y)(x - \overline{\tau} y).$$

Assume that $\tau \in \mathbb{H}^2$. This gives a map from the space of positive quadratic forms to $\mathbb{H}^2$, $Q \mapsto \tau$.

We can prove the following result:

Proposition 4.3. *The space of equivalence classes of positive quadratic forms of determinant* 1 *can be identified with* $\mathrm{SL}(2,\mathbb{Z})\backslash\mathbb{H}^2$.

The above identification can also be seen as follows: $\mathbb{H}^2 \cong \mathrm{SL}(2,\mathbb{R})/\mathrm{SO}(2)$, and for each $g \in \mathrm{SL}(2,\mathbb{R})$, gg^T is a positive definite matrix of determinant 1, and the quotient $\mathrm{SL}(2,\mathbb{Z})\backslash\mathrm{SL}(2,\mathbb{R})/\mathrm{SO}(2)$ can be identified with the space of equivalence classes of positive quadratic forms of determinant 1.

Since there are infinitely many quadratic forms in each equivalence, one natural question is how to pick or find the "*best representative*" in each equivalence class? This is called the *reduction theory*, related to finding "*good*" fundamental domains of the action of $\mathrm{SL}(2,\mathbb{Z})$ on $\mathbb{H}^2$.

Finding suitable representatives in equivalence classes of quadratic forms can help to solve the above problem of representation of integers by quadratic forms. See the book [16] for some examples and references.

This problem on fundamental domains for $\mathrm{SL}(2,\mathbb{Z})$ and other arithmetic subgroups of semisimple Lie groups was studied by Lagrange, Gauss, and in more general contexts by Minkowski, Siegel, Borel, et al. It is still not completely solved. See [40] [41] for some references.

4.3.3 The little Picard theorem

In complex analysis, one basic theorem is Liouville's Theorem which states that *every bounded holomorphic function on* $\mathbb{C}$ *is constant.*

A more difficult and amazing generalization is Picard's little theorem: *if* $f:\mathbb{C}\to\mathbb{C}$ *is a holomorphic function and the image* $f(\mathbb{C})$ *misses two points, then* f *is constant.*

The condition on f says that the complement $\mathbb{C}-f(\mathbb{C})$ contains at least two points. This is optimal since a nonconstant holomorphic function can miss one point. For example, $f(z)=e^z$ is holomorphic and is never equal to 0.

Of course, when f is bounded, its image $f(\mathbb{C})$ misses many points of $\mathbb{C}$.

There are several proofs of Picard's little Theorem. Picard's original proof uses modular forms with respect to a suitable subgroup of finite index of $\mathrm{SL}(2,\mathbb{Z})$. We will explain it now.

As mentioned before, $\mathrm{SL}(2,\mathbb{Z})$ has a rich collection of subgroups, and this is one reason for the importance of $\mathrm{SL}(2,\mathbb{Z})$. (Note that the plane geometry is rich because of large collections of geometric figures: lines, circles, triangles, and squares, etc.)

For each positive integer N, we have a principal congruence subgroup: $\Gamma(N)$. When $N=2$, the quotient $\Gamma(2)\backslash\mathbb{H}^2$ is biholomorphic to $\mathbb{C}-\{0,1\}$, which is also biholomorhic to the thrice punctured sphere $\mathbb{C}P^1-\{0,1,\infty\}$.

Note that $\Gamma(2)$ is generated by $\begin{pmatrix}1 & 2\\ 0 & 1\end{pmatrix}$, $\begin{pmatrix}1 & 0\\ 2 & 1\end{pmatrix}$, and $-I_2$. Then the above identification can be seen from a suitable fundamental domain of $\Gamma(2)$ (see [38, p. 154]). There is an explicit biholomorphic function

$$\lambda:\Gamma(2)\backslash\mathbb{H}^2\to\mathbb{C}-\{0,1\}.$$

The function λ is called *modular lambda function*, a holomorphic function on $\mathbb{H}^2$ invariant under $\Gamma(2)$, and there are explicit formulas for it. See [14, Chapter 7].

Now suppose $f:\mathbb{C}\to\mathbb{C}$ is holomorphic and misses two points. We can assume that $f:\ C\to\mathbb{C}-\{0,1\}$, after composing with suitable fractional linear transformations.

We note that the map $\lambda^{-1}(f):\mathbb{C}\to\mathbb{H}^2$ is well-defined, by the monodromy theorem since $\mathbb{C}$ is simply connected and the map $\mathbb{H}^2\to\Gamma(2)\backslash\mathbb{H}^2$ is a covering space. (Note that $\Gamma(2)$ is not torsion-free, but its image in $\mathrm{PSL}(2,\mathbb{Z})$ is, and hence the quotient $\Gamma(2)\backslash\mathbb{H}^2$ is a manifold.)

Let $\varphi:\mathbb{H}^2\cong D=\{z\in\mathbb{C}\mid |z|<1\}$ be an isomorphism. Then we obtain a bounded holomorphic map $\varphi(\lambda^{-1}(f)):\mathbb{C}\to D$. By Liouville's Theorem, it must be constant and hence f is constant. This finishes the proof.

The key point in the above proof is the modular function λ which uniformizes the Riemann surface $\mathbb{C}P^1-\{0,1,\infty\}$.

4.4 Modular forms and modularity

Besides modular functions such as λ before, we can also consider modular forms. They are holomorphic functions on $\mathbb{H}^2$ which are not invariant under $\mathrm{SL}(2, \mathbb{Z})$ or special subgroups such as the congruence subgroups $\Gamma(N)$ but satisfy simple rules under transformations coming from these groups.

For example, a modular form of weight k with respect to $\mathrm{SL}(2, \mathbb{Z})$ is a holomorphic function on $\mathbb{H}^2$ such that

$$f\left(\begin{pmatrix} a & b \\ c & d \end{pmatrix} z\right) = (cz + d)^k f(z)$$

for all $\begin{pmatrix} a & b \\ c & d \end{pmatrix} \in \mathrm{SL}(2, \mathbb{Z})$.

It can be checked that $f(z)$ is a modular form of weight $2k$ if and only if $f(z)dz^k$ is invariant under the action of $\mathrm{SL}(2, \mathbb{Z})$. This is the reason why f is called a modular *form*.

We note that if $f(z)$ is a modular form, then $f(z + 1) = f(z)$. Let $q = \exp 2\pi iz$. Then $f(z)$ admits a Fourier expansion (or a power expansion in q):

$$f(z) = \sum_{n=n_0}^{\infty} a_n q^n$$

for some integer n_0 and $a_n \in \mathbb{C}$.

It is known that $\mathrm{SL}(2, \mathbb{Z})\backslash\mathbb{H}^2$ is homeomorphic to $\mathbb{C}$ and the homeomorphism is given by the j-invariant,

$$j : \mathrm{SL}(2, \mathbb{Z})\backslash\mathbb{H}^2 \to \mathbb{C}.$$

The function j is a modular form of weight 0 (hence an automotphic function). Its q-expansion is given by

$$j(z) = \frac{1}{q} + 744 + 196884q + 21493760q^2 + 864299970q^3 + \cdots.$$

Usually modular forms are required to satisfy a polynomial growth condition when $\mathrm{Im}(z) \to +\infty$. Then its q-expansion starts with a_0.

There are many applications of modular forms, and one can easily spend one's whole life on studying them. The basic reason is that the transformation rules for $f(z)$ imply some symmetry conditions on the Fourier coefficients a_n. Conversely, when some natural sequences of numbers from different contexts can be realized as the Fourier coefficients a_n of a modular form f, often major problems can be solved.

Let me mention only two examples:

1. The solution of the Last Fermat Theorem by Andrew Wiles, as a special case of the Langlands program. Briefly, the modularity (or Taniyama-Shimura-Weil) conjecture says that every elliptic curve defined over $\mathbb{Q}$ is modular.

Briefly, for such an elliptic curve, there is an associated L-function constructed from counting of number of points over finite fields,

$$L(s) = \sum_{n=1} \frac{a_n}{n^s}.$$

The modularity conjecture says that the coffficients a_n are the Fourier coefficients of a modular form. If Fermat's last theorem were not true, an elliptic curve can be constructed to violate this conjecture. See [66] [19] for more details.

2. The Monstrous Moonshine, relating a modular function j for $\mathrm{SL}(2,\mathbb{Z})$ to an action of the biggest sporadic finite simple group, called *the Monster*, i.e., by relating the Fourier coefficients of j to graded dimensions of a module of the Monster group. For example, the first graded dimension is equal to 196884. For more details, see [11] [26]. Borcherds got a Fields medal in 1998 for proving this Monstrous Moonshine conjecture.

Note that there are several group actions here, and some new concepts were motivated by them. For example, $j : \mathrm{SL}(2,\mathbb{Z})\backslash\mathbb{H}^2 \cong \mathbb{C}$, but the projection map $\mathbb{H}^2 \to \mathbb{C}$ is not a covering map. The quotient $\mathrm{SL}(2,\mathbb{Z})\backslash\mathbb{H}^2$ is not smooth but an orbifold. The notion of orbifolds was essentially motivated by such quotients and first proposed by Satake.

There are other modularity applications. See [40, §3.4.16] for some references.

4.5 Closed geodesics of $\mathrm{SL}(2,\mathbb{Z})$, continued fractions and quadratic irrational numbers

Now I will briefly mention several other applications in geometry and analysis.

The quotient $\mathrm{SL}(2,\mathbb{Z})\backslash\mathbb{H}^2$ is a hyperbolic orbifold. We can study its geometry, for example, structures of geodesics, their distributions, their lengths, etc.

Closed geodesics of $\mathrm{SL}(2,\mathbb{Z})\backslash\mathbb{H}^2$ correspond to conjugacy classes of hyperbolic elements of $\mathrm{SL}(2,\mathbb{Z})$, and they correspond to indefinite binary quadratic forms. Now $\mathrm{SL}(2,\mathbb{Z})$ acts on both $\mathbb{H}^2$ and its boundary points, and the space of quadratic forms. The correspondence between the conjugacy classes of hyperbolic elements and the indefinite binary quadratic forms can be seen through the automorphism groups of indefinite binary quadratic forms, which in turn can be described via solutions of Pell's equations.

More specifically, every geodesic in $\mathbb{H}^2$ is determined by its end points. Note that any hyperbolic element $\begin{pmatrix} a & b \\ c & d \end{pmatrix}$ in $\mathrm{SL}(2,\mathbb{Z})$ fixes a geodesic in $\mathbb{H}^2$, whose limit (or end) points in $\mathbb{R}$ are roots of the quadratic equation $cz^2 + (d-a)z - b = 0$. The reason is that since $\begin{pmatrix} a & b \\ c & d \end{pmatrix}$ is hyperbolic, its trace satisfies $(a+d)^2 > 4$, and hence the quadratic form is indefinite. Clearly, the hyperbolic element fixes these end points.

Conversely, given any indefinite quadratic form Q, it has two real roots. These two points are fixed under the automorphism group of Q in $\mathrm{SL}(2,\mathbb{Z})$, and hence the unique geodesic between them is also invariant under the automorphism group, and the quotient is a closed geodesic on the modular surface. The automorphism group of the indefinite quadratic form Q can be described by solutions of Pell's equations associated with it. For more detail, see [56, §1].

There are also close relations between continued fractions and special matrices in $\mathrm{SL}(2,\mathbb{Z})$. Briefly, a continued fraction is eventually periodic if and only if it is a quadratic irrational number. Two quadratic irrational numbers are the end points of a closed geodesic (or a lift of a closed geodesic) if and only if they are conjugate in a fixed quadratic field. The length of the associated closed geodesic can be computed by the norm of the smallest unit, which also corresponds to a minimal solution of Pell's equation. See [1]. There are also close relations between quadratic forms and quadratic fields. See [13] [6]. So the geometry of $\mathrm{SL}(2,\mathbb{Z})\backslash\mathbb{H}^2$ is really about number theory.

4.6 Analysis of $\mathrm{SL}(2,\mathbb{Z})$

Since $\mathrm{SL}(2,\mathbb{Z})\backslash\mathbb{H}^2$ is a noncompact complete hyperbolic orbifold, we can also study analysis on it, such as eigenvalues and eigenfunctions of the Laplace operator. Specifically, its Laplace operator has a natural extension as a self-adjoint operator on the Hilbert space $L^2(\mathrm{SL}(2,\mathbb{Z})\backslash\mathbb{H}^2)$.

It has a continuous spectrum $[\frac{1}{4},+\infty)$, whose generalized eigenfunctions are given by Eisenstein series. See [46] for details and references.

It also has a discrete spectrum, consisting of eigenvalues of square integrable eigenfunctions. For example, 0 is an eigenvalue since the area of $\mathrm{SL}(2,\mathbb{Z})\backslash\mathbb{H}^2$ is finite. Though there are infinitely many eigenvalues in the discrete spectrum, it is very difficult to determine their values.

For example, a famous conjecture of Selberg says that for all $N \geqslant 1$, the first positive eigenvalue of the surface $\Gamma(N)\backslash\mathbb{H}^2$ is greater than or equal to $\frac{1}{4}$, i.e., it is embedded into the continuous spectrum. Some partial results are known. For example, uniform positive lower bounds with respect to N are known and have several important applications including expander graphs. See [57] for more details and references.

4.7 Expander graphs and other related works

One important problem in graph theory is to construct explicitly expander graphs, which are graphs whose subsets of at most half of the total size have large boundaries (i.e., having a large isoperimetric constant). These graphs allow people to transit information over networks quickly.

A positive low bound on the isoperimetric inequality is related to a positive lower bound on the first positive eigenvalue of a discrete Laplacian.

Partial results on the first positive eigenvalue of the hyperbolic surfaces $\Gamma(N)\backslash\mathbb{H}^2$ allow one to construct some good expander graphs. See [47] [8] for details and references.

4.8 Selberg trace formula

Eigenvalues of a Riemannian manifold are determined by and also reflect the geometry of the manifold. One important question is: *How to understand relations between geometry and spectral theory?* This is the subject of special geometry, motivated by M. Kac's famous paper [44] or question "Can one hear the shape of a drum?"

One more concrete problem is to describe and understand some explicit relations between eigenvalues of the Laplace operator and lengths of closed geodesics. This is in general very difficult or impossible for general manifolds.

For hyperbolic surfaces such as $\mathrm{SL}(2,\mathbb{Z})\backslash\mathbb{H}^2$ and $\Gamma(N)\backslash\mathbb{H}^2$, one answer is *The Selberg trace formula*, which relates eigenvalues and length spectrums and allow people to understand both sides of the trace formula. See the books [65] [58] for more details and references.

4.9 Classifying spaces in topology

The quotient $\mathrm{SL}(2,\mathbb{Z})\backslash\mathbb{H}^2$ is a classifying space for the group $\mathrm{SL}(2,\mathbb{Z})$.

In topology, given a discrete group Γ, there are special spaces called classifying spaces of Γ, such as $B\Gamma$, $E\Gamma$, first constructed by Milnor.

We recall that $B\Gamma$ is a CW-complex unique up to homotopy equivalence such that $\pi_1(B\Gamma)=\Gamma$, and for $i \geqslant 2$, $\pi_2(B\Gamma)=\{1\}$. Its universal covering space is $E\Gamma$ which is characterized by the conditions: Γ acts on $E\Gamma$ properly and fixed point freely, and $E\Gamma$ is contractible.

These models of Milnor are all infinite dimensional. A general result says that for any non-torsion-free group Γ such as $\mathrm{SL}(2,\mathbb{Z})$, every model of $B\Gamma$ and $E\Gamma$ is infinite dimensional.

For such groups, another classifying space is $\underline{E}\Gamma$, a universal space for proper actions Γ, which is characterized the conditions: Γ acts properly on $\underline{E}\Gamma$ such that for every finite subgroup $F \subset \Gamma$, the fixed point set $\underline{E}\Gamma^F$ is nonempty and contractible.

When Γ is torsion-free, $\underline{E}\Gamma$ is reduced to $E\Gamma$.

A basic problem is to find "small" models of these classifying space. It can be shown that $\mathbb{H}^2$ is a finite dimensional model of $\underline{E}\Gamma$ for $\Gamma=\mathrm{SL}(2,\mathbb{Z})$.

4.10 Optimal models of classifying spaces and an explicit presentation

A general result says that every model of $\underline{E}\Gamma$ has a lower bound on dimension given by the virtual cohomology dimension of Γ.

Since $\mathrm{SL}(2,\mathbb{Z})$ contains a free group on two generators as a subgroup of finite index, for example. the subgroup generated by $\begin{pmatrix}1 & 2\\ 0 & 1\end{pmatrix}$ and $\begin{pmatrix}1 & 0\\ 2 & 1\end{pmatrix}$, its virtual cohomology dimension is equal to 1.

An optimal model of $\underline{E}\Gamma$ for $\Gamma=\mathrm{SL}(2,\mathbb{Z})$ can be obtained by a trivalent tree T embedded in $\mathbb{H}^2$ which is an equivariant deformation retract.

This tree T can be obtained either from the classical fundamental domain of $\mathrm{SL}(2,\mathbb{Z})$ (or rather the associated tessellation) or the equivariant ideal triangulation determined by the Farey graph. In the first case, for each translate of the fundamental domain, take the geodesic boundary side which is bounded. In the second case, take the dual 1-complex.

One consequence of the action of $\mathrm{SL}(2,\mathbb{Z})$ on this tree T is a decomposition (or presentation) of $\mathrm{SL}(2,\mathbb{Z})$:

$$\mathrm{PSL}(2,\mathbb{Z}) = \mathbb{Z}/2\mathbb{Z} * \mathbb{Z}/3\mathbb{Z},$$

and $\mathrm{SL}(2,\mathbb{Z})$ is an extension of $\mathrm{PSL}(2,\mathbb{Z})$ by its center of order 2.

4.11 Actions on trees

The action of $\mathrm{SL}(2,\mathbb{Z})$ on the trivalent tree $T \subset \mathbb{H}^2$ in the previous subsection is one important example of actions of discrete groups on trees, i.e., the Bass-Serre theory. One of the motivations for Serre to develop this theory was to understand better lattices of $\mathrm{SL}(2,\mathbb{Q}_p)$. See the book [61] for detail.

Generalizations of the Bass-Serre theory such as cube complexes (for example, products of trees) and group actions on them have played a crucial role in the recent solutions of several conjectures of Thurston in three dimensional topology. See [67] and references there.

4.12 Geodesic flow and symbolic dynamics

As mentioned before, closed geodesics of the hyperbolic surface $\mathrm{SL}(2,\mathbb{Z})\backslash\mathbb{H}^2$ are related to several important concepts in number theory such as quadratic forms and Pell's equations. Nonclosed geodesics are also related to number theory and symbolic dynamics. For example, one consequence of such relations is the existence of dense geodesics $\mathrm{SL}(2,\mathbb{Z})\backslash\mathbb{H}^2$. Markoff numbers which appear in continued fractions can also be explained by geodesics of some hyperbolic surfaces related to $\mathrm{SL}(2,\mathbb{Z})\backslash\mathbb{H}^2$. See the papers [59] [60] [45] [17] and the book [18] for details. The book [1] contains a detailed discussion of Markov numbers and related topics.

4.13 The unit circle bundle of the modular surface as the complement of trefoil knot

The unit circle bundle of $\mathbb{H}^2$ is equal to $\mathrm{SL}(2,\mathbb{R})/\mathrm{SO}(2)$, and the unit circle bundle of the modular surface $\mathrm{SL}(2,\mathbb{Z})\backslash\mathbb{H}^2$ can be identified with $\mathrm{SL}(2,\mathbb{Z})\backslash\mathrm{SL}(2,\mathbb{R})$. Using modular functions, it can be shown that $\mathrm{SL}(2,\mathbb{Z})\backslash\mathrm{SL}(2,\mathbb{R})$ can be identified with the complement of trefoil knot in S^3, which is simplest example of a nontrivial knot. For detail, see [27].

4.14 Arithmetic subgroups of Lie groups

For probably obvious reasons, the group $\mathrm{SL}(2,\mathbb{Z})$ is called an arithmetic subgroup. In some sense, it is one of the most important arithmetic subgroups. For example, for every $n \geqslant 2$, $\mathrm{SL}(n, \mathbb{Z})$ and its subgroups of finite index are arithmetic groups. The study of $\mathrm{SL}(n, \mathbb{Z})$ is closely related to the geometry of numbers since it acts on the space of lattices in $\mathbb{R}^n$ and leaves the standard lattice $\mathbb{Z}^n$ invariant.

More generally, for every linear algebraic group $\mathbf{G} \subset \mathrm{GL}(n, \mathbb{C})$ defined over $\mathbb{Q}$, there is a class of arithmetic subgroups, i.e., subgroups of $\mathbf{G}(\mathbb{Q})$ which are commensurable with $\mathbf{G}(\mathbb{Z})$. They are basic objects in modular forms and automorphic representations. They also provide one important class of groups in geometric group theory.

Many questions and results about arithmetic groups are motivated by $\mathrm{SL}(2, \mathbb{Z})$ and its action on $\mathbb{H}^2$. See [40] for some references.

4.15 Automorphic forms and automorphic representations

Modular forms and the Selberg trace formula can be generalized to other groups besides $\mathrm{SL}(2, \mathbb{R})$ and $\mathrm{SL}(2, \mathbb{Z})$.

For example, for a semisimple Lie group G and an arithmetic subgroup Γ, Γ acts on a symmetric space $X = G/K$ associated with G, where K is a maximal compact subgroup of G. The corresponding trace formula is usually called the Authur-Selberg trace formula, and its geometric meaning is less clear. On the other hand, it can be used to understand automorphic forms, for example, to transfer automorphic representations from one group to another.

For such a purpose, suitable forms of the trace formula are needed, and one basic difficulty is called *the Fundamental Lemma*, which was conjectured by Langlands (maybe because it was difficult, crucial and technical step in his program on stabilization of the trace formula, he called it the Fundamental Lemma) and stayed open until Ngo proved it in 2010. He received a Fields Medal for this work. See [3] a summary.

4.16 A role mode of geometric group theory

Geometric group theory studies groups by considering groups as geometric objects or considering their actions on metric spaces. One of the main contributors was Gromov. See [31] for one of the early papers by him on this subject.

The group $\mathrm{SL}(2, \mathbb{Z})$ and its action on $\mathbb{H}^2$ is the most basic and important example in geometric group theory. Many questions and problems and solutions are motivated by it. I will mention two examples.

4.16.1 Mapping class groups

Let S_g be a compact oriented surface of genus g. Let $\mathrm{Diff}^+(S_g)$ be the group of orientation preserving diffeomorphisms of S_g, and $\mathrm{Diff}^0(S_g)$ its identity component. Then the mapping class group Mod_g of a compact oriented surface S_g of genus g is the quotient group $\mathrm{Mod}_g = \mathrm{Diff}^+(S_g)/\mathrm{Diff}^0(S_g)$.

When $g = 1$, $\mathrm{Mod}_g = \mathrm{SL}(2, \mathbb{Z})$. Therefore, for $g \geqslant 2$, it is a generalization of $\mathrm{SL}(2, \mathbb{Z})$.

The group Mod_g has played an important role in complex analysis, algebraic geometry and topology. For example, one basic problem introduced by Riemann was to understand the moduli space $\mathcal{M}_g$ of compact Riemann surfaces of genus g. Teichmüller rigidified Riemann surfaces by adding markings and introduced Teichmüller space $\mathcal{T}_g$ of Riemann surfaces and proved that it is a complex manifold diffeomorphic to $\mathbb{R}^{6g-6}$. Now Mod_g acts on $\mathcal{T}_g$ properly and holomorhically by changing markings, and the quotient $\mathrm{Mod}_g\backslash\mathcal{T}_g$ is equal to the moduli space $\mathcal{M}_g$. One immediate consequence is that $\mathcal{M}_g$ is a complex orbifold. It turns out the action of Mod_g on $\mathcal{T}_g$ is one of the most effective ways to study $\mathcal{M}_g$, and it is a direct generalization of the action of $\mathrm{SL}(2,\mathbb{Z})$ on $\mathbb{H}^2$. See [23] for a detailed introduction to mapping class groups and [41] for some other references.

Since many three dimensional manifolds can be realized as surface bundles or constructed from surfaces in other ways, Mod_g has also played a crucial role in understanding better the geometry and topology of three dimensional manifolds, for example, in the formulation and partial solutions of Thurston's geometrization program, which was finally solved by Perelman. The basic reason is that we need to understand maps between surfaces.

4.16.2 Outer automorphism groups of free groups

Free groups are basic objects in combinatorial group theory which describe groups by generators and relations. Automorphisms of free groups are also important for this purpose. Let F_n be the free group on n generators, and $Out(F_n) = Aut(F_n)/Inn(F_n)$ be its outer automorphism group. This is the most important group in combinatorial group theory, and many questions and results about it are motivated by arithmetic subgroups of Lie groups and mapping class groups.

In fact, when $n = 2$, $Out(F_n) = \mathrm{GL}(2,\mathbb{Z})$. (Note that $\mathrm{SL}(2,\mathbb{Z})$ is a subgroup of index 2 of $\mathrm{GL}(2,\mathbb{Z})$.) For $n \geqslant 3$, there is a surjective map $Out(F_n) \to \mathrm{GL}(n,\mathbb{Z})$ with infinite kernel. Therefore, it can be viewed as a generalization of $\mathrm{SL}(2,\mathbb{Z})$.

Based on the above discussion, we have three different generalizations of the basic group $\mathrm{SL}(2,\mathbb{Z})$: (1) $\mathrm{SL}(n,\mathbb{Z})$, $n \geqslant 2$ and more general arithmetic subgroups of semisimple Lie groups, (2) Mod_g, $g \geqslant 1$, (3) $Out(F_n)$, $n \geqslant 2$. See [41] for some references.

4.17 Is it a good question to ask for the most beautiful group?

There are many other applications of $\mathrm{SL}(2,\mathbb{Z})$ and related groups. See [40] for some references.

After seeing the above discussion, have you been convinced that $\mathrm{SL}(2,\mathbb{Z})$ is a beautiful group? or the most beautiful group? We hope that that the answer is *Yes!*

Probably we should have asked a more basic question before we started to answer the above question. *Is it a good question to ask which group is the most beautiful?*

The *most famous living mathematician* in the world *J. P. Serre* said that I have had asked a wrong question!

On July 10, 2013, in Taiwan, I asked Serre this question. His immediate answer was A_5, the alternative group on five elements. A minute later, he said that there is a saying in France: *One should not love only one girl, but many!* So there should be many beautiful groups.

5 What other groups are beautiful as well?

In this section, we give some reasons why some other groups are also beautiful. They often share two things: they are special (or exceptional) and have connections with other topics, in particular appear as automorphism (or symmetry) groups of some spaces or structures.

5.1 $\mathbb{Z}_2$

This is a very special group. For example, it is the smallest nontrivial group. It is also the smallest nontrivial cyclic group. It can also be realized as follows:

1. The general linear group $\mathrm{GL}(1, F_3)$, or equivalently, the multiplicative group of the field F_3 of 3 elements.
2. The group of units in the ring $\mathbb{Z}$.
3. The group $S^0(\mathbb{R})$, i.e., the group of unit vectors in $\mathbb{R}^1$.

To be fair, as a group, its structure is not so rich and hence does not make it very interesting. But it has many applications. In applications to symmetry, $\mathbb{Z}_2$ gives the mirror (or reflection) symmetry and hence is crucial in the sciences and humanities. It also appears in duality of many things, for example, in the Golden Rule in life.

5.2 Cyclic groups of prime order

Cyclic groups are basic groups in finite group theory. For example, every finite group contains cyclic subgroups. Their natural generalizations include Sylow subgroups, which are essential in finite group theory.

In Galois theory, field extensions with cyclic Galois groups are special. Closely related cyclotomic fields are obtained by adjoining a complex primitive root of unity to $\mathbb{Q}$. Their Galois group is isomorphic to the multiplicative group of the ring $\mathbb{Z}/n\mathbb{Z}$.

One important application of cyclic group actions is the construction of lens spaces, which were the first known examples of three dimensional manifolds not determined by their homology and fundamental group alone, and the simplest examples of closed manifolds whose homeomorphism type is not determined by their homotopy type. See [33].

In equivariant theories in geometric topology, cyclic groups are also special. According to Weinberger, "Actually there are a lot of cyclic groups that are lovely to me (it's like having a favorite integer, of course).

The reason I like $\mathbb{Z}/163$ so much is that it is the first example where there is a nontrivial restriction on possible equivariant signatures, but it can't be blamed on finiteness obstructions as a certain Tate cohomology group vanishes.

On the other hand, $\mathbb{Z}/17$ is fascinating because of its connection to construct ability of polygons."

According to Sullivan, "I like the class of groups whose orders are some power of a prime. When you break such a group into conjugacy classes you see class which is larger than one element has prime power order because it is the quotient of two Prime powers with one smaller than the other. Since the identity element is not such, we deduce there are other elements where the conjugacy class is one element, because the sum of all cardinalities of conjugacy classes adds up to the order of the group which is a multiple of p. Thus the group must have a nontrivial center. We can divide the group by this center to get a group of smaller prime power order. We repeat the argument to see this group has a nontrivial center. And continue down to the trivial group. This shows any group of prime power order is nilpotent.

There are very many such groups with a rich structure. This is good mathematics–rich but simple."

5.3 The Klein four-group

The Klein four-group $\mathbb{Z}_2 \times \mathbb{Z}_2$ is the smallest non-cyclic group. Every non-identity element is of order 2. It is abelian, and isomorphic to the smallest dihedral group D_2. We note that the dihedral group D_n is the symmetry group of the regular n-gon.

The Klein four-group $\mathbb{Z}_2 \times \mathbb{Z}_2$ is the symmetry group of a rhombus and of a rectangle which are not squares.

In Galois theory, a representation (or homomorphism) of the Klein four-group in the symmetric group S_4 explains the existence of the formula for calculating the roots of quartic equations in terms of radicals.

5.4 The symmetric group S_n

The symmetric group is a special group and has applications in many subjects such as Galois theory, invariant theory, the representation theory of Lie groups, enumerative algebraic geometry, etc.

For example, symmetric polynomials are really basic. S_n is the Weyl group of the simple Lie group $\mathrm{SL}(n+1)$. In some sense, it is a very special Coxeter group. Because of this, it figures prominently in the study of flag varieties and Schubert varieties.

It was observed by Cayley that every finite group G can be realized as a subgroup of some symmetric group S_n. This does not seem to have a direct impact on understanding G, but the point that we can understand G by studying its action on itself is profound. See the books [55] [25] for details and references.

Among all symmetric groups S_n, the group S_6 is special in that it has a nontrivial outer automorphism for all $n \geqslant 3$, as discovered by Holder in 1895. See the book [54, pp. 156-161] and the paper [36].

5.5 The symmetric group S_3 and triality

Among symmetric groups S_n, S_3 is special. This it the Coxeter group of the Dynkin diagram D_4 and is responsible for the triality in $Spin(8)$, which has the outer automorphism group isomorphic to S_3. Briefly, triality is a relationship among three vector spaces, analogous to the duality relation between dual vector spaces. See the book [4] for more detail and references.

S_3 appears in the fact that given every 4-tuples of points, we have 6 cross-ratio. They are λ, $\frac{1}{\lambda}$, $1-\lambda$, $\frac{1}{1-\lambda}$, $\frac{\lambda}{\lambda-1}$, $\frac{\lambda-1}{\lambda}$. They are generated by two transformation $\lambda \to \frac{1}{\lambda}$, $\lambda \to 1-\lambda$.

5.6 The alternating group A_5 and the (icosahedral) dodecahedral group

The alternative group A_5 is special in many ways. It is the smallest non-abelian simple group, and the smallest non-solvable group.

It is the icosahedral group, i.e., the group of orientation-preserving symmetries of the regular icosahedron (or equivalently, regular dodecahedron). Hence it is a finite triangle group with parameters $(2,3,5)$.

It is also isomorphic to the projective special linear group of degree two over the field of four elements $\mathrm{PSL}(2,4)$, the special linear group $\mathrm{SL}(2,4)$, the projective general linear group $P\mathrm{GL}(2,4)$, and the projective special linear group $\mathrm{PSL}(2,5)$.

The icosahedron group appears in Klein's book on its relation to algebraic equations of degree 5. This symmetry is needed to solve principal equations of degree 5.

There are also close relation between finite subgroups of $SU(2)$ and algebraic equations.

For $n \geqslant 5$, the alternative group A_n gives a family of infinitely many non-abelian simple groups.

5.7 $\mathrm{PSL}(2,7)$, the symmetry group of Klein's quartic curve

Another finite simple group is $\mathrm{PSL}(2,7)$. It is the automorphism group of the Klein quartic and the symmetry group of the Fano plane. It is the second-smallest nonabelian simple group after the alternating group A_5, which is isomorphic to $\mathrm{PSL}(2,5)$.

Note that the Klein quartic is the projective curve over the complex numbers defined by the quartic polynomial $x^3y + y^3z + z^3x = 0$. It is a compact Riemann surface of genus $g = 3$, and is the only such surface having the maximal automorphism group, i.e., of order $84(g-1)$.

The group $\mathrm{PSL}(2,7)$ is a maximal subgroup of the Mathieu group M_{21}.

For a history of $\mathrm{PSL}(2,7)$ and its role in the geometric works of Dedekind, Riemann, Clebsch, Weber and others, especially of Klein, see the paper [28].

5.8 The Mathieu group M_{24}

The Mathieu group M_{24} is a finite simple group introduced by Mathieu. It is a 5-transitive permutation group on 24 objects. It is the automorphism group of a Steiner system, a type of block design, and the permutation automorphism group of the binary Golay code. It can also be constructed as the symmetry group of the Klein quartic PSL(2, 7) by adding an additional permutation.

There are five Mathieu groups $M_{11}, M_{12}, M_{22}, M_{23}, M_{24}$, which are sporadic finite simple groups. The groups M_{23}, M_{22}, M_{12}, and M_{11} can be realized as stabilizer subgroups of M_{24} of a suitable action of M_{24}. See the book [29] for detail and references.

5.9 The monster group

The monster group is special in several ways. The most obvious is that it is the largest sporadic finite simple group. There are 26 sporadic finite simple groups. The monster group contains all but six of the other sporadic groups as subquotient groups.

It is also famous for its role in the Monstrous moonshine conjecture, which was mentioned before. See [11] [26].

5.10 The binary icosahedral group

The binary icosahedral group is the preimage in SU(2) of the icosahedral group in SO(3) and has order 120. It is a subgroup of the unit quaternions, under the isomorphism SU(2) ≅ Sp(1).

The binary icosahedral group corresponds to the exceptional Lie group E_8 under the McKay correspondence, whose Weyl group is a large finite group. According to Vogan, "One that's very nice is the binary icosahedral group: the preimage in SU(2) of the icosahedral group in SO(3). Of course the icosahedral group itself is wonderful for a lot of reasons: smallest nonabelian simple group, isomorphic to PSL$(2, F_5)$ or to SL$(2, F_4)$. But I am very fond of the McKay correspondence relating finite subgroups of SU(2) to simple algebraic groups; so the binary icosahedral group has E_8 hiding inside. This little 120-element group tells you how to get first the Weyl group of E_8 (696729600 elements), and then E_8 over any field. Over F_2, the cardinality is

$$337804753143634806261388190614085595079991692242467651576160 = 3.3810^{74}$$

WAY bigger than the monster group, which is less than 10^{54}.

So what's amazing is that all that beautiful complexity is encoded by the icosahedron."

5.11 All finite subgroups of SO(3) and SU(2), and ADE classification

Finite subgroups of SU(2) can be classified via Dynkin diagrams ADE and correspond to Klein singularities, simple algebraic groups via the so-called McKay correspondence, and other exceptional objects.

The ADE classification occurrs at many places which seem to be unrelated. One open problem promoted by Arnold is to understand the *real reason*, if there is one, for all these connections through ADE pattern. See [34] for more details and references.

5.12 The Hessian group inside $\mathrm{PGL}(3, \mathbb{C})$

The Hessian group is a finite group associated to the inflection points of a general family of cubic curves in $P^2(\mathbb{C})$, and is the symmetry group of a certain line configuration on the complex projective plane. See [32]. It was first introduced by Jordan as the group of determinant 1 affine transformations of the affine plane over the field of 3 elements. It was named after Hesse since it acts on the Hesse pencil and the Hesse configuration. It is also closely related to complex reflection groups.

5.13 $\mathbb{Z}$, the group of integers

Among infinite discrete groups, $\mathbb{Z}$ is very special in several ways. Every torsion-free infinite group is built up from copies of $\mathbb{Z}$, and every discrete dynamical system corresponds to an action of $\mathbb{Z}$.

According to W. Lück, "The group $\mathbb{Z}$ has the nice property that the set of group homomorphisms from $\mathbb{Z}$ to any other group is in bijective correspondence to the group G considered just as a set.

Its complex group ring is the ring of Laurent series in one variable which plays a basic role in complex analysis in one variable.

Algebraically the group ring is very nice. For example, it is an ideal principal domain and $\mathbb{Z}$ is the only group except for the trivial group for which the complex group ring has this property.

For von Neumann algebras of groups and their basic problems, perhaps culminating in the extension of the Murray-Von Neumann dimension from finitely generated projective modules to arbitrary modules, the von Neumann algebra of $\mathbb{Z}$ which is $L^\infty(S^1)$ was the main and very illuminating example to Lück.

Do we know everything about the complex group ring of Z? Oh no, think about the prominent and open Lehmer's problem and its implications to topology and number theory."

5.14 The Leech Lattice in $\mathbb{R}^{24}$

One natural generalization of $\mathbb{Z}$ consists of lattices of $\mathbb{R}^n$. The most obvious lattice is $\mathbb{Z}^n$. But from various of points of view such as sphere packing, this is not a very good lattice.

It should be emphasized that as groups, all lattices in $\mathbb{R}^n$ are isomorphic. But the key point is that there is a norm on the lattice, and its geometry is the crucial aspect.

Another natural class of lattices consists of root and weight lattices of simple Lie groups. For example, the root lattice of $SL(3,\mathbb{R})$ gives the densest lattice packing in dimension 2.

In high dimension, the Leech lattice in $\mathbb{R}^{24}$ is a special lattice with many good properties. For example, it has high density, and is highly symmetric. Its automorphism group is the double cover of a Conway group, a sporadic simple group. It is also connected to the monster vertex algebra which was used to prove the monstrous moonshine conjectures. It also enjoys many other properties. See [15] for detail and references.

5.15 Free group F_2

The free group on one generator is $\mathbb{Z}$ and its structure is not so rich. On the other hand, the free group F_2 on two generators contains many subgroups and subquotient groups. For example, all finitely generated groups are subquotient groups of F_2. The reason is that every free group F_n on finitely many generators can be realized as a subgroup of F_2.

The Cayley graph of F_2 with the standard choice of generators is a regular tree with valence 4. It is probably the most basic example people use for the class of Gromov hyperbolic groups.

5.16 Bianchi group

Let $K = \mathbb{Q}(\sqrt{-d})$ be an imaginary quadratic field, where d is a square free positive integer, and $\mathcal{O}_d$ its ring of integers. Then $SL(2,\mathcal{O}_d)$ is a Bianchi group. The most basic one is when $d = 1$.

One important feature is that some knot and link complements are hyperbolic three-manifolds which are quotients of the real hyperbolic space of dimension three by Bianchi groups and finite index subgroups. See [53] for details. Combinatorial properties of Bianchi groups have also been systematically studied. See [24].

5.17 $G_2(\mathbb{Z})$

Among Dynkin diagrams of rank 2, G_2 is very special. The compact form of G_2 is the automorphism group of the octonion algebra. It is the simplest reductive group over $\mathbb{Z}$which is compact over $\mathbb{R}$.

Its subgroup of integer points $G_2(\mathbb{Z})$ is a finite simple group and is equal to the automorphism group of Coxeter's order in the octonion algebra, and is isomorphic to $G_2(\mathbb{Z}/2\mathbb{Z})$ under reduction $\mod 2$, and that is how it appears in the ATLAS.

5.18 Thompson group

The Thompson group is an exotic group and has a collection of unusual properties and is a counterexamples to many general conjectures in group theory. It is an infinite but finitely presented group. There are two related groups. See [12] for an introduction and survey on these groups.

5.19 The circle group S^1

Among compact Lie groups, S^1 is probably the simplest one. In the theory of transformation groups, S^1-actions are also rather special. Extensive works on different aspects of circle actions have been done.

Harmonic analysis on S^1 is also classical. The Poisson formula relating the geometry and analysis of S^1 is basic and motivated the Selberg trace formula.

5.20 $\mathrm{SL}(2,\mathbb{R})$, $\mathrm{SL}(2,\mathbb{C})$, and Lie groups

Among simple Lie groups, $\mathrm{SL}(2,\mathbb{R})$, $\mathrm{SL}(2,\mathbb{C})$ and their Lie algebras are the most basic. General semisimple Lie groups can be constructed from them and their representations via root and weight systems.

Many properties of semisimple Lie groups of higher rank can be motivated by studying product groups $\mathrm{SL}(2,\mathbb{R}) \times \mathrm{SL}(2,\mathbb{R})$, and $\mathrm{SL}(2,\mathbb{C}) \times \mathrm{SL}(2,\mathbb{C})$. They are simple enough to be handled but yet reflect complicated phenomenon associated with higher rank Lie groups and symmetric spaces.

They are also special. For example, Mostow strong rigidity for locally symmetric spaces fail for those associated with $\mathrm{SL}(2,\mathbb{R})$, i.e., for hyperbolic surfaces of finite area.

A closely related Lie group is $\mathrm{SU}(1,1)$, the automorphism group of the unit disk in $\mathbb{C}$, and is ismomorphic to $\mathrm{SL}(2,\mathbb{R})$. It is the simplest group in the family of $\mathrm{SU}(n,1)$, $n \geqslant 1$. They are important in several complex variables and harmonic analysis.

5.21 GL(2) as a group scheme

We have mentioned several related groups such as $\mathrm{SL}(2,\mathbb{R})$, $\mathrm{SL}(2,\mathbb{C})$, $\mathrm{SL}(2,\mathbb{Q}_p)$ $\mathrm{SL}(2,\mathbb{Z})$, $\mathrm{SL}(2,\mathbb{Z}_p)$, and $\mathrm{GL}(2,\mathbb{R})$, $\mathrm{GL}(2,\mathbb{C})$, $\mathrm{GL}(2,\mathbb{Q}_p)$ $\mathrm{GL}(2,\mathbb{Z})$, $\mathrm{GL}(2,\mathbb{Z}_p)$, etc. They all come from one source GL(2) as a group scheme. This group scheme shows its power and explains subtle (or obvious) connections between these groups.

5.22 Classical Lie groups

Among semisimple Lie groups, classical Lie groups such as $\mathrm{SO}(n)$, $\mathrm{SU}(n)$, $\mathrm{Sp}(n)$ are directly related to geometries on vector spaces, i.e., they are isometry groups of bilinear or sesquilinear forms on vector spaces.

These classical groups can also be defined over general fields and rings. Analogous constructions for exceptional simple Lie groups are not so direct or easy.

Among classical groups, D_4 stands out due to its large symmetry, or outer automorphism group.

5.23 E_8, an exceptional simple Lie group

Among exceptional simple Lie groups, E_8 is special. For example, it has the largest rank.

The root lattice of E_8, i.e., the integral span of the E_8 root system forms a lattice in $\mathbb{R}^8$ which is the only even, unimodular lattice with rank less than 16.

The compact real form of E_8 is the isometry group of the 128-dimensional exceptional compact Riemannian symmetric space EVIII (in Cartan's classification), known as the "octooctonionic projective plane".

Since it is the largest exceptional simple Lie group, it is also important in string theory. For example, there is a possibility that E_8 might be the basis of some unified field theory.

E_8 lattice is important in sphere packing. See [15].

It is also useful in 4-dimensional manifolds. The reason is that for a connected compact oriented four-dimensional manifold M, the Poincare duality induces a quadratic form $H^2(M,\mathbb{Z}) \times H^2(M,\mathbb{Z}) \to H^4(M,\mathbb{Z}) \cong \mathbb{Z}$. The work of Donaldson imposes some restriction of this quadratic form when the manifold is smooth, and E_8 lattice provides a good lattice in the study of four dimensional manifolds. See [22]. For an introduction to quadratic forms and their applications in topology, see [51].

5.24 G_2, an exceptional simple Lie group

Among exceptional simple Lie groups, G_2 has the smallest rank.

The compact form of G_2 is the automorphism group of the octonions and one of the possible special groups that can appear as the holonomy group of a Riemannian metric. Manifolds of G_2 holonomy are also called G_2-manifolds and are of seven dimension. They are important in string theory.

The noncompact split G_2 group is important in automorphic representations and the Langlands program. According to Langlands, "I do not have a candidate for a "most beautiful group". On the other hand, the most interesting group-specific calculations I ever made were those for semisimple groups of type G in, I think, the 3rd appendix of the notes on Eisenstein series....

This example is, however, probably not of the kind you were looking for. Otherwise, functoriality for the L-functions for the tetrahedral and octahedral group figuring as they did in the proof of the Fermat theorem makes them perhaps the most interesting groups with which I was personally involved."

5.25 The Heisenberg group H_3

Among nonsemisimple and nonreductive Lie groups, the Heisenberg group H_3 in dimension 3 is special. It plays an important role in quantum mechanics. It is also essential in harmonic analysis. See [37] for a summary of its role in harmonic analysis. In complex geometry, it appears in the boundary of the unit ball in $\mathbb{C}^2$. See [35].

5.26 Coxeter groups via presentation

In general, when a group is given by a presentation, it is difficult to derive properties of the group from the presentation. This is the reason for the formulation of the so-called word problems in combinatorial group theory.

On the other hand, Coxeter groups can be effectively and cleanly described by presentations. They often provide important examples in different situations. See [20] for details and references.

Functions which are invariant under finite reflection groups are very rich. We note that finite reflection groups are closely related to root systems in the Lie theory. Classical symmetric functions are functions which are invariant under the symmetric group S_n. For various invariant theories of finite reflection groups and root systems, see [48] [49].

5.27 Simple, simply connected, complex groups

Simple, simply connected, complex groups and their structures (such as subgroups and representations and associated homogeneous spaces) can be described elegantly and effectively by Coxeter-Dynkin diagrams. They are complicated objects, yet they can be cleanly described and classified. They are not a single group and hence should not be a candidate for the most beautiful group, but "for some purposes, they seem to behave as one." There are many books on this topic already, and more will surely appear in the future as well.

5.28 Braid groups

Braid groups are special geometric groups which can be defined by simple presentations and have many applications in knot theory. They are also related to other groups such as mapping class groups. Their presentations have important generalizations to other algebras. See [10] for a survey of some applications.

5.29 Mapping class group of surfaces

Mapping class groups of surfaces provide important classes of groups in topology and geometric group theory. They are concrete but fairly complicated. Since surfaces are crucial ingredients in understanding the topology and geometry of higher dimensional manifolds, especially in dimensions 3 and 4, according to Farb, mapping class groups of surfaces "encode much of 2-dimensional, 3-dimensional and 4-dimensional topology, as well as algebraic geometry, and thus provide a (still very mysterious) link between all of these topics." See [23] for an extensive introduction.

References

[1] M. Aigner, *Markov's theorem and 100 years of the uniqueness conjecture. A mathematical journey from irrational numbers to perfect matchings.* Springer, Cham, 2013. x+257 pp.

[2] E. Arbarello, M. Cornalba, P. Griffiths, *Geometry of algebraic curves. Volume II. With a contribution by Joseph Daniel Harris.* Grundlehren der Mathematischen Wissenschaften, 268. Springer, Heidelberg, 2011. xxx+963 pp.

[3] J. Arthur, The work of Ngo Bao Chau. *Proceedings of the International Congress of Mathematicians.* Volume I, 57–70, Hindustan Book Agency, New Delhi, 2010.

[4] J.F. Adams, *Lectures on exceptional Lie groups.* Chicago Lectures in Mathematics. University of Chicago Press, Chicago, IL, 1996. xiv+122 pp.

[5] P. Bérard, From vanishing theorems to estimating theorems: the Bochner technique revisited. Bull. Amer. Math. Soc. (N.S.) 19 (1988), no. 2, 371–406.

[6] A.I. Borevich, I.R. Shafarevich, *Number theory.* Translated from the Russian by Newcomb Greenleaf. Pure and Applied Mathematics, Vol. 20 Academic Press, New York-London, 1966, x+435 pp.

[7] D. Brannan, M. Esplen, J. Gray, *Geometry.* Cambridge University Press, Cambridge, 1999, xii+497 pp.

[8] F. Bien, Constructions of telephone networks by group representations. Notices Amer. Math. Soc. 36 (1989), no. 1, 5–22.

[9] R.H. Bing, Some aspects of the topology of 3-manifolds related to the Poincaré conjecture, in *Lectures on Modern Mathematics II* (T. L. Saaty, ed.), Wiley, New York, 1964.

[10] J. Birman, T. Brendle, Braids: a survey. *Handbook of knot theory,* 19–103, Elsevier B. V., Amsterdam, 2005.

[11] R. Borcherds, What is Moonshine? *Proceedings of the International Congress of Mathematicians,* Vol. I (Berlin, 1998). Doc. Math. 1998, Extra Vol. I, 607–615

[12] J. Cannon, W. Floyd, W. Parry, Introductory notes on Richard Thompson's groups. Enseign. Math. (2) 42 (1996), no. 3-4, 215-256.

[13] J. W. S. Cassels, Rational Quadratic Forms, Dover Publications, 2008, 432 pages.

[14] K. Chandrasekharan, *Elliptic functions.* Grundlehren der Mathematischen Wissenschaften, 281. Springer-Verlag, Berlin, 1985. xi+189 pp.

[15] J. Conway, N. Sloane, *Sphere packings, lattices and groups.* Third edition. With additional contributions by E. Bannai, R. E. Borcherds, J. Leech, S. P. Norton, A. M. Odlyzko, R. A. Parker, L. Queen and B. B. Venkov. Grundlehren der Mathematischen Wissenschaften, 290. Springer-Verlag, New York, 1999. lxxiv+703 pp.

[16] D. Cox, *Primes of the form* $x^2 + ny^2$. *Fermat, class field theory and complex multiplication.* A Wiley-Interscience Publication. John Wiley & Sons, Inc., New York, 1989. xiv+351 pp.

[17] F. Dal'Bo, Crossroads between hyperbolic geometry and number theory. *Strasbourg master class on geometry,* pp. 183–232, IRMA Lect. Math. Theor. Phys., 18, Eur. Math. Soc., Zürich, 2012.

[18] F. Dal'Bo, *Geodesic and horocyclic trajectories.* Translated from the 2007 French original. Universitext. Springer-Verlag London, Ltd., London; EDP Sciences, Les Ulis, 2011. xii+176 pp.

[19] H. Darmon, A proof of the full Shimura-Taniyama-Weil conjecture is announced. Notices Amer. Math. Soc. 46 (1999), no. 11, 1397–1401.

[20] M. Davis, *The geometry and topology of Coxeter groups.* London Mathematical Society Monographs Series, 32. Princeton University Press, Princeton, NJ, 2008. xvi+584 pp.

[21] P. Davis, R. Hersh, *The mathematical experience. With an introduction by Gian-Carlo Rota.* Birkhäuser, Boston, Mass., 1980. xv+440 pp.

[22] S. Donaldson, Self-dual connections and the topology of smooth 4-manifolds. Bull. Amer. Math. Soc. (N.S.) 8 (1983), no. 1, 81–83.

[23] B. Farb, D. Margalit, *A primer on mapping class groups.* Princeton Mathematical Series, 49. Princeton University Press, Princeton, NJ, 2012. xiv+472 pp.

[24] B. Fine, *Algebraic theory of the Bianchi groups.* Monographs and Textbooks in Pure and Applied Mathematics, 129. Marcel Dekker, Inc., New York, 1989. viii+249 pp.

[25] W. Fulton, *Young tableaux. With applications to representation theory and geometry.* London Mathematical Society Student Texts, 35. Cambridge University Press, Cambridge, 1997. x+260 pp.

[26] T. Gannon, Monstrous moonshine: the first twenty-five years. Bull. London Math. Soc. 38 (2006), no. 1, 1–33.

[27] E. Ghys, Knots and dynamics. *International Congress of Mathematicians.* Vol. I, 247–277, Eur. Math. Soc., Zürich, 2007.

[28] J. Gray, From the history of a simple group, in *The eightfold way,* 115–131, Math. Sci. Res. Inst. Publ., 35, Cambridge Univ. Press, Cambridge, 1999.

[29] R. Griess, *Twelve sporadic groups.* Springer Monographs in Mathematics. Springer-Verlag, Berlin, 1998. iv+169 pp.

[30] M. Gromov, Rigid transformations groups. *Géométrie différentielle* (Paris, 1986), 65–139, Travaux en Cours, 33, Hermann, Paris, 1988.

[31] M. Gromov, Infinite groups as geometric objects. *Proceedings of the International Congress of Mathematicians*, Vol. 1, 2 (Warsaw, 1983), 385–392, PWN, Warsaw, 1984.

[32] I. Hambleton, R. Lee, Finite group actions on $P^2(\mathbb{C})$. J. Algebra 116 (1988), no. 1, 227–242.

[33] A. Hatcher, *Algebraic topology.* Cambridge University Press, Cambridge, 2002. xii+544 pp.

[34] M. Hazewinkel, W. Hesselink, D.Siersma, F. Veldkamp, *The ubiquity of Coxeter-Dynkin diagrams (an introduction to the A-D-E problem).* Nieuw Arch. Wisk. (3) 25 (1977), no. 3, 257–307.

[35] L. Capogna, D. Danielli, S. Pauls, J. Tyson, *An Introduction to the Heisenberg Group and the Sub-Riemannian Isoperimetric Problem,* Progress in Mathematics, vol. 259, Birkhäuser, 224 pages.

[36] B. Howard, J. Millson, A. Snowden, R. Vakil, *A description of the outer automorphism of S_6, and the invariants of six points in projective space.* J. Combin. Theory Ser. A 115 (2008), no. 7, 1296–1303.

[37] R. Howe, *On the role of the Heisenberg group in harmonic analysis.* Bull. Amer. Math. Soc. (N.S.) 3 (1980), no. 2, 821–843.

[38] B. Iversen, *Hyperbolic geometry.* London Mathematical Society Student Texts, 25. Cambridge University Press, Cambridge, 1992. xiv+298 pp.

[39] L. Ji, A summary of topics on group actions, preprint, 2014.

[40] L. Ji, *Arithmetic groups and their generalizations. What, why, and how.* AMS/IP Studies in Advanced Mathematics, 43. American Mathematical So-

ciety, Providence, RI; International Press, Cambridge, MA, 2008. xviii+259 pp.

[41] L. Ji, A tale of two groups: arithmetic groups and mapping class groups. *Handbook of Teichmüller theory.* Volume III, 157–295, IRMA Lect. Math. Theor. Phys., 17, Eur. Math. Soc., Zürich, 2012.

[42] L. Ji, Buildings and their applications in geometry and topology. *Differential geometry*, 89–210, Adv. Lect. Math. (ALM), 22, Int. Press, Somerville, MA, 2012.

[43] L. Ji, A. Papadopoulos, *Lie and Klein: The Erlangen program and its impact on mathematics and physics*, European Mathematical Society Publishing House, to appear in 2014.

[44] M. Kac, Can one hear the shape of a drum? Amer. Math. Monthly 73 1966 no. 4, part II, 1–23.

[45] S. Katok, I. Ugarcovici, Symbolic dynamics for the modular surface and beyond. Bull. Amer. Math. Soc. (N.S.) 44 (2007), no. 1, 87–132.

[46] T. Kubota, *Elementary theory of Eisenstein series.* Kodansha Ltd., Tokyo; Halsted Press, New York-London-Sydney, 1973. xi+110 pp.

[47] A. Lubotzky, *Discrete groups, expanding graphs and invariant measures.* With an appendix by Jonathan D. Rogawski. Progress in Mathematics, 125. Birkhäuser Verlag, Basel, 1994. xii+195 pp.

[48] I. Macdonald, *Symmetric functions and orthogonal polynomials.* Dean Jacqueline B. Lewis Memorial Lectures presented at Rutgers University, New Brunswick, NJ. University Lecture Series, 12. American Mathematical Society, Providence, RI, 1998. xvi+53 pp.

[49] I. Macdonald, *Symmetric functions and Hall polynomials.* Second edition. With contributions by A. Zelevinsky. Oxford Mathematical Monographs. Oxford University Press, 1995. x+475 pp.

[50] J. Milnor, The Poincaré conjecture. The millennium prize problems, 71–83, Clay Math. Inst., Cambridge, MA, 2006.

[51] J. Milnor, D. Husemoller, Dale *Symmetric bilinear forms.* Ergebnisse der Mathematik und ihrer Grenzgebiete, Band 73. Springer-Verlag, New York-Heidelberg, 1973. viii+147 pp.

[52] M. Newman, *Integral matrices.* Pure and Applied Mathematics, Vol. 45. Academic Press, New York-London, 1972. xvii+224 pp.

[53] C. Maclachlan, A. Reid, *The arithmetic of hyperbolic 3-manifolds.* Graduate Texts in Mathematics, 219. Springer-Verlag, New York, 2003. xiv+463 pp.

[54] J. Rotman, *An introduction to the theory of groups.* Fourth edition. Graduate Texts in Mathematics, 148. Springer-Verlag, New York, 1995. xvi+513 pp.

[55] B. Sagan, *The symmetric group. Representations, combinatorial algorithms, and symmetric functions.* Second edition. Graduate Texts in Mathematics, 203. Springer-Verlag, New York, 2001. xvi+238 pp.

[56] P. Sarnak, Class numbers of indefinite binary quadratic forms. J. Number Theory 15 (1982), no. 2, 229–247.

[57] P. Sarnak, Spectra of hyperbolic surfaces. Bull. Amer. Math. Soc. (N.S.) 40 (2003), no. 4, 441–478.

[58] A. Selberg, *Collected papers.* Vol. I. With a foreword by K. Chandrasekharan. Springer-Verlag, Berlin, 1989. vi+711 pp.

[59] C. Series, The modular surface and continued fractions. J. London Math. Soc. (2) 31 (1985), no. 1, 69–80.
[60] C. Series, The geometry of Markoff numbers. Math. Intelligencer 7 (1985), no. 3, 20–29.
[61] J.P. Serre, *Trees.* Corrected 2nd printing of the 1980 English translation. Springer Monographs in Mathematics. Springer-Verlag, Berlin, 2003. x+142 pp.
[62] G. Shimura, *Introduction to the arithmetic theory of automorphic functions.* Reprint of the 1971 original. Publications of the Mathematical Society of Japan, 11. Kano Memorial Lectures, 1. Princeton University Press, Princeton, NJ, 1994. xiv+271 pp.
[63] A. Sossinsky, *Geometries.* Student Mathematical Library, 64. American Mathematical Society, Providence, RI, 2012. xvi+301 pp.
[64] B. Sury, *The congruence subgroup problem. An elementary approach aimed at applications.* Texts and Readings in Mathematics, 24. Hindustan Book Agency, New Delhi, 2003. xvi+301 pp.
[65] A. Terras, *Harmonic analysis on symmetric spaces–Euclidean space, the sphere, and the Poincaré upper half-plane.* Second edition. Springer, New York, 2013. xviii+413 pp.
[66] A. Wiles, Twenty years of number theory. *Mathematics: frontiers and perspectives,* 329–342, Amer. Math. Soc., Providence, RI, 2000.
[67] D. Wise, *From riches to raags: 3-manifolds, right-angled Artin groups, and cubical geometry.* CBMS Regional Conference Series in Mathematics, 117. Published for the Conference Board of the Mathematical Sciences, Washington, DC; by the American Mathematical Society, Providence, RI, 2012. xiv+141 pp.
[68] H. Wu, The Bochner technique. *Proceedings of the 1980 Beijing Symposium on Differential Geometry and Differential Equations,* Vol. 1, 2, 3 (Beijing, 1980), 929–1071, Science Press, Beijing, 1982.
[69] H. Wu, *The Bochner technique in differential geometry. II.* Adv. in Math. (Beijing) 11 (1982), no. 1, 19-61.
[70] H. Wu, *The Bochner technique in differential geometry.* Math. Rep. 3 (1988), no. 2, ixii and 289–538.

Introduction to
Modern Mathematics
ALM 33, pp. 285–321

von Neumann Algebras in Mathematics and Physics*

Vaughan Jones†

Abstract

These are the notes for the Chern lectures that I gave at Tsinghua University in late December 2012. The notes were prepared in haste as I was preparing the lectures. As such everything in them must be regarded with suspicion—all definitions are incomplete, all theorems are wrong as stated and attributions are likely to be misleading. I have tried to include enough references to the literature for the interested readers to track down the truth, while trying to retain some readabilty to reflect the somewhat informal nature of the talks. I apologise to any mathematicians or physicists whose work is not given the correct recognition. It is also true of course that the choice and emphasis of topics is largely influenced by my own work. The story told by another worker in von Neumann algebras would no doubt be completely different.

2000 AMS subject classification: 46L10, 46L37, 46L54, 57M25, 57M27, 81R10, 81R15, 81T40.

Keywords and Phrases: von Neumann algebra, algebraic quantum field theory, knots and links, factors, subfactors, planar algebra, foliation random matrices, free probability.

1 Lecture 1: What is ... a von Neumann algebra?

1.1 Definition

Before starting let me give some examples of what are and are not von Neumann algebras.

(1) The 2×2 complex matrices $\left\{\begin{pmatrix} a & b \\ c & d \end{pmatrix}\right\}$ are a von Neuman algebra.

(2) The diagonal matrices $\left\{\begin{pmatrix} a & 0 \\ 0 & d \end{pmatrix}\right\}$ are a von Neumann algebra.

*Support from NSF grant DMS-0856316 and DARPA grant HR0011-12-1-0009.

†Department of Mathematics, Vanderbilt University, 1326 Stevenson Center, Nashville, TN 37240, USA, and Math. department, UC Berkeley, Berkeley, CA 94720, USA. Email: vaughan.f.jones@vanderbilt.edu

(3) The upper triangular matrices $\left\{\begin{pmatrix} a & b \\ 0 & d \end{pmatrix}\right\}$ are NOT a von Neumann algebra.
(4) If D is a linear differential operator operator realised as a self-adjoint operator on some Hilbert space $\mathfrak{H}$ (think id/dx on $L^2(\mathbb{R})$) then the set of all bounded operators that commute with D is a von Neumann algebra.
(5) The algebra of ALL bounded operators on a Hilbert space $\mathfrak{H}$, written $B(\mathfrak{H})$, is a von Neumann algebra.
(6) The algebra of all compact operators on a Hilbert space $\mathfrak{H}$ is NOT a von Neumann algebra (in infinite dimensions).
(7) If $g \to u_g$ is a unitary group representation on $\mathfrak{H}$ then its *commutant* or algebra of intertwining operators, $\{x \in B(\mathfrak{H}) | xu_g = u_g x, \forall g\}$ is a von Neumann algebra.
(8) If (X, μ) is a measure space then $L^\infty(X, \mu)$ is a von Neumann algebra.

The following features are common to all these "von Neumann" algebras:

a) They act on (complex) Hilbert space $\mathfrak{H}$.
b) They are closed under the * operatiton (conjugate transpose).
c) They contain the identity operator.
d) They are closed in pointwise convergence on $\mathfrak{H}$.

This then is the deceptively simple definition of a von Neumann algebra:

A von Neumann algebra M is an algebra of bounded operators on $\mathfrak{H}$ satisfying a), b), c), d).

Von Neumann had several reasons for introducing these algebras. Let us list some of these.

(i) An infinite dimensional version of the theory of semisimple associative algebras.
(ii) An infinite dimensional generalisation of projective geometry.
(iii) A mathematical framework for quantum physics.
(iv) A method of studying unbounded operators on Hilbert space by bounded operators.
(v) An abstract decomposition of unitary group representations.
(vi) The then very new subject of ergodic theory.

1.2 Factors

In the course of these lectures we will address most of the motivations but for this lecture let us restrict ourselves to the most obvious: (i). Indeed the Wedderburn theory immediately gives us the full classification of finite dimensional von Neumann algebras. They are all abstractly isomorphic to direct sums of matrix algebras, e.g.

$$M_2(\mathbb{C}) \oplus M_6(\mathbb{C}) \oplus \mathbb{C} \oplus \mathbb{C},$$

and the action on a given Hilbert space is determined up to an obvious notion of "multiplicity" (the action of $M_2(\mathbb{C})$ on $\mathbb{C}^2$ is multiplicity one).

This result generalises perfectly. The new phenomenon in infinite dimensions is that one must allow "direct integrals" to replace direct sums, in such a way that $L^\infty(X,\mu)$ becomes the direct integral $\int_X^\oplus \mathbb{C} d\mu$. A version of the spectral theorem says that any abelian von Neuman algebra is abstractly isomorophic to some $L^\infty(X,\mu)$. Then one uses the centre $Z(M)$ to decompose M as a direct integral of von Neumann algebras with trivial centre. (We will leave the discussion of multiplicity till later.) All of this was done by von Neumann himself in [61]. .

Already central in the theory was von Neumann's "bicommutant theorem". Given a subset $S \subseteq B(\mathfrak{H})$ we will denote its commutant ($\{x \in B(\mathfrak{H}) | xs = sx, \forall s \in S\}$) by S'. The bicommutant theorem asserts that for a von Neumann algebra $M = M''$ and that this condition could replace d) in the definition of a von Neumann algebra, making them in fact purely algebraic objects.

After the direct integral classification interest obviously shifts to von Neumann algebras with trivial centre which are called *factors*. If it were true that any factor is abstractly isomorphic to some $B(\mathfrak{H})$ for some $\mathfrak{H}$, the classification would be over and the infinite dimensional Wedderburn theory would be almost as simple as the finite dimensional one. But factors have turned out to be enormously complicated and enormously rich in structure. It is our first job in these lectures to describe the extent to which factors have been understood.

Let us agree to say a factor is type I if it is abstractly isomorphic to $B(\mathfrak{H})$ for some $\mathfrak{H}$.

The first example of a factor not of type I appeared in [42]. It already involved erdogic theory and an infinite dimensional version of the "crossed product" construction from abstract algebra. Recall that the group algebra of a group G is the algebra of all formal linear combinations $\sum_{g\in G} c_g g$ with $c_g \in \mathbb{C}$. Multiplication in this algebra is given by the group rule and the * operation $g^* = g^{-1}$ gives us a *-algebra. Now if the group acts by automorphisms on some other algebra N, we can change the coefficients of the group algebra from $\mathbb{C}$ to N and combine the multiplications of N and G using the rule $gng^{-1} = g(n)$ just as in the semidirect product of groups. Thus one obtains a new algebra (a *-algbera if N is) called the "crossed product" of N by G, written

$$M = N \rtimes G.$$

One has to be a bit careful defining this crossed product for von Neumann algebras but it works, even for groups G that are locally compact, and is an extremely useful construction.

The momentous discovery of Murray and von Neumann was that the crossed product can be used to give factors not of type I. Call an action of a group G by *-automorphisms of a von Neumann algebra N *ergodic* if the fixed point algebra $N^G = \{x \in N | g(x) = x, \forall g \in G\}$ is just the scalar multiples of the identity. Then it is morally true that

$N \rtimes G$ is a factor if the action of G is ergodic.

One needs to be careful especially if G is not discrete, but examples of appropriate ergodic actions abound. The easiest non-trivial example is to take N abelian so that the action is just the action of a group on a measure space and the notion of ergodicity is the now familiar one that the action of G leaves no subsets invariant unless they or their complents have zero measure. The example of [42] was to take $G = \mathbb{Z}$ and the action to be any ergodic measure preserving action such as the irrational rotation of the circle. We thus obtain the factor

$$M = L^\infty(S^1) \rtimes \mathbb{Z}.$$

What is so magical about this factor M? We may use the integral $\int_X f d\mu$ to define a linear functional tr on $N = L^\infty(S^1)$ and extend it to all of M by decreeing that $\mathrm{tr}(\sum_g x_g g) = \mathrm{tr}(x_e)$, e being the identity of the group. Because the action of the group preserves the measure, it preserves the integral and we have the tracial property

$$\mathrm{tr}(ab) = \mathrm{tr}(ba)$$

for all $a, b \in M$.

(Note that the condition $\mu(X) = 1$ is equivalent to the normalisation $\mathrm{tr}(1) = 1$.)

And here's the rub: no linear functional tr on $B(\mathfrak{H})$ can have this tracial property. Hence our M is not type I!!

By taking infinite matrices over this M (i.e. taking the tensor product with some infinite dimensional $B(\mathfrak{H})$) one obtains another factor, without a trace which is also not type I because it has a "corner" which is not type I. By corner of a von Neumann algebra Mwe mean a von Neumann algebra of the form pMp where p is some projection in M, i.e. an element with $p = p^2 = p^*$.

We will call a factor a type II_1 factor if it has a trace (and is infinite dimensional) and a type II_∞ factor if it has a corner which is of type II_1 (without being type II_1 itself).

Sooner or later we had to run into the notion of projections in a von Neumann algebra. They are fundamental to any technical ingredient of the theory. Two projections p and q in a von Neumann algebra are called "equivalent" relative to M if there is a $v \in M$ with $vv^* = e$ and $v^*v = f$. Clearly equivalent projections in a II_1 factor have the same trace and it is shown in [42] that projections with the same trace are equivalent. In fact the same result is elementary in $B(\mathfrak{H})$ where projections are orthogonal projections onto subspaces of $\mathfrak{H}$ and two such are equivalent iff the subspaces have the same dimension. The trace (=sum of diagonal elements) of a projection in $B(\mathfrak{H})$ is that dimension.

This leads us to the really stunning thing about II_1 factors-that the trace of projections is a notion of dimension of subspaces (relative to the factor of course) and that this dimension *varies continously*! For note that in our example $L^\infty(S^1) \rtimes \mathbb{Z}$ we can obtain a projection simply by taking the characteristic function of some measurable subset of S^1, whose measure is its trace. Thus

$$\{\mathrm{tr}(p) | p \text{ is a projection in } M\} = [0, 1].$$

One should compare and contrast the situation where M is the $n \times n$ matrices where if we normalise the trace similarly,

$$\{\mathrm{tr}(p) | p \text{ is a projection in } M\} = \left\{0, \frac{1}{n}, \frac{2}{n}, \frac{3}{n}, \cdots, 1\right\}$$

We see the continuous dimensionality that fascinated von Neumann and led him to give an abstract lattice-theoretic version of it known as continous geometry.

But the von Neumann algebra story is only just starting to warm up. We have at our disposal factors of types I and II but is that all? In a real tour de force Murray and von Neumann in [43] showed that this is not the case, that there are factors which do not even have a corner that admits a trace. Such factors are said to be of type III and have the property that any two non-zero projections are equivalent. They can be obtained from the crossed product by choosing an ergodic action of a discrete group on a measure space for which there is no equivalent (σ-finite) invariant measure. A concrete example is given by the action of $SL_2(\mathbb{Z})$ on the real line by linear fractional transformations.

Let us take leave of the Murray-von Neumann papers for a while to follow the evolution of the theory of type III factors.

A II_1 factor M actually contains the information for a canonical action on a Hilbert space. If we define the inner product $\langle x, y\rangle = \mathrm{tr}(y^*x)$ then M becomes a pre-Hilbert space for which left multiplication by $x \in M$ is a bounded operator. Thus the completion, which is called $L^2(M, \mathrm{tr})$ admits an action of M by left multiplication. The abstract *-algebra M is canonically represented on $L^2(M, \mathrm{tr})$. Moreover the extension of the * operation to $L^2(M, \mathrm{tr})$ is a conjugate linear isometry, written J. Murray and von Neumann showed that $JMJ = M' =$ {right multiplication by elements of M}.

If a factor M is of type III it still admits linear functionals ϕ (at worst defined on a dense subspace) with sufficient positivity and continuity for the above construction, called the GNS construction, to work, setting $\langle x, y\rangle = \phi(y^*x)$. However there is a fundamental snag. Although left multiplication by an element of M is bounded, the same is not necessarily true of right multiplication, or equivalently, the other side of the coin, the * operation is not an isometry and there is no immediate relation between M and M'.

In order to solve a relatively technical question, Tomita proposed a way around the non-isometry of *, namely to view it as an unbounded operator and take its polar decomposition. It is not a closed operator but it is pre-closed (the closure of its graph on $L^2(M, \phi)$ is the graph of an operator) so this is actually kosher. If we call S the closure of *, we have

$$S = J\Delta$$

as its polar decomposition, J being a conjugate linear isometry and Δ an injective positive self-adjoint operator. Tomita's original proof contained many difficulties but the theory was proved sound by Takesaki in [57] and we have the Tomita Takesaki theorem which states that

(i) $JMJ = M'$,
(ii) $\Delta^{it} M \Delta^{-it} = M$ for $t \in \mathbb{R}$.

While it was (i) that Tomita needed to solve his problem, (ii) was a massive spinoff as it shows that type III factors are in fact dynamic objects, possessing, for each ϕ a 1-parameter automorphism group of automorphisms, called the *modular group* σ_t^ϕ with

$$\sigma_t^\phi(x) = \Delta^{it} x \Delta^{-it}.$$

Of course σ_t^ϕ depends on ϕ. It was Connes in [4] who revealed the canonical nature of the modular group by showing that σ^ϕ depends only on ϕ up to inner automorphisms which are reflected in inessential perturbations of Δ. He thus defined two invariants $S(M)$ and $T(M)$ of a type III factor as

$$S(M) = \cap_\phi \{\text{spectrum of } \Delta_\phi\} \text{ and}$$
$$T(M) = \ker(\sigma_\phi) \text{ moduloinnerautomorphismsof } M.$$

He showed with van Daele in [13] that $S(M)$ is a closed multiplicative subgroup of $\mathbb{R}^+$ (and zero) and thus of a form according to which he organised type III factors:

$$\text{type III}_0: \ S(M) = \{0, 1\},$$
$$\text{type III}_\lambda: \ S(M) = \{\lambda^n | n \in \mathbb{Z}\} \cup \{0\} \text{ for } 0 < \lambda < 1 \text{ and},$$
$$\text{type III}_1: \ S(M) = \mathbb{R}^+ \cup \{0\}.$$

The one example we gave of a type III factor, from $SL_2(\mathbb{Z})$ on the circle, is of type III_1. Examples of type III_λ for $0 < \lambda < 1$ had been previously obtained (and shown to be mutually non-isomorphic) by Powers in [51].

Further to this, Connes gave in [4] a structural theorem for type III_λ factors with $0 \leqslant \lambda < 1$ as a crossed product of a type II_∞ von Neumann algebra by $\mathbb{Z}$. The remaining case, the type III_1 factor, was solved by Connes and Takesaki in [12] where they showed that a III_1 factor is actually a crossed product of a type II_∞ factor by an action of $\mathbb{R}$.

There are in this story significant connections with quantum statistical mechanics which I will take up in my lecture on von Neumann algebras and physics.

Although it is true that a factor will generically be of type III, Connes-Takesaki in principle lets us reduce type III problems to type II ones and it is fair to say that most of the emphasis in the structure theory of von Neumann algebras since Connes-Takesaki has been in the type II arena (not so of applications!).

So let us now return to the Murray von Neumann papers and II_1 factors.

One question that was hanging unanswered was the question of the isomorphism of II_1 factors. There is only one type I infinite dimensional factor per Hilbert space but what of type II_1's? Von Neumann recognised, as was later proved by Dye in [18] , that if the group G in $L^\infty(X, \mu) \rtimes G$ is $\mathbb{Z}$, one always obtains the same II_1 factor up to isomorphism. It is the hyperfinite II_1 factor R characterised in [43] by the fact that it contains a dense sequence of finite dimensional *-subalgebras. In a remarkable tour de force, Connes in [6] gave several other more flexible characterisations including "injectivity" which means that there is a rather discontinous partial trace from $B(\mathfrak{H})$ onto R. Among the many applications of Connes' result was: any subfactor of R factor is either finite dimensional or isomorphic to R.

But is any II_1 factor isomorhic to R? In their last paper Murray and von Neumann showed that this is not the case. To obtain a non-hyperfinite II_1 factor they used the crossed product construction with $N = \mathbb{C}$, i.e. nothing but the completed group algebra. This gives a factor, denoted LG, whenever all the non-trivial conjugacy classes of G are infinite, which happens quite often. They showed that if G is F_2, the free group on 2 generators, then LG has an even stronger factor property-if an element commutes up to ϵ with the group generators, then it is within 14ϵ of $\mathbb{C}id$. In R one can exhibit sequences called "central sequences" which asymptotically commute with anything but stay well away from $\mathbb{C}id$.

Variations on this central sequence theme eventually produced infinitely many mutually non-isomorphic II_1 factors ([41],[56]) but much easier methods now exist. In [5] Connes defined an invariant $\chi(M)$ which is an abelian group and he showed that it can be pretty much any abelian group within the limits of cardinality. But even $\chi(M)$ revolves around central sequences. Connes introduced an entirely new concept and obtained other new infinite families but to fully appreciate this new idea we need to go back and discuss multiplicity for II_1 factors.

A representation of the matrix algebra $M_n(\mathbb{C})$ on $\mathfrak{H}$ is determined by a single number, its multiplicity, namely the number of copies of $\mathbb{C}^n$ that constitute $\mathfrak{H}$. Because of the absence of irreducible representations, for a II_1 factor we need to change points of view and compare a given representation to $L^2(M)$ to which we will assign multiplicity equal to 1. From $L^2(M)$ one may obtain another representation by right multiplication by a projection p in M to which it is natural to assign a multiplicity of $\mathrm{tr}(p)$. By general theory any representation of the II_1 factor M on $\mathfrak{H}$ is a direct sum of an integral number n of copies of $L^2(M)$ and a "remainder" term which is of the form $L^2(M)p$ as above. We see that it is natural to use the terminolgy $\dim_M(\mathfrak{H})$ and for the $\mathfrak{H}$ we have just discussed, we have

$$\dim_M(\mathfrak{H}) = n + \mathrm{tr}(p).$$

(The way we have done it is a little messy-one can easily make it quite natural and less comprehensible ...)

The number $\dim_M(\mathfrak{H})$ was called the coupling constant by Murray and von Neumann as it measures the relative size of M and M'. It provides the answer to a relatively easy question in the theory-if two II_1 factors are abstractly isomorphic, are they isomorphic as algebras of operators on their Hilbert spaces? The answer-iff they have the same coupling constants.

And in this circle of ideas one cannot help but ask the question: When is M abstractly isomorphic to a corner subalgebra pMp? Clearly this depends only on the trace of p and by extending to corners of matrix algberas over M the trace values of projections with $M \cong pMp$ become a multiplicative subgroup of $\mathbb{R}^+$ called the *fundamental group* of M. (Nothing to do with π_1.) By [43] the fundamental group of R is $\mathbb{R}^+$ but it remained open for a long time if the fundamental group of a II_1 factor could be anything else. It is this problem that Connes solved with the introduction of Kazhdan's property T. (Remarkably, it is still not known if the fundamental group of LF_2 is $\mathbb{R}^+$ or $\{1\}$!) A group G has property T if the trivial representation is isolated in the space of all unitary representations of G and Connes showed in [8] that the fundamental group of LG is countable if G has property T. (Kazhdan had shown that lattices in Lie groups can have property T, in particular $PSL_n(\mathbb{Z})$ for $n \geqslant 3$.)

This use of property T gives uncountably many II_1 factors immediately. On the other hand it begs the important question of what the fundamental group actually is. For this another 20 years were to go by before Popa introduced a whole new circle of ideas into von Neumann algebras, his "deformation rigidity" program, the first fruit of which was to show that the crossed product algebra $L^\infty(\mathbb{T}^2) \rtimes \mathbb{Z}^2$ has trivial fundamental group ([49]). Deformation rigidity involves the position of the subalgebra in the crossed product which leads us for the first time in this talk to the relative theory of von Neumann algebras, namely the consideration of von Neumann subalgebras of factors.

1.3 The relative theory, subfactors and Cartan subalgebras

Let us reverse historical order and begin with subfactors since we will need them more in the subsequent lectures. A subfactor is nothing but that-a factor inside another one (we assume they both have the same identity so a corner is not considered a subfactor). In this generality factors themselves are subfactors-of $B(\mathfrak{H})$, but we will be interested in the non type I theory, starting with type II.

A natural source of examples of subfactors comes from group actions. If G acts on the II_1 factor N then the fixed point algebra N^G and the crossed product $N \rtimes G$ have a chance of being factors. The sweetest situation occurs when G is finite, Then both are factors if the action of the group is by outer automorphisms. We then have two subfactors

$$N^G \subseteq N \text{ and } N \subseteq N \rtimes G$$

which are in some sense dual to each other. One cannot help thinking of Galois theory and indeed in the late 1950's the Japanese school established a Galois theory with many of the ingredients of field extensions ([44]). The most obvious is the degree of the extension and we observe that this can be expressed in purely von Neumann algebraic language. Thus given a subfactor $N \subseteq M$ we define the index of N in M to be

$$[M : N] = \dim_N(L^2(M, \mathrm{tr})).$$

For both the group examples the index is $|G|$ so all positive integers occur as the index of a subfactor. But being a von Neumann coupling constant the index is a priori allowed to be any real $\geqslant 1$. The theory of subfactors got under way with the result that the index actually has to be in the set

$$\{4\cos^2 \pi/n | n = 3, 4, 5, \ldots\} \cup [4, \infty]$$

and all these values are realised by subfactors of R ([30]). The original proof involved a deeper knowledge of certain towers of finite dimensional algebras discovered in statistical mechanics by Temperley and Lieb ([59]). These algebras further yielded unitary representations of the braid groups and a polynomial invariant of knots that we will discuss further in the lecture on connections with topology. A subsequent proof of the restrictions on index values goes via the "principal graph" which Ocneanu defined as that connected component of the graph of induction/restriction of $N - N$ and $M - N$ bimodules containing the $N - N$ bimodule $L^2(N)$. If the index is less than 4 the principal graph has to be one of the Coxeter graphs A_n, $n \geqslant 3$, D_{2n}, $n \geqslant 2$, E_6 and E_8 all of which are realised in R. The first finite graph to occur for index > 4 was discovered by Haagerup. The square of the norm of the adjacency matrix of the graph is the index. For the Haagerup it is $\frac{5+\sqrt{13}}{2}$ ([28]). All principal graphs are known in index up to and including 5 ([34]).

The simplest known proof for the exisitence of the $\{4\cos^2\pi/n|n = 3, 4, 5, \ldots\}$ subfactors remains the original one though an independent construction was discovered by Wassermann in the theory of loop group rerpesentations and conformal field theory. We will discuss this further in our second lecture. This method actually constructs type III subfactors but the index theory had been extended to type III by Kosaki in [37] and the type II subfactors can be recovered if necessary. Algebraic quantum field theory provided an alternative interpretation of the index of a subfactor as the square of the statistical dimension of an endomorphism. Connes had observed the equivalence between endomorphisms and bimodules for type III factors ([38],[39]).

Now we pass to Cartan subalgebras. In the original construction of II_1 factors as $M = N \rtimes G$ with N abelian, N is a maximal abelian subalgebra (MASA) whose normaliser generates M (by definition). Such a MASA is called a *Cartan subalgebra* by analogy with Lie theory. Dixmier in [16] studied MASA's and found many that are not Cartan subalgebras. Indeed unlike in finite dimensions the normaliser of a MASA of a type II factor can be trivial. If $N = L^\infty(X, \mu)$ then in the II_1 case the normaliser defines an equivalence relation on X given by the orbits of G. Feldman-Moore showed in [21] how to construct the crossed product using only the equivalence relation, but that all equivalence relations do come from group actions, perhaps with fixed points for some group elements.

The question arises immediately of the uniqueness of the Cartan of a crossed product. Connes showed that R has a *unique* Cartan subalgebra up to automorphisms of R. But by [11] there are II_1 factors with more than one Cartan subalgebra.

Popa's first big breakthrough in this area was to show that Cartan subalgebras with two properties—a relative property T and a relative version of an approximation property due to Haagerup—are unique in their II_1 factors. (The non-relative versions of these properties can never coexist in a II_1 factor.) Combining this with ideas of Gaboriau on L^2 Betti numbers, Popa was able to exhibit II_1 factors with trivial fundamental group. Importantly, the kind of groups involved are not ad hoc but can be as mundane as $SL_2(\mathbb{Z})$. See [48].

This began the deformation rigidity program of Popa which sets about trying to recover information about the group and its action from nothing but the isomorphism class of the crossed product II_1 factor. It would be a great start if the Cartan subalgebra were unique and this is what Popa shows under favourable circumstances-[49]. Then in many cases one can go on to recover not just the equivalence relation but in fact the group itself and its action on $L^\infty(X, \mu)$. It is remarkable that Popa obtains uniqueness up to *inner* automorphisms though this becomes a bit less shocking when the deformation rigidity program produces II_1

factors with no outer automorphisms whatsoever! ([50]). The kinds of groups for which favourable circumstances occur are by no means obscure. Vaes and Popa show in [PV] that if one uses a non-abelian free group in the crossed product, the resulting Cartan subalgebra is unique up to an inner automorphism. This area of von Neumann algebra research has been highly active for over 10 years and remains so.

It should be noted that Connes proposed a rigidity for the group inside the group von Neumann algbegra for groups like $SL_3(\mathbb{Z})$ as long ago as 1980. This remains open.

But must a II_1 factor have a Cartan subalgebra? This was open for many years. We give the answer in the next section.

1.4 Free probability

We will return to this topic in our last lecture but we would like to end this one with a short preview of Voiculescu's theory.

If a crossed product von Neumann algebra is a II_1 factor, the non-identity group elements themselves all have trace equal to zero-this is how we exended the trace from N to $N \rtimes G$. Now consider the case $N = \mathbb{C}$ and $G = F_2$ so we are looking at LF_2. This II_1 factor comes equipped with two (non Cartan!) MASA's-the von Neumann algebras A and B generated by the two generators a and b of F_2. By freeness of the group and our observation about the trace of a group element it follows that A and B enjoy the following property:

$$\mathrm{tr}(x_1 x_2 \ldots x_n) = 0 \text{ whenever } x_i \text{ are all in either } A \text{ or } B,$$
$$\mathrm{tr}(x_i) = 0, \text{ and } x_i \in A \iff x_{i\pm 1} \in B.$$

If G is a group generated by a and b this purely von Neumann algebraic property is also obviously equivalent to G being the free product $\langle a \rangle * \langle b \rangle$.

It is also easy to see that if this property holds for two subalgebras A and B of a II_1 factor then one may calculate the trace of any element of the algebra generated by A and B by reducing to words on trace zero elements. In this situation we say that the von Neumann algebra generated by A and B is the *free product* of A and B. See [60],[3]. Moreover the property can be used to construct a free product of two (or more) von Neumann algebras with traces. One will almost always obtain a II_1 factor in this way.

Voiculescu used the free product to create his "free Gaussian functor" which works by analogy with the fermionic and bosonic functors of second quantisation. It takes Hilbert spaces to von Neumann algebras in a way that the direct sum of Hilbert spaces goes to the free product of algebras with traces. The fermionic and bosonic Fock spaces are replaced by the " full Fock space" of a Hilbert space which is the ℓ^2 completion of the tensor algebra. Annihiliation and creation operators act irreducibly on this Fock space as usual but when one takes just their real parts one obtains a II_1 factor. By the free product observation above this II_1 factor is isomorphic to $L(F_n)$, n being the dimension of the underlying Hilbert space.

Elements of these II_1 factors are to be thought of as "free" random variables and Voiculescu proves a free central limit theorem for which the Gaussian distribution is replaced by the semicircular law. For more, including a connection with random matrices, see our last lecture.

All of this would just be some pretty formalism and new work for underemployed probabilists but Voiculescu used it to make a major breakthrough in II_1 factors. He observed that the large N limit of the traces of words on n suitably independent random matrices defines a trace on the algebra of non-commuting polynomials in n variables which can then be completed to give a II_1 factor again isomorphic to LF_n! This new model for LF_n is flexible enough to give isomorphisms between certain corners of matrix algebras over LF_n's for certain n's. He combined them to show that the fundamental group of LF_∞ contains all of $\mathbb{Q}^+$, which was improved to all of $\mathbb{R}^+$ by Radulescu. ([54]).

A way to interpret the connection with random matrices is to say that, while not hyperfinite, the free group von Neumann algebras do admit some kind of finite dimensional approximation at least in terms of moments. By rethinking the concept of entropy in the free case Voiculescu came up with the idea of "microstates" and defined an entropy for a set of generators of a II_1 factor. Although this entropy has not been shown to be an invariant of the von Neumann algebra, Voiculescu was able to use it to show that LF_n cannot have a Cartan subalgebra. One does feel that MASA's should be important structural elements in von Neumann algebras so one wonders if some other property of a MASA could replace the normaliser in a stucture theorem.

With the work of Tomita, Takesaki, Connes, Popa, Voiculescu and others a whole host of problems raised implicitly by the work of Murray and von Neumann have been solved. I have only mentioned a few of them. Let me end by the one that remains and has resisted all attempts at solution:

$$\text{Is } LF_n \cong LF_m \text{ for } m \neq n?$$

Some progress has been made. Dykema and Radulescu have embedded LF_n in a continuous family LF_d and it is known that if any two of the family are isomorphic, they all are. ([19])

2 Lecture 2: von Neumann algebras and physics

2.1 Recall definition of von Neumann algebras

Let us recall the definition of a von Neumann algebra M. It is a subalgebra of bounded operators on some Hilbert space $\mathfrak{H}$ satisfying the three conditions

(a) M is closed under the * operation.
(b) M contains the identity operator.
(c) M is closed under the topology of pointwise convergence.

It turns out that (b) is not very important but (a) and (c) are,. To see how they are related to physics we first observe that (c) is equivalent to M being *weakly closed* i.e closed for the topology given by the seminorms

$$|a| = |\langle a\xi, \eta\rangle|$$

as ξ and η vary through $\mathfrak{H}$. (This is a consequence of von Neumann's bicommutant theorem since the weak topology is weaker than the topology of pointwise convergence.)

2.2 Quantum mechanics, the gospel

Now let us remind the audience of the basic mathematical structure of quantum mechanics. (This was most clearly understood by von Neumann himself who wrote a major book on the subject.) What I say is contained in some form in any serious textbook on quantum mechanics.

The pure states of a quantum system are given by one dimensional subspaces of some complex Hilbert space $\mathfrak{H}$. The observables of the system, such as position, momentum or spin are given by self adjoint operators on the Hilbert space which may be bounded as in the case of spin components or unbounded as in the case of momentum. If a is an observable and ξ is a unit vector representing a pure state then

$$\langle a\xi, \xi\rangle$$

is the average value of repeated measurements of a if the system is prepared in the state given by ξ. If an observable, or commuting family of observables is given the spectral theorem allows one to express $\mathfrak{H}$ concretely as

$$L^2(X, \mathfrak{K}),$$

so that the given observables become multiplication operators (they are "diagonalised"). $\mathfrak{K}$ is a Hilbert space expressing other degrees of freedom such as isospin, whatever that is.

Observables a and b may only be simultaneously observed with arbitrary accuracy if they commute, $ab = ba$. Fourier analysis gives the Heisenberg uncertainty principle restricting the standard deviations of momentum and position observables, or others in similar duality. In operator language it is given by

$$PQ - QP = i\hbar,$$

where P and Q are momentum and position and $\hbar$ is a fundamental physical constant due to Planck.

Pure states may be "superposed" by adding unit vectors in the corresponding one dimensional subspaces. This is a source of great mystery and confusion as the resulting state depends on the phases of the representing vectors. Mixed states are a simpler idea. They are states that yield convex combinations of pure states on observables. They are given by trace class operators t of trace 1 so that the average of a in the mixed state t is $\mathrm{tr}(ta)$.

If two systems are given by the Hilbert spaces $\mathfrak{H}$ and $\mathfrak{K}$, the Hilbert space of the joint system is given by the tensor product Hilbert space $\mathfrak{H}\otimes\mathfrak{K}$. What is meant by "joint system" should no doubt be considered more carefully but suffice it to say that one system might consist of one electron, the other of another electron and the joint system would consist of two electrons. In fact there are fermions and bosons where only the antisymmetric and symmetric parts of the tensor product occur when combining two identical particles.

Symmetries of the system are given by unitary operators on $\mathfrak{H}$ which are determined up to a phase so that a group of symmetries is given by a projective representation of that group on $\mathfrak{H}$. Time evolution is a privileged symmetry group given by the unitary group representation

$$t \mapsto e^{itH}$$

for some observable H called the energy or Hamiltonian, a *positive* self-adjoint operator.

2.3 Back to von Neumann algebras

We see that von Neumann algebras are hovering in the background throughout the discussion. As soon as we want to think about subsystems we are smack in von Neumann algebras. For very basic reasons-the weak topology is precisely the topology defined by physical measurements, and the adjoint is necessary to even talk about observables! It is true that the product of two operators never occurred but their commutators did and Lie von Neumann algebras are just about the same as von Neumann algebras. If a subsystem is described by some algebra of observables, the requirement that it be weakly closed is morally the same as saying that classical mechanics should be built on the real numbers. I believe it is von Neumann who said that it makes no sense to say that the angular momentum of the moon is some rational number, but it does make sense to say it is between two numbers. Thus physics does make necessary the real numbers-and von Neumann algebras ...

In the gospel, the von Neumann algebra of the whole system was $B(\mathfrak{H})$. If we are talking about a subsystem we should be open to the possibilty that its von Neumann algebra "of observables" might not be of type I. Thus the term *state* evolved in von Neumann algebras so that $\varphi(a)$ is the average value of measurements of a if the system is in the state a. The subsystem may not have any pure states.

2.4 Quantum statistical mechanics

Statistical mechanics deals with systems with a vast number of particles by taking a limit in which that number becomes infinite. According to the gospel that means taking an infinite tensor product of Hilbert space. Von Neumann studied the infinite tensor product in [62]. It is a subtle business and for the most part a better approach is to take tensor products of the von Neumann algebras defining the individual subsystems then to use the state/GNS construction to recover the Hilbert space. There is an intimate connection between these states and the Tomita-Takesaki theory. If H is a Hamiltonian then the state

$$\varphi(x) = \frac{\mathrm{tr}(e^{-\beta H}x)}{\mathrm{tr}(e^{-\beta H})}$$

gives the average of x where states are occupied with probability decreasing exponentially with their energy. It is easy to see that

$$\varphi(xy) = \varphi(y\sigma_{i\beta}(x))$$

where σ_t is the time evolution for H. We say that φ is β-KMS (Kubo-Martin-Schwinger) for σ_t. In Tomita Takesaki theory we have exactly the relation

$$\varphi(xy) = \varphi(y\sigma_i^\varphi(x))$$

whenever the modular group can be extended to imaginary t on x.

Note that this property is what morally replaces the tracial property of a state. For arbitrary elements x and y to which σ_i^φ may not be applied there is a characterisation in terms of bounded analytic functions on a strip.

For more on the connections between von Neumann algebras and quantum statistical mechanics, we refer the reader to [2].

2.5 The Haag-Kastler axioms and DHR superselection sectors

To give a model-independent and rigorous framework for quantum field theory Haag and Kastler proposed axioms ([27]) that derive from the gospel and special relativity. There will be a Hilbert space $\mathcal{H}$ for the entire universe which is a bit out of sight for ordinary mortals so one cuts things down to size by considering, for suitable bounded regions $\mathcal{O}$ of spacetime, the von Neumann algebra $\mathfrak{A}(\mathcal{O})$ of (i.e. generated by) observables *localised* in $\mathcal{O}$.

The assignment $\mathcal{O} \to \mathfrak{A}(\mathcal{O})$ satisfies certain axioms derived from the gospel. Here are the most important ones:

(1) (Isotony) If $\mathcal{O}_1 \subseteq \mathcal{O}_2$ then

$$\mathfrak{A}(\mathcal{O}_1) \subseteq \mathfrak{A}(\mathcal{O}_2)$$

(2) (Locality/causality) If $\mathcal{O}_1$ and $\mathcal{O}_2$ are space-like separated, meaning that no light ray connects them, then

$$\mathfrak{A}(\mathcal{O}_1) \subseteq \mathfrak{A}(\mathcal{O}_2)'$$

(3) (Poincaré Invariance) There is a unitary (projective) representation $g \mapsto u_g$ of the Poincaré group G on the Hilbert space with the property:

$$\mathfrak{A}(g(\mathcal{O})) = u_g\mathfrak{A}(\mathcal{O})u_g^{-1} \text{ for } g \in G$$

(4) (Positive Energy) The spectra of the generators of the spacetime translation subgroups of G are positive.

(5) (Vacuum) There is a vector, the vacuum $\Omega \in \mathcal{H}$ such that $\mathfrak{A}(\mathcal{O})\Omega$ is dense in the Hilbert space for all $\mathcal{O}$ and Ω is fixed by the Poincaré group.

These properties are very general and one may ask if the theory is capable of producing numbers. We have seen numbers in von Neumann algebras all ultimately originating with von Neumann's continuous dimension. With more conditions it follows from the HK axioms that the local algebras $\mathfrak{A}(\mathcal{O})$ are all factors. One may then ask of what type. The answer appears to be that $\mathfrak{A}(\mathcal{O})$ is always a (hyperfinite) type III_1 factor ([26]). There remains the modular group associated with the vacuum and this should interact with the observables of the theory. But another source of numbers was the index for subfactors and this is what I want to talk about for the rest of the lecture.

Doplicher Haag and Roberts in [17] had the fascinating idea that the entire universe may actually have a disequilibrium in, say, charge, but that one could not know that from local observations (what difference would an extra electron behind the moon make?) see also [26]. Thus they proposed "superselection sectors" which are representations π of the family of $\mathfrak{A}(\mathcal{O})$'s as abstract von Neumann algebras on some Hilbert space $\mathfrak{H}_\pi$ with all the axioms but the vacuum one. And these π should be indistinguishable from each other in the sense that if π and ψ are two superselection sectors and $\mathcal{O}$ a region then there should be a unitary $u : \mathfrak{H}_\pi \to \mathfrak{H}_\psi$ intertwining the actions of $\mathfrak{A}(\mathcal{O})$. In particular they are all equivalent to the vacuum representation and should thus be describable by the vacuum sector with additional data. Now Araki had proved that in the vacuum sector there is a Tomita-Takesaki duality for appropriate space-time regions:

$$\mathfrak{A}(\mathcal{O})' = \mathfrak{A}(\mathcal{O}')$$

where $\mathcal{O}'$ is the causal complement of $\mathcal{O}$. It follows immediately that a unitary intertwining the action of $\mathfrak{A}(\mathcal{O})$ in the vacuum sector with that in another sector defines an *endomorphism* ρ of $\mathfrak{A}(\mathcal{O})$ (by conjugation). These endomorphisms are studied in detail in [17] and exhibit a truly rich structure relating notions of gauge group, statistics, symmetric group representations and dimension.

For us the point is that an endomorphism need not be surjective and so one has a subfactor alternately described as $\rho(\mathfrak{A}(\mathcal{O})) \subseteq \mathfrak{A}(\mathcal{O})$ in the vacuum sector or $\mathfrak{A}(\mathcal{O}') \subseteq \mathfrak{A}(\mathcal{O})'$ in the other sector. So we have available a *numerical parameter*, namely the index of a subfactor. DHR showed that indeed, in the free theory, the endomorphisms do define proper subfactors. They worked in space-time dimension bigger than two so the subfactors they obtained had integer index. With the advent of the $4\cos^2 \pi/n$ subfactors and the braid group representations it was predicted by Fredenhagen Rehren and Schroer, and Longo, that in 2 spacetime dimensions one would get subfactors of these indices ([22],[38])

But from the subfactor point of view there is a difference of emphasis. Algebraic quantum field theory emphasizes the model-free aspects and assumes that physics will provide the examples, whereas subfactors are more traditionally rooted in mathematics and would like to see a careful construction of the subfactors in question. In higher dimensions this is not too hard because of group representation theory but the non-integer index subfactors do not come from locally compact groups. So the rigorous proof of the existence of these factors was required. This was accomplished by Wassermann in [63] in the context of loop group representations and we end with a brief description of this theory.

2.6 A word about second quantisation

Ordinary quantum mechanics with positivity of the energy operator is not compatible with relativity but Dirac found a way to turn the negative energy part of the spectrum of a Hamiltonian into positive spectrum by the concept of a sea of antiparticles, the annihilation of one of which creates a positive energy state. One is forced to deal with systems of arbitrarily many particles so if one starts with a one-particle system on a Hilbert space $\mathfrak{H}$ the Hilbert space of the "second quantised" system will be the so-called Fock space

$$\mathfrak{F}(\mathfrak{H}) = \bigoplus_{n=0}^{\infty} \otimes^n \mathfrak{H}$$

In fact of course the tensor product to be used will be either the symmetric or antisymmetric one for bosons or fermions respectively. For each vector $\xi \in \mathfrak{H}$ there is a *creation* operator $a(\xi)$ which creates a particle in state ξ:

$$a(\xi)(\otimes_{i=1}^n \eta_i) = \xi \otimes (\otimes_{i=1}^n \eta_i)$$

and the corresponding annihilation operator $a(\xi)^*$. These operators satisy the CCR (bosons)

$$[a(\xi), a(\eta)^*] = \langle \xi, \eta \rangle \text{ and } [a(\xi), a(\eta)] = 0$$

and the CAR (fermions)

$$\{a(\xi), a(\eta)^*\} = \langle \xi, \eta \rangle \text{ and } \{a(\xi), a(\eta)\} = 0$$

(where $[,]$ is the commutator and $\{,\}$ the anticommutator.)

The spectrum reversing trick is to split $\mathfrak{H}$ up into two orthogonal $\mathfrak{H}_+$ and $\mathfrak{H}_-$. (These will typically be the positive and negative spectral subspaces respectively for some selfadjoint operator h, the "pre-quantized" hamiltonian of the system.) We then consider

$$\mathfrak{F}((\mathfrak{H}_+) \oplus (\text{dual}(\mathfrak{H}_-)))$$

The CCR and CAR still hold on this new Fock space provided we replace $a(\xi)$ by $a(\xi)^*$ for $\xi \in \mathfrak{H}_-$ as required by linearity to obtain a new representation $A(\xi)$ of the CCR/CAR.

Of course this process is not at all innocent. The new representation of the CAR/CCR is not at all equivalent to the original one so symmetries of the one-particle quantum system do not necessarily pass to the second quantised one. However if a unitary u is not too far from the identity the map $\xi \mapsto A(u\xi)$ will give an equivalent representation of the CCR/CAR. All these representations being irreducible we obtain the following result (for a proof see [52]):

Theorem 2.1. *Let p be the projection onto $\mathfrak{H}_+$. If a unitary u on $\mathfrak{H}$ is such that $p - upu^*$ is a Hilbert-Schmidt operator, then there is a unitary U on $\mathfrak{F}((\mathfrak{H}_+) \oplus (\text{dual}(\mathfrak{H}_-)))$ with*

$$U(A(\xi))U^* = A(u(\xi)) \textit{ for all } \xi.$$

The U is unique up to a scalar.

Thus given a projection $p \in B(\mathfrak{H})$ one defines

$$U_{\text{res}}(\mathfrak{H})$$

to be the group of all unitaries u on $\mathfrak{H}$ with $u - upu^*$ a Hilbert-Schmidt operator.

The theorem says there is a projective unitary representation of $U_{\text{res}}(\mathfrak{H})$ on the Fock space of $p\mathfrak{H} \oplus \text{dual}((1-p)\mathfrak{H})$.

If one applies the previous construction to the situation where $\mathfrak{H}_-$ is the negative spectral subspace for a self-adjoint operator h one obtains a time evoloution on Fock space since e^{ith} actually commutes with p and is thus canonically implemented on Fock space by e^{itH} for some positive self adjoint H which is the second quantised Hamiltonian.

The gospel of quantum theory requires that time evolution be a one-parameter unitary group u_t with the state at any time obtained by applying u_t to the state at time zero. So if the Hamiltonian is a differential operator in space variables, the differential equations for the evolution of wave functions (e.g. the Schrodinger equation) must be first order in t. Unfortunately the Klein Gordon equation which is the relativistically invariant version of the Schrodinger equation, is second order in t. Dirac hit on the idea of factoring it as a product of first order operators, necessarily first order in both space and time, and taking the first order operator as the evolution equation. This operator necessarily has spectrum unbounded in both the positive and negative directions. So second quantization as above was the only way out.

In higher dimensions the theory is complicated by the necessity of multicomponent wave functions to factorise the Klein Gordon equation but in 1 dimension things are very simple, especially if we compactify space to obtain the circle S^1. Then in the massless free situation $\mathfrak{H}$ is just $L^2(S^1)$ and the relativistically invariant Hamiltonian is just $h = i\frac{d}{d\theta}$. The decompostion of $\mathfrak{H}$ into positive and negative spectral subspaces for h is just the decomposition of $L^2(S^1)$ into Hardy space $H^2(S^1)$ (boundary values of holomorphic functions in the disc) and its orthogonal complement.

2.7 Loop groups, subfactors and a 1-dimensional algebraic QFT

Let the Fock space constructed from $L^2(S^1)$ and Hardy space be called $\mathcal{F}$.

From a mathematical point of view we are interested in unitaries on $L^2(S^1)$ which leave holomorphic functions invariant up to a Hilbert-Schmidt perturbation. Each such unitary will give a unitary on $\mathcal{F}$ defined up to a constant by the commutation relations with the fermionic annihilation and creation operators.

There are two natural candidates.

(1) Multiplication operators M_f by measurable functions $f : S^1 \to \mathbb{C}$;

and

(2) Measurable transformations of S^1 which preserve the measure class of Lebesgue measure. (These can be made into unitaries on $L^2(S^1)$ by multiplying by a Radon-Nikodym derivative.)

Let us consider both in turn.

Using Fourier series it is easy to calculate the Hilbert-Schmidt norm of the commutator of M_f with projection onto H^2. One runs into the Sobolev 1/2 norm of f, $\sum_{\in\mathbb{Z}} n|a_n|^2$ where $|a_n|$ are the Fourier coefficients of f. So if f is C^∞ one is in luck and obtains a projective unitary representation of the loop group $C^\infty(S^1, S^1)$ by Theorem 6.1.

The Mobius group of linear fractional transformations preserving the unit disc, $SU(1,1)$, acting on S^1 obviously preserves H^2 so is canonically implemented on $\mathcal{F}$ (there is a holomorphic square root of the Radon-Nikodym derivative). But it is also true that arbitrary diffeomorphisms of S^1 commute up to a Hilbert-Schmidt with projection onto H^2 (see [53]).

Altogether we are equipped with a projective unitary represenation of the semidirect product $C^\infty(S^1, S^1) \rtimes \mathrm{Diff}(S^1)$ on $\mathcal{F}$. These considerations extend directly to the Hilbert space $L^(S^1)\otimes\mathbb{C}^n$ with the loop group now being $C^\infty(S^1, U(n))$.

It is not a priori obvious that this projective unitary representation is not an ordinary unitary representation, i.e. the second cohomology class of the representation might be zero. This question can be resolved by differentiation, for the Lie algebra of the loop group will be some version of the Lie algebra of functions from S^1 to the Lie algebra u_n. Restricting to $SU(n)$ one obtains after this differentiation exactly the affine Kac-Moody Lie algebras where the cohomology is well understood ([35]) and the representation is indeed properly projective. One may obtain more projective representations with"large" cohomology classes simply by taking tensor powers of the one we have just constructed.

For the diffeomorphism group, differentiation gives us the Virasoro algebra where the cohomology is also well understood (see[53]).

In this way one may actually obtain all the so-called positive energy representations of $LSU(n) = C^\infty(S^1, U(n))$ though the irreducible ones will occur in general with infinite multiplicity. Each irreducible representation is determined by its restriction to constant loops which is an irreducible representation of $SU(n)$. From the point of view of subfactors the most suggestive thing is that there is a restriction on the irreducibles of $SU(n)$ that may occur in this way if we fix the cohomology class of the central extension. For $SU(2)$ the cohomology class is given by a positive integer, the level ℓ, and the only allowed irreducible representations of $SU(2)$ that can occur are those whose dimension is less than or equal to the level.

Now we can define our algebraic quantum field theory. Space-time is the circle S^1, the Poincaré group is the group of Mobius transformations fixing the unit disk $SU(1,1)$, and the sets $\mathcal{O}$ are just intervals. The causal complement I' of the closed interval I is just the interval which is the closure of its complement. And the algebra $\mathfrak{A}(I)$ is just the von Neumann algebra generated by the loops supported in I. Given the level of the theory the different superselection sectors are the different representations of the loop group with that level and the vacuum sector is the one given by the trivial representation of the compact group. One can check the Haag Kastler axioms and thus conclude that the $\mathfrak{A}(I)$ are type III_1 factors. The modular group in the vacuum sector is given by the one parameter group of Mobius transformations having two fixed points at the ends of I. A strong version of local equivalence is true, namely the loop group representations restricted to the subgroup of loops supported in I are equivalent for all the representations at a fixed level.

In [63], Wassermann executes all of the above steps in detail for $SU(n)$ and shows that for $LSU(2)$ at level ℓ and spin j (dimension $= 2j+1$)

$$[\mathfrak{A}(I)' : \mathfrak{A}(I')] = \frac{\sin^2 \frac{(2j+1)\pi}{\ell+2}}{\sin^2 \frac{\pi}{\ell+2}}$$

This gives a complete construction of subfactors of irrational index completely independent of the construction of [30].

3 Lecture 3: von Neumann algebras and topology

3.1 Recall examples of II_1 factors

If G is a discrete group, the group von Neumann algebra LG is a II_1 factor (on the Hilbert space $\ell^2 G$) iff all conjugacy classes are infinite except of course that of the identity. More generally if N is a von Neumann algebra with a faithful trace tr (i.e. the scalar product $\langle x, y\rangle = \mathrm{tr}(y^* x)$ is positive definite), and G acts on N by automorphisms, then the crossed product $N \rtimes G$ is often a factor. Call an automorphism α "free" when $\alpha(x)y = yx, \forall x$ implies $y = 0$. If α arises from a measure preserving transformation of a probability space, freeness means that the set of fixed points of the transformation has measure zero. Then it is not hard to show that $N \rtimes G$ is a factor if each non-identity element of G acts freely and the action is ergodic. Thus ergodic theory produces many examples of factors. If N is a factor then free is the same as outer and $N \rtimes G$ is a factor if its action is free.

3.2 Group based applications

There are 3 main kinds of application of von Neumann algebras to topology. (If we were talking about C*-algebras, which are *-algebras of bounded operators on Hilbert space closed in the *norm* toopology, there would be many more and we would consider Connes' huge non-commutative geometry program [9].) The first kind of application is the most obvious-consideration of $L\pi_1(X)$.

We will say very little about this but we should mention the L^2 index theory of Atiyah which concerns differential operators on an (infinite) Galois covering space $\tilde{X}$ of the compact manifold X with Galois group G. An elliptic differential operator on X lifts to one on $\tilde{X}$. The kernel and cokernel of the operator on L^2 will no longer be finite dimensional but the projections onto those subspaces will be in the commutant of the action of G on the space of L^2 sections. This commutant will be a II_∞ factor if G is icc and the kernel and cokernel projections will have finite traces. The difference between these two traces is Atiyah's L^2 index. From its definition it is a real number but there are no known examples where it is irrational.

In fact given a triangulation of X, there is one of $\tilde{X}$ on which G acts. One may consider the boundary and coboundary operators on ℓ^2 chains and cochains. It will commute with the action of G and one may obtain finer invariants. I refer to the work of Lueck for all the above [40].

3.3 Foliations-Connes' non-commutative integration

If G is a compact Lie group acting on a manifold there may be orbits of different types so that the orbit space has singularities (e.g. fixed points). But the orbit space has only mild singularities which may not even be seen by topolgy, certainly not by measure theory. On the other hand if the group is noncompact all hell can break loose. Even in the case of $\mathbb{R}$: if one takes the flow of a linear vector field on $\mathbb{T}^2$ of irrational slope, then every orbit is dense so it is impossible to make sense of the quotient space. The crossed product serves as a non-commutative replacement for the quotient space. Though this should all be done in the context of C*-algebras, Connes began his investigation of very singular spaces with von Neumann algebras where he called his theory non-commutative integration [7].

The singular quotient space may arise in the absence of any group action. Connes' main examples occurred in the theory of foliations. A foliation of a manifold X is a decomposition into "leaves", which are manifolds of some fixed lower dimension p, in such a way that every point of X has a neighbourhood diffeomorphic to $\mathbb{R}^p \times \mathbb{R}^q$ so that the sets $\mathbb{R}^p \times \{x\}$ for each $x \in \mathbb{R}^q$ are all parts of leaves of the foliation. Thus the local structure of a foliation is trivial.(p is called the dimension of the foliation and q the codimension). But the point is that a leaf, once it has left one of the trivially foliated neighbourhoods, may wander all over the manifold and come back to a *different* leaf in the trivially foliated chart.

The simplest foliations are of dimension 1. They are basically the orbits of the flows of vector fields, for instance the irrational flow on the torus which we have considered above. We see from this example that the space of leaves may be very singular since any neighbourhood of a single point on one leaf will contain points on every other leaf. Foliations may be kinder and gentler of course. Everyone's favourite is the (codimension 1) Reeb foliation of the three sphere which begins with a simple foliation of the solid torus by non-compact leaves which wrap round and round inside the torus and come arbitrarily close to the boundary. This is called a Reeb component. Viewing the three-sphere as the union of two solid tori one obtains a fascinating picture by placing one Reeb component in one of the solid tori and another in the other.

Connes constructed, for any smooth foliation, a canonical von Neumann algebra. For the Reeb foliation it is not at all a factor and the only factors in its central decomposition are type I. But for the irrational rotation of the torus one obtains a type II_∞ factor. Connes' construction is quite general and uses the concept of a "measured groupoid" (the Atiyah L^2 theory is a very simple example where one takes the fundamental groupoid.). The groupoid in question for a foliation is the space of smooth paths along leaves where two paths are declared to be the same if the holonomies of those paths are the same.

Holonomy is one of the main concepts in the theory of foliations. Suppose we are given a codimension q foliation of a manifold X. Given a point x on a leaf and a path on the leaf to another point y, one may choose an open foliating chart containing x, and cover the rest of the path with a finite number of foliating charts. If we choose an open interval I in $\mathbb{R}^q$ and consider the subset $(0, I)$ of X with $(0,0) = x$, then provided I is small enough we can uniquely follow points in I along paths parallel to the given one but staying in their leaves, until we get to y. Thus we obtain a mapping from the groupoid of such paths to the germs of diffeomorphisms of $\mathbb{R}^q$. This map only depends on the path up to homotopy (in the leaf). In particular holonomy determines a homomorphism from $\pi_1(leaf)$ (based at x) to the germs of diffeomorphisms of $\mathbb{R}^q$ preserving the origin, thus yielding a "holonomy" covering space of each leaf. The holonomy of the irrational slope lines on the torus is trivial but the holonomy of the Reeb foliation is more fun, at least for the compact leaf. We leave the reader to calculate it and conclude that the Reeb foliation is not analytic.

Connes' von Neumann algebra construction takes for each leaf ℓ the manifold F_ℓ which is the holonomy covering space of ℓ and then the Hilbert space L^2_ℓ of half-densities on F_ℓ. Using Winkelnkemper's construction of the holomy groupoid which desingularises the leaf space this gives a perfectly well defined field of Hilbert spaces in the sense of direct integral theory and thus a field of type I factors $B(F_\ell)$. The von Neumann algebra M_X is the algebra of holonomy invariant sections of the field of type I factors so an element is "one operator per leaf". But the main requirement is that the operators be transversely measurable. It is possible to construct such operators-e.g. take a smooth vector field tangent to the leaves and flow along it to get a unitary. But to get an element of the centre one could be in trouble since that would mean choosing a scalar for each leaf which is transversally measurable. If all the leaves are dense as in our irrational slope example then this will not be possible (thinking continous rather than measurable for convenience) unless the scalar is the same for all leaves.

One sees that factors appear quite naturally.

The "integration" part of the theory comes when one considers Lebesgue measure on the transversals. If this measure is invariant under holonomy we say we have a measured foliation. Then the transverse measure gives traces on corners of the Connes von Neumann algebra by considering foliating charts. (In the case of the irrational slope foliation one may chose a rather global transversal and the corner becomes the crossed product algebra $L^\infty(S^1) \rtimes \mathbb{Z}$, $\mathbb{Z}$ acting by powers of the irrational rotation.) Thus the factors in this case will be of type II_∞. The real gift here is that geometrically defined operators like vector fields tangential to the leaves will give elements in the von Neumann algebra whose trace may be evaluated. Using this idea Connes was able to give a definition of the Betti numbers of a measured foliation and prove a Gauss Bonnet theorem. He proved a result asserting that vector fields tangent to the leaves of certain codimension one foliations of three manifolds must have lots of zeros. A theorem whose statement does not use von Neumann algebras at all.

I have of course only described the tip of the tip of the iceberg. For more see [7] and [9].

3.4 Knot theory and TQFT

Recall the theorem on the index of a subfactor [30]:

Theorem 3.1. *If* $[M:N]<4$ *it is* $4\cos^2\pi/n$ *for* $n=3,4,5,6,\ldots$ *and all these values occur. If the index is* $\geqslant 4$ *any real number is possible.*

One may prove this theorem by iterating a basic construction: let $\langle M, e_N\rangle$ be the von Neumann algebra on $L^2(M,\mathrm{tr})$ generated by M (left multiplication operators) and e_N, the orthogonal projection onto the subspace $L^2(N,\mathrm{tr})$.Then $[M:N]<\infty$ is equivalent to $<M,e_N>$ being a II_1 factor and $[<M,e_N>:M]=[M:N]$. Set $N=M_0$ and $M=M_1$ and define the tower $M_i\subseteq M_{i+1}$ by

$$M_{i+1}=<M_i,e_{M_{i-1}}>$$

Now set $e_i=e_{M_{i-1}}$ to obtain a sequence of projections satisfying

$$e_i^2=e_i=e_i^*$$
$$e_ie_j=e_je_i \text{ for } |i-j|\geqslant 2$$
$$e_ie_{i\pm1}e_i=\tau e_i \text{ with } \tau=[M:N]^{-1}$$

and a trace tr on the algebra generated by the $e_i's$ completely defined by the relation

$$\mathrm{tr}(xe_{i+1})=\tau\mathrm{tr}(x) \text{ if } x \text{ is a word on } e_1,e_2,\ldots,e_i$$

The "Jones-Wenzl" idempotents f_n are defined by $1-f_n$ being the identity of the algebra generated by $e_1,e_2,\cdots,e_n$. In language that von Neumann would have liked,

$$f_n=(e_1\vee e_2\vee\ldots\vee e_n)^\perp$$

Obviously it is an orthogonal projection with $f_ne_i=0$ for $1\leqslant i\leqslant n$. One can show using the above properties that

$$\text{If } \mathrm{tr}(f_n)\neq 0 \text{ then } \mathrm{tr}(f_{n+1})=\mathrm{tr}(f_n)-\tau\mathrm{tr}(f_{n-1})$$

For $1/4<\tau<1$ this means that $\mathrm{tr}(f_n)$ will eventually be negative, which is not so good for the trace of a projection, unless $\mathrm{tr}(f_k)$ is zero for some k. This happens only if τ is one of the allowed values. (For an inductive formula for f_n see [64].)

All we have used in the above proof is the e_i's. But there is much more to be had. For each i the algebra $N'\cap M_i$ is a finite dimensional von Neumann algebra and the first thing we saw in these lectures was that such algebras are given by some integers. Moreover each e_k for $1\leqslant k\leqslant i$ is canonically in $N'\cap M_i$ so these algebras actually have a lot of structure.

The $N'\cap M_i$ form what is called the standard invariant of the subfactor $N\subseteq M$.

As algebras the $N'\cap M_i$ can be realised as linear combinations of based loops on a pointed bipartite graph known as the *principal graph* of the subfactor. Subfactors of index <4 have one of the Coxeter graphs A_n, D_{2n}, E_6 or E_8 as their principal graph and these graphs all occur.

To say that the principal graph is A_n is the same as saying that $N' \cap M_k$ is generated by the e_i's and the identity for all k

Analysis of the above linear difference equation is simplest if one makes the change of variables $\tau^{-1} = 2 + t + t^{-1}$. Under this change of variables $4\cos^2 \pi/n$ corresponds to

$$t = e^{2\pi i/n}.$$

The n-string braid group has a presentation on generators $\sigma_1, \sigma_2, \cdots \sigma_{n-1}$ with the relations

$$\sigma_i \sigma_j = \sigma_j \sigma_i \text{ if } |i - j| \geqslant 2 \text{ and}$$

$$\sigma_i \sigma_{i+1} \sigma_i = \sigma_{i+1} \sigma_i \sigma_{i+1}$$

so sending σ_i to $te_i - (1 - e_i)$ gives a linear representation β_t of the braid group, unitary when $t = e^{2\pi i/n}$.

The faithfulness of β_t for any real $t \neq 0, 1$ remains open, especially for the braid group B_4.

Braids can be closed by tying the top to the bottom thus:

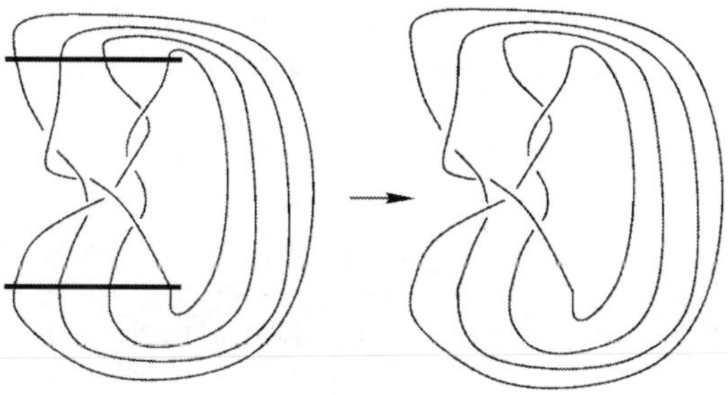

Denote by $\hat{\alpha}$ the closure of the braid α. Taking the trace of $\beta_t(\alpha)$ and renormalising by $\mathrm{tr}(\sigma_1)^{n-1}$ gives a polynomial invariant of knots and links in $\mathbb{R}^3$ known as the Jones polynomial. If L is a link this polynomial is written

$$V_L(t).$$

For knots it is a Laurent polynomial and for links it is $\sqrt{t}$ times a Laurent polynomial. It can be defined and calculated without reference to braids by the Conway-like scheme (skein relation):

(i) V_L is an invariant of knots.
(ii) $V_{unknot} = 1$
(iii) If L_+, L_- and L_0 are three diagrams of links differing only near one crossing as below:

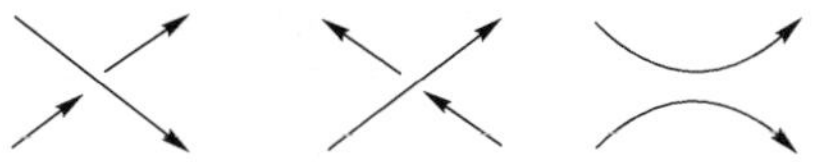

L_+ L_- L_0

then

$$\frac{1}{t}V_{L_+} - tV_{L_-} = (\sqrt{t} - \frac{1}{\sqrt{t}})V_{L_0}$$

For the trefoil knot $K =$ 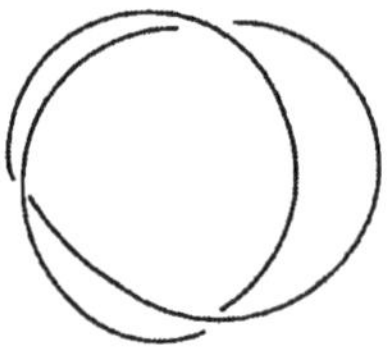

$$V_K(t) = t + t^3 - t^4$$

It is an open question whether there is a knot different from the unknot whose polynomial is identically equal to 1, though there are links different from the unkot with trivial polynomial ([58]).

This polynomial was interpreted in terms of statistical mechanical models in [32] but this led only to alot more polynomials using the machinery of quantum groups. Witten gave an interpretation in terms of Feynmann path integrals in [65]. Witten's version is only a posteriori a polynomial. Quantisation criteria give him the values $V_K(e^{2\pi i/n})$, exactly the values coming from the discrete series of subfactors. Witten's theory, though not rigorous, had the huge advantage of showing that the invariant can actually be defined for links in an arbitrary 3-manifold. This includes the empty link so one obtains an invariant of 3-manifolds. The method of calculation of these invariants is Topological Quantum Field Theory which describes how the invariant behaves under cutting and pasting. Various proofs were given making Witten's invariants rigourous. The most transparent rely heavily on the properties of the Jones-Wenzl projections we have described. See [55],[1].

3.5 Planar algebras

Kauffman gave an even simpler skein theory to calculate the polynomial $V_L(t)$. Start with *unoriented* links L and suppose that three are related as below:

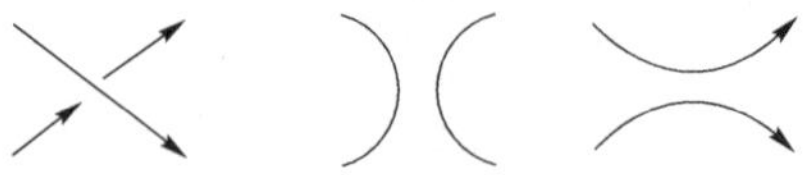

L_+ L_0 L_∞

then $< L_+ >= A < L_0 > +A^{-1} < L_\infty >$ where $< L >$ is a Laurent polynomial in A called the Kauffman bracket. Renormalising the Kauffman bracket when an orientation is imposed on L gives $V_L(A^4)$-see [36].

Along with the Kauffman bracket goes a theory of links with boundary called "tangles" by Conway in [14]. One uses the Kauffman skein relation to reduce all tangles to linear combinations of tangles with no crossings which have come to be called Temperley Lieb diagrams in this context though they are also known as non-crossing pairings. By splitting the boundary points into top and bottom these diagrams can be made into an algebra by multiplying just like braids. If we define E_i by the following diagram:

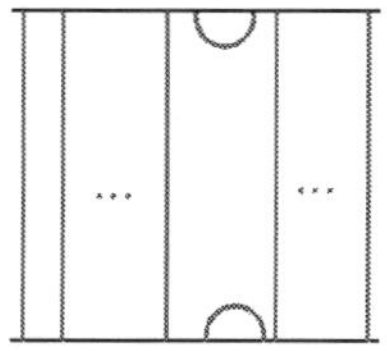

with the cups and caps between the ith and $(i+1)^{th}$ strings, then renormalising by $e_i = \delta^{-1}E_i$ then we get the same relations as in subfactor theory. (Here δ is the parameter $-A^2 - A^{-2}$, related to the index of the subfactor by $\delta^2 = [M : N]$. A closed curve in the middle of a Kauffman diagram can be removed by multiplying the diagram by δ.)

We thus get a diagrammatic basis and diagrammatic description of the algebra generated by the e_i's. One of the motivations of the study of planar algebras was to see to what extent Kauffman's diagrammatics could be extened to the entire standard invariant $\bigcup_k (N' \cap M_k)$.

Planar tangles are the essential structural ingredient of planar algebras. A planar tangle is a subset of $\mathbb{C}$ consisiting of some internal discs inside a larger external disc, all the discs being connected by non-crossing strings. Here is an example:

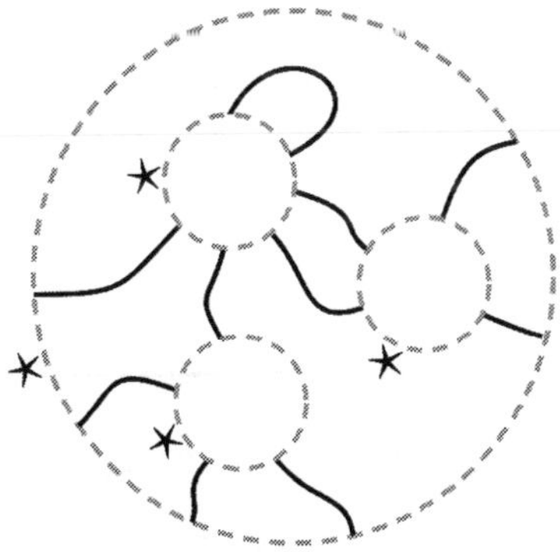

The Kauffman diagrams are exactly the planar tangles with no input discs, only strings. The fact that the $N' \cap M_i$ contain the e_i's suggests that there might be ways to manipulate *all* the elements of the standard invariant using tangles. This is indeed the case.

Tangles may be glued in the obvious way and we say that a *planar algebra* is a graded vector space P_n where the grading n corresponds to the number of boundary points of the discs in the tangle. A given tangle accepts inputs in each of its internal discs, from the vector space of the appropriate grading, and produces an element of P_n where n is the number of boundary points (intersections of discs with strings) of the external disc. These insertions of elements in planar tangles are multilinear and have to be compatible with the gluing. We say that tangles *act* on a planar algebra.

Theorem 3.2. *There is a canonical planar algebra structure on* $P_n = N' \cap M_{n+1}$ *which is entirely determined by the subfactor* $N \subseteq M$.

Note that $\dim P_0 = 1$ since N is a factor. Whenever $\dim P_0 = 1$ is true we may equip all the P_n's with bilinear forms according to the rule:

$[a, b] =$ 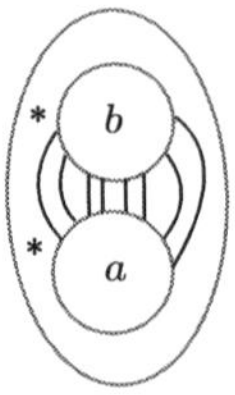

Subfactor planar algebras have *-structure compatible with taking the complex conjugate of a tangle. So the bilinear form can be made into a sesquilinear one which is in fact just the one coming from the GNS construction using the ambient trace. It is remarkable that any *-planar algebra with $\dim P_0 = 1$, finite dimensional P_n and for which the above sesquilinear form is positive definite, is in fact a subfactor planar algebra. This follows from a result of Popa in [47] but we will sketch a more direct proof in our next lecture on random matrices.

This gives a way to construct and analyse subfactors using the topology of the plane.

4 Lecture 4: von Neumann algebras and random matrices

4.1 Full Fock space

We discussed fermionic and bosonic Fock spaces in the second lecture. Let us now consider the "full" Fock space. Let $\mathfrak{H}$ be a real Hilbert space and define

$$\mathcal{F}(\mathfrak{H}) = \oplus_{n=0}^{\infty} \otimes^n \mathfrak{H}$$

which we complexify to make into a complex Hilbert space. The case $\dim \mathfrak{H} < \infty$ is already highly interesting and to simplify statements let us assume $\dim \mathfrak{H} = k < \infty$

The elements f of $\mathfrak{H}$ give left and right creation operators:

$$\ell_f(\xi_1 \otimes \xi_2 \otimes \cdots \otimes \xi_n) = f \otimes \xi_1 \otimes \xi_2 \otimes \cdots \otimes \xi_n$$

and

$$r_f(\xi_1 \otimes \xi_2 \otimes \cdots \otimes \xi_n) = \xi_1 \otimes \xi_2 \otimes \cdots \otimes \xi_n \otimes f$$

Since we have not quotiented by actions of the symmetric group the commutation relations between the ℓ's become rather simple:

$$\ell_f^* \ell_g = \langle f, g \rangle id$$

Thus in particular if v_i is an orthonomal basis then if we define $\mathcal{S}_i = \ell_{v_i}$ then

$$\mathcal{S}_i^* \mathcal{S}_j = \delta_{i,j}$$

The $\mathcal{S}_i \mathcal{S}_i^*$ are mutually orthogonal projections. Note that they all kill the vacuum vector so they cannot add up to 1, but they do add up to the projection onto the orthogonal complement of the vacuum vector.

If we call S_i the image of $\mathcal{S}_i$ in the quotient algebra of the *-algebra generated by the $\mathcal{S}_i$ by imposing the relation $\sum_{i=1}^k \mathcal{S}_i \mathcal{S}_i^* = 1$ we obtain the Cuntz algebra O_n ([15]) which has been highly influential in the theory of C*-algebras.

Left and right multiplication commute but the same is not true of ℓ_f and r_f^*. In fact it is not hard to show that the ℓ_f act irreducibly on $\mathcal{F}(\mathfrak{H})$ so there is no real von Neumann algebraic interest in looking at the algebra they generate. The interest occurs when one looks at the real parts of the ℓ_f. This was done first by Voiculescu - see [60].

Theorem 4.1. *Let $F(\mathfrak{H})$ be the von Neumann algebra on $\mathcal{F}(\mathfrak{H})$ generated by the elements $\ell_f + \ell_f^*$. It is a II_1 factor whose trace is defined by the vacuum vector. It is isomorphic to LF_k.*

Voiculescu envisioned $\mathfrak{H} \mapsto F(\mathfrak{H})$ as a functor on the same footing as the fermionic and bosonic second quantisation functors, with the direct sum of Hilbert spaces being sent to the free product of von Neumann algebras. By itself this theory did not lead to any structural revelations about LF_k but in combination with Voiculescu's connection with random matrices it has proved a powerful tool.

4.2 Wigner's semicircle law

Let A be an $N \times N$ random hermitian matrix. That is to say the entries $A_{i,j}$ will be complex Gaussian random variables with mean zero, mutually indendent apart from the constraints $A_{i,j} = \overline{A_{j,i}}$ which mean that the $A_{i,i}$ are real. The variances $E(A_{i,j}A_{j,i})$ of the off diagonal elements will all be the same, equal to v, as will the variances of the diagonal elements. We will adjust those variances later to make certain limits exist.

Recall Wick's theorem which says that if X_i are mutually Gaussian centred random variables then $E(X_1 X_2 \cdots X_n)$ is the sum over all pairings $\{i, \pi(i)\}$ of the set $\{1, 2, \ldots, n\}$ of the product of the covariances $E(X_i X_{\pi(i)})$. (A pairing π is just a fixed-point free involution of a set dividing it up into orbits $\{i, \pi(i)\}$.)

$$E(X_1 X_2 \cdots X_n) = \sum_{\pi} \prod_{\{i,\pi(i)\}} E(X_i X_{\pi(i)})$$

In particular this expected value is zero if n is odd. To study the distribution of the eigenvalues of A we are interested in its *moments*, i.e. the expected values of the traces of its powers. (This is because the trace of the nth power of a matrix is the sum of the nth powers of the eigenvalues.) Write explicitly

$$E(\mathrm{tr}(A^n)) = \sum_{i_1, i_2, \cdots, i_n} E(A_{i_1,i_2} A_{i_2,i_3} \cdots A_{i_n,i_1})$$

By Wick we obtain a sum over pairings of $\{(i_1,i_2),(i_2,i_3),\ldots,(i_n,i_1)\}$. We will depicted these pairings by arranging $\{(i_1,i_2),(i_2,i_3),\ldots,(i_n,i_1)\}$ on a circle in a way we hope is self-explanatory:

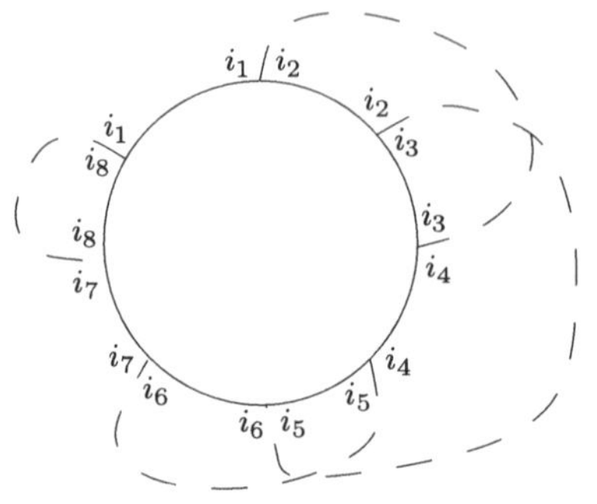

By the independence of the $A_{i,j}$ the pairing we have depicted will only contribute something non-zero if $i_2 = i_3 = i_5 = i_6$ and $i_1 = i_4 = 1_7$ and then for fixed i_8 it will contribute a factor v^4. Now thicken the dotted lines joining the pairs of indices to bands glued to a central disc so as to form a surface with boundary. We see that pairings contributing non-zero terms to the trace can be arranged as below on a surface with boundary in such a way that the indices are actually constant on the boundary components:

Moreover any way of assigning the numbers from 1 to N to the boundary components of such a surface gives a pairing which contributes to the sum. If the number of boundary components is b, the contribution will be $N^b v^{n/2}$.

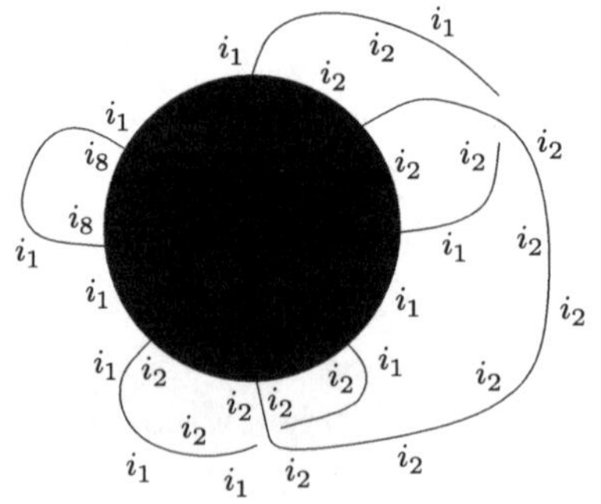

If one caps off the boundary components one obtains a closed orientable surface of Euler characeristic $1-n/2+b$, in other words of genus $g = n/4-b/2+1/2$. We see that

$$\begin{aligned}\frac{1}{N}E(\mathrm{tr}(A^n)) &= \sum_{\text{pairings of } \{1,2,\ldots,n\}} N^{n/2-2g}v^n \\ &= \sum_{\text{pairings of } \{1,2,\ldots,n\}} (\sqrt{N}v)^2 N^{-2g}\end{aligned}$$

where g is the genus of the surface obtained as we have described.

We now want to examine the behaviour of this as $N \to \infty$. Remember that n and g stay bounded so if this sum is to have a limit then obviously terms with $g > 0$ will not contribute, and if we set $v = N^{-1/2}$ we find that

$$\lim_{N\to\infty} \frac{1}{N} E(\mathrm{tr}(A^n)) = \text{ number of pairings giving genus 0 surfaces.}$$

It is easy to see that the surface is of genus 0 iff the pairing is planar, or, drawing it on the inside of the circle rather than the outside, exactly what we called a Temperley-Lieb diagram in the last lecture!!

The number of such pairings is well known to be the Catalan number

$$\frac{1}{n/2+1}\binom{n}{n/2},$$

so finally we get

$$\lim_{N\to\infty} \frac{1}{N} E(\mathrm{tr}(A^n)) = \begin{cases} 0, & \text{if } n \text{ is odd,} \\ \dfrac{1}{n/2+1}\dbinom{n}{n/2}, & \text{if } n \text{ is even.} \end{cases}$$

The probability distribution having this as its moments is the semicircle law with support between -2 and $+2$ whose density function looks like:

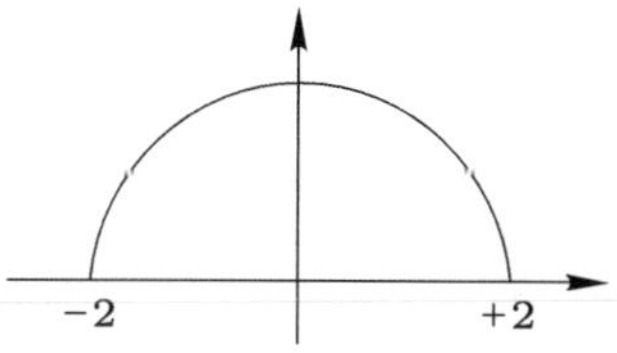

So we have derived Wigner's semicircle law for the density of eigenvaules of a large Hermitian matrices.

Suggestive exercise: Show that $\mathrm{tr}(\ell_f + \ell_f^*)^n$ (from the previous section) has the same value ...

4.3 Voiculescu land

There are many directions one could go from here and we will follow a few of them. The first, due to Voiculescu is to consider several hermitian random matrices $A_1, A_n, \ldots, A_k$. Their entries will be Gaussian as before but all the entries of A_i will be independent of all those of A_j for $i \neq j$. The obvious thing to calculate is the trace of a word in the A_i's. For concreteness let us choose one: $A_1A_2A_2A_1A_1A_1$ and repeat the calculation of the previous section to calculate

$$\lim_{N\to\infty} \frac{1}{N} E(\mathrm{tr}(A_1A_2A_2A_1A_1A_1)).$$

Wick's theorem may be used in exactly the same way and all non-planar pairings will disappear in the limit. The only difference will be that pairings that connect an A_i to A_j with $i \neq j$ will contribute zero to the trace. So the answer will be

$$\lim_{N\to\infty} \frac{1}{N} E(\mathrm{tr}(A_1 A_2 A_2 A_1 A_1 A_1)) = \text{number of planar pairings connecting like} A_i\text{'s}.$$

Of the 5 planar pairings in this case where the length of the word is 6, only two satisfy the condition so the limit of the trace of our word is 2.

This is a very nice generalisation of Wigner's semicircle law but what is one to do with it? It could certainly be used to get the moments of self-adjoint words in the A_i's but obviously it has a much more noncommutative context. A natural reaction of the von Neumann algebraist would be to use this limit to define a trace on an algebra and complete to get a von Neumann algebra. The algebra is simply the *-algebra of non-commuting polynomials in k self-adjoint variables $X_1, X_2, \ldots, X_k$,written $\mathbb{C} < X_1, X_2, \ldots, X_k >$. Define the Voiculescu trace $tr : \mathbb{C} < X_1, X_2, \ldots, X_k > \to \mathbb{C}$ by

$$tr(w) = \lim_{N\to\infty} \frac{1}{N} E(\mathrm{tr}(w)),$$

where w denotes a word in the X's and the same word in the A's. This is clearly a positive semidefinite trace by properties of the trace on a matrix algebra. So we may perform the GNS construction to obtain a von Neumann algebra with a trace. But in fact by a generalisation of the exercise at the end of the previous section we already know the answer! For the same trace would be obtatined by taking the word w as a word on the $\ell(f_i) + \ell(f_i)^*$ (for an orthonormal basis $f_1, f_2, \ldots, f_k$ of $\mathfrak{H}$) from the first section. Thus by Voiculescu's free functor we know we obtain LF_k, our friend the II_1 factor of the free group on k generators.

This shows that although the LF_n do not contain dense sequences of finite dimensional algebras, they can actually be approximated by matrix algebras in the sense of traces of words on a set of generators. Another way of saying this is that the von Neumann algebra LF_n provides an algebra for working directly in the infinite N limit of random hermitian matrices.

4.4 Potentials

We could have gone in another direction from our calculation of Wigner's law, staying with a single matrix, a direction of considerably more interest to physics and with nontrivial mathematical applications.

One can realise a random matrix by taking the space of all determinisitc Hermitian $N \times N$ matrices with Gaussian measure. More explicitly the probability density as a function of a matrix A is a suitably normalised form of

$$e^{-\mathrm{tr}(A^2)} dA,$$

dA being Lebesgue measure. A random matrix is then just one drawn from this distribution. A physicist on seeing this will say that the quadratic term $\mathrm{tr}(A^2)$

means that we are dealing with a free theory and that more interesting theories will be obtained if we change the probability density to

$$e^{-\mathrm{tr}(A^2)+\beta V(A)}dA,$$

where V is a"potentia" such as $\mathrm{tr}(A^4)$. At least a potential should be a function of A which is invariant so expressible as a function of the traces of powers of A. β will be called a coupling constant.

Let us take the special case of $\beta\mathrm{tr}(A^4)$ and attempt to repeat the calculation we did in the free case. (This method is due to 't Hooft [29].)

Expanding the exponential as a power series in $\beta\mathrm{tr}(A^4)$ we see we will need to calculate terms such as (here the expectation E is taken with respect to the Gaussian distribution)

$$\lim_{N\to\infty}\frac{1}{N}E(\mathrm{tr}(A^n)\mathrm{tr}(A^4)^r),$$

and the large N limit of the average of the trace, with the potential defining the probabiltiy distribution, will be a power series in β with this term as the coefficient of β^r.

Wick's theorem may again be applied but now we see that in order to accomodate the $\mathrm{tr}(A^4)$ terms, we have to add to the diagrams (besides the central circle with n boundary points) r other circles with 4 boundary points each. Such a diagram would look like (for $n = 8$ and $r = 4$):

As before the terms in the Wick sum can be organised as functions from the boundary of this surface to the set $\{1, 2, \ldots, N\}$ and the same formula obtained for the coefficient of β^n in terms of the genus of the surface. Not surprisingly it is once again only the genus zero terms that contribute to the large N limit and one ends up counting the number of figures like the one above that can be drawn on a genus zero surface, i.e. the 2-sphere. So the large N limit of the average value is the generating function for the numbers of various combinatorial planar pictures.

In the above discussion we ignored the normalisation constant for the probability distribution with potential. This is just the average value of the identity so it counts the number of pictures containing discs with only 4 boundary points. Shrinking these discs down to points one sees in fact the projection of a knot or link. Since there is no crossing data the best one can hope for is to be counting alternating link diagrams but Zinn-Justin and Zuber did just that in [66] and obtained Thistlethwaite's asymptotics for the number of alternating links by evaluating the integral with the quartic potential by totally different means.

For any potential the average values of the traces of high powers should give the moments of a measure, thus one may be able to use differential equations or other techniques of classical analysis to find it. The answer is known for very general potentials.

4.5 Several random matrices with a potential

It is obvious what the title of this section means, given the previous discussion. We interpret the large N limit of the average value of the trace of a word on matrices $A_1, A_2, \dots, A_k$ (now with a non-Gaussian measure) as defining a trace on $\mathbb{C} < X_1, X_2, \dots, X_k >$. The trace will be positive semidefinite as before so the GNS construction should be available to build a von Neumann algebra M with a trace.

But in order to obtain general results for arbitrary potentials we meet two serious snags. The first is that the integrals one is able to do by explicit integration are few and far between, but perhaps more seriously, the second is the interpretation of the result! For we are no longer dealing with a measure on the line but a trace on a highly non-commutative algebra. Worse, the von Neumann algebras we obtain by the GNS construction are not very well understood even in the Gaussian case- remember that it is still an open question as to whether $LF_n \cong LF_m$ for $n \neq m$!!

So we can expect progress to be slow. The first step is to get some control of the isomorphism class of M. Is it a factor for instance? The best results so far have been obtained by Guionnet and Shlyakhtenko in [24] where they show that in certain cases M is indeed a factor and in fact isomorphic to a free group factor.

4.6 Planar algebra constructions

The fact that large N limits are governed by planar diagrams is highly suggestive of a connection with the planar algebras we defined in the third lecture. We never gave any explicit examples of planar algebras so let us immediately remedy that situation. To give an example we just have to give a graded vector space and an action of planar tangles on it.

The Temperley-Lieb planar algebra:

For each n let P_n be the vector space having as a basis the set of Temperley Lieb diagrams with n boundary points (and no strings that do not connect boundary points, i.e. no closed loops).

This is the graded vector space. Tangles act in an obvious way. Temperley-Lieb tangles can be glued into the internal discs of a planar tangle to form another basis element (in fact closed loops are inevitably formed and must be removed, multiplying by δ each time.

This Temperley Lieb planar algebra is clearly a planar subalgebra of any other planar algebra, hence its central importance in the theory.

Another planar algebra is directly relevant to random matrices. It is called the *tensor planar algebra.* Choose a vector space $\mathcal{V}$ of dimension k for which we will choose a basis for simplicity. Then the graded vector space is equal to the nth tensor power of $\mathcal{V}$:

$$P_n = \otimes^n \mathcal{V}.$$

A basis for P_n consists of tensors in a very elementary sense - numbers $R_{i_1,i_2,\ldots,i_n}$ with the i's running between 1 and k.

Penrose gave a diagrammatic formalism for the contraction of tensors according to rather arbitrary schemes ([46]) and we can just borrow that formalism to define the action of planar tangles on tensors. If tensors $R_1, R_2, \cdots, R_m$ are inserted into the internal discs of a planar tangle, one assigns an i between 1 and k to each boundary point on the external disc and contracts the tensors according the pairing of tensor indices given by the tangle. (One has to add that if two external boundary points are connected by a string then the answer is zero unless the indices assigned to the ends of that string are the same, in which case the answer is just the contraction of the rest of the tangle.)

As an easy exercise to get used to the formalism, check that the tangle below defines matrix multiplication on the space of tensors with two indices:

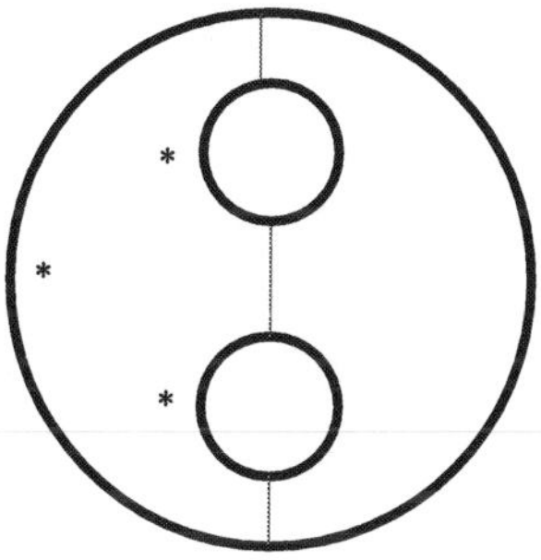

Various tangles define algebra structures on a planar algebra. The one that concerns us here will be called graded multiplication and makes $\oplus_n P_n$ into a graded *-algebra. Here is the tangle that defines graded multiplication, illustrated as the map from $P_3 \times P_5$ to P_8:

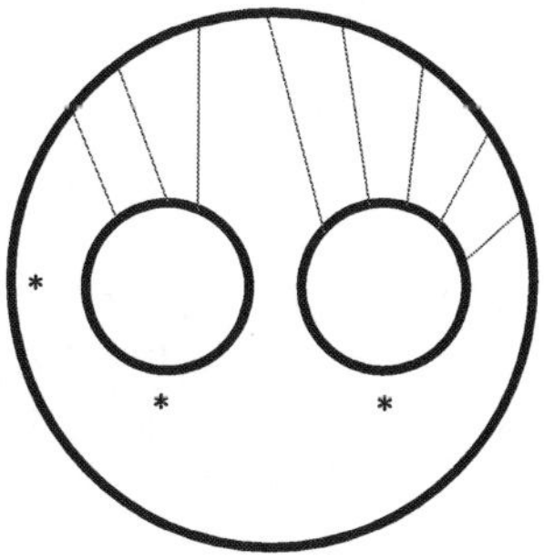

To obtain the product ab for $a \in P_3$ and $b \in P_5$ one inserts a into the left input disc and b into the right. The action of the tangle produces the element ab - this is the definition of multiplication. It is easy to check associativity and the *-algebra property.

The relevance to our random matrix story is this: the algebra given by this graded multiplication on the tensor planar algebra is nothing but the algebra of non-commuting polynomials in k self-adjoint variables. Moreover the Voiculescu trace on this algebra defined by the large N limit of k random matrices can be described in purely planar algebraic terms. We can define an element $\sum TL$ of P_{2n} as the sum of the Temperley-Lieb basis elements. Then

$$\text{Voiculescu tr}(x) = [\sum TL, x],$$

where $[,]$ is the bilinear pairing defined at the end of the last lecture.

The point is that all these ingredients make sense in any planar algebra and if it is one of the subfactor kind the trace on the graded algebra will be positive definite. In [25] it is shown that this process produces a II_1 factor with a subfactor whose standard invariant is the given planar algebra. More interestingly in [25] it is shown that there is a matrix model (somewhat more complicated than just hermitian matrices) whose large N limit produces this trace we have defined on the planar algebra.

What is most intriguing about these matrix models appears when one considers the meaning of the loop parameter. For the tensor planar algebra (non-commuting polynomials) this loop parameter is exactly the number k of random matrices. But for a general planar algebra it can be any number in the set

$$\{2\cos\pi/n : n = 3, 4, 5, \ldots\} \bigcup [2, \infty).$$

Thus we have produced matrix models with a continuously varying number of matrices. And the ultimate source of this continuum was von Neumann's continuous dimension in a II_1 factor, an outcome he would have loved but surely did not foresee.

References

[1] Blanchet, C., Habegger, N., Masbaum, G., and Vogel, P. (1995). Topological quantum field theories derived from the Kauffman bracket. *Topology*, **34**, 883–927.

[2] Bratelli, O. and Robinson, D. (1997). Operator Algebras and Quantum Statistical Mechanics 2, 2nd Edition. Springer-Verlag, New York.

[3] Ching, Wai-Mee (1973). Free products of von Neumann algebras. *Transactions AMS*, **178**.

[4] Connes, A. (1973). Une classification des facteurs de type III. *Annales Scientifiques de l'École Normale Supérieure*, **6**, 133–252.

[5] Connes, A. (1975). Sur la classification des facteurs de type II. *C.R. Acad. Sci. Paris*, **281**, 13–15.

[6] Connes, A. (1976). Classification of injective factors. *Annals of Mathematics*, **104**, 73–115.

[7] Connes, A. (1979). Sur la théorie non commutative de l'integration. *Springer Lecture Notes in Math.*, **725**, 19–143.

[8] Connes, A. (1982). A factor of type II_1 with countable fundamental group. *J. Operator Theory*, **4**, 151–153.

[9] Connes, A. (1985). Non-commutative differential geometry I–II. *Publication Mathématiques IHES*, **62**, 41–144.

[10] Connes, A. (1994). Noncommutative geometry. *Academic Press.*

[11] Connes, A. and Jones, V. (1982). A II1 factor with two non-conjugate Cartan subalgebras. *Bull. Amer. Math. Soc.*, **6**, 211–212.

[12] Connes, A. and Takesaki, M. (1977). The flow of weights on factors of type III. *Tohoku Mathematical Journal*, **29**, 73–555.

[13] Connes, A. and Van Daele, A. (1973). The Group Property of the Invariant S of von Neuman Algebras. *Math. Scand.*, **32**, 187–192.

[14] J.H. Conway (1970). An enumeration of knots and links, and some of their algebraic properties. *Computational Problems in Abstract Algebra* (Proc. Conf., Oxford, 1967), 329–358.

[15] Cuntz, J. (1977). Simple C^*-algebras generated by isometries. *Communications in Mathematical Physics*, **57**, 173–185.

[16] Dixmier, J. (1953). Sous-anneaux abeliens maximaux dans les facteurs de type fini. *Annals of Math.*, **59**, 279–286.

[17] Doplicher, S., Haag, R. and Roberts, J. E. (1969). Fields, observables and gauge transformations II. *Communications in Mathematical Physics*, **15**, 173–200.

[18] Dye H., (1959). On groups of measure preserving transformation. I, *Amer. J. Math.*, **81**, 119–159.

[19] Dykema, K. (1992). *Interpolated free group factors.* **arXiv:funct-an/9211012v1**

[20] Evans, D. and Kawahigashi, Y. (1998). Quantum symmetries on operator algebras. Oxford University Press.

[21] Feldman J. and Moore, C.C. (1975). Ergodic equivalence relations, cohomology, and von Neumann algebras. *Bulletins AMS*, **81**, 921–924.

[22] Fredenhagen, K., Rehren, K.-H. and Schroer, B. (1989). Superselection sectors with braid group statistics and exchange algebras. *Communications in Mathematical Physics*, **125**, 201–226.

[23] Goodman, F.M. Harpe, P. de la and Jones, V.F.R. (1989). Coxeter graphs and towers of algebras, *Publ. MSRI* **62**, Springer.

[24] Guionnet, A. and Shlyakhtenko, D. (2012), Free monotone transport. *arXiv: 1204.2182*

[25] Guionnet, A., Jones, V.F.R., Shlyakhtenko, D. (2010). Random matrices, free probability, planar algebras and subfactors, in "Quanta of maths", pp. 201–239, Clay Math. Proc., 11, Amer. Math. Soc., Providence, RI, arXiv.org/0712.2904

[26] Haag, R. (1996). Local Quantum Physics. Springer-Verlag, Berlin-Heidelberg-New York.

[27] Haap, R. and Kastler, D. (1964). An Algebraic Approach to Quantum Field Theory. *Journal of Mathematical Physics*, **5**, 848–861.

[28] Haagerup, U. (1994). Principal graphs of subfactors in the index range $4 < 3+\sqrt{2}$. in *Subfactors — Proceedings of the Taniguchi Symposium, Katata —*, (ed. H. Araki, et al.), World Scientific, 1–38.
[29] Hooft G. (1974). A planar diagram theory for strong interactions. *Nuclear Phys. B*, **72**, 461–473.
[30] Jones, V.F.R. (1983). Index for subfactors. *Invent. Math.*, **72**, 1–25.
[31] Jones, V.F.R. Planar Algebras I, preprint.
[32] Jones, V. F. R. (1989). On knot invariants related to some statistical mechanical models. *Pacific Journal of Mathematics*, **137**, 311–334.
[33] Jones, V.F.R. (2001). The annular structure of subfactors, in "Essays on geometry and related topics", *Monogr. Enseign. Math.*, **38**, 401–463.
[34] Jones, V. Morrison, S. and Snyder, N. (2013). The classification of subfactors of index $\leqslant 5$. To appear.
[35] Kac, V. (1990). Infinite Dimensional Lie Algebras, 3rd Edition. Cambridge University Press.
[36] Kauffman, L. (1987). State models and the Jones polynomial. *Topology*, **26**, 395–407.
[37] Kosaki, H. (1986). Extension of Jones' theory on index to arbitrary factors. *Journal of Functional Analysis*, **66**, 123–140.
[38] Longo, R, (1989). Index of subfactors and statistics of quantum fields, I. *Comm. Math. Phys.*, **126**, 217–247.
[39] Longo, R, (1990). Index of subfactors and statistics of quantum fields, II. *Comm. Math. Phys.*, **130**, 285–309.
[40] Lueck, W. L^2-Invariants from the Algebraic point of view. (2008) *arXiv:math/0310489v1[MATH GT]*
[41] McDuff, D. (1969). Uncountably many II_1 factors. *Annals of Mathematics*, **90**, 372–377.
[42] Murray, F. J. and von Neumann, J. (1936). On rings of operators. *Annals of Mathematics*, **37**, 116–229.
[43] Murray, F. J. and von Neumann, J. (1943). On rings of operators IV. *Annals of Mathematics*, **44**, 716–808.
[44] Nakamura, M. and Takeda, Z. (1960). A Galois theory for finite factors. *Proc. Japan Acad.*, **36**, 258–260.
[45] Ocneanu, A. (1988). Quantized groups, string algebras and Galois theory for algebras, in "Operator algebras and applications", Vol. 2, 119–172, London Math. Soc. Lecture Note Ser., 136, Cambridge Univ. Press.
[46] Penrose, R. (1971). Applications of negative dimensional tensors. *Applications of Combinatorial Mathematics*, Academic Press, 221–244.
[47] Popa, S. (1995). An axiomatization of the lattice of higher relative commutants of a subfactor. *Inventiones Mathematicae*, **120**, 427–446.
[48] Popa, S. (2004). On the fundamental group of type II_1 factors. *Proceedings of the National Academy of Sciences of the United States of America*, **101**, 723–726. math.OA/0210467.
[49] Popa, S. (2006). On a class of type II_1 factors with Betti numbers invariants. *Annals of Mathematics*, **163**, 809–899. math.OA/0209130.

[50] Popa, S. and Vaes, S. (2008). Strong rigidity of generalized Bernoulli actions and computations of their symmetry groups. *Adv. Math.*, **217**, no. 2, 833–872.

[PV] Popa, S. and Vaes, S. *Unique Cartan decomposition for II_1 factors arising from arbitrary actions of free groups* Preprint, arXiv:1111.6951

[51] Powers, R. T. (1967). Representations of uniformly hyperfinite algebras and their associated von Neumann rings. *Annals of Mathematics*, **86**, 138–171.

[52] Powers, R. T. and Størmer, E. (1970). Free states of the canonical anti-commutation relations. *Communications in Mathematical Physics*, **16**, 1–33.

[53] Pressley, A. and Segal, G. (1986). Loop groups. Oxford University Press.

[54] Radulescu, Florin (1992). The fundamental group of the von Neumann algebra of a free group with infinitely many generators is $\mathbb{R}+\{0\}$. *Journal of the AMS* **5**, 517–532.

[55] Reshetikhin, N. Yu. and Turaev, V. G. (1991). Invariants of 3-manifolds via link polynomials and quantum groups. *Inventiones Mathematicae*, **103**, 547–598.

[56] Sakai, S. (1971). C^*-algebras and W^*-algebras. Springer-Verlag, Berlin-Heidelberg-New York.

[57] Takesaki, M. (1970). Tomita's theory of modular Hilbert algebras and its apllications. *Lecture Notes in Mathematics*, Springer, **128**.

[58] Thistlethwaite, M. (2001). Links with trivial Jones Polynomial. *Journal of Knot Theory and its Ramifications*, **10**, 641–643.

[59] Temperley, H. N. V. and Lieb. E. H. (1971). Relations between the "percolation" and "colouring" problem and other graph-theoretical problems associated with regular planar lattices: some exact results for the "percolation" problem. *Proceedings of the Royal Society A*, **322**, 251–280.

[60] Voiculescu, D., Dykema, K., and Nica, A. (1993). Free random variables. *CRM Monogr. Series* **1**, AMS.

[61] von Neumann, J. (1949). On rings of operators. Reduction theory. *Ann. Math.* **50**, 401–485.

[62] von Neumann, J. (1939). On infinite direct products. *Compositio Mathematica*, **6**, 1–77.

[63] Wassermann, A. (1998). Operator algebras and conformal field theory III: Fusion of positive energy representations of $LSU(N)$ using bounded operators. *Inventiones Mathematicae*, **133**, 467–538.

[64] Wenzl, H. (1987). On sequences of projections. *Comptes Rendus Mathématiques, La Société Royale du Canada, L'Academie des Sciences*, **9**, 5–9.

[65] Witten, E. (1989). Quantum field theory and Jones polynomial. *Communications in Mathematical Physics*, **121**, 351–399.

[66] Zinn-Justin P. and Zuber J. B. (2002). Matrix integrals and the counting of tangles and links. *Discrete Math.*, **246**, no. 1–3, 343–360, Formal power series and algebraic combinatorics (Barcelona, 1999).

Introduction to
Modern Mathematics
ALM 33, pp. 323–347

Formules de Poisson non Linéaires et Principe de Fonctorialité de Langlands*

Laurent Lafforgue

Résumé

On introduit un énoncé général de formules de Poisson relatives à des opérateurs de transformation de Fourier sur des groupes réductifsadéliques. Ces formules de Poisson sont équivalentes au principe de fonctorialité de Langlands.

Mots-clefs: Groupes réductifs, adèles, transformation de Fourier, formule de Poisson, principe de fonctorialité.

Classification (AMS): 11R56, 43A30, 22E57.

On connaît la transformation de Fourier sur $\mathbb{R}$. Elle définit un automorphisme unitaire de l'espace des fonctions de carré intégrable $\mathbb{R} \to \mathbb{C}$. De plus, cet automorphisme et son inverse respectent le sous-espace des fonctions de Schwartz, c'est-à-dire des fonctions de classe C^∞ à décroissance rapide. Enfin, une telle fonction de Schwartz f et sa transformée de Fourier $\widehat{f}$ sont reliées par la formule de Poisson

$$\sum_{n\in\mathbb{Z}} f(n) = \sum_{n\in\mathbb{Z}} \widehat{f}(n).$$

Étant donné un "corps global" F, c'est-à-dire une extension finie de $\mathbb{Q}$ ou du corps $\mathbb{F}_q(X)$ des fonctions rationnelles sur un corps fini $\mathbb{F}_q$, la théorie de l'anneau des adèles $\mathbb{A} = \prod_{x\in|F|} F_x$ de F a permis à Tate, en 1950 dans sa fameuse thèse, de transporter et d'exploiter transformation de Fourier et formule de Poisson dans un cadre arithmétique. Il choisit pour cela un caractère additif continu unitaire non trivial $\psi = \prod_{x\in|F|} \psi_x : \mathbb{A} = \prod_{x\in|F|} F_x \to U(1) = \{z \in \mathbb{C}^\times \mid |z| = 1\}$ invariant par le sous-groupe discret F de $\mathbb{A}$.

*Ces notes ont servi de base à un double exposé donné au séminaire Takagi, à Tokyo, les 25 et 26 mai 2013, ainsi qu'à un exposé à l'université Tsinghua de Pékin, au titre de "Loo-Keng Hua Distinguished Lecture", le 8 novembre 2013.

L'auteur remercie chaleureusement les organisateurs de ces séminaires qui l'ont invité. Il exprime également sa grande reconnaissance à Cécile Gourgues, qui a réalisé la frappe du manuscrit avec son efficacité et sa disponibilité toujours parfaites, ainsi qu'au traducteur du texte en langue chinoise.

En toute "place" x de F, c'est-à-dire pour toute complétion F_x de F relativement à une certaine norme, que celle-ci soit "archimédienne" (avec alors $F_x \cong \mathbb{R}$ ou $\mathbb{C}$) ou "ultramétrique" (si bien que F_x est une extension finie d'un corps p-adique $\mathbb{Q}_p$ ou d'un corps $\mathbb{F}_q((X))$ de séries de Laurent), le caractère ψ_x définit un opérateur de transformation de Fourier $f_x \mapsto \widehat{f_x}$. Cet opérateur est un automorphisme unitaire de l'espace des fonctions de carré intégrable sur F_x.

De plus, cet opérateur et son inverse respectent un certain sous-espace de fonctions de Schwartz qui, lorsque x est une place ultramétrique, est simplement l'espace des fonctions localement constantes à support compact sur F_x. Le sous-espace des fonctions de Schwartz en les places ultramétriques est stable par translations multiplicatives si bien qu'il admet une caractérisation spectrale. Il s'avère que cette caractérisation peut s'exprimer en termes de certaines fractions rationnelles

$$L_x(\chi, Z)$$

associées aux caractères multiplicatifs $\chi : F_x^\times \to \mathbb{C}^\times$.

D'autre part, la transformation de Fourier est compatible avec les translations multiplicatives, et son action sur l'espace des fonctions de Schwartz admet une caractérisation spectrale. Il s'avère que cette caractérisation peut s'exprimer en termes de certains monômes

$$\varepsilon_x(\chi, Z) = \varepsilon_x(\chi, \psi_x, Z)$$

également associés aux caractères multiplicatifs $\chi : F_x^\times \to \mathbb{C}^\times$.

En les places archimédiennes, la décomposition spectrale des espaces de fonctions de Schwartz et de leur automorphisme de transformation de Fourier conduit à définir de manière analogue des facteurs

$$L_x(\chi, s) \quad \text{et} \quad \varepsilon_x(\chi, s) = \varepsilon_x(\chi, \psi_x, s)$$

associés aux caractères multiplicatifs $\chi : F_x^\times \to \mathbb{C}^\times$ et qui sont cette fois des fonctions analytiques de $s \in \mathbb{C}$.

Le produit des opérateurs de transformation de Fourier "locale" sur les F_x associés aux composantes ψ_x de ψ définit un opérateur de transformation de Fourier "globale" sur $\mathbb{A} = \prod_{x \in |F|} F_x$ associé à ψ. Il envoie chaque fonction produit $f = \bigotimes_{x \in |F|} f_x$ sur $\widehat{f} = \bigotimes_{x \in |F|} \widehat{f_x}$. C'est un automorphisme unitaire de l'espace des fonctions de carré intégrable sur $\mathbb{A}$. Il préserve le sous-espace des fonctions de Schwartz globales, défini comme sous-espace engendré par les produits de fonctions de Schwartz locales sur les complétions F_x.

Tate a alors démontré que toute fonction de Schwartz globale $f : \mathbb{A} \to \mathbb{C}$ satisfait la formule de Poisson

$$\sum_{\gamma \in F} f(\gamma) = \sum_{\gamma \in F} \widehat{f}(\gamma).$$

Puis, en décomposant spectralement cette identité sous l'action du groupe multiplicatif $\mathbb{A}^\times$, il en a déduit les propriétés globales des fonctions L

$$\mathbb{C} \ni s \mapsto L(\chi, s) = \left[\prod_{x\in|F| \text{ ultramétrique}} L_x(\chi_x, q_x^{-s})\right] \cdot \left[\prod_{x\in|F| \text{ archimédienne}} L_x(\chi_x, s)\right]$$

associées aux caractères multiplicatifs

$$\chi = \prod_{x\in|F|} \chi_x : \mathbb{A}^\times \to \mathbb{C}^\times$$

qui sont "automorphes", c'est-à-dire invariants par le sous-groupe discret $F^\times$ de $\mathbb{A}^\times$. Ces propriétés sont la convergence absolue pour $\mathrm{Re}\,(s)$ assez grande, le prolongement analytique à $\mathbb{C}$ tout entier et l'équation fonctionnelle

$$L(\chi^{-1}, 1-s) = L(\chi, s) \cdot \varepsilon(\chi, s)$$

avec

$$\varepsilon(\chi, s) = \left[\prod_{x\in|F| \text{ ultramétrique}} \varepsilon_x(\chi_x, q_x^{-s})\right] \cdot \left[\prod_{x\in|F| \text{ archimédienne}} \varepsilon_x(\chi, s)\right].$$

Au début des années 1970, Godement et Jacquet ont généralisé la théorie de Tate aux groupes linéaires GL_r de rang arbitraire plongés dans les espaces de matrices M_r.

Ils ont défini une transformation de Fourier et un espace de fonctions de Schwartz sur $M_r(F_x)$ pour toute place x. La décomposition spectrale de ces espaces et de ces opérateurs leur a permis d'associer des facteurs locaux

$$L_x(\pi, \bullet) \quad \text{et} \quad \varepsilon_x(\pi, \bullet)$$

à toute représentation irréductible π du groupe $\mathrm{GL}_r(F_x)$.

Puis, en faisant le produit sur toutes les places x de F, ils ont défini une transformation de Fourier globale sur $M_r(\mathbb{A})$, ainsi qu'un espace de fonctions de Schwartz globales $f : M_r(\mathbb{A}) \to \mathbb{C}$ satisfaisant la formule de Poisson

$$\sum_{\gamma\in M_r(F)} f(\gamma) = \sum_{\gamma\in M_r(F)} \widehat{f}(\gamma).$$

La décomposition spectrale de cette identité sous l'action du groupe multiplicatif $\mathrm{GL}_r(\mathbb{A})$ a permis à Godement et Jacquet d'établir les propriétés globales (convergence absolue dans un demi-plan, prolongement analytique et équation fonctionnelle) des fonctions L globales

$$L(\pi, \bullet) = \prod_{x\in|F|} L_x(\pi_x, \bullet)$$

des représentations irréductibles $\pi = \bigotimes_{x\in|F|} \pi_x$ de $\mathrm{GL}_r(\mathbb{A}) = \prod_{x\in|F|} \mathrm{GL}_r(F_x)$ qui sont "automorphes", c'est-à-dire admettent une réalisation dans l'espace des fonctions périodiques

$$\mathrm{GL}_r(F)\backslash\mathrm{GL}_r(\mathbb{A}) \to \mathbb{C}.$$

Si G est un groupe réductif (déployé) sur F, on dispose plus généralement de la notion de représentation irréductible "automorphe" de $G(\mathbb{A}) = \prod_{x\in|F|} G(F_x)$, c'est-à-dire qui admet une réalisation dans l'espace des fonctions périodiques

$$G(F)\backslash G(\mathbb{A}) \to \mathbb{C}.$$

Dans le cas général, on ne dispose sur G d'aucune structure linéaire qui permette de reproduire les constructions et les démonstrations de Tate généralisées par Godement et Jacquet.

Cependant, on dispose du "principe de fonctorialité" conjecturé par Langlands en 1967.

Selon Langlands, on doit pouvoir associer à toute représentation

$$\rho : \widehat{G} \to \mathrm{GL}_r(\mathbb{C})$$

du groupe réductif complexe $\widehat{G}$ dual de G, une application

$$\pi = \bigotimes_{x\in|F|} \pi_x \longmapsto \pi' = \bigotimes_{x\in|F|} \pi'_x$$

de l'ensemble des représentations automorphes de $G(\mathbb{A})$ vers l'ensemble des représentations automorphes de $\mathrm{GL}_r(\mathbb{A})$. Cette application doit être essentiellement compatible avec une famille d'applications entre ensembles de représentations irréductibles locales de $G(F_x)$ et $\mathrm{GL}_r(F_x)$

$$\pi_x \mapsto \pi'_x$$

qui, lorsque les π_x sont "non ramifiées", sont définies par une règle très simple associée à ρ.

Si ces applications $\pi_x \mapsto \pi'_x$ existent, elles permettent donc d'associer aux représentations irréductibles π_x des $G(F_x)$ des facteurs

$$L_x(\rho, \pi_x, \bullet) = L_x(\pi'_x, \bullet) \quad \text{et} \quad \varepsilon_x(\rho, \pi_x, \bullet) = \varepsilon_x(\pi'_x, \bullet).$$

À toute représentation automorphe $\pi = \bigotimes_x \pi_x$ de $G(\mathbb{A})$ est alors associée une fonction L globale relative à ρ

$$L(\rho, \pi, \bullet) = \prod_{x\in|F|} L_x(\rho, \pi_x, \bullet)$$

qui possède les propriétés globales usuelles des fonctions L si le principe de fonctorialité de Langlands est connu.

Bien sûr, les arguments de Tate, Godement et Jacquet pourraient être reproduits sur $G(\mathbb{A})$, et conduire à une démonstration directe des propriétés globales des fonctions L des représentations automorphes de $G(\mathbb{A})$, si l'on disposait sur les $G(F_x)$ et sur $G(\mathbb{A})$ de transformations de Fourier naturellement définies à partir de ρ, d'espaces de fonctions de Schwartz respectés par ces transformations et d'une formule de Poisson convenable sur l'espace des fonctions de Schwartz globales. Ce vœu a été exprimé pour la première fois dans la littérature dans un article de Braverman et Kazhdan publié en l'an 2000.

Le but du présent texte d'exposition est d'expliquer qu'il y a en fait équivalence entre le principe de fonctorialité de Langlands et le problème d'associer à toute représentation $\rho : \widehat{G} \to \mathrm{GL}_r(\mathbb{C})$ des opérateurs de transformation de Fourier locaux, des espaces de fonctions de Schwartz locales et une formule de Poisson globale.

Plus précisément, la connaissance de facteurs locaux $L_x(\rho, \pi_x, \bullet)$ permet de construire en chaque place un espace de fonctions de Schwartz. Celle de facteurs locaux $\varepsilon_x(\rho, \pi_x, \bullet)$ permet de définir un opérateur de transformation de Fourier agissant sur cet espace. Enfin, les propriétés globales des fonctions $L(\rho, \pi, \bullet) = \prod_{x \in |F|} L_x(\rho, \pi_x, \bullet)$ associées aux représentations automorphes $\pi = \bigotimes_{x \in |F|} \pi_x$ de $G(\mathbb{A})$ permettent de définir sur l'espace des fonctions de Schwartz globales une certaine forme linéaire fixée par la transformation de Fourier.

1 Transformation de Fourier sur $M_r(\mathbb{R})$

Munissons $\mathbb{R}$ d'un caractère additif continu unitaire non trivial

$$\psi_\infty : \mathbb{R} \to U(1) \subset \mathbb{C}^\times,$$

par exemple

$$t \mapsto e^{2\pi i t}.$$

Ce choix détermine celui de l'unique mesure invariante de $\mathbb{R}$ qui attribue le volume 1 au quotient compact de $\mathbb{R}$ par son sous-groupe discret $\mathrm{Ker}\, \psi_\infty$, et donc de la mesure invariante produit de $M_r(\mathbb{R}) = \mathbb{R}^{r^2}$.

On a:

Proposition 1. *La ψ_∞-transformation de Fourier*

$$f \mapsto \widehat{f} = \left[m' \mapsto \int_{M_r(\mathbb{R})} dm \cdot f(m) \cdot \psi_\infty(\mathrm{Tr}(mm')) \right]$$

définit un automorphisme de l'espace des fonctions de Schwartz (c'est-à-dire C^∞ à décroissance rapide)

$$f : M_r(\mathbb{R}) \to \mathbb{C}.$$

Son automorphisme réciproque est la $\overline{\psi}_\infty = \psi_\infty^{-1}$-transformation de Fourier.

Le groupe des éléments $g \in \mathrm{GL}_r(\mathbb{R})$ agit par translation à droite $f \mapsto f^g = f(\bullet\, g)$ et par translation à gauche $f \mapsto {}^g f = f(g\, \bullet)$ sur l'espace des fonctions de Schwartz. Ces actions sont compatibles avec la transformation de Fourier au sens suivant:

Lemme 2. *Pour toute fonction de Schwartz*

$$f : M_r(\mathbb{R}) \to \mathbb{C}$$

et pour tout élément $g \in \mathrm{GL}_r(\mathbb{R})$, *on a*

$$\begin{aligned} \widehat{f^g} &= |\det(g)|^{-r} \cdot {}^{g^{-1}}\widehat{f}, \\ \widehat{{}^g f} &= |\det(g)|^{-r} \cdot \widehat{f}^{g^{-1}}. \end{aligned}$$

2 Transformation de Fourier sur $M_r(F_x)$

$\mathbb{R}$ est une complétion de $\mathbb{Q}$. Or $\mathbb{Q}$ possède d'autres complétions: les corps p-adiques $\mathbb{Q}_p$.

On appelle "corps locaux" F_x les éléments de la liste suivante:

- $\mathbb{R}$ et son unique extension finie $\mathbb{C}$,
- les corps $\mathbb{Q}_p$ et leurs extensions finies,
- les corps de séries de Laurent $\mathbb{F}_{q_x}((X))$ sur un corps fini $\mathbb{F}_{q_x}$.

Tout corps local F_x est muni d'une unique norme $|\bullet|_x$ telle que tout élément $\gamma \in F_x^\times$ agisse sur les mesures additives de F_x par $|\gamma|_x$.

Si $F_x = \mathbb{R}$ ou $\mathbb{C}$, F_x est dit "archimédien".

Sinon, il est dit "ultramétrique" car sa norme $|\bullet|_x$ satisfait l'inégalité

$$|a+b|_x \leq \max\{|a|_x, |b|_x\}, \quad \forall a, b \in F_x.$$

Dans ce cas, F_x possède le sous-anneau ouvert compact

$$O_x = \{a \in F_x \mid |a|_x \leq 1\}$$

dont le quotient par l'idéal maximal ouvert

$$m_x = \{a \in F_x \mid |a|_x < 1\}$$

est un corps fini $\mathbb{F}_{q_x}$.

Si F_x est un corps local, une fonction de Schwartz $M_r(F_x) \to \mathbb{C}$ est une fonction

- C^∞ à décroissance rapide si F_x est archimédien,
- localement constante à support compact si F_x est ultramétrique.

Choisissant un caractère additif continu unitaire non trivial

$$\psi_x : F_x \to U(1) \subset \mathbb{C}^\times$$

puis une mesure invariante dm_x de $M_r(F_x)$, on a

Proposition 3. (i) *La ψ_x-transformation de Fourier*

$$f_x \mapsto \widehat{f_x} = \left[m'_x \mapsto \int_{M_r(F_x)} dm_x \cdot f_x(m_x) \cdot \psi_x(\mathrm{Tr}(m_x m'_x)) \right]$$

définit un automorphisme de l'espace des fonctions de Schwartz.

(ii) *Il existe une unique mesure invariante dm_x (dite autoduale) pour laquelle l'automorphisme réciproque est la $\overline{\psi}_x$-transformation de Fourier.*

(iii) *Pour toute fonction de Schwartz $f_x : M_r(F_x) \to \mathbb{C}$ et tout élément $g_x \in \mathrm{GL}_r(F_x)$, on a*

$$\widehat{f_x^{g_x}} = |\det(g_x)|_x^{-r} \cdot {}^{g_x^{-1}}\widehat{f_x},$$
$$\widehat{{}^{g_x}f_x} = |\det(g_x)|_x^{-r} \cdot \widehat{f_x}^{g_x^{-1}}.$$

3 L'espace des représentations lisses admissibles irréductibles de $G(F_x)$

Pour tout corps local F_x, il est naturel de chercher à décomposer l'espace des fonctions de Schwartz sous la double action de $\mathrm{GL}_r(F_x)$.

Dans ce but, il faut d'abord introduire un ensemble de représentations irréductibles de $\mathrm{GL}_r(F_x)$ assez riche pour permettre une telle décomposition, et montrer que cet ensemble a une structure naturelle.

Traitons le cas où le corps local F_x est ultramétrique.

Définition 4. *Soit G un groupe linéaire GL_r ou, plus généralement, un groupe réductif sur un corps local ultramétrique F_x.*

Une représentation complexe π de $G(F_x)$ est dite "lisse admissible" si

- *pour tout sous-groupe ouvert compact $K \subset G(F_x)$,*

$$\pi_K = \{v \in \pi \mid K \cdot v = v\}$$

est de dimension finie sur $\mathbb{C}$,
- *π est la réunion filtrante de ses sous-espaces π_K.*

Le groupe topologique localement compact $G(F_x)$ est unimodulaire et peut être muni d'une mesure bi-invariante dg_x. Soient alors $\mathcal{H}_x^G$ l'algèbre de convolution des fonctions localement constantes à support compact

$$h_x : G(F_x) \to \mathbb{C}$$

et, pour tout $K \subset G(F_x)$, $\mathcal{H}_{x,K}^G$ sa sous-algèbre des fonctions à support compact

$$K\backslash G(F_x)/K \to \mathbb{C}.$$

Pour toute représentation lisse admissible π de $G(F_x)$, chaque π_K est une représentation de dimension finie de $\mathcal{H}_{x,K}^G$ et $\pi = \varinjlim \pi_K$ peut être vue comme une représentation de $\mathcal{H}_x^G = \varinjlim \mathcal{H}_{x,K}^G$.

Notons $\{\pi\}_x^G$ l'ensemble des classes d'isomorphie de représentations lisses admissibles irréductibles de $G(F_x)$ et, pour tout $K \subset G(F_x)$, $\{\pi\}_{x,K}^G$ le sous-ensemble des π telles que $\pi_K \neq 0$.

Pour tout $K \subset G(F_x)$, les éléments $h_x \in \mathcal{H}_{x,K}^G$ définissent des fonctions

$$\{\pi\}_{x,K}^G \ni \pi \mapsto \mathrm{Tr}_\pi(h_x),$$

et on peut noter $A_{x,K}^G$ l'algèbre de fonctions

$$\{\pi\}_{x,K}^G \to \mathbb{C}$$

qu'elles engendrent. On a:

Théorème 5. *Soit toujours un groupe réductif G sur un corps local ultramétrique F_x. Alors:*

(i) *Pour tout $K \subset G(O_x)$, l'algèbre $A_{x,K}^G$ est un produit fini d'algèbres de type fini et intègres sur $\mathbb{C}$. Autrement dit,* $\mathrm{Spec}\, A_{x,K}^G$ *est réunion disjointe finie de variétés algébriques affines intègres sur $\mathbb{C}$.*

(ii) *Toute $\pi \in \{\pi\}_{x,K}^G$ définit un caractère $A_{x,K}^G \to \mathbb{C}$ de $A_{x,K}^G$ qui le caractérise, si bien que $\{\pi\}_{x,K}^G$ se plonge dans le spectre maximal de $A_{x,K}^G$. C'est un ouvert de Zariski.*

(iii) *La réunion filtrante $\{\pi\}_x^G = \varinjlim \{\pi\}_{x,K}^G$ s'écrit comme une réunion disjointe de variétés algébriques intègres sur $\mathbb{C}$, de telle façon que chaque $\{\pi\}_{x,K}^G$ est une réunion finie de composantes connexes.*

Remarque 6. Dans le cas d'un corps local archimédien $F_x = \mathbb{R}$ ou $\mathbb{C}$, on a un théorème analogue: cette fois, l'ensemble des représentations irréductibles a une structure naturelle de variété analytique complexe.

Pour tout $K \subset G(F_x)$, soit $\mathrm{Im}\,\{\pi\}_{x,K}^G \subset \{\pi\}_{x,K}^G$ le sous-ensemble des représentations unitaires. On a:

Théorème 7. *Soit toujours un groupe réductif G sur un corps local ultramétrique F_x. Alors, pour tout $K \subset G(F_x)$, on a:*

(i) *Le sous-ensemble* $\mathrm{Im}\,\{\pi\}_{x,K}^G$ *des représentations unitaires est une sous-variété algébrique réelle de la variété algébrique complexe $\{\pi\}_{x,K}^G$.*

(ii) *Il existe sur* $\mathrm{Im}\,\{\pi\}_{x,K}^G$ *une unique mesure $d\pi$, appelée mesure de Plancherel, telle que, pour tout $h_x \in \mathcal{H}_{x,K}^G$, on ait*

$$h_x(1) = \int_{\mathrm{Im}\,\{\pi\}_{x,K}^G} d\pi \cdot \mathrm{Tr}_\pi(h_x).$$

Disons qu'une fonction

$$\{\pi\}_{x,K}^G \to \mathbb{C}$$

est "polynomiale" si elle est élément de l'algèbre $A_{x,K}^G$ engendrée par les fonctions traces

$$\pi \mapsto \mathrm{Tr}_\pi(h_x), \quad h_x \in \mathcal{H}_{x,K}^G.$$

En remplaçant les fonctions h_x par leurs translatées h_x^g ou ${}^g h_x$, on déduit du théorème ci-dessus:

Corollaire 8. *Pour tout $K \subset G(F_x)$ et si $d\pi$ désigne la mesure de Plancherel sur $\mathrm{Im}\,\{\pi\}_{x,K}^G$, on a:*

(i) *Toute fonction $h_x \in \mathcal{H}_{x,K}^G$ s'écrit sous la forme*

$$h_x(g) = \int_{\mathrm{Im}\,\{\pi\}_{x,K}^G} d\pi \cdot h_{x,\pi}(g), \quad g \in G(F_x),$$

où:

- *pour tout $g \in G(F_x)$,*
$$\pi \mapsto h_{x,\pi}(g)$$
est une fonction polynomiale sur $\{\pi\}_{x,K}^G$,
- *pour toute $\pi \in \{\pi\}_{x,K}^G$, la fonction sur $G(F_x)$*
$$g \mapsto h_{x,\pi}(g)$$
est élément de l'espace $\pi_K^\vee \boxtimes \pi_K$ engendré par les "coefficients matriciels" $v^\vee \boxtimes v : g \mapsto \langle v^\vee, g \cdot v\rangle = \langle g^{-1} \cdot v^\vee, v\rangle$ associés à un vecteur $v \in \pi_K$ et une forme linéaire $v^\vee : \pi \to \mathbb{C}$ invariante par K.

(ii) Cette décomposition spectrale de toute fonction $h_x \in \mathcal{H}_{x,K}^G$ est unique, et on a

$$h_{x,\pi}(g) = \mathrm{Tr}_\pi(h_x^g) = \mathrm{Tr}_\pi({}^g h_x), \quad \forall \pi, \quad \forall g \in G(F_x).$$

4 Décomposition spectrale de la transformation de Fourier linéaire

Revenons aux groupes linéaires GL_r sur un corps local ultramétrique F_x.

Comme les fonctions de Schwartz (c'est-à-dire localement constantes à support compact) $M_r(F_x) \to \mathbb{C}$ sont déterminées par leurs restrictions à $\mathrm{GL}_r(F_x)$, et que leur espace est stable par translation à gauche ou à droite, on s'attend à pouvoir caractériser cet espace en termes spectraux. Effectivement, on a:

Théorème 9. *Pour tout $K \subset \mathrm{GL}_r(F_x)$, il est possible d'associer à toute représentation $\pi \in \{\pi\}_{x,K}^{\mathrm{GL_r}} = \{\pi\}_{x,K}^r$ une fraction rationnelle*

$$L_x(\pi, Z)$$

telle que:

(1) *L'inverse $L_x(\pi, Z)^{-1}$ est un polynôme en $\pi \in \{\pi\}_{x,K}^r$ et Z qui vaut 1 pour $Z = 0$.*

(2) *On a*
$$L_x(\pi \otimes |\det(\bullet)|^s, Z) = L_x(\pi, q_x^{-s} \cdot Z), \quad \forall \pi, \quad \forall s \in \mathbb{C}.$$

(3) *Une fonction*
$$f_x : K\backslash\mathrm{GL}_r(F_x)/K \to \mathbb{C}$$

se prolonge en une fonction de Schwartz sur $M_r(F_x)$ si et seulement si ses restrictions aux fibres de l'homomorphisme

$$|\det(\bullet)|_x : \mathrm{GL}_r(F_x) \to q_x^{\mathbb{Z}}$$

sont à support compact, et qu'elle se décompose spectralement sous la forme

$$f_x(g) = |\det(g)|_x^{-\frac{r}{2}} \cdot \int_{\mathrm{Im}\,\{\pi\}^r_{x,K}} d\pi \cdot f_{x,\pi}(g) \cdot L_x\left(\pi^\vee, q_x^{-\frac{1}{2}}\right)$$

où

- *pour tout $g \in \mathrm{GL}_r(F_x)$*

$$\pi \mapsto f_{x,\pi}(g)$$

  est une fonction polynomiale sur $\{\pi\}^r_{x,K}$,
- *pour toute $\pi \in \{\pi\}^r_{x,K}$, la fonction sur $\mathrm{GL}_r(F_x)$*

$$g \mapsto f_{x,\pi}(g)$$

  est élément de l'espace $\pi_K^\vee \boxtimes \pi_K$,
- *$\pi^\vee = \varinjlim \pi_K^\vee$ désigne la représentation contragrédiente de toute $\pi \in \{\pi\}^r_x$.*

Comme la transformation de Fourier est compatible avec les translations à gauche et à droite, on obtient:

Corollaire 10. *Le caractère non trivial $\psi_x : F_x \to U(1) \subset \mathbb{C}^\times$ étant choisi, il est possible d'associer à toute représentation $\pi \in \{\pi\}^r_{x,K}$ un monôme (c'est-à-dire un polynôme inversible) en π et $Z^{\pm 1}$*

$$\varepsilon_x(\pi, \psi_x, Z) = \varepsilon_x(\pi \otimes |\det(\bullet)|^s, q_x^s \cdot Z), \quad \forall s \in \mathbb{C},$$

tel que, pour toute fonction de Schwartz sur $M_r(F_x)$

$$f_x : K\backslash \mathrm{GL}_r(F_x)/K \to \mathbb{C}$$

décomposée spectralement sous la forme du théorème

$$f_x(g) = |\det(g)|_x^{-\frac{r}{2}} \cdot \int_{\mathrm{Im}\,\{\pi\}^r_{x,K}} d\pi \cdot f_{x,\pi}(g) \cdot L_x\left(\pi^\vee, q_x^{-\frac{1}{2}}\right), \quad \forall g,$$

sa ψ_x-transformée de Fourier $\widehat{f}_x$ soit donnée par la décomposition spectrale

$$\widehat{f}_x(g) = |\det(g)|_x^{-\frac{r}{2}} \cdot \int_{\mathrm{Im}\,\{\pi\}^r_{x,K}} d\pi \cdot f_{x,\pi}(g^{-1}) \cdot L_x\left(\pi, q_x^{-\frac{1}{2}}\right) \cdot \varepsilon_x\left(\pi, \psi_x, q_x^{-\frac{1}{2}}\right), \quad \forall g.$$

Remarque 11. On utilise le fait que si $\varphi_\pi : \mathrm{GL}_r(F_x) \to \mathbb{C}$ est élément de l'espace $\pi_K^\vee \boxtimes \pi_K$ des coefficients matriciels de π_K, alors $g \mapsto \varphi_\pi(g^{-1})$ est élément de l'espace $\pi_K \boxtimes \pi_K^\vee$ des coefficients matriciels de $\pi_K^\vee$.

Le calcul des facteurs L_x et ε_x des représentations $\pi \in \{\pi\}_x^r$ de $\mathrm{GL}_r(F_x)$ se ramène à celui des facteurs correspondants des caractères $\chi \in \{\pi\}_x^1$ de $F_x^\times$ dans au moins deux cas extrêmes:

- le cas des représentations non ramifiées, c'est-à-dire des $\pi \in \{\pi\}_{x,\emptyset}^r = \{\pi\}_{x,K}^r$ où $K = \mathrm{GL}_r(O_x)$,
- le cas des représentations de la forme

$$\pi \cdot \omega = \pi \otimes (\omega \circ \det(\bullet))$$

où $\pi \in \{\pi\}_{x,K}^r$ pour un certain $K \subset \mathrm{GL}_r(F_x)$ et $\omega : F_x^\times \to \mathbb{C}^\times$ est un caractère assez ramifié en fonction de K.

Les représentations non ramifiées sont connues grâce au théorème suivant:

Théorème 12 (Satake). *Si $K = \mathrm{GL}_r(O_x)$, l'algèbre de convolution "non ramifiée"*

$$\mathcal{H}_{x,\emptyset}^r = \mathcal{H}_{x,K}^{\mathrm{GL}_r} = C_c(\mathrm{GL}_r(O_x)\backslash \mathrm{GL}_r(F_x)/\mathrm{GL}_r(O_x))$$

est commutative et canoniquement isomorphe à

$$(\mathcal{H}_{x,\emptyset}^{T_r})^{\mathfrak{S}_r} = C_c(T_r(F_x)/T_r(O_x))^{\mathfrak{S}_r} = \mathbb{C}[\widehat{T}_r]^{\mathfrak{S}_r}$$

où $T_r = \mathbb{G}_m^r$ désigne le tore diagonal de GL_r et $\widehat{T}_r = (\mathbb{C}^\times)^r$ son tore complexe dual.

Remarque 13. Comme $\mathcal{H}_{x,\emptyset}^r$ est commutative, elle s'identifie aussi à l'algèbre $A_{x,\emptyset}^r = A_{x,\mathrm{GL}_r(O_x)}^{\mathrm{GL}_r}$.

Il résulte de ce théorème que se donner une représentation non ramifiée $\pi \in \{\pi\}_{x,\emptyset}^r$ équivaut à se donner une famille de r caractères non ramifiés

$$\chi_\pi^1, \dots, \chi_\pi^r : F_x^\times / O_x^\times \to \mathbb{C}^\times$$

bien définie à l'ordre près.

On a:

Théorème 14. (i) (Godement, Jacquet) *Pour toute représentation non ramifiée $\pi \in \{\pi\}_{x,\emptyset}^r$, on a*

$$L_x(\pi, Z) = \prod_{1 \le i \le r} L_x(\chi_\pi^i, Z)$$

et

$$\varepsilon_x(\pi, \psi_x, Z) = \prod_{1 \le i \le r} \varepsilon_x(\chi_\pi^i, \psi_x, Z).$$

(ii) (Jacquet, Shalika) *Si $\pi \in \{\pi\}_{x,K}^r$ pour un certain $K \subset \mathrm{GL}_r(F_x)$ et $\omega : F_x^\times \to \mathbb{C}^\times$ est un caractère assez ramifié en fonction de K, on a*

$$L_x(\pi \cdot \omega, Z) = 1$$

et

$$\varepsilon_x(\pi \cdot \omega, \psi_x, Z) = \varepsilon_x(\chi_\pi \cdot \omega, \psi_x, Z) \cdot \varepsilon_x(\omega, \psi_x, Z)^{r-1}$$

où $\chi_\pi : F_x^\times \to \mathbb{C}^\times$ désigne le caractère central de π.

En dehors de ces cas, les facteurs L_x et surtout ε_x des représentations $\pi \in \{\pi\}_x^r$ sont beaucoup plus complexes et ne peuvent en général être ramenés au cas du rang $r = 1$.

5 Transformations de Fourier sur $G(F_x)$

La décomposition spectrale sur $\mathrm{GL}_r(F_x)$ de la ψ_x-transformation de Fourier sur $M_r(F_x)$ amène à poser la définition suivante:

Définition 15. *Soit F_x un corps local ultramétrique.*

Soit G un groupe réductif sur F_x muni d'un caractère non trivial

$$\det_G : G \to \mathbb{G}_m.$$

Soit ρ la donnée d'un caractère

$$\det_\rho : G \to \mathbb{G}_m$$

et d'une manière d'associer à toute $\pi \in \{\pi\}_x^G$ une fraction rationnelle en Z

$$L_x(\rho, \pi, Z) = L_x(\rho, \pi \otimes |\det_G(\bullet)|_x^s, q_x^s \cdot Z), \quad \forall s,$$

et un monôme en $Z^{\pm 1}$

$$\varepsilon_x(\rho, \pi, Z) = \varepsilon_x(\rho, \pi \otimes |\det_G(\bullet)|_x^s, q_x^s \cdot Z), \quad \forall s,$$

tels que:

- *l'inverse $L_x(\rho, \pi, Z)^{-1}$ est un polynôme en Z et π sur chaque $\{\pi\}_{x,K}^G$ qui vaut 1 pour $Z = 0$,*
- *le monôme $\varepsilon_x(\rho, \pi, Z)$ est un polynôme inversible en Z et π sur chaque $\{\pi\}_{x,K}^G$.*

Alors:

(i) *On appelle ρ-fonctions sur $G(F_x)$ les fonctions*

$$f_x : K \backslash G(F_x)/K \to \mathbb{C}$$

dont les restrictions aux fibres de l'homomorphisme

$$|\det_G(\bullet)|_x : G(F_x) \to q_x^{\mathbb{Z}}$$

sont à support compact, et qui se décomposent spectralement sous la forme

$$f_x(g) = |\det_\rho(g)|_x^{-\frac{1}{2}} \cdot \int_{\mathrm{Im}\, \{\pi\}_K^G} d\pi \cdot f_{x,\pi}(g) \cdot L(\rho, \pi^\vee, q_x^{-\frac{1}{2}})$$

où

- *pour tout $g \in G(F_x)$,*

$$\pi \mapsto f_{x,\pi}(g)$$

est une fonction polynomiale sur $\{\pi\}_{x,K}^G$,

- *pour toute* $\pi \in \{\pi\}_{x,K}^G$, *la fonction sur* $G(F_x)$

$$g \mapsto f_{x,\pi}(g)$$

est élément de l'espace $\pi_K^\vee \boxtimes \pi_K$,

- $\pi^\vee = \varinjlim \pi_K^\vee$ *désigne la représentation contragrédiente de chaque* $\pi \in \{\pi\}_x^G$.

(ii) On appelle ρ*-transformation de Fourier l'opérateur qui associe à toute* ρ*-fonction* f_x *décomposée spectralement comme ci-dessus la fonction*

$$\widehat{f_x}(g) = |\det_\rho(g)|_x^{-\frac{1}{2}} \cdot \int_{\mathrm{Im}\,\{\pi\}_{x,K}^G} d\pi \cdot f_{x,\pi}(g^{-1}) \cdot L(\rho, \pi, q_x^{-\frac{1}{2}}) \cdot \varepsilon_x(\rho, \pi, q_x^{-\frac{1}{2}}).$$

Remarque 16. Il résulte de la définition que l'espace des ρ-fonctions f_x est invariant par translation à gauche ou à droite par les éléments $g \in G(F_x)$ et que

$$\widehat{f_x^g} = |\det_\rho(g)|_x^{-1} \cdot {}^{g^{-1}}\widehat{f_x},$$
$$\widehat{{}^g f_x} = |\det_\rho(g)|_x^{-1} \cdot \widehat{f_x}^{g^{-1}}.$$

Si $G = T$ est un tore déployé (c'est-à-dire isomorphe à une puissance de $\mathbb{G}_m$) sur F_x, on note $\widehat{T}$ le tore complexe "dual" de T défini en échangeant les réseaux de caractères et de cocaractères

$$X_{\widehat{T}} = X_T^\vee, \quad X_{\widehat{T}}^\vee = X_T.$$

Voici une manière naturelle de définir sur un tel $G = T$ une donnée ρ comme ci-dessus:

Proposition 17. *Soit* $G = T$ *un tore déployé sur* F_x *muni d'un caractère*

$$\det_G = \det_T : T \to \mathbb{G}_m.$$

Soit

$$\rho_T = (\rho_T^1, \ldots, \rho_T^r) : \widehat{T} \to (\mathbb{C}^\times)^r = \widehat{T}_r$$

une collection de caractères de $\widehat{T}$ *telle que:*

- ρ_T *est injectif, si bien que l'homomorphisme dual*

$$\rho_T^\vee = (\rho_T^{1\vee}, \ldots, \rho_T^{r\vee}) : T_r = \mathbb{G}_m^r \to T$$

est surjectif,

- *pour* $1 \leq i \leq r$, *le composé du caractère* $\rho_T^i : \widehat{T} \to \mathbb{C}^\times$ *et du cocaractère* $\widehat{\det}_T : \mathbb{C}^\times \to \widehat{T}$ *dual de* $\det_T$ *est l'identité de* $\mathbb{C}^\times$.

Posons

$$\det_{\rho_T} = \det_T$$

et, pour tout caractère $\chi \in \{\pi\}_x^T$ *de* $T(F_x)$,

$$L_x(\rho_T, \chi, Z) = \prod_{1 \leq i \leq r} L_x(\chi \circ \rho_T^{i\vee}, Z),$$
$$\varepsilon_x(\rho_T, \chi, Z) = \prod_{1 \leq i \leq r} \varepsilon_x(\chi \circ \rho_T^{i\vee}, \psi_x, Z).$$

Alors:

(i) *Les ρ_T-fonctions sur $T(F_x)$ sont exactement les fonctions déduites des fonctions localement constantes à support compact*

$$F_x^r \to \mathbb{C}$$

par intégration le long des fibres de l'homomorphisme surjectif

$$\rho_T^\vee : (F_x^\times)^r \to T(F_x).$$

(ii) *La ρ_T-transformation de Fourier sur $T(F_x)$ est induite par la ψ_x-transformation de Fourier sur F_x^r via $\rho_T^\vee$.*

Si G est un groupe réductif déployé sur F_x, il possède un tore maximal T déployé sur F_x, muni d'une action du groupe de Weyl W_G. Si de plus $\rho_T = (\rho_T^1, \ldots, \rho_T^r) : \widehat{T} \to (\mathbb{C}^\times)^r$ est une famille de caractères de $\widehat{T}$ comme ci-dessus, stable par l'action de W_G, les facteurs

$$L_x(\rho_T, \chi, Z) \quad \text{et} \quad \varepsilon_x(\rho_T, \chi, Z)$$

des caractères $\chi \in \{\pi\}_x^T$ sont invariants par l'action induite de W_G sur $\{\pi\}_x^T$.

Or on a:

Théorème 18. *Soit G un groupe réductif déployé sur F_x, de tore maximal T. Alors:*

(i) *G a une structure O_x-rationnelle naturelle, et $G(O_x)$ est un sous-groupe ouvert compact de $G(F_x)$.*

(ii) *Si $K = G(O_x)$, l'algèbre de convolution "non ramifiée"*

$$\mathcal{H}_{x,\emptyset}^G = \mathcal{H}_{x,K}^G = C_c(G(O_x)\backslash G(F_x)/G(O_x))$$

est commutative et canoniquement isomorphe à

$$(\mathcal{H}_{x,\emptyset}^T)^{W_G} = C_c(T(F_x)/T(O_x))^{W_G} = \mathbb{C}[\widehat{T}]^{W_G}.$$

Il résulte de ce théorème que se donner une représentation non ramifiée $\pi \in \{\pi\}_{x,\emptyset}^G = \{\pi\}_{x,G(O_x)}^G$ équivaut à se donner un caractère

$$z_\pi : T(F_x)/T(O_x) \to \mathbb{C}^\times$$

bien défini modulo l'action de W_G.

Si donc la famille $\rho_T = (\rho_T^1, \ldots, \rho_T^r)$ de caractères ρ_T^i, $1 \le i \le r$, de $\widehat{T}$ est stable par l'action de W_G, on peut poser

$$L_x(\rho, \pi, Z) = L_x(\rho_T, z_\pi, Z)$$
$$\varepsilon_x(\rho, \pi, Z) = \varepsilon_x(\rho_T, z_\pi, Z)$$

pour toute représentation non ramifiée $\pi \in \{\pi\}_{x,\emptyset}^G$.

Sous la même hypothèse, le caractère

$$\widehat{\mu}_G = \prod_{1 \le i \le r} \rho_T^i : \widehat{T} \to \mathbb{C}^\times$$

est fixé par l'action de W_G, et il lui correspond un cocaractère central

$$\mu_G : \mathbb{G}_m \to G.$$

Pour toute $\pi \in \{\pi\}_x^G$, $F_x^\times$ agit sur π via μ_G par un caractère $\chi_\pi : F_x^\times \to \mathbb{C}^\times$ que l'on peut appeler le caractère central de π.

Pour tout $K \subset G(F_x)$, toute $\pi \in \{\pi\}_{x,K}^G$ et tout caractère $\omega : F_x^\times \to \mathbb{C}^\times$, notons

$$\pi \cdot \omega = \pi \otimes (\omega \circ \det_G(\bullet)).$$

Si ω est assez ramifié en fonction de K, il est naturel de poser

$$\begin{aligned} L_x(\rho, \pi \cdot \omega, Z) &= 1, \\ \varepsilon_x(\rho, \pi \cdot \omega, Z) &= \varepsilon_x(\chi_\pi \omega, \psi_x, Z) \cdot \varepsilon_x(\omega, \psi_x, Z)^{r-1}. \end{aligned}$$

Restent, bien sûr, bien d'autres représentations $\pi \in \{\pi\}_x^G$ pour lesquelles on ne saurait définir naturellement des facteurs

$$L_x(\rho, \pi, Z) \quad \text{et} \quad \varepsilon_x(\rho, \pi, Z)$$

à partir d'une donnée

$$\rho_T = (\rho_T^1, \ldots, \rho_T^r) : \widehat{T} \to (\mathbb{C}^\times)^r$$

stable par W_G comme ci-dessus.

6 Formule de Poisson sur $M_r(\mathbb{R})$ puis $M_r(\mathbb{A})$

Revenons au corps $\mathbb{R}$ muni du caractère

$$\psi_\infty : \mathbb{R} \ni t \mapsto e^{2\pi i t}$$

et aux opérateurs de ψ_∞-transformation de Fourier

$$f \mapsto \widehat{f} = \left[m' \mapsto \int_{M_r(\mathbb{R})} dm \cdot f(m) \cdot \psi_\infty(\mathrm{Tr}(mm')) \right]$$

des espaces de fonctions de Schwartz $f : M_r(\mathbb{R}) \to \mathbb{C}$.

Le groupe additif localement compact $\mathbb{R}$ contient $\mathbb{Z}$ comme sous-groupe discret. Le quotient $\mathbb{R}/\mathbb{Z}$ est compact et de volume 1 pour la mesure autoduale. Les caractères de $\mathbb{R}$ triviaux sur $\mathbb{Z}$ sont exactement les

$$\mathbb{R} \ni t \mapsto \psi_\infty(n \cdot t), \quad n \in \mathbb{Z},$$

d'où l'on déduit la formule de Poisson:

Théorème 19. *Pour toute fonction de Schwartz*

$$f : M_n(\mathbb{R}) \to \mathbb{C},$$

on a

$$\sum_{\gamma \in M_n(\mathbb{Z})} f(\gamma) = \sum_{\gamma \in M_n(\mathbb{Z})} \widehat{f}(\gamma).$$

Nous allons rappeler la généralisation arithmétique de cette formule démontrée par Tate dans le cadre de la théorie des adèles.

Considérons pour cela un "corps global" F, c'est-à-dire

- $\mathbb{Q}$ ou une extension finie de $\mathbb{Q}$, appelée "corps de nombres",
- $\mathbb{F}_q(X)$ ou une extension finie de $\mathbb{F}_q(X)$, appelée "corps de fonctions".

Notons $|F|$ l'ensemble des corps locaux $(F_x, |\bullet|_x)$ qui s'obtiennent en complétant F par une norme $|\bullet|_x$. Cet ensemble est toujours dénombrable; ses éléments x s'appellent les places. Si F est un corps de fonctions, toutes ses places sont ultramétriques. Si F est un corps de nombres, il a un nombre fini de places archimédiennes – telles que $F_x \cong \mathbb{R}$ ou $\mathbb{C}$ – et les autres sont ultramétriques.

L'anneau topologique des "adèles" de F est défini comme le sous-anneau

$$\mathbb{A}_F \subset \prod_{x\in|F|} F_x$$

constitué des familles $(a_x \in F_x)_{x\in|F|}$ telles que

$$a_x \in O_x, \quad \text{soit } |a_x|_x \leq 1,$$

en "presque toute" place $x \in |F|$ (ce qui signifie "toute place sauf un ensemble fini").

Le sous-groupe diagonal $F \hookrightarrow \prod_{x\in|F|} F_x$ est contenu dans $\mathbb{A}$, ce qui signifie que pour tout $\gamma \in F^\times$, on a

$$|\gamma|_x = 1 \text{ en presque toute place } x \in |F|.$$

On a:

Proposition 20. (i) ("Formule du produit") *Pour tout $\gamma \in F^\times$, on a*

$$\prod_{x\in|F|} |\gamma|_x = 1.$$

(ii) *F est un sous-groupe discret du groupe topologique localement compact $\mathbb{A}$.*
(iii) *Le quotient $\mathbb{A}/F$ est compact.*

Choisissons une fois pour toutes un caractère additif continu non trivial

$$\psi = \prod_{x\in|F|} \psi_x : \mathbb{A} \to \mathbb{C}^\times$$

qui est trivial sur le sous-groupe discret cocompact F. Ce caractère est nécessairement unitaire, ainsi que ses composantes

$$\psi_x : F_x \to \mathbb{C}^\times.$$

En toute place $x \in |F|$, munissons F_x et $M_r(F_x)$ de la mesure additive "autoduale" pour laquelle l'automorphisme réciproque de la ψ_x-transformation de Fourier des fonctions de Schwartz

$$f_x \mapsto \widehat{f}_x = \left[m' \mapsto \int_{M_r(F_x)} dm \cdot f_x(m) \cdot \psi_x(\mathrm{Tr}(mm'))\right]$$

est la $\overline{\psi}_x = \psi_x^{-1}$-transformation de Fourier.

Munissons $\mathbb{A}$ et $M_r(\mathbb{A})$ de la mesure produit, et appelons fonctions de Schwartz sur $M_r(\mathbb{A})$ les combinaisons linéaires de produits

$$f = \bigotimes_{x \in |F|} f_x$$

de fonctions de Schwartz $f_x : M_r(F_x) \to \mathbb{C}$ égales à la fonction caractéristique $\mathbb{I}_{M_r(O_x)}$ en presque toute place ultramétrique x.

La ψ-transformation de Fourier globale

$$f \mapsto \widehat{f} = \left[m' \mapsto \int_{M_r(\mathbb{A})} dm \cdot f(m) \cdot \psi(\mathrm{Tr}(mm')) \right]$$

envoie chaque $\bigotimes_{x \in |F|} f_x$ sur $\bigotimes_{x \in |F|} \widehat{f_x}$.

On montre que les caractères

$$\mathbb{A}/F \to \mathbb{C}^\times$$

sont exactement les

$$\mathbb{A} \ni a \mapsto \psi(\gamma \cdot a), \quad \gamma \in F,$$

et que le volume du quotient compact $\mathbb{A}/F$ est 1.

On en déduit la formule de Poisson adélique de Tate:

Théorème 21. *Pour toute fonction de Schwartz*

$$f : M_r(\mathbb{A}) \to \mathbb{C},$$

on a

$$\sum_{\gamma \in M_r(F)} f(\gamma) = \sum_{\gamma \in M_r(F)} \widehat{f}(\gamma).$$

7 Nouvelle expression de la formule de Poisson sur $\mathrm{GL}_r(\mathbb{A})$

De la même façon que nous avons ramené les transformations de Fourier des $M_r(F_x)$ aux $\mathrm{GL}_r(F_x)$ et, à partir de là, proposé une forme très générale de transformation de Fourier sur les groupes réductifs locaux $G(F_x)$, nous voudrions ramener les formules de Poisson des $M_r(\mathbb{A})$ aux $\mathrm{GL}_r(\mathbb{A})$ afin de proposer une forme plus générale de formule de Poisson sur les groupes réductifs adéliques $G(\mathbb{A})$.

Toute la difficulté réside dans la contribution des matrices non-inversibles $\gamma \in M_r(F) - \mathrm{GL}_r(F)$ à la fonctionnelle de Poisson $f \mapsto \sum_{\gamma \in M_r(F)} f(\gamma)$.

Voici une manière simple de faire disparaître les termes de bord associés à ces matrices dans le cas d'une fonction de Schwartz $f = \bigotimes_{x \in |F|} f_x$ très ramifiée en au moins une place:

Proposition 22. *Soit* $f = \bigotimes_{x \in |F|} f_x : M_r(\mathbb{A}) \to \mathbb{C}$ *une fonction de Schwartz telle que, en au moins une place ultramétrique* x_0, *le facteur* f_{x_0} *est le produit*

$$f_{x_0} = f'_{x_0} \cdot \omega_{x_0} \circ \det(\bullet)$$

d'une fonction f'_{x_0} *invariante par un certain sous-groupe ouvert compact* $K \subset \mathrm{GL}_r(F_{x_0})$ *et d'un caractère* $\omega_{x_0} : F_{x_0}^\times \to \mathbb{C}^\times$ *très ramifié en fonction de* K.

Alors le support de f_{x_0} *est une partie compacte de* $\mathrm{GL}_r(F_{x_0})$, *de même que celui de* $\widehat{f}_{x_0}$, *et on a*

$$\sum_{\gamma \in \mathrm{GL}_r(F)} f(\gamma) = \sum_{\gamma \in \mathrm{GL}_r(F)} \widehat{f}(\gamma).$$

Dans le cas général d'une fonction de Schwartz arbitraire

$$f : M_r(\mathbb{A}) \to \mathbb{C},$$

on a besoin de la définition suivante:

Définition 23. *Soit* x_0 *une place ultramétrique de* F.

Soit $f_{x_0} : M_r(F_{x_0}) \to \mathbb{C}$ *une fonction de Schwartz qui est non-ramifiée, c'est-à-dire invariante à gauche et à droite par* $\mathrm{GL}_r(O_{x_0})$, *et dont la décomposition spectrale a la forme*

$$f_{x_0}(g) = |\det(g)|_{x_0}^{-\frac{r}{2}} \cdot \int_{\mathrm{Im}\,\{\pi\}_{x,\emptyset}^r} d\pi \cdot f_{x_0,\pi}(g) \cdot L_{x_0}\left(\pi^\vee, q_{x_0}^{-\frac{1}{2}}\right), \quad \forall g.$$

Alors, pour tous $N, N' \in \mathbb{N}$, *on note*

$$f_{x_0}^{N,N'} : M_r(F_{x_0}) \to \mathbb{C}$$

la fonction de Schwartz définie par l'expression spectrale

$$f_{x_0}^{N,N'}(g) = |\det(g)|_{x_0}^{-\frac{r}{2}} \cdot \int_{\mathrm{Im}\,\{\pi\}_{x,\emptyset}^r} d\pi$$
$$\cdot f_{x_0,\pi}(g) \cdot L_{x_0}(\pi^\vee, q_{x_0}^{-\frac{1}{2}}) \cdot I_{x_0}^N(\pi, q_{x_0}^{-\frac{1}{2}}) \cdot I_{x_0}^{N'}(\pi^\vee, q_{x_0}^{-\frac{1}{2}})$$

où $I_{x_0}^N(\pi, Z)$ *désigne le polynôme en* Z *et* π *produit de*

$$L_{x_0}(\pi, Z)^{-1}$$

et du monôme de degré N *en* Z *qui apparaît dans le développement en série formelle en* Z *de l'inverse*

$$L_{x_0}(\pi, Z).$$

Remarque 24. On remarque que chaque fonction $f_{x_0}^{N,N'}$ et sa transformée de Fourier $\widehat{f_{x_0}^{N,N'}}$ sont à support compact dans $\mathrm{GL}_r(F_{x_0})$.

D'autre part, on a dans l'anneau des séries formelles en Z l'égalité

$$\sum_{N\in\mathbb{N}} I_{x_0}^N(\pi, Z) = 1$$

d'où l'on déduit

$$\sum_{N,N'\in\mathbb{N}} f_{x_0}^{N,N'}(g) = f_{x_0}(g), \quad \forall g \in \mathrm{GL}_r(F_{x_0}).$$

Cette définition permet de formuler le théorème suivant qui ramène l'expression de la formule de Poisson de $M_r(\mathbb{A})$ à $\mathrm{GL}_r(\mathbb{A})$:

Théorème 25. *Soit* $f = \bigotimes_{x\in|F|} f_x : \mathrm{GL}_r(\mathbb{A}) \to \mathbb{C}$ *une fonction de l'espace de Schwartz de* $M_r(\mathbb{A})$.

Soit $x_0 \in |F|$ *une place ultramétrique en laquelle le facteur* f_{x_0} *de* f *est non ramifié.*

Alors:

(i) *La série formelle*

$$\sum_{N,N'\in\mathbb{N}} Z^{N+N'} \cdot \sum_{\gamma\in\mathrm{GL}_r(F)} \left(f_{x_0}^{N,N'} \otimes \left(\bigotimes_{x\neq x_0} f_x \right) \right)(\gamma)$$

est une fraction rationnelle en Z.

(ii) *Sa "valeur régularisée" en* $Z = 1$ *(définie en soustrayant l'unique expression de la forme* $\sum_{1\leq i\leq k} a_i \cdot (Z-1)^{-i}$ *qui fait disparaître l'éventuel pôle en* $Z = 1$*), notée* $S(f)$*, ne dépend pas du choix de la place* x_0.

(iii) *On a*

$$\sum_{\gamma\in M_r(F)} f(\gamma) = \left(\sum_{\gamma\in\mathrm{GL}_r(F)} f(\gamma) \right) + \left(\sum_{\gamma\in\mathrm{GL}_r(F)} \widehat{f}(\gamma) \right) - S(f).$$

Remarque 26. La démonstration de ce théorème utilise:

- le théorème de décomposition spectrale de Langlands,
- les propriétés globales des fonctions L démontrées par Godement et Jacquet.

8 Forme cherchée de formules de Poisson sur $G(\mathbb{A})$

Considérons maintenant un groupe réductif G sur un corps global F, muni d'un caractère non trivial

$$\det_G : G \to \mathbb{G}_m.$$

Soit ρ la donnée de:

- un caractère $\det_\rho : G \to \mathbb{G}_m$,

- en toute place ultramétrique $x \in |F|$, une manière d'associer à toute $\pi \in \{\pi\}_x^G$ une fraction rationnelle en Z

$$L_x(\rho, \pi, Z) = L_x(\rho, \pi \otimes |\det_G(\bullet)|_x^s, q_x^s \cdot Z), \quad \forall s,$$

et un monôme en $Z^{\pm 1}$

$$\varepsilon_x(\rho, \pi, Z) = \varepsilon_x(\rho, \pi \otimes |\det_G(\bullet)|_x^s, q_x^s \cdot Z), \quad \forall s,$$

vérifiant les propriétés du paragraphe 5,
- en toute place archimédienne $x \in |F|$, des facteurs spectraux analogues.

Comme on a vu, ρ permet de définir en toute place ultramétrique x de F l'espace des ρ-fonctions

$$f_x(\bullet) = |\det_\rho(\bullet)|_x^{-\frac{1}{2}} \cdot \int_{\mathrm{Im}\,\{\pi\}_{x,K}^G} d\pi \cdot f_{x,\pi}(\bullet) \cdot L_x\left(\rho, \pi^\vee, q_x^{-\frac{1}{2}}\right)$$

puis la ρ-transformation de Fourier de ces fonctions

$$f_x \mapsto \widehat{f}_x(\bullet) = |\det_\rho(\bullet)|_x^{-\frac{1}{2}} \cdot \int_{\mathrm{Im}\,\{\pi\}_{x,K}^G} d\pi \cdot f_{x,\pi}((\bullet)^{-1}) \cdot L_x\left(\rho, \pi, q_x^{-\frac{1}{2}}\right) \cdot \varepsilon_x\left(\rho, \pi, q_x^{-\frac{1}{2}}\right).$$

Et de même en les places archimédiennes de F.

En presque toute place ultramétrique x de F, le groupe réductif G est "non ramifié", donc admet une structure O_x-rationnelle, et le groupe $G(F_x)$ contient le sous-groupe ouvert compact $G(O_x)$. En une telle place, on appelle "ρ-fonction standard" l'unique ρ-fonction non-ramifiée

$$f_x : G(O_x)\backslash G(F_x)/G(O_x) \to \mathbb{C}$$

dont les raies spectrales $f_{x,\pi}$, $\pi \in \{\pi\}_{x,\emptyset}^G = \{\pi\}_{x,G(O_x)}^G$ vérifient

$$f_{x,\pi}(1) = 1, \quad \forall \pi.$$

On appelle ρ-fonction globale

$$f : G(\mathbb{A}) \to \mathbb{C}$$

les combinaisons linéaires de produits

$$\bigotimes_{x \in |F|} f_x : G(\mathbb{A}) \to \mathbb{C}$$

de ρ-fonctions locales

$$f_x : G(F_x) \to \mathbb{C}$$

presque toutes égales aux ρ-fonctions standard.

Par analogie avec le cas de GL_r, posons:

Définition 27. *Soit x_0 une place ultramétrique de F en laquelle le groupe réductif G est non ramifié.*

Soit $f_{x_0} : G(F_{x_0}) \to \mathbb{C}$ une ρ-fonction non ramifiée, c'est-à-dire invariante à gauche et à droite par $G(O_x)$, et dont la décomposition spectrale a la forme

$$f_{x_0}(\bullet) = |\det_\rho(\bullet)|_{x_0}^{-\frac{1}{2}} \cdot \int_{\mathrm{Im}\,\{\pi\}_{x,\emptyset}^G} d\pi \cdot f_{x_0,\pi}(\bullet) \cdot L_{x_0}\left(\rho, \pi^\vee, q_{x_0}^{-\frac{1}{2}}\right).$$

Alors, pour tous $N, N' \in \mathbb{N}$, on note

$$f_{x_0}^{N,N'} : G(F_{x_0}) \to \mathbb{C}$$

la ρ-fonction définie par l'expression spectrale

$$f_{x_0}^{N,N'}(\bullet) = |\det_\rho(\bullet)|_{x_0}^{-\frac{1}{2}} \cdot \int_{\mathrm{Im}\,\{\pi\}_{x,\emptyset}^G} d\pi$$
$$\cdot f_{x_0,\pi}(\bullet) \cdot L_{x_0}\left(\rho, \pi^\vee, q_{x_0}^{-\frac{1}{2}}\right) \cdot I_{x_0}^N\left(\rho, \pi, q_{x_0}^{-\frac{1}{2}}\right) \cdot I_{x_0}^{N'}\left(\rho, \pi^\vee, q_{x_0}^{-\frac{1}{2}}\right)$$

où $I_{x_0}^N(\rho, \pi, Z)$ désigne le polynôme en Z et π produit de

$$L_{x_0}(\rho, \pi, Z)^{-1}$$

et du monôme de degré N en Z qui apparaît dans le développement en série formelle en Z de l'inverse

$$L_{x_0}(\rho, \pi, Z).$$

Remarque 28. Comme dans le cas linéaire, on remarque que chaque fonction $f_{x_0}^{N,N'}$ et sa ρ-transformée de Fourier $\widehat{f_{x_0}^{N,N'}}$ sont à support compact dans $G(F_{x_0})$.

D'autre part, on a dans l'anneau des séries formelles en Z l'égalité

$$\sum_{N \in \mathbb{N}} I_{x_0}^N(\rho, \pi, Z) = 1$$

d'où l'on déduit

$$\sum_{N,N' \in \mathbb{N}} f_{x_0}^{N,N'}(g) = f_{x_0}(g), \quad \forall g \in G(F_{x_0}).$$

Le cas de GL_r et de la transformation de Fourier linéaire sur $M_r(\mathbb{A})$ amène à envisager la forme générale suivante pour d'éventuelles autres formules de Poisson:

Définition 29. *Soit G un groupe réductif sur un corps global F, muni d'un caractère non trivial* $\det_G : G \to \mathbb{G}_m$.

Soit ρ une donnée globale comme ci-dessus.

(i) *On dira que ρ satisfait la formule de Poisson restreinte si, pour toute ρ-fonction*

$$f = \bigotimes_{x \in |F|} f_x : G(\mathbb{A}) \to \mathbb{C}$$

dont le facteur f_{x_0} en au moins une place ultramétrique x_0 est le produit

$$f_{x_0} = f'_{x_0} \cdot \omega \circ \det_G(\bullet)$$

d'une fonction f'_{x_0} invariante par un certain sous-groupe ouvert compact $K \subset G(F_{x_0})$ et d'un caractère $\omega_{x_0} : F_{x_0}^\times \to \mathbb{C}^\times$ très ramifié en fonction de K, on a

$$\sum_{\gamma \in G(F)} f(\gamma) = \sum_{\gamma \in G(F)} \widehat{f}(\gamma).$$

(ii) *On dira que ρ satisfait la formule de Poisson générale si, pour toute ρ-fonction*

$$f = \bigotimes_{x \in |F|} f_x : G(\mathbb{A}) \to \mathbb{C}$$

et pour toute place ultramétrique $x_0 \in |F|$ en laquelle G et f_{x_0} sont non-ramifiés, la série formelle

$$\sum_{N,N' \in \mathbb{N}} Z^{N+N'} \cdot \sum_{\gamma \in G(F)} \left(f_{x_0}^{N,N'} \otimes \left(\bigotimes_{x \neq x_0} f_x \right) \right) (\gamma)$$

est une fraction rationnelle en Z et que sa "valeur régularisée" en $Z = 1$, notée $S(f)$, ne dépend pas du choix de x_0 et vérifie

$$S(f) = S(\widehat{f}).$$

Les propriétés connues des facteurs L_x et ε_x linéaires amènent à supposer:

Conjecture 30. *Soit toujours G un groupe réductif sur F, muni d'un caractère non trivial $\det_G : G \to \mathbb{G}_m$.*

Si ρ est une donnée globale qui satisfait la formule de Poisson (générale ou seulement restreinte), elle est uniquement déterminée par la connaissance des facteurs

$$L_x(\rho, \pi, Z) \quad \textit{et} \quad \varepsilon_x(\rho, \pi, Z)$$

des représentations non ramifiées $\pi \in \{\pi\}_{x,\emptyset}^G$ en presque toute place ultramétrique x de F où G est non ramifié.

9 Le cas des tores déployés

On se place toujours sur un corps global F.

On considère le cas où $G = T$ est un tore déployé sur F, muni d'un caractère non trivial

$$\det_T : T \to \mathbb{G}_m.$$

On rappelle que $\widehat{T}$ désigne le tore complexe dual de T.

Comme au paragraphe 5, on considère une famille de caractères de $\widehat{T}$

$$\rho_T = (\rho_T^1, \dots, \rho_T^r) : \widehat{T} \to (\mathbb{C}^\times)^r = \widehat{T}_r$$

telle que:

- ρ_T est injectif, si bien que l'homomorphisme dual

$$\rho_T^\vee = (\rho_T^{1\vee}, \dots, \rho_T^{r\vee}) : T_r = \mathbb{G}_m^r \to T$$

  et surjectif,
- pour $1 \leq i \leq r$, le composé du caractère $\rho_T^i : \widehat{T} \to \mathbb{C}^\times$ et du cocaractère $\widehat{\det}_T : \mathbb{C}^\times \to \widehat{T}$ dual de $\det_T$ est l'identité de $\mathbb{C}^\times$.

On pose

$$\det_{\rho_T} = \det_T.$$

Comme on a vu, ρ_T permet de définir en toute place ultramétrique x de F des facteurs

$$L_x(\rho_T, \chi, Z) = \prod_{1 \leq i \leq r} L_x(\chi \circ \rho_T^{i\vee}, Z),$$
$$\varepsilon_x(\rho_T, \chi, Z) = \prod_{1 \leq i \leq r} \varepsilon_x(\chi \circ \rho_T^{i\vee}, \psi_x, Z)$$

des caractères $\chi \in \{\pi\}_x^T$ et donc un espace des ρ_T-fonctions sur $T(F_x)$ ainsi qu'un opérateur de ρ_T-transformation de Fourier de ces fonctions.

Et de même en les places archimédiennes de F.

D'où une notion de ρ_T-fonctions sur $T(\mathbb{A})$ et un opérateur de ρ_T-transformation de Fourier de ces fonctions globales.

On a:

Théorème 31. *Dans les conditions ci-dessus, la donnée ρ_T satisfait la formule de Poisson (restreinte ou générale) au sens du paragraphe précédent.*

Principe de démonstration: C'est déjà connu si $T = \mathbb{G}_m$ est muni de la ψ-transformation de Fourier linéaire de $\mathbb{A}$ et donc, plus généralement, si

$$\rho_T : \widehat{T} \to \widehat{T}_r = (\mathbb{C}^\times)^r$$

est un isomorphisme.

Dans le cas général, la ρ_T-formule de Poisson sur $T(\mathbb{A})$ se déduit de la formule de Poisson standard sur $(\mathbb{A}^\times)^r$ par intégration le long des fibres de l'homomorphisme surjectif

$$\rho_T^\vee : (\mathbb{A}^\times)^r/(F^\times)^r \to T(\mathbb{A})/T(F).$$

10 Le cas des groupes réductifs déployés

On considère maintenant un groupe réductif G déployé sur un corps global F. Il possède un tore maximal T défini et déployé sur F.

En toute place ultramétrique x de F, le groupe réductif déployé G est non ramifié sur F_x, donc muni d'une structure O_x-rationnelle naturelle et d'un sous-groupe ouvert compact $G(O_x)$ de $G(F_x)$.

On suppose G muni d'un caractère non trivial

$$\det_G : G \to \mathbb{G}_m$$

dont la restriction à T est notée $\det_T$.

Enfin, on considère une famille de caractères

$$\rho_T = (\rho_T^1, \ldots, \rho_T^r) : \widehat{T} \to (\mathbb{C}^\times)^r = \widehat{T}_r$$

qui vérifie les mêmes propriétés qu'au paragraphe précédent.

Supposons de plus que cette famille de caractères ρ_T^i, $1 \leq i \leq r$, est stable par l'action naturelle du groupe de Weyl W_G de G sur le tore T et son dual $\widehat{T}$.

Comme on a vu au paragraphe 5, il en résulte que, en toute place ultramétrique x de F, les facteurs

$$L_x(\rho_T, \chi, Z) \quad \text{et} \quad \varepsilon_x(\rho_T, \chi, Z)$$

des caractères $\chi \in \{\pi\}_x^T$ sont invariants par l'action de W_G.

Or, en toute telle place x, les représentations non ramifiées de $G(F_x)$

$$\pi \in \{\pi\}_{x,\emptyset}^G$$

sont paramétrées par des caractères non ramifiés

$$z_\pi \in \{\pi\}_{x,\emptyset}^T$$

bien définis modulo l'action de W_G, et on peut poser

$$L_x(\rho, \pi, Z) = L_x(\rho_T, z_\pi, Z),$$
$$\varepsilon_x(\rho, \pi, Z) = \varepsilon_x(\rho_T, z_\pi, Z).$$

La conjecture de transfert automorphe de Langlands des groupes réductifs déployés sur F vers les groupes linéaires est équivalente à l'énoncé suivant:

Conjecture 32. *Dans la situation ci-dessus, la famille des facteurs*

$$L_x(\rho, \pi, Z) \quad et \quad \varepsilon_x(\rho, \pi, Z)$$

des représentations non ramifiées $\pi \in \{\pi\}_{x,\emptyset}^G$ en les places ultramétriques x de F peut être complétée (de façon nécessairement unique) en une donnée globale ρ qui satisfasse la formule de Poisson (restreinte ou générale) si et seulement si la famille de caractères stable par W_G

$$\rho_T = (\rho_T^1, \ldots, \rho_T^r) : \widehat{T} \to (\mathbb{C}^\times)^r = \widehat{T}_r$$

provient d'une représentation algébrique

$$\widehat{G} \to \mathrm{GL}_r(\mathbb{C})$$

du groupe réductif complexe $\widehat{G}$ dual de G.

Remarque 33. On rappelle que le groupe réductif $\widehat{G}$ sur $\mathbb{C}$ dual de G est déduit de G par une construction purement combinatoire: On demande que sa donnée radicielle $(X_{\widehat{G}}, \Delta_{\widehat{G}}, X_{\widehat{G}}^\vee, \Delta_{\widehat{G}}^\vee)$ s'identifie à la duale $(X_G^\vee, \Delta_G^\vee, X_G, \Delta_G)$ de la donnée radicielle $(X_G, \Delta_G, X_G^\vee, \Delta_G^\vee)$ de G.

Il admet pour tore maximal le dual $\widehat{T}$ du tore maximal T de G.

Référence

[1] A. Braverman et D. Kazhdan, "γ-functions of representations and lifting". With an appendix by V. Vologodsky, in "GAFA 2000, Visions in Mathematics", Geom. Funct. Anal. 2000, Special Volume, Part I, Birkhäuser Verlag, 2000, p. 237–278.

[2] R. Godement et H. Jacquet, "Zeta functions of simple algebras", LNM 260, Springer-Verlag, 1972.

[3] L. Lafforgue, "Noyaux du transfert automorphe de Langlands et formules de Poisson non linéaires", Japan J. Math. 9, p. 1–68, 2014.

[4] L. Lafforgue, "Du transfert automorphe de Langlands aux formules de Poisson non linéaires", prépublication de l'IHES numéro M/14/05, 2014. (http://preprints.ihes.fr/2014/M/M-14-05.pdf)

[5] J. Tate, "Fourier analysis in number fields and Hecke's zeta-functions", Ph. D. Thesis, Princeton Univ. 1950, in : "Algebraic Number Theory" (éditeurs: J. Cassels et A. Fröhlich), Academic Press, 1967, p. 305–347.

Introduction to
Modern Mathematics
ALM 33, pp. 349–361

A Static Uniqueness Theorem for Higher Genus Kottler Metrics

Dan A. Lee

Abstract

A (generalized) Kottler metric is the analog of a Schwarzschild metric in the asymptotically *locally* hyperbolic setting. When the conformal boundary of such a manifold has genus at least two, it is possible for these Kottler metrics to have negative mass. We prove a static uniqueness theorem for negative mass Kottler metrics. The result comes from combining a previous result of P. Chruściel and W. Simon with a new Penrose inequality for asymptotically locally hyperbolic manifolds of *negative* mass.

1 Introduction

This is a report on joint work with André Neves of Imperial College, the details of which will appear in a forthcoming paper [9]. We consider the vacuum Einstein equations with cosmological constant:

$$\mathrm{Ric}(h) - \tfrac{1}{2}R(h)h + \Lambda h = 0,$$

where h is a Lorentzian metric, and Λ is the cosmological constant. It is natural to formulate questions in the $\Lambda < 0$ setting that are analogous to questions that have been settled in the $\Lambda = 0$ setting. One well-known theorem in the $\Lambda = 0$ setting is the static uniqueness of Schwarzschild space.

Definition. We say that (M^3, g, V) is a *complete vacuum static data set with cosmological constant* Λ if and only if M is a smooth manifold (possibly with boundary) equipped with a complete Riemannian metric g and a nonnegative function V (called the *static potential*) such that $\partial M = \{V = 0\}$, and

$$\Delta_g V = -\Lambda V \tag{1}$$

$$\mathrm{Ric}(g) = \frac{1}{V}\,\mathrm{Hess}_g V + \Lambda g. \tag{2}$$

This definition is motivated by the fact that if a metric g satisfies (1) and (2), then the Lorentzian metric $h = -dt^2 + g$ on $\mathbb{R} \times M$ is a solution of the vacuum Einstein equations with cosmological constant Λ, and ∂_t is a Killing field of h.

In this paper, *Schwarzschild space* will be an exterior region of a Riemannian Schwarzschild manifold (which is just a standard $t = 0$ spacelike slice of a Schwarschild spacetime), whose boundary is the horizon. Explicitly, for $m > 0$, define

$$V(r) = \sqrt{1 - \frac{2m}{r}},$$

and set $M = [2m, \infty) \times S^2$ with

$$g = V^{-2}dr^2 + r^2 g_{S^2},$$

where g_{S^2} is the standard round unit sphere metric. This (M, g) is what we will call the *Schwarzschild space of mass* m. Written in this way, it is easy to verify that (M, g, V) is a complete vacuum static data set with cosmological constant $\Lambda = 0$.

Indeed, if one searches for spherically symmetric solutions of the $\Lambda = 0$ vacuum static equations (1) and (2), one can easily derive the above formula for V. In fact, one need only solve for V using the necessary condition $R = 0$, which easily follows from (1) and (2) when $\Lambda = 0$. Similarly, if one searches for spherically symmetric solutions of the vacuum static equations for general Λ, then solving for V using the necessary condition of constant scalar curvature will yield the so-called Kottler metrics (which are called Schwarzschild–anti-de Sitter for $\Lambda < 0$ and Schwarzschild–de Sitter for $\Lambda > 0$). As is the case for Schwarzschild, the Kottler metrics of positive mass turn out to define complete vacuum static data sets with cosmological constant Λ in the region where $V \geq 0$. In the $\Lambda < 0$ case, one can go further and search for warped product solutions with any compact surface of constant curvature, rather than just looking at the standard round unit sphere. For $\Lambda = -3$, this construction leads to the following definition.

Definition. Let $(\hat{\Sigma}, \hat{g})$ be a surface with constant curvature $\hat{k}$ equal to 1, 0, or -1. Let $m \in \mathbb{R}$ be large enough so that the function

$$V(r) = \sqrt{r^2 + \hat{k} - \frac{2m}{r}}$$

has a positive zero. Let r_m be the largest positive zero of V, and let $M = [r_m, \infty) \times \hat{\Sigma}$ with metric

$$g = V^{-2}dr^2 + r^2\hat{g}$$

We say that (M, g) is a *(generalized* $\Lambda = -3$*) Kottler space with conformal infinity* $(\hat{\Sigma}, \hat{g})$ *and mass* m.

We will usually drop the "generalized $\Lambda = -3$" part of the description, because the context will be clear throughout this article. Note that these metrics have $R = -6$. While they are asymptotically hyperbolic when $\hat{\Sigma}$ is a sphere, they are only asymptotically *locally* hyperbolic in general. Also observe that if $\hat{k}$ is 1 or 0, then m must be positive in order to have $r_m > 0$, but if $\hat{k} = -1$, then m need only be greater than $m_{\text{crit}} = -\frac{1}{3\sqrt{3}}$.

Recall the following important characterization of Schwarzschild space by Müller zum Hagen-Robinson-Seifert [11], and Bunting and Masood-ul-Alam [5].

Theorem 1.1 (Static uniqueness theorem for Schwarzschild space.). *Let (M^3, g, V) be a complete vacuum static data set with zero cosmological constant, such that (M, g) is asymptotically flat. Then (M, g) is either Euclidean space, or it is a Schwarzschild space, with V equal to the usual static potential, up to a constant multiple.*

A natural question is whether we can prove an analogous theorem for the $\Lambda = -3$ Kottler spaces (regular or generalized). The general picture is far from fully understood. Boucher-Gibbons-Horowitz [2] and X. Wang [14] proved that the only *asymptotically hyperbolic* complete vacuum static data set with $\Lambda = -3$ and *no boundary* is hyperbolic space (also called anti-de Sitter space in this context). Horowitz and Myers discovered complete vacuum static data sets with $\Lambda = -3$ that are NOT generalized Kottler spaces [7]. They have *no boundaries* while their conformal boundaries at infinity have the topology of a torus (thus only asymptotically *locally* hyperbolic). So one can see that the most naive conjecture one could make about static uniqueness of generalized Kottler spaces fails. Galloway, Surya, and Woolgar proved a static uniqueness theorem for the Horowitz-Myers examples [6].

In order to state our main result, we must first introduce the concept of surface gravity. If (M, g, V) is a complete vacuum static data set, a standard argument shows that $|\nabla V|$ is constant on each component of ∂M, and we call this constant the *surface gravity* κ of that component. For the Kottler metrics with $\hat{k} = -1$, there is a bijection between possible surface gravities in $(0, \infty)$ and possible masses in $(-\frac{1}{3\sqrt{3}}, \infty)$, so we can define a function $m(\kappa)$ according to the fixed relationship between mass and surface gravity for Kottler metrics with $\hat{k} = -1$. We can now state our main theorem, which is a static uniqueness theorem for Kottler metrics of negative mass.

Theorem 1.2 (Lee-Neves). *Let (M^3, g, V) be an asymptotically locally hyperbolic, complete vacuum static data set with cosmological constant $\Lambda = -3$. If ∂M is homeomorphic to the conformal boundary of M, and if $m(\kappa) \leq 0$, then (M, g) must be isometric to a Kottler metric of mass $m(\kappa)$, while V is equal to the usual static potential of the Kottler metric, up to a constant multiple.*

2 Penrose inequalities

Recall the following theorem.

Theorem 2.1 (Riemannian Penrose inequality). *Let (M^3, g) be a complete asymptotically flat manifold with boundary. If (M, g) has nonnegative scalar curvature, and if ∂M is an outermost minimal surface (i.e., there are no compact minimal surfaces separating ∂M from infinity), then*

$$m \geq \sqrt{\frac{A}{16\pi}},$$

where A is the area of ∂M. Furthermore, equality occurs if and only if (M, g) is isometric to a Schwarzschild space.

Huisken and Ilmanen [8] first proved this inequality for A equal to the largest area of a connected component of ∂M. Bray [3] later proved the more general inequality described above using a different method, and this result was later extended to dimensions less than 8 by Bray and the author [4].

We are interested in proving an analog of this theorem for asymptotically locally hyperbolic manifolds, which we now define.

Definition. We say that a Riemannian metric g on a smooth manifold M^3 is *asymptotically locally hyperbolic* if there exists a compact set K and a surface $(\hat{\Sigma}, \hat{g})$ of constant curvature $\hat{k}$ (called the *conformal boundary* of (M, g)) such that $M \smallsetminus K$ is diffeomorphic to $(1, \infty) \times \hat{\Sigma}$ with the metric satisfying

$$g = (\hat{k} + \rho^2)^{-1} d\rho^2 + \rho^2 \hat{g} + \rho^{-1} h + Q,$$

where

- ρ is the coordinate on $(1, \infty)$;
- Q is a symmetric 2-tensor on M such that

$$|Q|_b + \rho|\bar{\nabla} Q|_b + \rho^2 |\bar{\nabla}^2 Q|_b = o(\rho^{-3}),$$

  where b is the hyperbolic metric $(\hat{k}+\rho^2)^{-1}d\rho^2+\rho^2\hat{g}$, and $\bar{\nabla}$ is the Levi-Civita connection of b;
- h is a symmetric 2-tensor on $\hat{\Sigma}$ depending on ρ in such a way that there exists a function μ on $\hat{\Sigma}$, called the *mass aspect function*, such that $\lim_{\rho\to\infty} \frac{3}{4} \operatorname{tr}_{\hat{g}} h = \mu$, where the convergence is in C^2.

The surface $\hat{\Sigma}$ may always be chosen to have the normalization that $\hat{k}$ is 1, 0, or -1, and in the $\hat{k} = 0$ case, we can normalize so that $|\hat{\Sigma}| = 4\pi$.

Finally, we define the *mass* to be

$$m = \fint_{\hat{\Sigma}} \mu \, d\hat{g} = \frac{1}{|\hat{\Sigma}|_{\hat{g}}} \int_{\hat{\Sigma}} \mu \, d\hat{g},$$

and we also define

$$\bar{m} = \sup_{\hat{\Sigma}} \mu.$$

In order to see that the Kottler metrics are asymptotically locally hyperbolic according to this definition, one must perform a change of coordinates, in which case the h in the definition above is just $\frac{2}{3} m\hat{g}$.

Huisken and Ilmanen proved the Riemannian Penrose inequality using Geroch monotonicity, which states that the Hawking mass of a surface moving by inverse mean curvature flow (IMCF) is monotone nondecreasing if the scalar curvature of the ambient space is nonnegative. In order to adapt their argument to the asymptotically locally hyperbolic setting, we can define the *Hawking mass* of a compact surface Σ in a manifold M^3 by fixing a nonnegative integer $\mathfrak{g}$ (which would ideally be the genus of Σ) and defining

$$m_H(\Sigma) := \sqrt{\frac{|\Sigma|}{16\pi}} \left(1 - \mathfrak{g} - \frac{1}{16\pi} \int_\Sigma (H^2 - 4) \right).$$

It is not hard to show that Geroch monotonicity of $m_H(\Sigma_t)$ holds for *smooth* inverse mean curvature flow if $R \geq -6$.

Proposition 2.2 (Geroch monotonicity). *Suppose that we have a family of compact, connected, two-sided surfaces Σ_t in a Riemannian manifold (M^3, g) that* smoothly *evolves via inverse mean curvature flow $\frac{dx}{dt} = H^{-1}\nu$, where ν is a normal vector of Σ_t. Also assume that the genus of Σ_t is at least $\mathfrak{g}$. If $R \geq -6$, then*

$$\frac{d}{dt} m_H(\Sigma_t) \geq 0.$$

The proof is nearly the same as the $R \geq 0$ version, but we provide the proof here for the sake of exposition.

Proof. We will compute the rate of change of the Hawking mass of Σ_t. Recall that for IMCF, $\frac{d}{dt} d\sigma_t = d\sigma_t$, where $d\sigma_t$ is the area measure on Σ_t, and that

$$\frac{dH}{dt} = -\Delta H^{-1} - (|A|^2 + \mathrm{Ric}(\nu,\nu))H^{-1}.$$

A standard computation using the Gauss-Codazzi equations shows that

$$|A|^2 + \mathrm{Ric}(\nu,\nu) = \frac{1}{2}(R - 2K + \tfrac{1}{2}|\mathring{A}|^2 + \tfrac{3}{2}H^2).$$

Therefore

$$\begin{aligned}
\frac{d}{dt}\int_{\Sigma_t}(H^2 - 4) &= \int_{\Sigma_t}\left(2H\frac{dH}{dt} + H^2 - 4\right) \\
&= \int_{\Sigma_t}(-2H\Delta H^{-1} - (R - 2K + \tfrac{1}{2}|\mathring{A}|^2 + \tfrac{3}{2}H^2) + H^2 - 4) \\
&= \int_{\Sigma_t}(2\nabla H \cdot \nabla H^{-1} - R + 2K - \tfrac{1}{2}|\mathring{A}|^2 - \tfrac{1}{2}H^2 - 4) \\
&= 4\pi\chi(\Sigma_t) + \int_{\Sigma_t}(-2H^{-2}|\nabla H|^2 - \tfrac{1}{2}|\mathring{A}|^2 - (R+6) - \tfrac{1}{2}(H^2 - 4)) \\
&= \frac{1}{2}\left(8\pi\chi(\Sigma_t) - \int_{\Sigma_t}[(H^2 - 4) + 2(R+6) + |\mathring{A}|^2 + 4H^{-2}|\nabla H|^2]\right).
\end{aligned}$$

If we define $\hat{\chi} := 2 - 2\mathfrak{g}$, then

$$\begin{aligned}
\frac{d}{dt} m_H(\Sigma_t) &= (16\pi)^{-3/2}\frac{d}{dt}\left(|\Sigma_t|^{1/2}\left[8\pi\hat{\chi} - \int_{\Sigma_t}(H^2 - 4)\right]\right) \\
&= (16\pi)^{-3/2}\left(\frac{1}{2}|\Sigma_t|^{-1/2}\frac{d}{dt}|\Sigma_t|\right)\left[8\pi\hat{\chi} - \int_{\Sigma}(H^2 - 4)\right] \\
&\quad - (16\pi)^{-3/2}|\Sigma_t|^{1/2}\frac{d}{dt}\int_{\Sigma}(H^2 - 4) \\
&= (16\pi)^{-3/2}|\Sigma_t|^{1/2}\cdot\frac{1}{2}\left[8\pi\hat{\chi} - \int_{\Sigma}(H^2 - 4)\right] \\
&\quad - (16\pi)^{-3/2}|\Sigma_t|^{1/2}\cdot\frac{1}{2}\Bigg[8\pi\chi(\Sigma_t) - \int_{\Sigma_t}((H^2 - 4)
\end{aligned}$$

$$+ 2(R+6) + |\mathring{A}|^2 + 4H^{-2}|\nabla H|^2)\Big]$$
$$= \frac{1}{2}(16\pi)^{-3/2}|\Sigma_t|^{1/2}\Big[8\pi(\hat{\chi} - \chi(\Sigma_t))$$
$$+ \int_{\Sigma_t} (2(R+6) + |\mathring{A}|^2 + 4H^{-2}|\nabla H|^2)\Big],$$

which is clearly nonnegative given our assumptions. □

Consider the IMCF with initial condition $\Sigma_0 = \partial M$. If the flow were smooth and behaved nicely as $t \to \infty$, we would have

$$\sqrt{\frac{|\partial M|}{16\pi}}\left(1 - \mathfrak{g} + \frac{|\partial M|}{4\pi}\right) = m_H(\partial M) \leq \lim_{t\to\infty} m_H(\Sigma_t) = m\left(\frac{|\hat{\Sigma}|}{4\pi}\right)^{3/2}$$
$$= m\left(\max\{1, \mathfrak{g} - 1\}\right)^{3/2},$$

where the last line just follows from our normalization of $\hat{\Sigma}$ and Gauss-Bonnet, and the line before that is the equality that *would* be true *if* Σ_t were a constant ρ surface. This non-rigorous computation suggests the following conjecture, originally proposed by X. Wang in the $\mathfrak{g} = 0$ case.

Conjecture 2.3 ($\Lambda = -3$ Penrose inequality). *Let (M^3, g) be a complete asymptotically locally hyperbolic manifold with boundary, and with conformal boundary $(\hat{\Sigma}, \hat{g})$ of genus $\mathfrak{g}$. If (M, g) has scalar curvature $R \geq -6$, and if ∂M is an outermost minimal surface, then*

$$m\left(\max\{1, \mathfrak{g} - 1\}\right)^{3/2} \geq \sqrt{\frac{|\partial M|}{16\pi}}\left(1 - \mathfrak{g} + \frac{|\partial M|}{4\pi}\right).$$

Furthermore, equality occurs if and only if (M, g) is isometric to the Kottler space with conformal boundary $(\hat{\Sigma}, \hat{g})$ and mass m.

There are two big problems with the proof sketch of the conjecture described above. The first is that we do not expect IMCF to be smooth. This is the main problem tackled by Huisken and Ilmanen in their proof of the Penrose inequality in the asymptotically flat case [8]. They gave a weak formulation of IMCF and were able to make the argument rigorous for the weak flow. The weak formulation of IMCF works just as well in the asymptotically locally hyperbolic setting, though we have to work harder to prove Geroch monotonicity for the weak flow. Instead, the main problem for us is that the long-time behavior of Σ_t is not well-behaved enough to conclude that

$$\lim_{t\to\infty} m_H(\Sigma_t) \leq m\left(\frac{|\hat{\Sigma}|}{4\pi}\right)^{3/2}.$$

In particular, Neves proved that there are counterexamples, even when $\hat{\Sigma}$ is a round sphere [12].

However, we are able to show that a version of this inequality holds when the mass aspect μ is nonpositive. We call this the "negative mass" Penrose inequality.

Theorem 2.4 (Lee-Neves). *Let (M^3, g) be an asymptotically locally hyperbolic manifold with conformal boundary $(\hat{\Sigma}, \hat{g})$ of genus $\mathfrak{g}$, and assume $\bar{m} \leq 0$.*

We also assume that $R \geq -6$, ∂M is an outermost minimal surface, and ∂M also has genus $\mathfrak{g}$. Then

$$\bar{m}\left(\max\{1, \mathfrak{g}-1\}\right)^{3/2} \geq \sqrt{\frac{A}{16\pi}}\left(1-\mathfrak{g}+\frac{A}{4\pi}\right),$$

where A is the area of ∂M.

Furthermore, equality occurs if and only if (M, g) is isometric to the Kottler space with infinity $(\hat{\Sigma}, \hat{g})$ and mass $\bar{m}$.

By minimizing the right side of the inequality, we obtain a simple corollary.

Corollary 2.5. *Let (M^3, g) be an asymptotically locally hyperbolic manifold with conformal boundary of genus $\mathfrak{g}$.*

Assume that $R \geq -6$, ∂M is an outermost minimal surface, and ∂M also has genus $\mathfrak{g}$. Then

$$\bar{m} > 0 \quad \text{for } \mathfrak{g} = 0 \text{ or } 1,$$
$$\bar{m} > -\frac{1}{3\sqrt{3}} \quad \text{for } \mathfrak{g} > 1.$$

While the assumption on the boundary is not desirable, note that the Horowitz-Myers examples described in the Introduction have $\mathfrak{g} = 1$, no boundaries, and *negative* masses. Of course, the $\mathfrak{g} = 0$ case of the corollary is essentially a special case of the positive mass theorem for asymptotically hyperbolic manifolds [13, 1]. We note that the inverse mean curvature flow technique allows us to a give a new proof of (a non-optimal version of) the positive mass theorem for asymptotically hyperbolic manifolds.

Theorem 2.6 (Positive mass theorem for asymptotically hyperbolic manifolds). *Let (M^3, g) be a complete manifold (with or without a minimal boundary) that is asymptotically hyperbolic (that is, asymptotically locally hyperbolic with spherical conformal boundary), with scalar curvature $R \geq -6$. Then $\bar{m} \geq 0$.*

The case with boundary is already covered by the corollary above, while the case with no boundary can be handled byconsidering the IMCF whose initial condition is effectively a single point, just as in Huisken and Ilmanen's proof of the positive mass theorem for asymptotically flat manifolds [8].

3 The work of Chruściel and Simon

P. Chruściel and W. Simon observed that our main result (Theorem 1.2) would follow from a Penrose inequality for asymptotically locally hyperbolic manifolds, combined with the following theorem of theirs.

Theorem 3.1 (Chruściel-Simon). *Let (M^3, g, V) be an asymptotically locally hyperbolic, complete vacuum static data set with cosmological constant $\Lambda = -3$. Assume that ∂M is homeomorphic to the conformal boundary, and let κ be its surface gravity.*

Suppose that $m(\kappa) < 0$, and let (M_0, g_0, V_0) denote the Kottler space with the same conformal boundary as M and mass $m(\kappa)$. Then

$$|\partial M| \geq |\partial M_0|$$
$$\bar{m} \leq m(\kappa).$$

We can now describe Chruściel and Simon's proof of how our main theorem follows from the theorem above and the negative mass Penrose inequality.

Proof of Theorem 1.2. Let (M, g, V) be as in the statement of the theorem, whose infinity is $\hat{\Sigma}$. In particular, all of the hypotheses of Theorem 3.1 are satisfied, and thus

$$A \geq A_0$$
$$\bar{m} \leq m_0 := m(\kappa),$$

where A is the area of ∂M and A_0 is the area of ∂M_0 in the reference solution. In particular, $\bar{m} \leq m(\kappa) \leq 0$. With some work we can show that ∂M is an outermost minimal surface, and therefore we can also apply our negative mass Penrose inequality (Theorem 2.4). Since $m(\kappa) < 0$, it must be the case that the conformal boundary has genus $\mathfrak{g}$ at least 2. Therefore the Penrose inequality tells us that

$$\bar{m}(\mathfrak{g}-1)^{3/2} \geq \sqrt{\frac{A}{16\pi}}\left(1 - \mathfrak{g} + \frac{A}{4\pi}\right). \tag{3}$$

It is convenient to define constants

$$\mathfrak{r} := \sqrt{\frac{A}{4\pi(\mathfrak{g}-1)}}$$
$$r_0 := \sqrt{\frac{A_0}{4\pi(\mathfrak{g}-1)}},$$

so that we have

$$\mathfrak{r} \geq r_0.$$

Inequality (3) then becomes

$$\bar{m} \geq \frac{1}{2}\mathfrak{r}(-1 + \mathfrak{r}^2)$$
$$2\bar{m} + \mathfrak{r} - \mathfrak{r}^3 \geq 0.$$

Meanwhile, on the reference space we know that r_0 is the largest root of

$$2m_0 + r_0 - r_0^3 = 0.$$

One can deduce that $r_0 \geq \frac{1}{\sqrt{3}}$ using elementary reasoning. Therefore

$$\begin{aligned} 0 &\leq 2\bar{m} + \mathfrak{r} - \mathfrak{r}^3 \\ &= (2\bar{m} + \mathfrak{r} - \mathfrak{r}^3) - (2m_0 + r_0 - r_0^3) \\ &= 2(\bar{m} - m_0) + (\mathfrak{r} - r_0) - (\mathfrak{r}^3 - r_0^3) \\ &= 2(\bar{m} - m_0) + (\mathfrak{r} - r_0)[1 - (\mathfrak{r}^2 + \mathfrak{r}r_0 + r_0^2)] \\ &\leq 2(\bar{m} - m_0) + (\mathfrak{r} - r_0)(1 - 3r_0^2) \\ &\leq 0. \end{aligned}$$

Therefore all of the inequalities must be equalities, so it follows from the rigidity part of Theorem 2.4 that (M, g) is isometric to a Kottler space of mass $m_0 = m(\kappa)$. □

4 Proof of the negative mass Penrose inequality

As mentioned earlier, Huisken and Ilmanen's weak formulation of IMCF works in essentially the same way in the asymptotically locally hyperbolic case. However, if one wishes to prove monotonicity for the weak flow, it is clear from the proof of Proposition 2.2 that one must show that the Euler characteristic of Σ_t does not increase. For this part of the argument, we use the seminal work of W. Meeks, L. Simon, and S.-T. Yau [10]. The argument can be roughly outlined as follows: The main results of Meeks, Simon, and Yau allow us to find some Σ' that is obtained from Σ_t by a sequence of isotopies and γ-reductions, with the property that each component of Σ' is either very small or is essentially a "multiple" of a minimal surface. The outermost assumption on ∂M, together with some simple topological arguments, allows us to conclude that one component of Σ' may be taken to be ∂M, and also that Σ_t must be connected. Since γ-reductions are topological surgeries that replace annuli by a pair of disks, one can conclude that the genus of Σ_t is at least as large as the genus of ∂M.

Another important feature of the weak IMCF is that it can be modified so that it jumps over obstacles. When the $\mathfrak{g} \geq 2$, a new argument is needed to explain why monotonicity still holds when we jump over obstacles, but we choose not to discuss this argument here. Given these arguments, we can conclude that Geroch monotonicity holds for the weak IMCF, possibly with obstacles.

We now turn to the main technical problem in our work, which is the problem of long-time behavior of Σ_t. As seen in earlier work of Neves [12], we do not have very good behavior as $t \to \infty$, but we are able to control some of the geometry of Σ_t as $t \to \infty$.

We consider a conformal compactification $\tilde{g} = \rho^{-2} g$ of the exterior region of (M, g), where ρ is the coordinate we used in the definition of asymptotically locally hyperbolic. Then if we set $s = \rho^{-1}$, then we have

$$\tilde{g} = ds^2 + \hat{g} + \tilde{Q},$$

on the space $(0, s_1) \times \hat{\Sigma}$ for small enough s_1, where

$$|\tilde{Q}| + s|\tilde{\nabla}\tilde{Q}| + s^2|\tilde{\nabla}\tilde{Q}| = O(s^2).$$

Using sub- and supersolutions of inverse mean curvature flow, we know that there is a constant C such that the s coordinate on Σ_t must lie in $(\frac{1}{C}e^{-t/2}, Ce^{-t/2})$. We will use $\tilde{\Sigma}_t$ to denote the surface Σ_t endowed with the induced metric from $\tilde{g}$. The bounds on s, together with the fact that $|\Sigma_t| = |\Sigma_0|\, e^t$, allow us to uniformly bound the area of $\tilde{\Sigma}_t$.

The asymptotically locally hyperbolic assumption gives us some knowledge of the asymptotics of the Ricci curvature. If $\frac{\nabla\rho}{|\nabla\rho|}, e_1, e_2$ is an orthonormal frame at a point in M, then

$$\begin{aligned} \mathrm{Ric}\left(\frac{\nabla\rho}{|\nabla\rho|}, \frac{\nabla\rho}{|\nabla\rho|}\right) &= -2 - 2\mu\rho^{-3} + o(\rho^{-3}) \\ \mathrm{Ric}(e_i, e_j) &= -2\delta_{ij} + O(\rho^{-3}) \\ \mathrm{Ric}\left(\frac{\nabla\rho}{|\nabla\rho|}, e_i\right) &= o(\rho^{-3}) \\ R &= -6 + o(\rho^{-3}). \end{aligned}$$

Using this information, we are able to derive the following estimates on Σ_t.

$$\left|\int_{\Sigma_t} (H^2 - 4)\right| \leq C, \tag{4}$$

$$\int_{\tilde{\Sigma}_t} |\tilde{\nabla}^T s|^2 = O(e^{-t/2}), \tag{5}$$

$$|\tilde{\Sigma}_t| = |\hat{\Sigma}| + O(e^{-t/2}), \tag{6}$$

$$\lim_{i\to\infty} \int_{\Sigma_{t_i}} |\mathring{A}|^2 = 0 \text{ for some sequence } t_i, \tag{7}$$

$$\chi(\Sigma_{t_i}) = \chi(\hat{\Sigma}) \text{ for sufficiently large } i. \tag{8}$$

If we think of the first factor in $(0, s_1) \times \hat{\Sigma}$ as "vertical" and the $\hat{\Sigma}$ factor as "horizontal," then estimate (5) tells us that the $\tilde{\Sigma}_t$ is mostly horizontal in an L^2 sense. This alone does not prevent $\tilde{\Sigma}$ from "folding over" on itself, but together with the knowledge that Σ_t is mean convex, we can conclude estimate (6).

Using the Gauss-Codazzi equations, the asymptotics of scalar curvature, and the Gauss-Bonnet Theorem, we have

$$\begin{aligned} m_H(\Sigma_t) &= \sqrt{\frac{|\Sigma_t|}{16\pi}}\left(1 - \mathfrak{g} - \frac{1}{16\pi}\int_{\Sigma_t}(H^2 - 4)\right) \\ &= \sqrt{\frac{|\Sigma_t|}{(16\pi)^3}}\left(16\pi(1-\mathfrak{g}) - \int_{\Sigma_t}(-2R + 4K_{\Sigma_t} + 4\mathrm{Ric}(\nu,\nu) + |\mathring{A}|^2 - 4)\right) \\ &= \sqrt{\frac{|\Sigma_t|}{(16\pi)^3}}\left(8\pi\chi(\hat{\Sigma}) - \int_{\Sigma_t}(4K_{\Sigma_t} + 4(\mathrm{Ric}(\nu,\nu) + 2) + |\mathring{A}|^2 + O(r^{-5}))\right) \end{aligned}$$

$$= \sqrt{\frac{|\Sigma_t|}{(16\pi)^3}} \left(8\pi\chi(\hat{\Sigma}) - 8\pi\chi(\Sigma_t) - \int_{\Sigma_t} |\mathring{A}|^2 \right.$$
$$\left. - 4 \int_{\Sigma_t} (\mathrm{Ric}(\nu,\nu) + 2) + O(e^{-3t/2}) \right).$$

If we choose t to be one of the times from the sequence described in (7), we can use (7) and (8) to conclude

$$m_H(\Sigma_t) \leq -\frac{1}{2}(4\pi)^{-3/2}\sqrt{|\Sigma_t|} \int_{\Sigma_t} (\mathrm{Ric}(\nu,\nu) + 2) + O(e^{-t}). \tag{9}$$

If we decompose the unit normal ν of Σ_t as

$$\nu = a\frac{\nabla\rho}{|\nabla\rho|} + v$$

so that v is orthogonal to $\nabla\rho$, then we can see that $|v| = \frac{\nabla^T\rho}{|\nabla\rho|}$ while $a^2 = 1 - |v|^2$. Then it follows from the asymptotics of Ricci that

$$\begin{aligned} \mathrm{Ric}(\nu,\nu) &= a^2(-2 - 2\mu\rho^{-3}) + |v|^2(-2 + O(\rho^{-3})) + o(\rho^{-3}) \\ &= -2 - 2\mu r^{-3} + |v|^2 O(\rho^{-3}) + o(\rho^{-3}). \end{aligned} \tag{10}$$

Note that

$$\begin{aligned} \int_{\Sigma_t} \mu|v|^2\rho^{-3} &= \int_{\Sigma_t} \mu \left(\frac{|\nabla^T\rho|}{|\nabla\rho|}\right)^2 \rho^{-3} \\ &= \int_{\Sigma_t} \mu \left(\frac{|\nabla^T s|}{|\nabla s|}\right)^2 s^3 \\ &= \int_{\tilde{\Sigma}_t} \mu \left(\frac{|\tilde{\nabla}^T s|}{|\tilde{\nabla} s|}\right)^2 s \\ &= O(e^{-t}), \end{aligned}$$

where we used (5) in the last line. Putting this together with formulas (9) and (10) we obtain

$$\begin{aligned} m_H(\Sigma_t) &\leq (4\pi)^{-3/2}\sqrt{|\Sigma_t|} \int_{\Sigma_t} \mu\rho^{-3} + O(e^{-t/2}) \\ &= (4\pi)^{-3/2} \left(\int_{\tilde{\Sigma}_t} s^{-2}\right)^{1/2} \left(\int_{\tilde{\Sigma}_t} \mu s\right) + O(e^{-t/2}) \end{aligned} \tag{11}$$

We use the Hölder inequality with $p = 3$ and $q = 3/2$ to see that

$$\begin{aligned} \int_{\tilde{\Sigma}_t} \mu^{2/3} &= \int_{\tilde{\Sigma}_t} s^{-2/3}(\mu s)^{2/3} \\ &\leq \left(\int_{\tilde{\Sigma}_t} s^{-2}\right)^{1/3} \left(\int_{\tilde{\Sigma}_t} |\mu| s\right)^{2/3} \end{aligned}$$

Now, if we know that the mass aspect $\mu \leq 0$, then by taking the 3/2 power of both sides of the above inequality, we conclude that

$$-\left(\int_{\tilde{\Sigma}_t} \mu^{2/3}\right)^{3/2} \geq \left(\int_{\tilde{\Sigma}_t} s^{-2}\right)^{1/2} \left(\int_{\tilde{\Sigma}_t} \mu s\right).$$

Combining this with inequality (11), we have

$$\begin{aligned} m_H(\Sigma_t) &\leq -\left(\frac{1}{4\pi}\int_{\tilde{\Sigma}_t} \mu^{2/3}\right)^{3/2} + O(e^{-t/2}) \\ &\leq \bar{m}|\Sigma_t|^{3/2} + O(e^{-t/2}), \end{aligned}$$

for any sequence of t's as in (7). Our negative mass Penrose inequality (Theorem 2.4) now follows from Geroch monotonicity of the weak flow, and by applying our normalization of $\hat{\Sigma}$ and Gauss-Bonnet to the right side. The rigidity follows in a similar manner as in Huisken and Ilmanen [8].

References

[1] Vincent Bonini and Jie Qing. A positive mass theorem on asymptotically hyperbolic manifolds with corners along a hypersurface. *Ann. Henri Poincaré*, 9(2):347–372, 2008.

[2] W. Boucher, G. W. Gibbons, and Gary T. Horowitz. Uniqueness theorem for anti-de Sitter spacetime. *Phys. Rev. D (3)*, 30(12):2447–2451, 1984.

[3] Hubert L. Bray. Proof of the Riemannian Penrose inequality using the positive mass theorem. *J. Differential Geom.*, 59(2):177–267, 2001.

[4] Hubert L. Bray and Dan A. Lee. On the Riemannian Penrose inequality in dimensions less than eight. *Duke Math. J.*, 148(1):81–106, 2009.

[5] Gary L. Bunting and A. K. M. Masood-ul Alam. Nonexistence of multiple black holes in asymptotically Euclidean static vacuum space-time. *Gen. Relativity Gravitation*, 19(2):147–154, 1987.

[6] G. J. Galloway, S. Surya, and E. Woolgar. On the geometry and mass of static, asymptotically AdS spacetimes, and the uniqueness of the AdS soliton. *Comm. Math. Phys.*, 241(1):1–25, 2003.

[7] Gary T. Horowitz and Robert C. Myers. AdS-CFT correspondence and a new positive energy conjecture for general relativity. *Phys. Rev. D (3)*, 59(2):026005, 12, 1999.

[8] Gerhard Huisken and Tom Ilmanen. The inverse mean curvature flow and the Riemannian Penrose inequality. *J. Differential Geom.*, 59(3):353–437, 2001.

[9] Dan A. Lee and André Neves. The penrose inequality for asymptotically locally hyperbolic spaces with nonpositive mass. 2013, arXiv:1310.3002.

[10] William Meeks, III, Leon Simon, and Shing Tung Yau. Embedded minimal surfaces, exotic spheres, and manifolds with positive Ricci curvature. *Ann. of Math. (2)*, 116(3):621–659, 1982.

[11] H. Müller zum Hagen, David C. Robinson, and H. J. Seifert. Black holes in static vacuum space-times. *General Relativity and Gravitation*, 4:53–78, 1973.

[12] André Neves. Insufficient convergence of inverse mean curvature flow on asymptotically hyperbolic manifolds. *J. Differential Geom.*, 84(1):191–229, 2010.

[13] Xiaodong Wang. The mass of asymptotically hyperbolic manifolds. *J. Differential Geom.*, 57(2):273–299, 2001.

[14] Xiaodong Wang. On the uniqueness of the AdS spacetime. *Acta Math. Sin. (Engl. Ser.)*, 21(4):917–922, 2005.

Introduction to
Modern Mathematics
ALM 33, pp. 363–390

Lectures on Finite Groups*

Ronald Solomon†

Abstract

This series of lectures is organized in the following way. Each of the first four lectures explains and illustrates one idea fundamental to the study of finite groups. The fifth lecture presents an overview of a proof of the Classification of Finite Simple Groups (CFSG), highlighting the roles played by these four fundamental concepts. Finally, the sixth lecture discusses some directions of current and future research.

We hope to inspire young mathematicians to achieve new and deeper insights into finite simple groups and related algebraic structures, in the spirit of the following quote from John G. Thompson in *Group Theory: Essays for Philip Hall* (1984):

"Those of us who have been engaged in this work [the Classification of Finite Simple Groups] are the intellectual confréres of Linnaeus ... I wonder if a future Darwin will conceptualize and unify our hard-won theorems. ... Possibly, *The Origin of Groups* remains to be written."

1 Lecture 1: Counting and characters

Finite group theory was born in the memoir, "Réflexions sur la résolution algébrique des équations," presented to the Berlin Academy of Sciences in 1770–71 by Joseph Louis Lagrange. Near the very end of this lengthy work appears the theorem which in modern language states:

Lagrange's Theorem. *Let G be a group which acts transitively on a set X. Then the left coset space of G_x in G is permutation isomorphic to X as G-set, where $G_x := \{g \in G : g(x) = x\}$, for any $x \in X$. In particular, if G is finite, then* $|G| = |X||G_x|$.

By transitive action on X, we mean that, for any $x, y \in X$, there exists $g \in G$ with $y = g(x)$. Approximately 100 years later, Ludwig Sylow proved

Sylow's Theorem. *Let G be a finite group with $|G| = p^a \cdot m$, where p is a prime not dividing m. Then G acts transitively by conjugation (i.e., $g(P) = gPg^{-1}$) on*

*Professor Emeritus, The Ohio State University, Columbus, Ohio, U. S. A. Email: solomon@math.ohiostate.edv

†Our sincere thanks to Professor Lizhen Li, Tsinghua University, and the Morningside Center for Mathematics, for their generous support and for affording the opportunity to present these lectures.

the set

$$Syl_p(G) := \{P \leq G : |P| = p^a\},$$

and $|Syl_p(G)| \equiv 1 \pmod{p}$.

Here is an example of an easy consequence of these two theorems.

Application. *Suppose* $|G| = p(p+1)$ *with* p *a prime and* $P \in Syl_p(G)$. *Then either*

(1) $P \lhd G$; *or*
(2) $p = 2$ *and* $G \cong S_3$; *or*
(3) $p + 1 = 2^a$ *and there exists* $K \lhd G$ *with* $|K| = 2^a$ *and* K *abelian of exponent* 2.

Proof. Assume that neither (1) nor (2) holds. Then p is odd and by Sylow, $|Syl_p(G)| = p + 1$, whence if $X := \{g \in G : g^p \neq 1\}$, then by counting we see that $|X| = p$. (Note: If $P, P_1 \in Syl_p(G)$, then $P \cap P_1 = P$ or $\{1\}$.) By Lagrange, $N_G(P) = P$, whence P centralizes no member of X, and hence P acts transitively by conjugation on X. As p is odd and $|G|$ is even, X contains elements of order 2. Hence,

$$K := X \cup \{1\} = \{k \in G : k^2 = 1\}.$$

In particular, every element of G has order 1, 2, or p. If K is not closed under multiplication, then for some $x, y \in K$, xy has order p and $y(xy)y^{-1} = yx = (xy)^{-1}$. But then y normalizes the p-group $\langle xy \rangle$, contrary to the fact that $N_G(P) = P$. Hence K is an abelian normal subgroup of G of exponent 2. It follows that $|K| = p + 1 = 2^a$, whence (3) holds. □

Note that a key fact in this proof is that $P \cap P^g = \{1\}$ or P, for all $P^g := gPg^{-1}$. This inspires the

Definition. *A subgroup* T *of* G *is a TI-subgroup if* $T \cap T^g = \{1\}$ *or* T *for all* $g \in G$.

For much of the 19th century, the theory of groups focused on finite permutation groups, although the theories of (infinite) discrete and continuous groups developed rapidly starting in the 1870s, thanks primarily to the work of Sophus Lie, Felix Klein, and Henri Poincaré. In 1861, Emile Mathieu discovered an interesting new subgroup of the symmetric group, S_{12}, which came to be known as M_{12}. It is highly transitive in the following sense.

Definition. *Suppose* G *acts on a set* X *with* $|X| \geq m$ *so that if* $(\alpha_1, \dots, \alpha_m)$ *and* $(\beta_1, \dots, \beta_m)$ *are any m-tuples of distinct points, there exists* $g \in G$ *such that* $g(\alpha_i) = \beta_i$ *for all i. Then we say that* G *acts m-transitively on* X. *If, moreover,* g *is unique, we say that* G *acts sharply m-transitively on* X.

Now, Mathieu's new group, M_{12}, acts sharply 5-transitively on $\{1, 2, \dots, 12\}$. Of course S_m acts sharply m-transitively on $\{1, 2, \dots, m\}$ and A_{m+2} acts sharply m-transitively on $\{1, 2, \dots, m+2\}$, but otherwise this property is quite rare for $m > 1$, and Mathieu's fellow Parisian, Camille Jordan, decided to investigate it.

He was able to show that for $m \geq 4$, the only examples besides A_{m+2} and S_m are M_{12} and M_{11}, the latter being the stabilizer of a point in M_{12}. Moreover, he proved:

Jordan's Theorem. *Suppose that G acts sharply 2-transitively on a set X with $|X| = n$. Then $n = p^a$ for some prime p and $a > 0$, and $G = KG_x$, where K is a normal Sylow p-subgroup of G with $|K| = p^n$ and K abelian of exponent p, and $K \cap G_x = \{1\}$.*

Exercises.

(1) *Prove Jordan's Theorem, using the ideas in the proof of the Application above. [Note that in cases (2) and (3) above, G must in fact be a sharply 2-transitive group of degree $p + 1 = 2^a$. You may also want to use Cauchy's Theorem: If r is a prime divisor of $|G|$, then G contains an element of order r. This is, of course, a corollary of Sylow's Theorem.]*

(2) *Prove: Every group G acts sharply 1-transitively on some set X.*

In the late 1890s, F. G. Frobenius began the development of the theory of group characters. Here, if $\rho : G \to GL(n, \mathbb{C})$ is a group homomorphism, then the function $\chi : G \to \mathbb{C}$ via $\chi(g) = Tr(\rho(g))$ is called the associated group character. In 1897, William Burnside wrote a beautiful book, "The Theory of Groups of Finite Order". He expressed appropriate skepticism concerning this new theory of group characters, which had not proved its value in any applications. But, by 1911, when Burnside published the second edition of his book, he had completely reversed his opinion and devoted half of his book to the theory of group characters. Why?

In 1901, Frobenius applied his theory to prove the following important generalization of Jordan's Theorem.

Frobenius' Theorem. *Let G act transitively on the finite set X with $|X| > 1$. Suppose that $G_\alpha \neq \{1\}$, but $G_{\alpha\beta} := G_\alpha \cap G_\beta = \{1\}$ for all distinct $\alpha, \beta \in X$. Then $G = KG_\alpha$ with K a normal subgroup of G and $K \cap G_\alpha = \{1\}$.*

We call G a **Frobenius group**. It is easy to guess what the **Frobenius kernel** K should be:

$$K = \{1\} \cup \{g \in G : g(x) \neq x \text{ for all } x \in X\}.$$

But, to this day, there is no known proof that K is closed under multiplication which does not use the theory of group characters to recognize K as the kernel of of a representation (homomorphism) $\rho : G \to GL(n, \mathbb{C})$. This marks the beginning of the theory of exceptional characters of finite groups, developed in the 1950s by Richard Brauer, Michio Suzuki, and Walter Feit.

We now move to a different application of counting – counting with involutions. This was first exploited in a paper of William Burnside in 1899, in which he determined all finite groups of even order in which every element has either odd order or order 2. This paper was ignored for 70 years. Burnside's method was independently discovered by Richard Brauer around 1950. It rests on the following elementary observation, part of which is implicit in our Application above.

Lemma. *Let* $s, t \in I(G) := \{x \in G : x^2 = 1 \neq x\}$. *(We call the members of* $I(G)$ *involutions.) If* $s \neq t$*, then* $\langle s, t \rangle = \langle r, t : r^n = t^2 = 1, trt^{-1} = r^{-1} \rangle$ *is a dihedral group (with* $r = st$ *of order* $n \geq 2$*). Moreover, if* $s \notin t^G := \{gtg^{-1} : g \in G\}$*, then* $n = 2m$ *is even and* $z := r^m$ *is an involution commuting with both* s *and* t.

We now consider the following commuting graph $\Gamma_1(G)$. This is an undirected graph without multiple edges whose vertex set is $I(G)$, the set of all involutions of G. A pair $\{s, t\}$ of distinct involutions is an edge of $\Gamma_1(G)$ if and only if $st = ts$, i.e., s and t commute.

Exercises.

(1) *Draw the graphs* $\Gamma_1(G)$ *for* $G = S_3$, S_4, A_5, *and* S_5. .
(2) *Draw the subgraph of* $\Gamma_1(S_5)$ *on the set* T *of transpositions* (a, b) *in* S_5. *This is a famous graph, the so-called Petersen graph.*
(3) *Prove:* G *acts as a group of graph automorphisms of* $\Gamma_1(G)$.
(4) *Prove: The Petersen graph can be realized as follows: Take the* 1*-skeleton of the regular dodecahedron and identify antipodal vertices and edges. The resulting graph is the Petersen graph. Using this, prove that the group of all rotational symmetries of the dodecahedron is isomorphic to* A_5.

The following is an easy corollary of the previous lemma. We leave it as an exercise.

Exercise. *Prove: Let* G *be a finite group having at least two distinct* G*-conjugacy classes of involutions. Then* $\Gamma_1(G)$ *is a connected graph. In fact, the diameter of* $\Gamma_1(G)$ *is at most* 3*, i.e., any two vertices are connected by a path of length at most* 3.

By the exercise, if $\Gamma_1(G)$ is a disconnected graph, then G has exactly only G-conjugacy class of involutions. Hence, G transitively permutes the vertices of $\Gamma_1(G)$, and the connected components of $\Gamma_1(G)$. Let Γ_0 be one of the connected components of $\Gamma_1(G)$, and let $C_0 = \{g \in G : g(\Gamma_0) = \Gamma_0\}$, the stabilizer in G of Γ_0.

Definition. *Suppose that* $\Gamma_1(G)$ *is a disconnected graph and let* M *be a proper subgroup of* G *containing* C_0. *We call* M *a* **strongly 2-embedded subgroup** *(or sometimes simply, a strongly embedded subgroup) of* G.

Note that by the Lemma, if $\Gamma_1(G)$ is disconnected, then G transitively permutes the vertex set of $\Gamma_1(G)$, and hence transitively permutes the connected components of Γ_0. In particular, G_0 is a proper subgroup of G and the stabilizers of connected components of $\Gamma_1(G)$ are the G-conjugates of G_0.

Example. *If* $G = S_3$*, then* $M \cong S_2$ *is a strongly* 2*-embedded subgroup of* G. *If* $G = A_5$*, then* $M \cong A_4$ *is a strongly* 2*-embedded subgroup of* G.

Exercise. *Prove that the following definitions of strongly* 2*-embedded subgroup are equivalent to the definition above:*

(1) *A proper subgroup* M *of* G *is strongly* 2*-embedded in* G *if* $|M|$ *is even but* $|M \cap M^g|$ *is odd for all* $g \in G - M$.

(2) *Let G be a finite group of even order which acts transitively on a set X with $|X| > 1$. A point stabilizer $M = G_\alpha$ is strongly 2-embedded in G if every involution of G fixes exactly one point of X.*

Exercise. *Prove: Let G be a Frobenius group acting on a finite set X, with $|G_\alpha|$ even. Then G_α is a strongly 2-embedded subgroup of G.*

Here is an easy counting argument.

Theorem. *Let G be a finite group of even order. Choose $x \in I(G)$ with $c := |C_G(x)|$ as large as possible. Suppose that $y \in I(G) - x^G$. Then $|G| < c^3$.*

Proof. Let $\Gamma = \Gamma_1(G)$ and let $\Delta_i(x) = \{w \in I(G) : d(x,w) = i\}$, where $d(x,w)$ denotes the length of the shortest path from x to w. As $y \notin x^G$, the Lemma implies that $d(x,y) = 1$ or $d(x,y) = 2$ (with z adjacent to both x and y). Thus $|y^G| \leq |\Delta_1(x)| + |\Delta_2(x)|$. Clearly, $|\Delta_1(x)| \leq c - 2$. Also, applying the same reasoning to each $w \in \Delta_1(x)$ and remembering that $|C_G(w)| \leq c$, we see that $|\Delta_2(x)| \leq (c-2)(c-3)$. Hence

$$|y^G| \leq |\Delta_1(x)| + |\Delta_2(x)| \leq (c-2) + (c-2)(c-3) = (c-2)^2.$$

By Lagrange's Theorem, $|G| = |y^G||C_G(y)| \leq (c-2)^2 \cdot c < c^3$, as claimed. □

Exercises. *Let G and c be as in the Theorem.*

(1) *Prove: $c = 4n$ for some $n \geq 1$.*
(2) *Determine the possibilities for G when $c = 4$ or 8.*
(3) *Let p be any odd prime. Give an example of a group $G = G(p)$ such that $c = 4p$ and $|G| = 4p^3$.*
(4) *Suppose G has a non-identity normal 2-subgroup N. Prove: $|G| < c^3$. (Hint: Consider $y \in Z(N)$ of order 2.)*

Notice that if G is a dihedral group of order $2n$ with n odd, then $|C_G(x)| = 2$ for every involution x in G. So, in general, $|G|$ is not bounded by any function of $|C_G(x)|$. On the other hand, the following fact is true.

Fact. *Let G be a finite group of even order. Choose $x \in I(G)$ with $c := |C_G(x)|$ as large as possible. Suppose that the graph $\Gamma_1(G)$ has at least one edge. Then $|G| < c^3$.*

However, I do not know any way to prove this statement without using the full strength of the Classification of Finite Simple Groups. Notice that if $G = A_5$, then $c = 4$ and $|G| = 60 < 4^3$. However, this is close. Indeed, if $G = SL(2, 2^n)$, the group of 2×2 matrices of determinant 1 with entries from the finite field of order 2^n, then $c = 2^n$ and $|G| = 2^n(2^{2n} - 1)$. So $\frac{|G|}{c^3} = \frac{2^{2n}-1}{2^{2n}} = 1 - \frac{1}{2^{2n}} \to 1$ as $n \to \infty$. (Note: $A_5 \cong SL(2,4)$.)

Although there is no known proof of the Fact without using CFSG, a similar but much weaker theorem was discovered around 1952 by Richard Brauer and his student Kenneth Fowler:

Brauer-Fowler Theorem. *Let G be a finite simple group of even order. Let $x \in I(G)$ and let $c = |C_G(x)|$. Then $|G| < (c^2)!$.*

Although the $(c^2)!$ upper bound is useless for practical purposes, the Brauer-Fowler Theorem was a great inspiration for the work on CFSG, especially after Feit and Thompson proved 10 years later that every non-abelian finite simple group has even order.

Around 1970, while preparing a survey article on finite group theory in the 20th century, Walter Feit discovered the 1899 Burnside paper mentioned earlier. In this paper, Burnside proved the following lovely theorem.

Theorem (Burnside). *Let G be a finite group of even order in which every element of even order is an involution. Let T be a Sylow 2-subgroup of G. Then either $T \lhd G$ or $|T| = 2$ (in which case, G is a generalized dihedral group) or $G \cong SL(2, 2^n)$ for some $n \geq 2$.*

Notice that it is easy to show that if T is not normal in G, then $\Gamma_1(G)$ is disconnected and $T - \{1\}$ is the vertex set of a connected component Γ_0 of $\Gamma_1(G)$. Thus $M := N_G(T)$ is a strongly 2-embedded subgroup of G. It is also easy to show that $|M| = (q-1)q$ with $q = |T|$. Finally, with a bit more careful but elementary argument, it is possible to show that $|G| = (q-1)q(q+1)$ and G is a transitive permutation group on the set $X = Syl_2(G)$ with $|X| = q+1$. It is then not difficult to prove that for each q, there is at most one such group G. Since $SL(2,q)$ satisfies the hypotheses of Burnside's Theorem, this completes the proof. Details may be found on pages 295–305 of Suzuki, *Group Theory III.*

One of the fundamental theorems in the work on CFSG is the following theorem of Helmut Bender, based on an important theorem of Michio Suzuki:

Theorem (Bender-Suzuki). *Let G be a finite simple group having a strongly 2-embedded subgroup. Then $G \cong SL(2, 2^n)$ or $PSU(3, 2^n)$ or $Sz(2^n)$.*

Here $SU(3, 2^n)$ is the group of 3×3 unitary matrices of determinant 1 defined over a field of order 2^n, and the groups $Sz(2^n)$ are a family of 4×4 matrix groups discovered by Suzuki.

2 Lecture 2: Architecture

A large finite group may be visualized as a large building constructed out of many bricks. A fundamental theorem about finite groups and their bricks is the Jordan-Hölder Theorem:

Jordan-Hölder Theorem. *Let G be a finite group. There is a chain $\mathcal{C}$ of subgroups:*

$$\mathcal{C} : \{1\} = G_0 \lhd G_1 \lhd \cdots \lhd G_{n-1} \lhd G_n = G,$$

with each quotient group $K_i := G_i/G_{i-1}$ simple. Let $\mathcal{K} = \{K_1, \ldots, K_n\}$. Then $\mathcal{K}$ is determined (up to isomorphism and a permutation of the set $\mathcal{K}$) by G, independent *of the choice of the chain $\mathcal{C}$.*

One may then imagine each finite group G as a tower with brick K_n at the bottom and brick K_1 at the top. For example, if $G = S_4$, then the tower would look like:

$$\begin{array}{c} C_2 \\ C_3 \\ C_2 \\ C_2 \end{array}$$

However, this is not a very good picture of S_4. Let's call the ordered family $(K_1, \ldots, K_n)$ an (ordered) Jordan-Hölder data set for G.

Exercises.

(1) *Prove: There are, up to isomorphism,* 15 *groups of order* 24, 14 *of which have the same ordered Jordan-Hölder data set as* S_4. *In fact, the only one which doesn't is the group* $SL(2,3)$ *of all* 2×2 *matrices of determinant* 1 *with entries from the field of order* 3.

(2) *Draw a Jordan-Hölder Tower for* $SL(2,3)$. *Prove that there are six non-isomorphic groups with the same Jordan-Hölder data set as* $SL(2,3)$.

Remark. *The ordered Jordan-Hölder data set for* S_4 *and for* $SL(2,3)$ *is unique. For most groups, this is not the case.*

A better picture for S_4 is the following, which I will call a **Fitting tower** for S_4:

$$\begin{array}{c} C_2 \\ C_3 \\ C_2 \times C_2 \end{array}$$

Here are the defining rules for a Fitting tower:

$$\{1\} = F_0 \lhd F_1 \lhd \cdots \lhd F_{n-1} \lhd F_n = G$$

is a Fitting tower for G if the following conditions hold:

(1) $F_i \lhd G$ for all i.
(2) For all $i \geq 1$, $C_{G/F_{i-1}}(F_i/F_{i-1}) \leq F_i/F_{i-1}$.
(3) F_i is minimal subject to the above.

The associated ordered Fitting data is the ordered n-tuple

$$\mathcal{F} = (F_1/F_0, F_2/F_1, \ldots, F_n/F_{n-1}).$$

Exercises.

(1) *Construct a Fitting tower for* $SL(2,3)$. *(This illustrates that the Fitting quotients for G need not be abelian groups, even if G is a solvable group.)*

(2) *Prove that S_4 and $SL(2,3)$ have unique Fitting towers.*
(3) *Prove that S_4 and $SL(2,3)$ are determined up to isomorphism by their Fitting data.*
(4) *Prove that D_8 does not have unique Fitting data.*
(5) *Prove that D_8 is not uniquely determined by the ordered Fitting data (C_4, C_2).*

The important feature of a Fitting tower is a consequence of the following observation:

Observation. *Let N be a normal subgroup of a group G. Then G-conjugation induces a homomorphism $c : G \to Aut(N)$ with kernel $C_G(N)$. We call N a* **large normal subgroup** *if $C_G(N) \leq N$. In this case, G/N is isomorphic to a section (quotient of a subgroup) of $Aut(N)$.*

Notice that if $\mathcal{F}$ is a Fitting tower for a group G, then, for all $i \geq 1$, F_i/F_{i-1} is a large normal subgroup of G/F_{i-1}.

Exercises.
(1) *Prove: If N is a large normal subgroup of the finite group G, then $|G| \leq |N|!$.*
(2) *Determine all cases when equality holds, and justify your claim.*

We use the term Fitting tower to honor the mathematician Hans Fitting, who proved the important theorem below. First we need some terminology.

Definition. *A group N is* **nilpotent** *if there is a chain of subgroups:*

$$N = N^0 > N^1 = [N,N] > \cdots > N^{r-1} > N^r = \{1\},$$

with $N^{i+1} = [N^i, N]$ for all $i \geq 0$.

The fundamental example of a nilpotent group is the subgroup $U(n,F)$ of $GL(n,F)$ consisting of upper triangular matrices with all diagonal entries equal to 1. The terminology comes from the fact that the associated Lie algebra $n(n,F)$ of upper triangular matrices with all diagonal entries 0 is an algebra of nilpotent matrices. If F is a finite field of characteristic p, then $U(n,F)$ is a finite p-group, and indeed is a Sylow p-subgroup of $GL(n,F)$. It follows easily from Cauchy's lemma (that the center of a non-trivial finite p-group is non-trivial) that every finite p-group is nilpotent. Also, direct products of nilpotent groups are nilpotent. In fact, the following important fact holds.

Proposition. *Let N and N_1 be normal nilpotent subgroups of a group G. Then NN_1 is a nilpotent subgroup of G. Hence, if G is finite, G has a unique maximal normal nilpotent subgroup, $F(G)$, the Fitting subgroup of G.*

The following theorem establishes the role of nilpotent subgroups in our architectural study of finite groups.

Fitting's Theorem. *Let G be a finite solvable group. Then $F(G)$ is a large normal subgroup of G.*

Sketch of Proof. Let $F = F(G)$ and $C = FC_G(F)$. Suppose that $C > F$. As G/F is solvable, there is a non-trivial characteristic abelian subgroup A/F of C/F.

Then $A \lhd G$ and it is easy to check that A is nilpotent, contrary to the fact that F is the maximal normal nilpotent subgroup of G.

In general, the Fitting subgroup is not a minimal large normal subgroup of G. However it does provide a tower

$$\{1\} \leq F(G) = F^1(G) \leq F^2(G) \leq \cdots \leq F^h(G) = G,$$

with $F^{i+1}(G)/F^i(G) = F(G/F^i(G))$ for all $i \geq 1$, which satisfies the conditions for a Fitting tower, except for the minimality condition, whenever G is a solvable group. We call h the Fitting height of G.

Definition. *Let G be a finite group. We call G* **constrained** *if $F(G)$ is a large normal subgroup of G.*

Note: Fitting worked in Germany in the early 1930s under the influence of Emmy Noether's development of a structural approach to algebra via the systematic study of homomorphic images. The Fitting subgroup is the analogue of the nil radical of a ring.

Fitting's Theorem asserts that every finite solvable group is constrained. However, there are many examples of constrained groups which are not solvable. Indeed, just as unipotent linear groups provide the fundamental examples of nilpotent groups, parabolic subgroups of reductive Lie groups or linear algebraic groups afford the fundamental examples of constrained groups. We will not develop the general theory of parabolic subgroups here, but only give the following class of important examples.

Example. *Let $G = GL(V)$ be the group of invertible linear operators on the n-dimensional F-vector space V, for some field F. A chain $\mathcal{F}$ of subspaces*

$$\{0\} = V_0 \subseteq V_1 \subseteq \cdots \subseteq V_r = V$$

is called a **flag** *of subspaces in V. The stabilizer $P(\mathcal{F})$ of $\mathcal{F}$ in G is $\{g \in G : g(V_i) = V_i \text{ for all } i\}$. Any such stabilizer of a non-trivial flag is called a* **parabolic subgroup** *of $GL(V)$. A* **maximal flag** *in V is a flag of length n, i.e., $dim(V_i) = i$ for all i. The stabilizer of a maximal flag is called a* **Borel subgroup** *of G.*

Each parabolic subgroup $P = P(\mathcal{F})$ has a Levi decomposition: $P = UL$, where U is the unipotent radical of P, i.e., the largest normal unipotent (and hence, nilpotent) subgroup of G, and L is a complement to U in P, called a Levi complement.

Exercise. *Let V be a 4-dimensional F-vector space with basis $\mathcal{E} = (e_1, e_2, e_3, e_4)$. Use $\mathcal{E}$ to identify $GL(V)$ with $GL(4, F)$. Let $\mathcal{F}$ be the maximal flag*

$$\{0\} = V_0 \subseteq V_1 = Fe_1 \subseteq V_2 = Fe_1 + Fe_2 \subseteq V_3 = Fe_1 + Fe_2 + Fe_3 \subseteq V_4 = V.$$

(1) *Prove: The associated Borel subgroup $B(\mathcal{F})$ is the group of upper triangular matrices.*

(2) *List all flags $\mathcal{F}'$ such that $B(\mathcal{F}) \leq P(\mathcal{F}')$, and identify $P(\mathcal{F}')$ as a subgroup of $GL(4, F)$.*
(3) *For each $P(\mathcal{F}')$, identify $U(\mathcal{F}')$ and $L(\mathcal{F}')$. Describe each of these, up to isomorphism. [Note: $L(\mathcal{F}')$ is not uniquely determined, but it is unique up to isomorphism and, given $\mathcal{E}$, there is a natural choice for it.]*

Although U is a normal nilpotent subgroup of P, it is not, in general, the Fitting subgroup of P. Let $Z = Z(GL(V))$, the group of invertible scalar operators, $z = \alpha I$, $\alpha \neq 0$. Then $Z \subseteq Z(P)$ and $F(P) = U \times Z$. Even when the field F is infinite, $U \times Z$ is the unique largest normal nilpotent subgroup of $P = P(\mathcal{F})$ and $C_P(U \times Z) \leq U \times Z$, i.e., $F(P) = U \times Z$ is a large normal subgroup of P. Thus P is constrained. Now,

$$L \cong GL(V_1/V_0) \times GL(V_2/V_1) \times \cdots \times GL(V/V_{r-1}).$$

If $n_i := dim(V_i/V_{i-1}) > 1$ for some i, then $GL(V_i/V_{i-1})$ is not a solvable group unless $n_i = 2$ and $|F| \leq 3$. Thus we have many examples of constrained parabolic subgroups which are not solvable groups.

It is reasonable to wonder if all finite groups are constrained. This is certainly not the case. If G is a non-abelian finite simple group, then $F(G) = \{1\}$, and G is certainly not constrained. Looking to linear groups again, we see that, just as parabolic subgroups are the standard examples of constrained groups, reductive or semisimple subgroups are the standard examples of non-constrained groups. By definition, a linear group G is **reductive** if the unipotent radical of G is trivial. Roughly speaking (for connected groups), this implies that $Z(G)$ is the largest normal nilpotent subgroup of a reductive group G. Hence, in general, G is reductive and constrained only if G is abelian. (This is false if $G = GL(2,2) \cong S_3$ or $G = GL(2,3)$ or $SL(2,3)$. These groups are both reductive and constrained, but not abelian.)

Example. *Let $G = GL(V)$ and let s be a diagonalizable element of G. If $V = V_{\lambda_1} \oplus \cdots \oplus V_{\lambda_r}$ is the decomposition of V into eigenspaces for s, then*

$$C_G(s) = GL(V_{\lambda_1}) \times \cdots \times GL(V_{\lambda_r}).$$

Thus, $C_G(s)$ is a reductive group and, generically, $F(C_G(s)) = Z_1 \times \cdots \times Z_r = Z(C_G(s))$, where $Z_i = Z(GL(V_{\lambda_i}))$. Hence, if $|F| > 3$, then $C_G(s)$ is not constrained unless each eigenspace is 1-dimensional. Thus we have many examples of non-constrained groups.

Around 1970, Helmut Bender discovered a generalization of the Fitting subgroup which provides a large normal subgroup for every finite group G. We need some definitions.

Definitions.

(1) *A finite group L is* **quasisimple** *if $L = [L, L]$ and $L/Z(L)$ is simple.*
(2) *A finite group E is* **semisimple** *if $E = L_1 L_2 \ldots L_r$, with each L_i quasisimple and with $[L_i, L_j] = 1$ for all $i \neq j$.*

(3) *If G is a finite group, then E(G) is defined to be the largest normal semisimple subgroup of G, and $F^*(G) = E(G)F(G)$ is called the* **generalized Fitting subgroup** *of G.*

The condition $L = [L, L]$ in the definition of a quasisimple group rules out the possibility that $L = S \times A$, where S is a nonabelian simple group and A is a nontrivial abelian group. The quintessential example of a quasisimple group is $SL(n, F)$, the group of $n \times n$ matrices of determinant 1 with entries from the field F, when $n > 1$ and $(n, |F|) \neq (2, 2)$ or $(2, 3)$. It is a nontrivial exercise to show that the product of normal semisimple subgroups of a group G is again a normal semisimple subgroup of G. With this in hand, it follows that there is a unique maximal normal semisimple subgroup of any finite group G, which we call $E(G)$. It is also a fact that $[E(G), F(G)] = 1$. [See GLS2, Chapter 3, for more information and proofs.]

Finally, we have Bender's fundamental theorem:

Theorem (Bender). *Let G be a finite group. Then $F^*(G)$ is a large normal subgroup of G.*

Sketch of proof. As in the proof of Fitting's Theorem, let $F = F(G)$ and $C = FC_G(F)$. As in that proof, we can see that the socle E/F of C/F is a direct product of nonabelian simple groups, and of course, $C_{C/F}(E/F) = 1$. Then, the full pre-image of that socle in G is $F^*(G)$ and $C_G(F^*(G)) = C_C(F^*(G)) \leq F^*(G)$, as claimed.

Thus, $F^*(G)$ is the sturdy foundation on which every finite group G is constructed. If $F^*(G)$ is known and $Aut(F^*(G))$ is known, then structure of the entire group G is at least "bounded", if not completely determined. We will look at an important special case of this in the next lecture.

Exercises.

(1) *Prove: Let $E = L_1 \times L_2 \times \cdots \times L_r$ be a direct product of nonabelian simple groups, and let α be an automorphism of E. Then α permutes the set $\{L_1, L_2, \ldots, L_r\}$. (Also: Show that this is (dramatically) false if the L_i are abelian simple groups.)*

(2) *Prove: Let $E = F^*(G)$ with $E = L_1 \times L_2 \times \cdots \times L_r$ with $L_1 \cong L_2 \cong \ldots \cong L_r$, and with L_i a nonabelian simple group for all i. Let $A_i = Aut(L_i)$. Then G is isomorphic to a subgroup of $A = (A_1 \times \cdots \times A_r) : S$, where $S \cong S_r$ acting faithfully by permutations on the set $\{A_1, \ldots, A_r\}$.*

(3) *Let $\Pi = \Pi_1 \cup \cdots \cup \Pi_r$ be a partition of $\{1, 2, \ldots, n\}$. Let G be the stabilizer in S_n of Π, i.e., if $g \in G$ and $1 \leq i \leq r$, then $g(\Pi_i) = \Pi_j$ for some j. Describe $F^*(G)$. Now describe G.*

(4) *Prove: Let $E = L_1 \times L_2 \times \cdots \times L_r \lhd G$ with $r > 1$ and with L_i nonabelian simple of even order for all i. Then G has more than one conjugacy class of involutions. [Hence, with c as in Lecture 1, $|G| < c^3$.]*

3 Lecture 3: Linear methods

It turns out that the most important structures of $F^*(M)$ which arise in the study of subgroups M of finite simple groups G are:

(1) $F^*(M) = F(M)$ is a p-group for some prime p (the p-Constrained Case); or
(2) $F^*(M) = E(M)Z(F^*(M))$ (the Reductive Case).

A clue as to why this should be so is that most finite simple groups are simple groups of Lie type, i.e., are similar to $PSL(n, F) = SL(n, F)/Z(SL(n, F))$. In this case, parabolic subgroups M are always p-constrained, where $p = char(F)$, while centralizers M of semisimple elements are reductive.

Exercises.

(1) *Prove: Let $G = S_9$ and let $t = (12)(34) \in G$. Then $C_G(t)$ is neither constrained nor reductive. However, if $s = (12)$, then $C_G(s)$ is reductive, while if $x = (12)(34)(56)(78)$, then $C_G(x)$ is 2-constrained.*
(2) *Let $G = S_n$ with $n \geq 7$, and let $s = (12)$. Then $C_G(s)$ is reductive.*
(3) *Let $G = S_n$ with $n = 2^m$ or $2^m + 1$, and let $x = (12)(34)\dots(2^m - 1, 2^m)$. Then $C_G(x)$ is 2-constrained.*

In this lecture we will introduce some important ideas for the study of the p-constrained case. Thus, throughout we will work under the following hypotheses:

Hypotheses.

(1) *p is a prime and M is a finite p-constrained group;*
(2) *$R = F(M) = F^*(M)$ is a p-group.*

Now, $Aut(R)$ may be quite difficult to determine. A very important subgroup is the **Frattini subgroup** $\Phi(R)$.

Definition. *Let R be a finite group. Then, $\Phi(R)$ is the intersection of all of the maximal subgroups of R.*

Note that $\Phi(R)$ is another type of radical subgroup, analogous to the Jacobson radical. It is easy to see that $\Phi(R)$ is the set of all non-generators of R, in the sense that if R is generated by $\Phi(R)$ and H, then R is generated by H alone.

Exercises.

(1) *Prove: If R is generated by $\Phi(R)$ and H, then R is generated by H alone.*
(2) *Prove: $\Phi(R)$ is a characteristic subgroup of R, i.e., if α is an automorphism of R, then $\alpha(\Phi(R)) = \Phi(R)$.*

When R is a finite p-group, it is easy to see that T is a maximal subgroup of R if and only if $|R : T| = p$. It follows that $R/\Phi(R)$ is abelian of exponent p (elementary abelian). The right way to think about an elementary abelian p-group is to think of it as a vector space over the field $\mathbb{F}_p$. We introduce the following additional notation:

$$R/\Phi(R) := V.$$

The Burnside Basis Theorem asserts that every minimal generating set for R has cardinality $dim_{\mathbb{F}_p}(V)$. The key fact for us is:

$$C_M(V) = R.$$

Here, $C_M(V) = \{g \in M : [R,g] \leq \Phi(R)\}$.

Exercise. *Prove: Let M be a finite group with $R = F(M) = F^*(M)$ a p-group. Then $C_M(R/\Phi(R)) = R$.*

Hint:

(1) *Prove that $C_R(R/\Phi(R)) = R$.*
(2) *Prove: If $g \in C_M(R/\Phi(R))$ and g has prime order q, then $q = p$. Hence $C_M(R/\Phi(R))$ is a p-group.*
(3) *Prove: $C_M(R/\Phi(R))$ is a normal subgroup of M.*

Thus, the conjugation action of M on R induces an injective homomorphism

$$c : M/R \to Aut(V) = GL(V).$$

Hence, we may study M/R as a group of linear operators (transformations) on the vector space V. Here are some of the most important consequences.

Co-prime action

The "easy" case is the study of the action of subgroups of M of order co-prime to p, so-called p'-subgroups, on R. The key fact, which is an extension of Hint 2 above, and is easy to verify by counting, is:

Lemma. *Let Q be a q-subgroup of M, where q is a prime different from p. Then $C_V(Q) = C_R(Q)\Phi(R)/\Phi(R)$.*

Because of this fact, we can reduce most questions about co-prime action on R to purely linear algebra questions about co-prime action on V.

Schur's Lemma. Let Q be a non-cyclic abelian q-subgroup of M of exponent q, where q is a prime different from p. Then $R = \langle C_R(x) : 1 \neq x \in Q\rangle$.

Sketch of Proof. Using the Lemma and the non-generation property of $\Phi(R)$, we reduce to the case $R = V$. Now, we can tensor V with an extension field E of $\mathbb{F}_p$ containing the qth roots of 1, to get a new field $\overline{V}$ on which Q acts diagonalizably. Thus $\overline{V} = V_1 \oplus \cdots \oplus V_n$ is a sum of 1-dimensional Q-invariant subspaces. If $Q_i = C_Q(V_i)$, then $|Q : Q_i| \leq q$. Hence, $Q_i \neq 1$ and, if $1 \neq x_i \in Q_i$, then $\overline{V} = \langle C_{\overline{V}}(x_i) : 1 \leq i \leq n\rangle$, proving the result.

Thompson's $P \times Q$ Lemma. Let $P \times Q \leq M$ with P a p-group and Q a q-group, with q a prime different from p. If Q centralizes $C_R(P)$, then $Q = 1$.

Sketch of Proof. Again, we reduce to the case $R = V$. Then $V = C_V(Q) \oplus (Q - I)(V)$ is a decomposition of V into P-invariant subspaces. [Here, we use additive notation in V, and $(Q-I)(V)$ denotes the linear span of $\{T(v) - v : T \in Q, v \in V\}$, i.e., the sum of all Q-invariant subspaces of V not centralized by Q. By hypothesis, $C_V(P) \leq C_V(Q)$. As P and Q commute, $(Q - I)(V)$ is P-invariant. But then, P is a unipotent group of linear operators on $(Q - I)(V)$ having no (non-trivial) eigenvector with eigenvalue 1. This is possible only if $(Q - I)(V) = \{0\}$, i.e., $V = C_V(Q)$. Thus, Q centralizes $V = R$. But $C_G(R) \leq R$. So $Q \leq R$, whence $Q = 1$, as claimed.

Exercise. *Let G, $\Gamma_1(G)$, x, and c be as in the Fact stated in Lecture 1. Assume that groups of odd order are solvable. [This is a famous and deep theorem of Feit and Thompson.] Prove: If G is a minimal counterexample to the statement that $|G| < c^3$, then $F^*(G)$ is a nonabelian simple group, and G has only one conjugacy class of involutions. [Hint: By the final exercises in Sections 1 and 2, this is true unless G has a non-identity normal subgroup N of odd order. Moreover, G has only one class of involutions. Since N is solvable, we may assume that N is an elementary abelian p-group for some odd prime p. (Why?) Now use Schur's Lemma to show that $|N| \leq |C_N(x)|^3$.]*

p-Group action

Much more difficult is the study of the action on R of p-subgroups of M. It is still true that if P is a Sylow p-subgroup of M, then $P/P \cap R$ acts faithfully on V. But this is usually not the most useful source of information. Instead, John Thompson taught us to study the action on a subgroup Z of $\Omega_1(Z(R))$ containing $\Omega_1(Z(P))$. [Note: As $C_M(R) = Z(R)$, $Z(P) \leq Z(R)$ for all $P \in Syl_p(M)$.]

Definition. *Let P be a p-group. Then $\Omega_1(P) = \langle x \in P : x^p = 1 \rangle$.*

We choose Z so that $M/C_M(Z)$ has no nontrivial normal p-subgroup. For example, we may choose Z to be the subgroup generated by all M-conjugates of $\Omega_1(Z(P))$.

Exercises. *Let $Z_1 = \Omega_1(Z(P)) \leq \Omega_1(Z(R))$. Let $Z = \langle gZ_1g^{-1} : g \in M \rangle$.*

(1) *Prove: $Z \leq \Omega_1(Z(R))$.*
(2) *Prove: $M/C_M(Z)$ has no nontrivial normal p-subgroup.*

Thus the conjugation action of M on Z induces a homomorphism $c : M \to GL(Z)$ and $c(M)$ has no nontrivial normal p-subgroup. This sets the stage for an application of a deep theorem of Philip Hall and Graham Higman (1956).

Theorem (Hall-Higman). *Let p be a prime and V a finite-dimensional F-vector space, where F is a field of characteristic p. Suppose that H is a constrained subgroup of $GL(V)$ having no nontrivial normal p-subgroup. Let $g \in H$ of order p. Then the minimum polynomial of g is $(x-1)^p$, unless p is a Fermat prime and $Q = [Q, g]$ is a nonabelian 2-subgroup of G, in which case the minimum polynomial of g is either $(x-1)^{p-1}$ or $(x-1)^p$.*

We say that g **acts quadratically** on V if the minimum polynomial of g is $(x-1)^2$. Thus, under the Hall-Higman hypotheses, g can only act quadratically on V if $p = 3$ and H has non-abelian Sylow 2-subgroups. In fact, a more elementary argument shows that quadratic action can only occur when $p = 3$ and H has a subgroup isomorphic to $SL(2,3)$.

Exercise. *Let* $V = C_3 \times C_3$ *and* $M = V : H$, *where* $H \cong SL(2,3)$ *and where if* $T = \begin{pmatrix} a & b \\ c & d \end{pmatrix}$ *and* $v = \begin{pmatrix} x \\ y \end{pmatrix} \in V$, *then*

$$TvT^{-1} = \begin{pmatrix} a & b \\ c & d \end{pmatrix} \begin{pmatrix} x \\ y \end{pmatrix} \in V.$$

Prove: M *is a* 3*-constrained group,* H *is a* 2*-constrained group with no non-trivial normal* 3*-subgroup, and if* $T \in H$ *has order* 3*, then the minimum polynomial of* T *is* $(x-1)^2$*, i.e.,* T *acts quadratically on* V.

If one drops the hypothesis that H is constrained, then $SL(2,3)$ proves to be the tip of an iceberg. Nevertheless, for p odd, quadratic action still gives a considerable amount of information about the subgroup H_0 of H generated by quadratic elements and the nature of V as an H_0-module. Indeed, for $p > 3$, Thompson's Quadratic Pairs Theorem gives a very detailed description of H_0 and V as H_0-module.

Thompson discovered a wonderful way to exploit this unlikelihood of quadratically acting p-elements (for p odd). The key idea was the Thompson subgroup, $J(P)$.

Definitions. *Let* P *be a finite* p*-group.*

(1) *The* **rank** *of* P, $m(P)$, *is the dimension of the largest elementary abelian* p*-subgroup of* P.
(2) $\mathcal{E}_i(P)$ *is the set of all elementary abelian subgroups of* P *of rank* i.
(3) $\mathcal{A}(P) = \mathcal{E}_{m(P)}(P)$.
(4) $J(P) = \langle A : A \in \mathcal{A}(P) \rangle$.

With M, R, and Z as before, and with $P \in Syl_p(M)$, Thompson's Replacement Theorem says that there exists $A \in \mathcal{A}(P)$ such that $[Z, A, A] = 0$, i.e., A acts trivially or quadratically on Z. In consequence, we get:

Thompson Factorization Theorem. *Let* M *be a constrained finite group with* $R = F(M)$ *a* p*-group for some odd prime* p. *Let* $P \in Syl_p(M)$. *Suppose that* Z *can be chosen with* $\Omega_1(Z(P)) \leq Z \leq \Omega_1(Z(R))$ *and such that* $M/C_M(Z)$ *is a constrained group having no nontrivial normal* p*-subgroup. Then one of the following holds:*

(1) $M = C_M(\Omega_1(Z(P)))N_M(J(P))$ *(the Thompson factorization); or*
(2) $p = 3$ *and* $M/C_M(Z)$ *has a subgroup isomorphic to* $SL(2,3)$.

Idea of Proof. If $J(P) \leq C_M(Z) \lhd M$ and $P_0 = P \cap C_M(Z)$, then $J(P) = J(P_0)$ and so

$$M = C_M(Z)N_G(P_0) = C_M(\Omega_1(Z(P)))N_G(J(P)),$$

as claimed. If not, then by Thompson Replacement, some $A \in \mathcal{A}(P)$ satisfies: $[Z, A, A] = 0 \neq [Z, A]$. So, if $\overline{M} = M/C_M(Z)$, then $(\overline{M}, Z)$ satisfies the hypotheses for (H, V) in the Hall-Higman Theorem, and if $\overline{g}$ is a non-identity element of $\overline{A}$, then $\overline{g}$ has minimum polynomial $(x-1)^2$ on Z, whence by Hall-Higman, $p = 3$ and $M/C_M(Z)$ has a subgroup isomorphic to $SL(2, 3)$, as claimed.

The power of this theorem emerges in the following context:

Definition. *Let p be a prime. A subgroup M of a finite group G is a p-local subgroup of G if M is the normalizer of a non-trivial p-subgroup of G.*

Now suppose that G is a finite simple group all of p-local subgroups are p-constrained. If Thompson factorization holds for every p-local subgroup M of G, then it is reasonable to assume that there are at most two maximal p-local subgroups of G–$C := C_G(\Omega_1(Z(P)))$ and $N := N_G(J(P))$ – containing a given Sylow p-subgroup P of G. If there are more than two – $M_1 = C$, $M_2 = N$, $M_3, \ldots, M_r$, then a Quadratic Pairs Theorem or analogue thereof will give detailed structural information about $M_i/F(M_i)$ for $i > 3$.

In particular, if G is a finite simple group of odd order all of whose proper subgroups are solvable, then the hypotheses can be shown to hold if $m(P) \geq 3$, and the second conclusion is impossible. So one may deduce that G has at most two conjugacy classes, C^G and N^G, of maximal p-local subgroups of G. Indeed, Glauberman was able to prove under these hypotheses that, in fact, G has exactly one conjugacy class of maximal p-local subgroups, $M^G := N_G(ZJ(P))^G$. From this it is not difficult to prove that M is a strongly p-embedded subgroup of G, i.e., $M \cap M^g$ is a p'-group for all $g \in G - M$. This was the first great application of these ideas, but they are marvelously robust and have applications throughout the classification of finite simple groups.

4 Lecture 4: Balance

Philip Hall and Graham Higman provided deep insights into the ways that Architecture and Linear Methods can be used to study the structure of an arbitrary finite group M. However, it is easy to see that an arbitrary finite group M can have arbitrarily complicated architecture. On the other hand, the architecture of finite groups M arising as maximal or maximal p-local subgroups of finite simple groups G is often much more restricted than one might expect. [Think of maximal parabolic subgroups and centralizers of semisimple elements in $GL(n, F)$.] If it were not possible to prove this SIMPLY FROM THE HYPOTHESIS THAT G IS SIMPLE, then there would be little hope of classifying the finite simple groups. Fortunately, it was possible.

One of the first positive hints came from a theorem of Walter Feit. First recall Frobenius' Theorem.

Frobenius' Theorem. *Let G be a transitive permutation group on a finite set X such that $G_\alpha \neq 1$, but $G_{\alpha\beta} = 1$ for all distinct $\alpha, \beta \in X$. Then G has a normal subgroup K such that $G = KG_\alpha$ and $K \cap G_\alpha = 1$.*

In his Ph.D. thesis (1958), Thompson proved that K must be a nilpotent group. It is easy to see that all abelian groups of odd order, as well as many other nilpotent groups, arise as Frobenius kernels K. In particular, K need not be a p-group. On the other hand, in 1960, Feit proved:

Feit's Theorem. *Let G be a Zassenhaus group, i.e., a 2-transitive permutation group having no regular normal subgroup and in which G_α is a Frobenius group. Then the Frobenius kernel of G_α must be a p-group for some prime p.*

Note that in the context of Feit's Theorem, G must be an (almost) simple group and $M = G_\alpha$ must be a maximal subgroup of G. Somehow the simplicity of G and the maximality of M severely restrict the structure of M.

Here is a trivial observation: Let M and N be distinct maximal subgroups of a simple group G. Suppose that X is a non-identity normal subgroup of M. Then X cannot also be a normal subgroup of N. Otherwise, X would be a proper normal subgroup of $G = \langle M, N\rangle$.

Suppose M and N are distinct maximal subgroups of a finite simple group G and X is a non-identity subgroup of $F(M)$. Then X is a **subnormal** subgroup of M, where subnormality is the transitive extension of the relation of normality. So, one might hope that X cannot also be contained in $F(N)$, i.e., subnormal in BOTH M and N. This is not true as the following example shows.

Example. *Let $G = GL(V)$, where V is a 3-dimensional vector space over $\mathbb{F}_p$. Let $\{0\} = V_0 < V_1 < V_2 < V_3 = V$ be a maximal flag of subspaces of V. Let P_i be the maximal parabolic subgroup of G which is the stabilizer of the subspace V_i, for $i = 1, 2$. Then P_1 and P_2 are distinct constrained maximal subgroups of G. Let $U = \{T \in G : (T - I)(V) \subseteq V_2, (T - I)(V_2) \subseteq V_1, (T - I)(V_1) = \{0\}\}$. Then U is a Sylow p-subgroup of G contained in $P_1 \cap P_2$. Moreover, $Z(U) \leq F(P_1) \cap F(P_2)$.*

Indeed, this example is the typical occurence for maximal parabolic subgroups in Lie-type groups. Surprisingly, Helmut Bender proved the following remarkable theorem.

Definition. *Let M and N be subgroups of a finite group G. We say that M infects N if there is a subgroup X of $F(M)$ such that $N_{F^*(M)}(X) \leq N$.*

Bender's Maximal Subgroup Theorem. *Let G be a finite simple group. Suppose that M and N are distinct maximal subgroups of G which infect each other. Then there is a prime p such that both $F^*(M)$ and $F^*(N)$ are p-groups.*

The key to the proof of this theorem is repeated application of Thompson's $P \times Q$ Lemma.

Let us now focus on a very important problem.

Definition. *Let G be a finite group, p a prime divisor of $|G|$, and $P \in Syl_p(G)$. The* ***p-local data for*** *G is*

$$\mathcal{L}_P(G) := \{N_G(R) : 1 \neq R \leq P\}.$$

If P is not specified, we shall write $\mathcal{L}_p(G)$. Up to isomorphism, this is independent of the choice of P.

QUESTION. *Can you decide, just by looking at $\mathcal{L}_p(G)$, whether G is a simple group?*

A fundamental special case is the following:

Suppose that G is a finite simple group, p and r are distinct primes, and V is an $\mathbb{F}_r$-vector space which is an irreducible G-module. Form the group $G^* = V : G$. Let $P \in Syl_p(G) \subseteq Syl_p(G^*)$.

QUESTION. *Is it possible, by looking only at $\mathcal{L}_P(G^*)$, to deduce that G^* is not a simple group?*

Since the only proper normal subgroup of G^* is V, you must find V and prove that it is normal in G^*. If G has p-rank 1, this may be impossible.

Exercise. *Let $G = GL(3,2)$, $V = C_2 \times C_2 \times C_2$, and $G^* = V : G$. Prove: $\mathcal{L}_7(G) = \mathcal{L}_7(G^*)$.*

But if G contains an elementary abelian p-subgroup A of rank at least 2, then by Schur's Lemma, $V = \langle C_V(a) : a \in A - \{1\}\rangle$. You can **see** $C_V(a)$ as a normal r-subgroup of $C_{G^*}(a)$. The trick is to reconstruct V from the $C_V(a)$ pieces which are visible inside the p-local subgroups of G^*, and to prove that V is a proper normal subgroup of G^*.

We shall discuss a strategy that accomplishes this in many cases. The following lemma illustrates how the $P \times Q$ Lemma is used.

Definition. *Let H be a finite group and p a prime. Then $O_{p'}(H)$ is the largest normal p'-subgroup of H.*

Note: The notation $O(H)$ (for $O_{2'}(H)$) was introduced by Richard Brauer. In this context O stands for odd.

Thompson's Balance Lemma. Let M be a solvable group and P a p-subgroup of M. Then $O_{p'}(N_M(P)) \leq O_{p'}(M)$.

Proof. We may assume that $O_{p'}(M) = 1$, and try to prove that $O_{p'}(N_M(P)) = 1$. Let $R = F(M)$, a p-group, and let Q be a q-subgroup of $X := O_{p'}(N_M(P))$. Then $P \times Q$ acts on R. Let $C = C_R(P) \leq N_M(P)$. Then, since $X \triangleleft N_M(P)$ and $R \triangleleft M$, we have $[Q, C] \leq X \cap R = 1$. Thus $C_R(P) \leq C_R(Q)$ and so, by Thompson's $P \times Q$ Lemma, Q centralizes $R = F(M)$. Thus $Q \leq R$ by Fitting's Theorem, and so $Q = 1$. As X is the product of its q-subgroups for different primes q, it follows that $X = 1$, as desired. □

The term balance arises from the following immediate corollary.

Theorem. *Let G be a finite group all of whose p-local subgroups are solvable. Let A and B be commuting abelian p-subgroups of G. Then the following Balance Equation holds:*

$$C_G(A) \cap O_{p'}(C_G(B)) = O_{p'}(C_G(A)) \cap C_G(B).$$

I conjecture that the symmetry of the above equation was the motivation for Danny Gorenstein to call it the Balance Equation. Unfortunately, the Balance Equation does not hold in general when p-local subgroups of G are not constrained. However, it is often possible to define a so-called signalizer functor Θ to replace the functor $O_{p'}$ for which the analogous Balance Equation does hold. For the remainder of the discussion, I will write $\Theta(a)$ for $\Theta(C_G(A))$. where $A = \langle a \rangle$. You may think of $\Theta(a)$ as $O_{p'}(C_G(a))$. I will also state results only for $p = 2$, although more general results hold. We let $\mathcal{E}_i(G)$ denote the set of all elementary abelian 2-subgroups of G of rank i. The fundamental theorem is:

Signalizer Functor Theorem (Gorenstein-Goldschmidt). *Let G be a finite group of 2-rank at least 3, and let Θ be a function (signalizer functor) defined on $I(G) = \mathcal{E}_1(G)$. Assume that for each $A \in \mathcal{E}_3(G)$ and each $a \in A - \{1\}$, $\Theta(a)$ is an A-invariant odd order subgroup of $C_G(a)$. Assume that the Balance Equation holds for Θ. Then $\Delta_\Theta(A) := \langle \Theta(a) : a \in A - \{1\} \rangle$ is a group of odd order. Moreover, $\Delta_\Theta(A) = \Delta_\Theta(B)$ for all $B \in \mathcal{E}_2(A)$.*

An easy consequence is the following result.

Theorem. *Let G be a finite group of 2-rank at least 3, and let Θ be a balanced G-equivariant signalizer functor on $\mathcal{E}_1(G)$, i.e., $\Theta(a^g) = \Theta(a)^g$ for all $a \in \mathcal{I}(G)$ and all $g \in G$. Let $\Gamma_2(G)$ be the commuting graph whose vertex set is $\mathcal{E}_2(G)$. The following conclusions hold.*

(1) *The function Δ_Θ is constant on connected components of $\Gamma_2(G)$.*
(2) *If $\Gamma_2(G)$ is connected, then $W := \Delta_\Theta(B)$ is a normal odd order subgroup of G, independent of the choice of $B \in \mathcal{E}_2(G)$.*
(3) *If G is a simple group and $\Gamma_2(G)$ is connected, then $\Theta(a) = 1$ for all $a \in \mathcal{I}(G)$.*

Proof. Let $B, C \in \mathcal{E}_2(G)$ with $[B, C] = 1$. Let $A = BC$. We may assume that $A \in \mathcal{E}_3(G)$. By the Signalizer Functor Theorem, $\Delta_\Theta(B) = \Delta_\Theta(A) = \Delta_\Theta(C)$. Thus, (1) holds. Now assume that $\Gamma_2(G)$ is connected and let $W = \Delta_\Theta(B)$, independent of the choice of B. By the Signalizer Functor Theorem, W has odd order. Let $g \in G$. As $\Theta(b^g) = \Theta(b)^g$ for all $b \in B - \{1\}$, it follows that

$$W^g = \Delta_\Theta(B)^g = \Delta_\Theta(B^g) = W.$$

Hence, $W \lhd G$, proving (2). If also G is simple (of even order), then as $|W|$ is odd and $W \lhd G$, $W = 1$, whence $\Theta(a) = 1$ for all $a \in \mathcal{I}(G)$, proving (3). □

In particular, we obtain the following theorem of John Thompson.

Theorem. *Let G be a finite simple group all of whose 2-local subgroups are solvable. Suppose that $\Gamma_2(G)$ is connected and non-empty. Then $O_{2'}(C_G(a)) = 1$ for all $a \in \mathcal{I}(G)$.*

Exercises.

(1) *Prove: Let G be a finite simple group. If $\Gamma_2(G)$ is connected and non-empty, then G has 2-rank at least 3.*
(2) *Prove this theorem of Thompson using the Signalizer Functor Theorem of Gorenstein and Goldschmidt.*

In fact, Thompson was able to establish this result whenever $\Gamma_1(G)$ is connected and G has 2-rank at least 3. When $\Gamma_1(G)$ is disconnected, G has a strongly 2-embedded subgroup, and hence is known by the Bender-Suzuki Theorem. Using this theorem, we get

Theorem. *Let G be a finite simple group of 2-rank at least 3, all of whose 2-local subgroups are solvable. Then $O_{2'}(C_G(a)) = 1$ for all involutions a of G. Hence $F^*(H) = O_2(H)$ for every 2-local subgroup H of G.*

Exercise. *Prove that the final statement of the theorem above follows from the previous statement.*

However, Thompson's method breaks down when G has 2-rank 2, and indeed, the theorem is generally false in this case. The principal example of a simple group of 2-rank 2 all of whose 2-local subgroups are solvable is $G = PSL(2,q)$ with q a power of an odd prime. In this case, if a is an involution in G, then $C_G(a)$ is a dihedral group of order $q-\epsilon$, where $\epsilon = \pm 1$ and $q \equiv \epsilon \pmod 4$. Moreover, $O_{2'}(C_G(a))$ is a cyclic group whose order is the odd part of $q-\epsilon$, which is typically not 1.

Exercise. *Let $G_1 = A_5$, $V = C_5 \times C_5 \times C_5$, $G^* = V : G_1$. Let $H = PSL(2,19)$. Prove: $\mathcal{L}_2(G^*) \cong \mathcal{L}_2(H)$. Thus, you cannot detect 2-locally that G^* is not a simple group.*

5 Lecture 5: Synthesis

We now illustrate how the ideas discussed in the first four lectures come together to achieve the Classification of the Finite Simple Groups. The outline below is designed for clarity. All theorems stated below are true. However, the actual flowchart of the classification proof is significantly oversimplified at several key points. First we state the Classification Theorem.

Classification Theorem (CFSG). *Let G be a non-abelian finite simple group. Then either*

(1) *G is a finite simple group of Lie type [prototype: $PSL(n,p^m)$, $n \geq 2$, $(n,p^m) \neq (2,2)$ or $(2,3)$]; or*
(2) *G is an alternating group, A_n, for some $n \geq 5$; or*
(3) *G is one of 26 sporadic simple groups.*

Step 1: Minimal simple groups

Work on CFSG began in earnest in the 1950s. The period from 1950 to 1970 was largely devoted to the minimal simple groups, i.e., non-abelian simple groups all of whose proper subgroups are solvable. First, let's generalize the notion of strong 2-embedding.

Definition. *A proper subgroup M of a finite group G is* **strongly p-embedded** *in G if p divides $|M|$ but p does not divide $|M \cap M^g|$ for any $g \in G - M$, i.e.,* *every non-identity p-subgroup of G lies in a* **unique** *conjugate of M. Often, M is called a* **p-uniqueness subgroup** *of G.*

Example. *Let* $F = \mathbb{F}_{p^m}$, $G = SL(2, F)$ *and* $M = \left\{ \begin{pmatrix} \alpha & t \\ 0 & \alpha^{-1} \end{pmatrix} : \alpha \in F - \{0\}, t \in F \right\}$.

Theorem. *Let G be a minimal non-abelian simple group. Let p be the smallest prime divisor of* $|G|$. *If p is odd or if* $p = 2$ *and G has 2-rank at least 3, then G has a strongly p-embedded subgroup M.*

Idea of Proof for p odd. The Thompson Balance Theorem may be used to show that p-locals are p-constrained, i.e., $F(M)$ is a p-group for every p-local subgroup M of G. Then, Thompson Factorization (or Glauberman's ZJ-Theorem) may be used to prove that there is a unqiue conjugacy class M^G of maximal p-local subgroups of G, from which it follows easily that M is a strongly p-embedded subgroup of G.

The theorem is much more difficult when $p = 2$, but is a corollary of Thompson's N-Group Theorem.

Theorem (Feit-Thompson, Bender-Suzuki). *Let G be a non-abelian simple group with a strongly p-embedded subgroup M, where p is the smallest prime divisor of* $|G|$. *Then* $p = 2$ *and* $G \cong SL(2, 2^n)$, $Sz(2^{2n-1})$, *or* $PSU(3, 2^n)$, *for some* $n \geq 2$.

Counting arguments and character theory play a very large role in the proof of this theorem. When $p = 2$, a key step is the proof that G acts 2-transitively on the left cosets of M in G. If this could be proved directly when p is odd, this would give an alternative proof of the Odd Order Theorem of Feit and Thompson.

Notice that these two theorems complete the classification of minimal simple groups, except for the case when G has even order but 2-rank 1 or 2. This case is completely handled by the following result, completed by 1970.

Theorem (Brauer, Suzuki, Gorenstein, Walter, Alperin, Lyons). *Let G be a non-abelian simple group of 2-rank at most 2. Then* $G \cong PSL(2, q)$, $PSL(3, q)$, *or* $PSU(3, q)$, *q odd, or* $G \cong A_7$, M_{11}, *or* $PSU(3, 4)$.

Corollary (Thompson). *Let G be a minimal non-abelian simple group. Then* $G \cong PSL(2, p)$,$p \geq 5$, $PSL(2, 2^p)$, $PSL(2, 3^p)$, $Sz(2^p)$, *or* $PSL(3, 3)$.

Step 2: The dichotomy

For the remainder of the discussion we may and shall assume that G is a non-abelian simple group of 2-rank ≥ 3. Having reached this point, the analysis splits in half thanks to the following theorem.

Strong Dichotomy Theorem. *Let G be a finite simple group of 2-rank at least 3. Then either*

(1) *For every 2-local subgroup H,* $F^*(H)$ *is a 2-group (G is of characteristic 2-type); or*

(2) *There exists an involution t of G such that $E(C_G(t)) \neq 1$ (G is of odd (or component) type).*

The proof of this theorem is quite lengthy. It is here where the Balance Theorem plays a critical role. If for every 2-local H, $H/O_{2'}(H)$ is constrained, then it is not hard to reach conclusion (1), just as in Thompson's theorem when all 2-locals H are solvable. However, the general case is much more difficult. In particular, the case where $\Gamma_2(G)$ is disconnected was handled by Aschbacher, who showed that either $\Gamma_1(G)$ is disconnected or $G \cong J_1$.

Step 3: The characteristic 2-type case

Here one expects to prove that G is a simple group of Lie type defined over a field $\mathbb{F}_{2^m}$, although some sporadic simple groups also occur.

Let $U \in Syl_2(G)$, $B = N_G(U)$, and $\mathcal{M}(B)$ be the set of all maximal 2-local subgroups of G containing B. In the Lie case, these will be the maximal parabolic subgroups of G. If $|\mathcal{M}(B)| = 1$, then Aschbacher proved that G has a strongly 2-embedded subgroup. Hence we may assume that $\mathcal{M}(B) = \{M_1, \ldots, M_r\}$ with $r \geq 2$. Let $L_i = M_i/F(M_i)$ and let Z_i be the L_i module contained in $\Omega_1(Z(F(M_i)))$ studied in the Thompson Factorization Method. Then linear methods give some control over the pairs (L_i, Z_i).

Define the Tits geometry Γ whose objects of type i are the left cosets of M_i in G, with incidence defined by:

$$xM_i \text{ is incident with } yM_j \text{ if and only if } xM_i \cap yM_j \neq \emptyset.$$

Exercise. *Prove: If $G \cong PSL(n, p^m)$, $n \geq 3$, then Γ is the (Desarguesian) projective geometry of dimension $n-1$ over $\mathbb{F}_{p^m}$.*

The goal is to prove that Γ is a Tits building of spherical type and G is a flag-transitive subgroup of $Aut(\Gamma)$. Sometimes this fails leading to the sporadic examples.

The most difficult case is the case when $r = 2$. This is essentially the case treated in the Quasithin Theorem of Aschbacher and Smith (2004).

Step 4: The odd type case

Here one expects to prove that either $G \cong A_n$ for some $n \geq 9$ or G is a simple group of Lie type defined over a field $\mathbb{F}_{p^m}$, for some odd prime p. Again, some sporadic groups arise.

The first key step is a refinement of the structure theorem for $C_G(t)$.

Aschbacher Component Theorem. *There exists an involution t such that $C_G(t)$ has a normal subgroup L (called a "component" of $C_G(t)$) with*

(1) *L either quasisimple or isomorphic to $O^+(4, q)' \cong SL(2, q) \circ SL(2, q)$, q odd; and*

(2) *$Q := C_G(L)$ having a Sylow 2-subgroup T with T either cyclic, generalized quaternion, or a Klein 4-group.*

Now, L is a very large subgroup of $C_G(t)$ and one expects the "type"of L to determine the "type"of G. Thus

(1) If $L \cong A_n$, then we expect $G \cong A_{n+4}$ with $t = (12)(34) \in G$; while
(2) If $L \cong SL(n, p^m)$, then we expect $G \cong PSL(n + r, p^m)$ with $r \in 1, 2$ and with $t = \text{diag}(1, -1, -1, \ldots, -1)$ or $t = \text{diag}(-1, -1, 1, \ldots, 1)$.

Indeed, generically it is possible to prove such a theorem. First, Glauberman's Z^*-Theorem guarantees the existence of a G-conjugate $s = t^g$ of t with $s \neq t$, but $st = ts$. The information about $C_G(L)$ supplied by Aschbacher's Component Theorem permits us to narrow down the possibilities for s.

For example, when $L \cong A_n$, one arrives at the case where $J = E(C_L(s)) \cong A_{n-4} \leq L^g \cong L$. When n is sufficiently large, it is then easy to read off a set of Coxeter generators and relations for $G_0 := \langle L, L^g \rangle \cong A_{n+4}$.

When L is of Lie type, there are Curtis-Tits and Phan presentations which can be used in a similar way to recognize G_0. Small cases present additional challenges. For example, if $L \cong SL(6, p^m)$, then G_0 may be either $PSL(8, p^m)$ or the exceptional Lie type group $E_6(p^m)$.

Finally, one shows that $G = G_0$, using Bender's strongly 2-embedded subgroup theorem.

Remark. *This simplified outline presents the logically most elementary outline of a proof of the Classification Theorem. This is not in fact an accurate outline of the existing proof, where the "generic"characteristic* 2*-type case is handled by switching primes and applying the analogue of the Odd Type Case analysis with the prime* 2 *replaced by a suitable odd prime* p. *There is a substantial body of work-in-progress aimed at a treatment of the Charateristic* 2*-Type Case along the lines indicated in my outline above, but this work is not yet completed.*

6 Lecture 6: Future directions

This section is of a speculative nature, mostly raising questions and suggesting directions for future study.

Question 1: What is a sporadic simple group?

The term "sporadic"in the context of simple groups was first used by William Burnside in the final line of Note N in his splendid group theory textbook. Having surveyed the known simple groups of order less than 1,000,000, he mentions the three smallest Mathieu groups, and then remarks: "These apparently sporadic simple groups would probably repay a closer examination than they have yet received.". Burnside is clearly using the word in the still current sense of: a simple group which belongs to none of the known families of finite simple groups, i.e., the alternating groups and the finite simple groups of Lie type.

It would be lovely to have an intrinsic definition of sporadic simple group; ideally, a definition such that one could then hope to prove the theorem:

Theorem. *There are only finitely many sporadic simple groups.*

without first proving the complete Classification Theorem. Here is one possibility.

Definition. *Let p be a prime. A finite group G is of* **parabolic characteristic** p *if every p-local subgroup H of G which contains a Sylow p-subgroup of G satisfies the condition $F^*(H) = O_p(H)$.*

Of course, all finite simple groups of Lie type in characteristic p are of parabolic characteristic p, and even of characteristic p-type, which is a more restrictive condition. However, many sporadic simple groups are of parabolic characteristic 2, though few are of characteristic 2-type. However, the property parabolic characteristic 2 is not, of itself, a good criterion for sporadicity. Let's try a more restrictive condition.

Definition. *Let p and r be distinct primes. A finite group G is of* **bicharacteristic** p, r *if the p-rank and r-rank of G are both at least 2, and G is of both parabolic characteristic p and parabolic characteristic r.*

Note that the condition on the p-rank and r-rank is necessitated by examples like: Suppose p and $r = p - 2$ are primes. Then A_p is of bicharacteristic p, r. As there are probably infinitely many twin primes, this would produce infinitely many "sporadic" alternating groups. Perhaps surprisingly, many of the sporadic simple groups are of bicharacteristic p, r for some distinct primes p and q, usually $2, 3$, specifically:

(1) $(2,3)$: M_{11}, M_{12}, J_3, Co_2 (also 5), Co_1 (also 5), Fi_{22}, Fi'_{24}, HN (also 5), Th (also 5), BM (also 5), M (also 5, 7, and 13)
(2) $(2,5)$: HS, Ru
(3) $(2,7)$: He
(4) $(2,11)$: J_4
(5) $(3,5)$: Mc, Co_3

Thus, 17 of the sporadic simple groups are of bicharacteristic type. A slight further weakening of the definition (which we won't specify here) will capture all of the sporadic simple groups except for J_1. But are there many non-sporadic (in the traditional sense) simple groups of bicharacteristic p, r? I don't think so. Certainly there are some. An obvious example is $PSp(4,3) = PSU(4,2)$, which is of bicharacteristic $2, 3$. Indeed the history of studying simple groups of bicharacteristic $2, p$ begins with John Thompson's characterization theorem for $PSp(4,3)$ and $G_2(3)$ in the N-Group Paper. Another example is A_9, which is also of bicharacteristic $2, 3$.

Considerable study of groups of bicharacteristic $2, p$ has been undertaken by Klinger-Mason, Gorenstein-Lyons, and Lyons-Solomon-Korchagina. This work is ongoing.

Question 2: Can we prove fundamental statements about finite groups and their representations without using CFSG?

There are numerous good examples in this context, but I will focus on my favorite one: the Alperin Weight Conjecture.

Definition. *For k a field and G a finite groups, $\ell(kG)$ is the number of isomorphism classes of simple (irreducible) kG-modules, while $\ell_0(kG)$ is the number of projective, simple kG-modules.*

Note: Suppose that $char(k) = p$. If p does not divide $|G|$, then every kG-module is projective. On the other hand, if G is a p-group, then no simple kG-module is projective. If G is a simple group of Lie type in characteristic p, then there is a unique $\overline{\mathbb{F}}_pG$-module which is both simple and projective, namely the Steinberg module of degree $|G|_p$.

Alperin's Weight Conjecture. *Let G be a finite group and $k = \overline{\mathbb{F}}_p$. Then*

$$\ell(kG) = \sum_{Q} \ell_0(kN_G(Q)/Q),$$

where Q runs over a set of representatives of the G-conjugacy classes of p-subgroups of G.

This conjecture generalizes the theory of weights in the representation theory (in the defining characteristic) for Lie groups and groups of Lie type. In this context, the set $\{N_G(Q)\}$ is the set of parabolic overgroups of a fixed Borel subgroup of G, and the conjecture is easily verified. The mystery is: why should this be true for all finite groups? Considerable progress has been made in the direction of reducing this conjecture to a related conjecture for simple groups G. (See work of Dade, Isaacs, Navarro, Malle, Späth, etc.) Also, many cases for G simple have been verified. But a deeper understanding is desirable.

Now, kG is the direct sum of indecomposable 2-sided ideals, called block algebras, B, of G. Knörr and Robinson formulated a block-by-block version of the Alperin Weight Conjecture as follows:

$$dim_k(Z(B)) = \sum_{\sigma} (-1)^{|\sigma|} dim_k(Z(B_\sigma)),$$

where σ runs over a set of representatives for the G-classes of chains of non-trivial p-subgroups:

$$\sigma : Q_0 < Q_1 < \cdots < Q_{|\sigma|},$$

and B_σ is a sum of blocks of $kN_G(\sigma)$.

The Knörr-Robinson formulation has the form of an Euler characteristic or Lefschetz number, and begs for a topological interpretation. We remark that $Z(B) \cong HH^0(B)$, the 0-th Hochschild cohomology group of the algebra B. Connections with the full Hochschild cohomology algebra $HH^*(B)$ have been explored recently by Kessar and Linckelmann.

It is natural to consider the following posets:

$S_p(G)$: the poset of nontrivial p-subgroups of a group G; and

$A_p(G)$: the subposet of nontrivial elementary abelian p-subgroups of G.

Quillen proved that these two posets are homotopy equivalent, and he formulated the following conjecture:

Quillen Conjecture. *$A_p(G)$ is contractible if and only if G has a nontrivial p-subgroup.*

The "if" direction is trivial. The best result to date in the "only if" direction is the following theorem of Aschbacher and Smith:

Theorem. *Quillen's Conjecture holds when $p > 5$ and G has no subnormal subgroup isomorphic to the unitary group $U_n(q)$, with q odd and $q \equiv -1 \pmod{p}$.*

A finite group G can be localized at a prime p in the following sense. Let $S \in Syl_p(G)$. Then $\mathcal{F}_S(G)$ is the category whose object set is the set of all subgroups of S, and whose morphism set is the set of all injective group homomorphisms $\phi : P \to Q$ induced by G-conjugations and inclusions. We call $\mathcal{F}_S(G)$ the p-fusion system of G. This category arose in the work of homotopy theorists in the following context.

Martino-Priddy Conjecture. *Let G and H be finite groups, and let p be a prime. Then the p-completed classifying spaces $BG^{\wedge}_p$ and $BH^{\wedge}_p$ are homotopy equivalent if and only if $\mathcal{F}_p(G)$ and $\mathcal{F}_p(H)$ are isomorphic categories.*

This conjecture has been proved by Bob Oliver, using CFSG. A crucial step is the construction of a related category, $\mathcal{L}_p(G)$, the associated centric linking system. A new proof of the existence and uniqueness of centric linking systems (in the more general context of abstract (saturated) fusion systems, defined next) has recently been given by Andy Chermak.

An abstract fusion system $\mathcal{F}$ over a finite p-group S is a category where $Obj(\mathcal{F})$ is the set of all subgroups of S, as before, and, for all $P, Q \leq S$,

$$Hom_S(P,Q) \subseteq Hom_{\mathcal{F}}(P,Q) \subseteq Inj(P,Q),$$

i.e., $Hom_{\mathcal{F}}(P,Q)$ is a set of injective group homomorphisms including all S-conjugations. There are two additional axioms which model elementary consequences of Sylow's Theorem when $\mathcal{F} = \mathcal{F}_S(G)$, with G a finite group having S as a Sylow p-subgroup.

Definition. *If $\phi \in Mor(\mathcal{F})$ with $\phi(P) = Q$, we say that P and Q are $\mathcal{F}$-conjugate. A subgroup P of S is fully centralized (reap., fully normalized) if $|C_S(P)|$ (resp., $|N_S(P)|$) is maximal in the $\mathcal{F}$-conjugacy class of P.*

Definition. *$\mathcal{F}$ is a (saturated) fusion system if the following conditions hold:*

(1) *Each fully normalized P is fully centralized and $Aut_S(P) \in Syl_p(Aut_{\mathcal{F}}(P))$; and*

(2) *If* $\phi \in Hom_{\mathcal{F}}(P, S)$ *with* $\phi(P)$ *fully normalized, then* ϕ *extends to a morphism* $\Phi : N_\phi \to S$, *where*

$$N_\phi = \{g \in N_S(P) : \phi \circ c_g \circ \phi^{-1} \in Aut_S(\phi(P))\}.$$

(Here, c_g denotes conjugation by the element g.)

The work of Broto-Levi-Oliver had been anticipated (in the early 1990s) by Lluis Puig's definition of a Frobenius category. His motivation was to unify the fusion category $\mathcal{F}_S(G)$ of a finite group with a fusion category implicit in the work of Alperin and Broué on Brauer pairs in the context of representation theory, where the objects are the subgroups of a defect group D of a p-block B of kG. When $B = B_0(G)$ is the principal p-block, a defect group is a Sylow p-subgroup of G and the fusion category for $B_0(G)$ is just $\mathcal{F}_S(G)$.

Meanwhile, in a parallel universe, Bill Dwyer and Clarence Wilkerson defined p-compact groups, as a homotopy-theoretic generalization of the p-completion of the classifying space of a compact Lie group. They also discovered an "exotic" 2-compact group, $DI(4)$, a 45-dimensional "overgroup" of the simply connected compact Lie group $Spin(7)$, the universal covering group of the simple 7-dimensional complex orthogonal group, B_3. The cohomology ring of $DI(4)$ is the ring of Dickson invariants in four variables, $\mathbb{F}_2[x_1, x_2, x_3, x_4]^{GL(4,2)}$, whence the name $DI(4)$.

Dave Benson suggested that certain "exotic" 2-fusion systems $\mathcal{F}_{Sol}(q)$ (which contain $\mathcal{F}_2(Spin_7(q))$) might arise as the fixed points of Adams operations on $DI(4)$. This was confirmed by Levi and Oliver, yielding a remarkable chain:

$$\mathcal{F}_2(Co_3) \subseteq \mathcal{F}_{Sol}(3) \subseteq \mathcal{F}_{Sol}(3^2) \subseteq \cdots \subseteq \mathcal{F}_{Sol}(3^\infty) \sim DI(4),$$

where Co_3 is the smallest sporadic Conway group, i.e., the stabilizer of a certain vector in the celebrated Leech lattice Λ, $\mathcal{F}_2(G)$ denotes the 2-fusion system of a group G, and $\sim$ denotes homotopy equivalence. The phrase "exotic fusion system" denotes a saturated fusion system $\mathcal{F}$ on a finite p-group P, which is not isomorphic to $\mathcal{F}_P(G)$ for any finite group G having P as a Sylow p-subgroup. In work related to my PhD dissertation in 1972, I had in effect studied the 2-fusion systems $\mathcal{F}_{Sol}(3^n)$ and proved that they were exotic. There is an intriguing analogy between the fusion systems $\mathcal{F}_{Sol}(q)$ and the Lie-type groups $F_4(q)$, which lie over the groups $Spin_9(q)$ in much the same way that $\mathcal{F}_{Sol}(q)$ lies over $Spin_7(q)$.

In 2007, Andersen and Grodal (and Moller and Viruel) completed the classification of p-compact groups for all primes p. They are associated with p-adic reflection groups in much the same way that Lie groups are associated with real reflection groups (Weyl groups). In particular, their classification establishes that $DI(4)$ is the only exotic simple 2-compact group. This inspires the following conjecture.

Conjecture. *The family* $\mathcal{F}_{Sol}(3^n)$, *for* n *a power of* 2, *is the only family of exotic simple* 2*-fusion systems.*

Of course this conjecture requires the definition of normal subsystems and simple p-fusion systems. These definitions have been supplied by Michael Aschbacher (and others). Indeed, Aschbacher has formulated a strategy for the classification of simple 2-fusion systems, which he is pursuing, along with others including Justin Lynd, Julianne Rainbolt, and Matt Welz.

Warnings.

(1) *$SL(2,2^n)$ is a simple group for all $n \geq 2$. But $\mathcal{F}_2(SL(2,2^n))$ is never a simple 2-fusion system for $n \geq 2$, since $\mathcal{F}_2(SL(2,2^n)) \cong \mathcal{F}_2(B(2,2^n))$, where $B(2,2^n)$ is the (solvable) Borel subgroup of upper triangular matrices in $SL(2,2^n)$.*

(2) *In general, there is at most one simple 2-fusion system on a given 2-group S, although there may be infinitely many simple groups with Sylow 2-subgroup S. For example, if $\mathcal{F}$ is the unique simple fusion system on $S = D_8$, the dihedral group of order 8, then $\mathcal{F} \cong \mathcal{F}_S(L_2(q))$ for all q such that $q \equiv \pm 7 \pmod{16}$, and $\mathcal{F} \cong \mathcal{F}_S(A_7)$.*

Finally, we note that Markus Linckelmann (J. Algebra 277) has a reformulation of the Alperin Weight Conjecture in the language of linking systems related to block algebras. This might afford a new line of attack on the Alperin Weight Conjecture, possibly independent of CSFG. There is work-in-progress of Chermak, Grodal, and Yalcin directed towards a solution of a relevant "gluing" problem.

In an even more speculative vein, we observe that p-compact groups are connected to complex reflection groups. But many sporadic quasisimple groups (e.g., Suz, Co_0, etc.) are quaternionic or octonionic in nature.

Question 3: Is there a more general class of topological objects connected to these sporadic groups?

In summary, *The Origin of Groups* has yet to be written, but new insights continue to emerge. Possibly, the Darwin of Group Theory is reading this article.

Introduction to
Modern Mathematics
ALM 33, pp. 391–408

What We Know and What We Don't Know about 4-dimensions

Cliff Taubes*

My plan for this lecture is to tell you a little about what we do know and something of what we don't know about 4-dimensional spaces.

1 Manifolds, dimensions and coordinates

I must digress first to tell you what I mean by a 4-dimensional space. To this end, let me start with the notion of a *manifold.* This is a space that at small scales looks flat and featureless. Imagine the surface of a ball. If you look at it at very small scales, you won't see that it is curved. To an ant, it seems to all intents and purposes like the surface of a table. Figure 1 below depicts a surface in space, an example of a manifold. A region that looks flat has been magnified.

Figure 1.

When I say featureless, I mean featureless in the sense that the straight line or a flat plane or any Euclidean space is featureless. There is nothing to distinguish one point from another or one direction from another. Figure 2 depicts the line and the plane, the two smallest Euclidean spaces.
A technical definition has a manifold being a space whose points have neighborhoods that can be identified with a Euclidean space.

The number of coordinates of the relevant Euclidean space is the *dimension* of the manifold. Said differently, the dimension a given manifold is the number of

*William Petschek Professor of Mathematics, Harvard University, USA. Email: chtaubes@math.harvard.edu

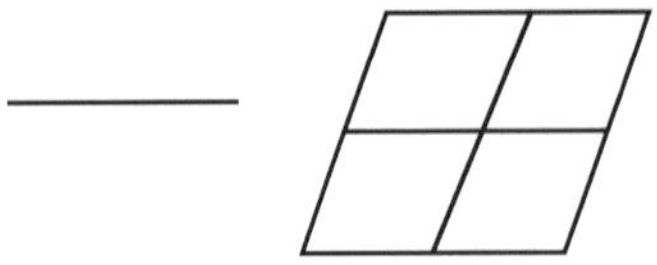

Figure 2.

local *coordinates* that you need to specify your position in a neighborhood of any one of its points. For example, if you were to live on a circle, then you would need to give but one coordinate to specify your position. You would need only to tell your friends the angle of your abode from some reference point. See Figure 3; the house is 138° counter clockwise from the reference point.

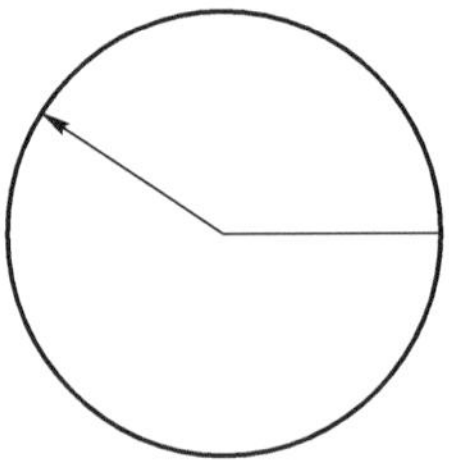

Figure 3.

The circle is an example of a 1-dimensional manifold. It is denoted by S^1; the S standing for sphere (not sircle) and the 1 being the dimension.

The surface of the earth is an example of a 2-dimensional manifold. You need to give locally 2 parameters–latitude and longitude–to completely specify the position of a point. For example, the position of this Math Forum building is $18°19'42''$ north latitude and $109°24'37''$ east longitude.

If I wanted to give someone directions as to where to find this lecture room, I would also have to give them a third coordinate, this being the height over the surface of the earth. According to Google earth, this lecture hall is 153 meters about sea level. So, we need not two, but three coordinates to describe our world because we can move in three directions, not two.

Now suppose that someone wanted to actually *see* my lecture. If this is the case, then they would have to also know a fourth coordinate, this being the time of the lecture, December 19, 2013 at 6:30 GMT which is 14:30 local time. This illustrates the fact that our observable universe is 4-dimensional, one needs 3 spatial coordinates and a fourth temporal coordinate to completely specify the position of any given point.

In general, a 4-dimensional manifold is characterized by the property that each of its points has a neighborhood that looks like 4-dimensional Euclidean space. This means that each point of this neighborhood is completely specified by four numbers and that any given set of four numbers specifies a point. These 4 numbers are the coordinates of the points. Note however that the coordinates that are used for a neighborhood of a chosen point can differ from those that are used near some other point. If there were just one coordinate system that worked everywhere, then the manifold would be either the whole of 4-dimensional Euclidean space, or a part of 4-dimensional Euclidean space.

This fact that one coordinate set may not work over the whole manifold signals the existence of structure at large scales. This is true in any dimension. The case of the circle is perhaps the simplest example whereby the Euclidean coordinate given by the counter-clockwise angle is only *locally* single valued. Go around the circle once and the angle increases by 360°. This is a manifestation of the fact that the circle looks like the 1-dimensional Euclidean space only at small scales. It has a non-Euclidean structure at large scales. By way of a second example, the coordinates of latitude and longitude for the surface of the earth don't work as local coordinates near the north and south poles. You will need to use a different set of local coordinates near these two points. This is a manifestation of the fact that the surface of the earth has non-Euclidean structure at large scales.

2 The large scale shapes of the earth and the universe

The earth is a solid sphere and its surface is just like the surface of a ball. Modulo relatively small bumps (i.e hills and mountains), the surface of the earth looks to our eyes like a flat surface, the 2-dimensional Euclidean space. Indeed, many people in antiquity thought the earth was *flat*. Even so, the ancient Greeks already knew that the earth was round. There is an annotated version of a lecture course given by a Greek savant by the name of Gemino. The annotated version is by James Evans and J. Lennart Berggren and it is called *Gemino's "Introduction to the Phenomena" A Translation and Study of a Hellenistic Survey of Astronomy* (Princeton University Press 2006). Gemino was the equivalent of a college professor and his lectures were meant to teach his students about the 'phenomena', the phenomena being astronomy and geometry. In Hellenistic times, astronomy and geometry were taught together as aspects the same thing. Gemino in his lectures gives three pieces of evidence for the earth being round. The first is the following fact: On the summer solstice, if you put a vertical stake in the ground at Alexandria in Egypt at noon local time and do the same thing due north in central Anatolia (modern Turkey) and then somewhat farther north, you will find that the shadows pointing north have different lengths. The distances between the stakes and the shadow lengths can be used to compute the curvature of the earth. See Figure 4.

The second piece of evidence can be seen when there is an eclipse of the moon. The shadow of the earth on the moon is curved. The Greeks knew that an eclipse is due to the earth getting between the sun and the moon.

The third piece of evidence concerns elephants. If you go west from Alexandria towards the Atlantic ocean, you will first get to lands where there are elephants. The famous army of Hannibal (from Carthage in what is now Lybia) crossed the Alps into Italy coming from Spain with African elephants. Meanwhile, if you go east from Alexandria, across Anatolia towards India, you will again get

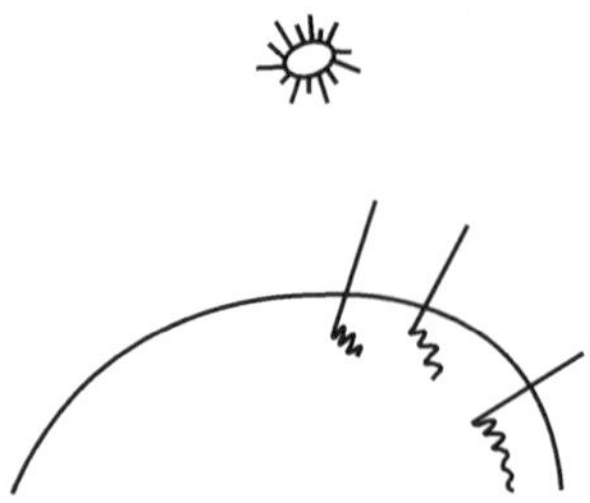

Figure 4.

to lands with elephants. For example, Alexander from Macedon fought against armies with Indian elephants during his campaigns in what is now Pakistan. If there are elephants to the west and elephants to the east and none in between, then the east and west must be connected *around the back*.

Gemino makes a rather amazing prediction in his lectures from the fact that the earth is round. There was much commerce between the far flung parts of Europe and Asia during Hellenistic times, and as a consequence the Greeks knew that the climate got ever colder as you went to the north. For example, they knew that there are lands in the far north where there is perpetual ice and snow. They also knew that the climate got ever hotter to the south. In particular, the climate in the Sahara Desert was such that it was a barrier to travel and thus to knowledge. Gemino makes the bold statement that if you were to cross the Sahara, you would again get to lands with temperate climate, and then eventually to lands with perpetual snow and cold as in the far north. He then conjectures that the hypothetical southern temperate lands could be populated with people, people whom we (the Greeks) know nothing about.

The surface of the earth is an example of a *manifold* that is curved in the large so as to have the shape of the boundary of a ball. This is to say that it is a 2-dimensional sphere. As I said, our universe is a 4-dimensional manifold, and this begs the question as to what is its shape in the large? This is a question for astronomers to sort out. Even so, it is a mathematical question as to the list of all *possible* 4-dimensional spaces.

3 Lists of manifolds

When talking about the listing manifolds, it is important to distinguish between the intrinsic structure and the shape that appears when a manifold is viewed as a subset of some larger, ambient space. By way of an example, the circle can be depicted in the plane in its standard round form and also with various bumps as in the left and right hand drawing of Figure 5. From an intrinsic point of view, both drawings are depictions of the same circle since in both cases, any given point is specified by an angular coordinate.

By the same token, the standard unknotted circle in 3-space and a knotted circle in the 3-dimensional Euclidean space are both standard circles. What differs is the identification as subspaces in the 3-dimensional Euclidean space. See

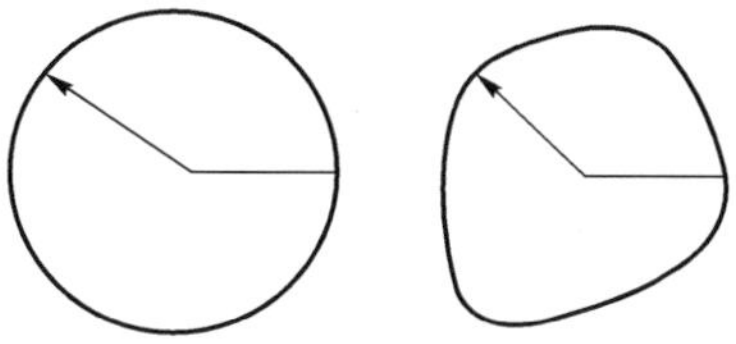

Figure 5.

Figure 6.

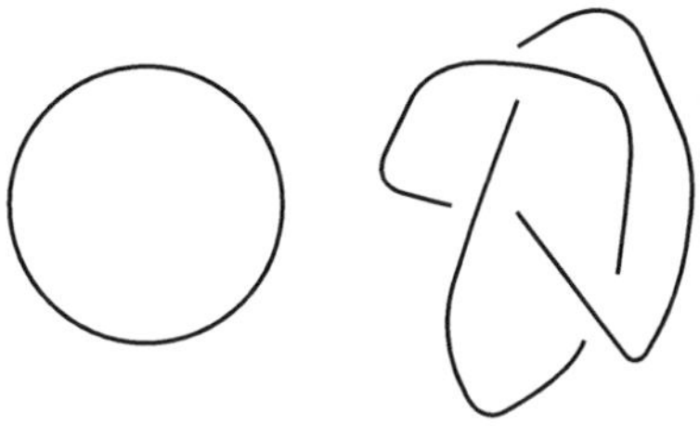

Figure 6.

By way of a second example, the surface of a perfectly round ball and the surface of the earth are both 2-dimensional spheres even as the latter has hills and valleys. The hills and valleys are not intrinsic to the sphere; they are the manifestation of sphere's appearance as a subspace of the larger 3-dimensional Euclidean space.

The key point here is that when I talk momentarily about a list for manifolds of a given dimension, I am not going to distinguish extrinsic features that arise from distinct appearances in some higher dimensional space. With this important point understood, consider now 'listing' the manifolds of a given dimension.

THE 0-DIMENSIONAL CASE: Figure 7 below depicts the 0 dimensional list.

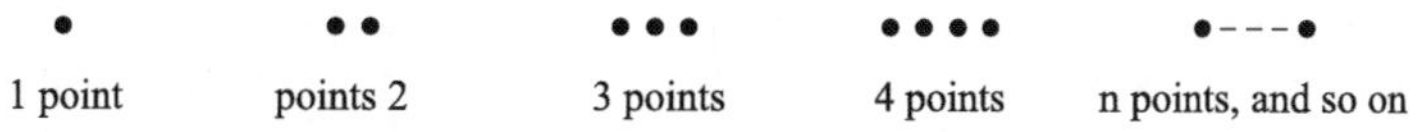

Figure 7.

Note in this regard that a point is 0-dimensional. If your universe were a point, then you wouldn't have to give any number to specify your neighbors. Everyone would be living at the same place.

Keep in mind with regards to Figure 7 that a manifold consisting of 2 points is also 0-dimensional since locally, no number is needed. Of course, this isn't the case globally since you do have to say which of the two points you live on. This example illustrates the distinction between a manifold with *connected* manifold and a manifold with multiple components. I will henceforth restrict to manifolds that are connected.

THE 1-DIMENSIONAL CASE: The list of 1-dimensional manifolds has two elements, these being the line and the circle. They are depicted in Figure 8.

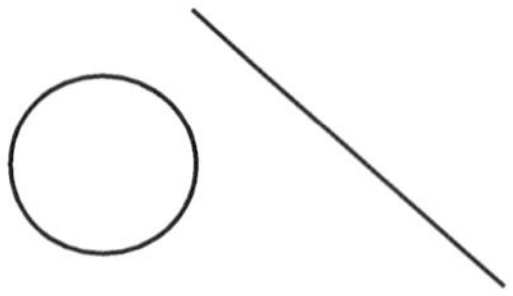

Figure 8.

Keep in mind that I am considering only intrinsic differences; thus a perfectly round circle and an ellipse, for example, are considered identical. By the same token, an infinite line and an open interval in the real numbers are the same. The proof that the line and the circle are the only examples of connected, 1-dimensional manifolds is little more than a calculus exercise.

The examples of the line and the circle illustrate the distinction between *compact* manifolds and *noncompact* manifolds. The circle is compact and the line is not. A manifold is compact if every sequence of points has a subsequence that limits to a point in the manifold. To keep things manageable, I will henceforth talk only about compact manifolds.

THE 2-DIMENSIONAL CASE: The list of the 2-dimensional (compact and connected) manifolds has two parts. The manifolds from the first part of the list are the surfaces in 3-dimensional Euclidean space that depicted below in Figure 9.

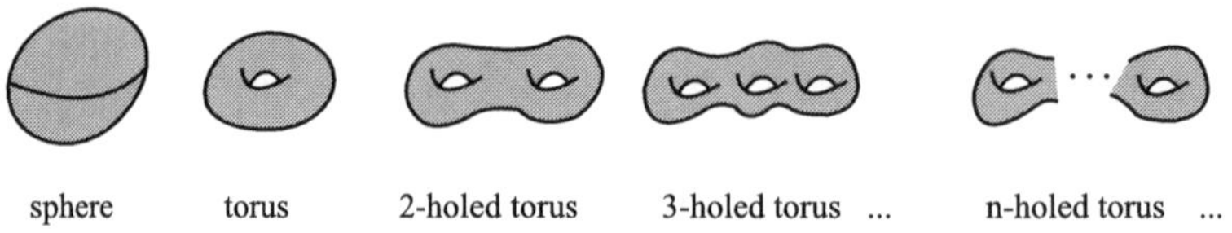

Figure 9.

This left-most surface on this list is the 2-dimensional sphere, the next one is called the 2-dimensional torus. It is the surface of solid tube. The next one after the torus is called the 2-holed torus. Then there are tori with more and more holes.

The torus can also be depicted as the set of pairs points from the circle. This is to say that each point in the torus is uniquely determined by two angular variables as depicted in Figure 10.

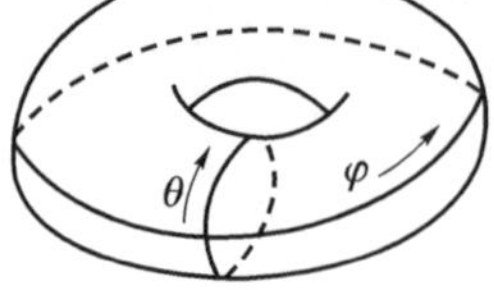

Figure 10.

The standard circle is denoted by S^1 and the torus is denoted by $S^1 \times S^1$. This illustrates the fact that manifolds of smaller dimension can be used as building blocks for manifolds of higher dimension by taking the larger dimensional manifold to be the space of pairs of points in a given smaller dimensional manifold just as the space of pairs of points in the circle define the 2-dimensional torus manifold.

The manifolds from the second of part of the list of 2-dimensional manifolds are harder to depict because they can not be drawn in a nice way as subsets of 3-dimensional Euclidean space. In any event, my description starts with the Möbius band which is depicted in 3-dimensional Euclidean space below in Figure 11.

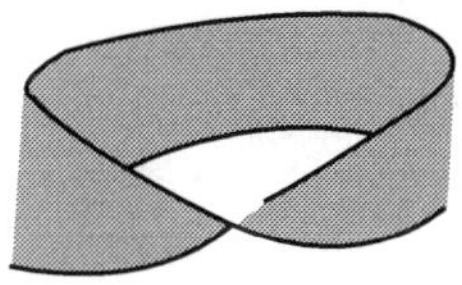

Figure 11.

Notice that the boundary of the Möbius band is a circle, albeit not the standard round depiction of the circle in 3-dimensional Euclidean space. See the next figure.

Figure 12.

As it turns out, the boundary of the 2-dimensional disk is also a circle. This being the case, the boundary of the disk can be *glued* to the boundary of the Möbius band with the result being a 2-dimensional, compact, connected manifold. This manifold is called the *real projective plan.* See the next figure. It is the first manifold on the second part of the list of 2-dimensional manifolds.

To obtain the second manifold on the second part of the 2-dimensional manifold list, glue two Möbius bands together by identifying their respective boundary circles. The resulting manifold is called the *Klein bottle.* The third manifold on the list is obtained as follows: Cut out a disk from the 2-dimensional torus. This is depicted below in Figure 14. You will notice that the boundary of the result is also a circle. This being the case, it can also be glued to the boundary of the

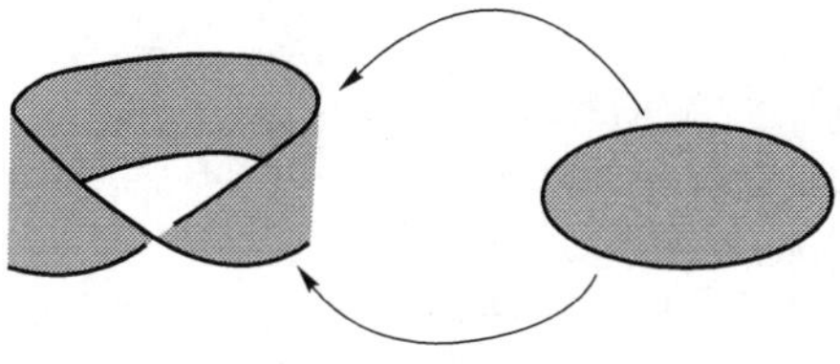

Figure 13.

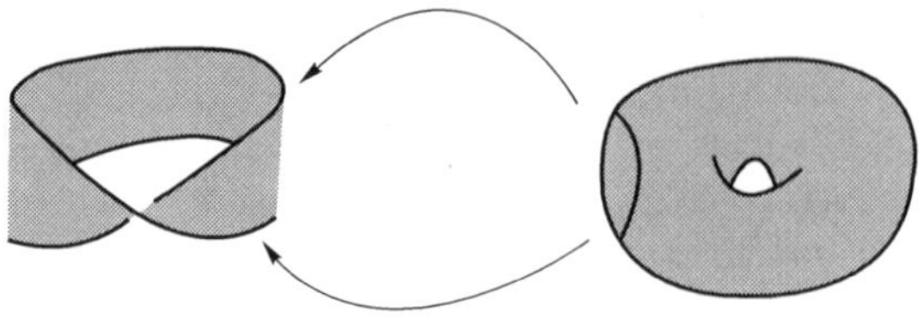

Figure 14.

Möbius band. The result of doing so is the second of the manifolds on Part 2 of the list of 2-dimensional manifolds. The remaining manifolds on the list are obtained from those on Part 1's list in Figure 9 by this same operation of cutting out a disk and replacing the latter by the Möbius band.

This operation of cutting and then gluing to construct new manifolds from old ones is called *surgery.* The surgery operation plays a central role in the subsequent discussion.

The proof that the list of 2-dimensional manifolds is as I just describe can be had using (in part) the theory of complex functions.

THE 3-DIMENSIONAL CASE: The list of compact, connected 3-dimensional manifolds was only recently determined. This was done by Grisha Pereleman ([P1-P3]. A conjecture for the list was made in 1982 by William Thurston [T]. The conjecture was called the *geometrization conjecture.* Perelman proved this conjecture using work of Richard Hamilton. See [MF] for an introduction and overview of all of this work. The now proved geometrization conjecture asserted that any given compact oriented 3-manifold could be constructed in an essentially unique way by gluing together parts from a list of eight basic kinds. The gluing was is to be done along certain specified two dimensional boundaries that are either 2-dimensional spheres or 2-dimensional tori. I won't say more about the details.

What follows instead depicts some of the simpler 3-dimensional manifolds. The purpose is to give you a way to see something of their large scale structure. Keep in mind in this regard that a neighborhood of each point in any one of these manifolds looks just like a neighborhood of the origin in 3-dimensional Euclidean space. Even so, none of these manifolds fits entirely into 3-dimensional space. For this reason, I depict these manifolds by cutting them into parts that *do* sit as subspaces in 3-dimensional Euclidean space. I then draw these parts and indicate how the parts fit together so as to make a single, 3-dimensional manifold.

The 3-dimensional sphere: Take two solid balls as depicted in the next figure. The boundary of each ball is a 2-dimensional sphere. Now glue the two boundaries together.
The result is depicted schematically in the next figure. The figure shows a path that exits the left most ball at the boundary 2-sphere but due to the gluing, the path then enters the right most sphere.
Likewise, a path the exits the right most solid ball through its boundary sphere must enter the left most solid ball. The manifold that is obtained by this gluing is called the 3-dimensional sphere. This depiction of the 3-dimensional sphere as

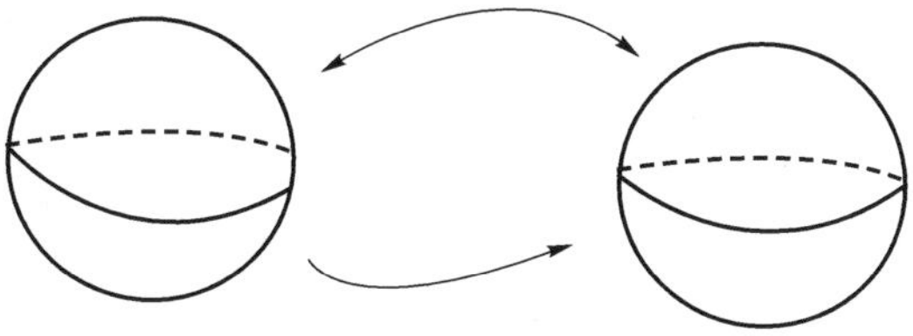

Figure 15.

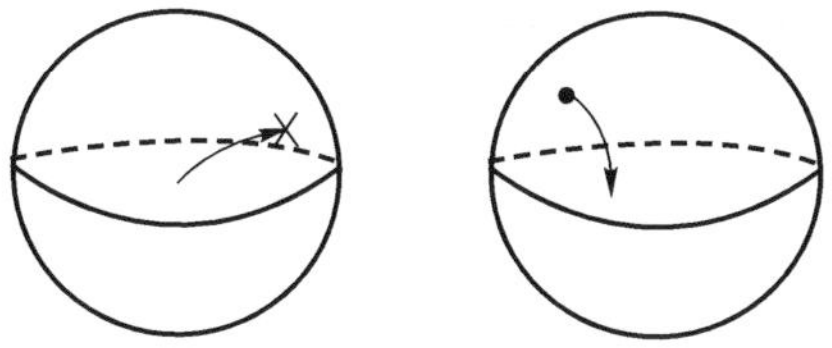

Figure 16.

two balls with their boundaries glued together is analogous to the depiction of the 2-dimensional sphere below in Figure 17 as the union of the upper hemisphere (a disk) and the lower hemisphere (another disk) with their boundary circles identified as the equatorial circle.

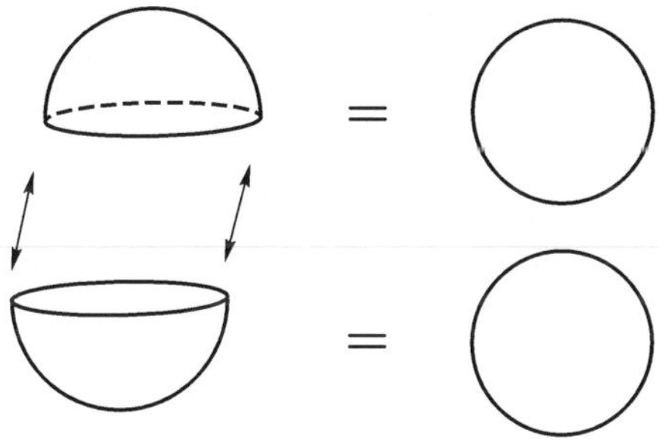

Figure 17.

The manifold $S^1 \times S^2$: This manifold is one of two direct 3-dimensional analogs of the two dimensional torus that is depicted in Figure 10. It can be viewed as the set of pairs of points with the first member of the pair being a point in the circle and the second member of the pair being a point in the 2-dimensional sphere. This manifold is depicted in the upcoming Figure 18. The inner and outer 2-dimensional spheres in the depiction are glued one to the other so that a path that leaves the region between the spheres by exiting through the outer sphere automatically re-enters the region through the inner sphere. This path is depicted in the figure. By the same token, a path that exists the region between the spheres through the inner sphere automatically re-enters through the outer sphere.

The analog to $S^1 \times S^1$ can be seen by depicting the latter space as in Figure 19 as a cylinder with the two ends glued together. The analog in Figure 18 of the cylinder in Figure 19 is the region between the inner and outer spheres.

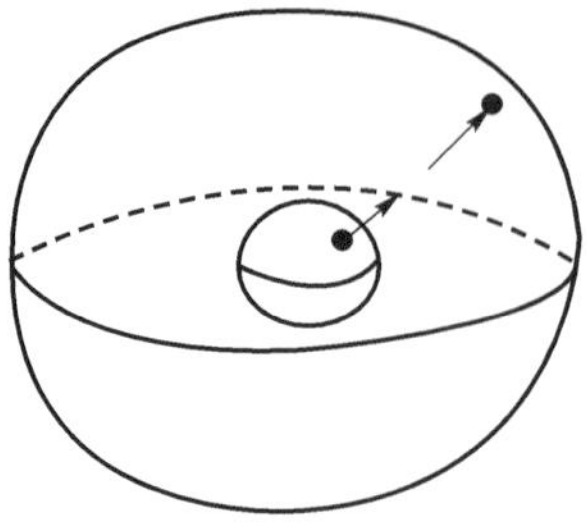

Figure 18.

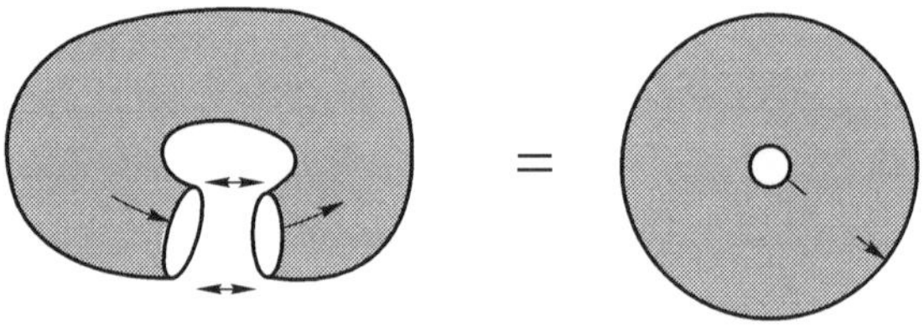

Figure 19.

The manifold $S^1 \times S^1 \times S^1$: This manifold consists of the set of triples of points with each point being a point in the circle. It is depicted in the next figure as the cube in Euclidean 3-space with the opposite sides identified. Granted this identification, then a path that leaves the interior of the cube by exiting the top side immediately re-enters the cube from the bottom. By the same token, a path the exists the cube through the right hand side immediately reenters through the left hand side; and a path that leaves through the front side immediately reenters through the back side. One of these paths is depicted in the figure.

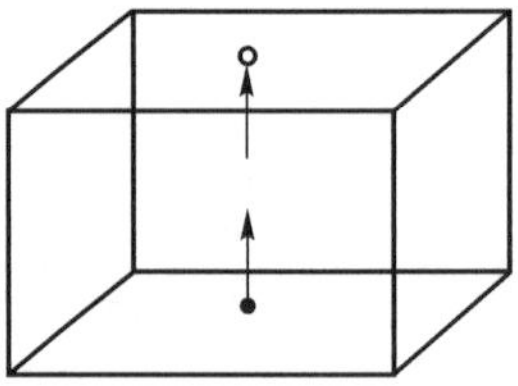

Figure 20.

An analogous depiction of the 2-dimensional torus $S^1 \times S^1$ as the square in the plane with opposite sides identified is given on the left below. The same square is depicted in the right most figure as a subset of the torus with the torus depicted as a surface in 3-dimensional Euclidean space.

Manifolds from knots: The description that follows explains how to construct a new 3-dimensional manifold from two old ones via a surgery. To start the construction, take a 3-dimensional sphere as depicted in Figures 15 and 16 as two balls with their boundaries identified. A solid knotted tube in one of the

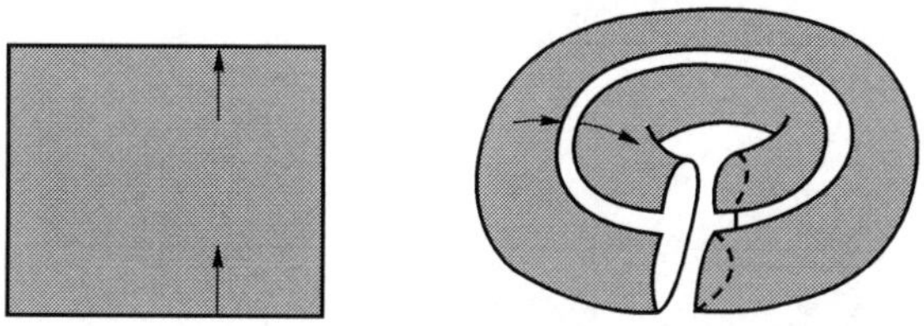

Figure 21.

ball components of the 3-sphere is chosen. The two balls that comprise the 3-dimensional sphere and the solid knotted tube are depicted on the left in the figure that follows.

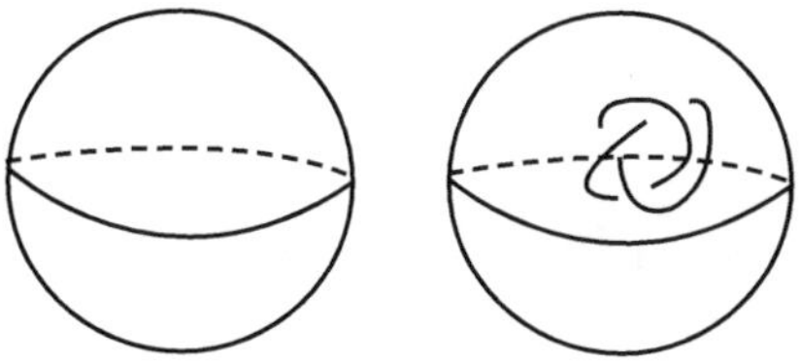

Figure 22.

Note that the boundary of the solid knotted tube is a knotted realization in the 3-dimensional Euclidean space of the 2-dimensional torus that is depicted in Figure 10. Figure 23 shows the boundary of the knot with the two angular parameters that identify it with the standard 2-dimensional torus as depicted in Figure 10.

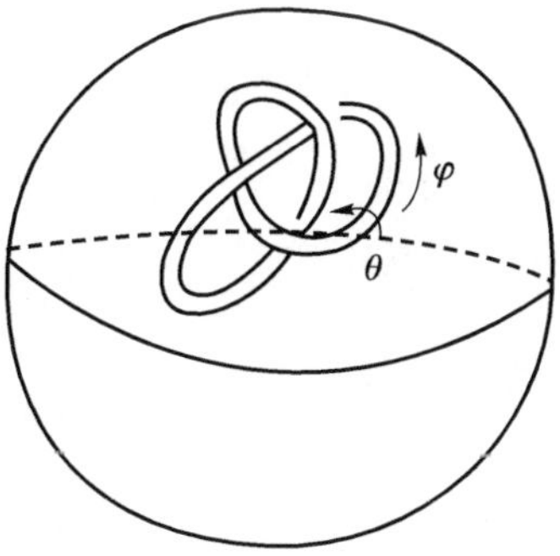

Figure 23.

The next figure again shows the two balls that comprise the 3-dimensional sphere with another solid knotted tube depicted.
Take out the solid knotted tube in Figure 24. The boundary of the result is the 2-dimensional torus. Likewise take out the solid knotted tube in Figure 22. The result has a 2-dimensional torus for a boundary also. Identify these two

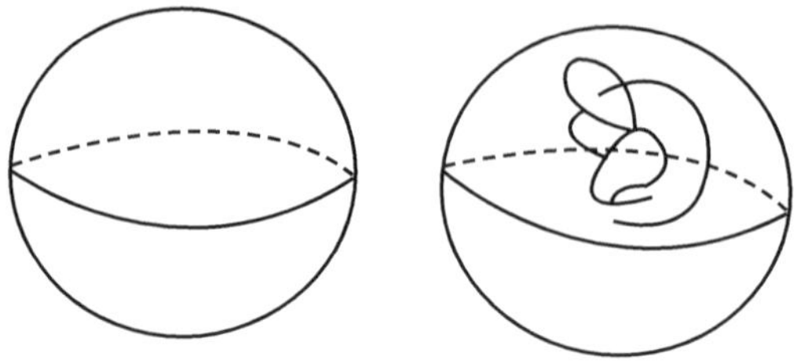

Figure 24.

torus boundaries and use the identification to glue them together. The result is a compact, connected, 3-dimensional manifold.

THE CASE OF MANIFOLDS OF DIMENSION 5 OR GREATER: As it turns out, the question of a list of sorts for manifolds of dimension strictly greater than 4 has also been sorted out. This was work done in the 1960's by Smale, Stallings, Milnor, Kirby, Siebenmann, and others (see for example [M] and [KS]). I won't say more about dimensions greater than 4 but for the remark that distinctions between manifolds involve certain subtle sorts of algebraic data.

THE CASE OF 4-DIMENSIONS: I left this case for last because 4 is the only dimension where the question of a list is as yet unsettled. It is known that the analog in dimension 4 of the algebraic data that is needed for the cases of dimension greater than 4 is necessary, but it is also known that the algebraic data is by no means sufficient. There is no compelling conjecture as to what else is involved.

4 Visualizing 4-dimensional manifolds

Manifolds of dimension 4 are hard to picture. This is so even though we live in a four dimensional universe. Yet, some people are extremely good at visualizing and hence exploiting the intrincally 4-dimensional aspects of our universe. Magicians in particular are adept at this. Many standard magic tricks are done by exploiting the fact that the universe is 4-dimensional.

A simple example involves a common trick whereby the magician asks someone from the audience to pick a card from a shuffled deck. The card is then shown to the audience but not to the magician. After the audience sees the card, the magician inserts the card in the deck and then has another member of the audience shuffle the deck many times. The magician then takes the deck and 'magically' picks out the chosen card.

The trick is done as follows: In the morning before the performance, the magician travels in the 4'th dimension forward in time at an accelerated rate so as to appear in the back of the performance hall a split second before the trick is performed before the audience. The incarnation of the magician in the back of the hall then sees the chosen card when it is shown to the audience. The incarnation of the magician in the back of the hall then travels backwards in time to when he started, and then travels forward in time at the normal rate knowing *before the trick is even begun* what the chosen card will be.

If you are at a magic show where this trick is performed and if you listen closely just after the chosen card is shown to the audience, you will hear a popping sound coming from the back of the performance hall. This popping sound is the sound that the air makes when it refills the void that is left when the incarnation of the magician leaves what was for him the future to return to the 'past' with the knowledge of the chosen card. This popping sound is the give-away. If you are not specifically listening for it, you won't notice it amongst the hub-bub and misdirection on the stage.

5 Smooth manifolds and topological manifolds

To say what we know and what we don't know about 4-dimensional manifolds, I have to digress momentarily to talk about the notion of a *smooth* manifold. A manifold is said to be *smooth* if there is a consistent definition across the whole manifold of which functions have derivatives and which do not. This is a distinguished set of manifolds.

As it turns out, there are manifolds that lack a consistent notion of a derivative. Of course, near any given point, the manifold looks like Euclidean space and we know from multi-variable calculus what it means to take derivatives of functions in Euclidean space. The issue appears because neighborhoods of different points will have, in general, different identifications with Euclidean spaces and the functions that appear differentiable with respect to one identification need not be differentiable with respect to the other (and vice-versa). This issue is not present in dimensions 1, 2 and 3. It is however present in dimensions 4 and greater. Two manifolds that are equivalent with no regard for differentiation are said to define the same *topological manifold.*

If differentiation is of no interest, then the question of a 'list' for 4-dimensional manifolds has been determined for cases when the higher dimensional analog of what I called algebraic data is not too complicated. This was done by Michael Freedman in the early 1980's (see [FQ]). Freedman proved that the algebraic data in these cases is all that is needed to determine a 4-dimensional manifold as a topological manifold.

If you care about differentiation, then the story is far more complicated and, as I said previously, not at all sorted out. That this is so follows from the pioneering work of Simon Donaldson (see [DK]). I say more about this in what follows.

The notion of a smooth 4-dimensional manifolds is in any event, very natural and this is one motivation for listing them. In particular, the notion of differentiation is central to much of modern physics. By way of an example, classical mechanics in the guise of Newton's laws equates the second *derivative* with respect to time of the position of a particle with the force acting on it. By way of a second example, the time evolution of the wave function in quantum mechanics is determined by the *Schrödinger equation* which is an equation that involves both time and spatial derivatives.

With the preceding understood, I henceforth restrict my attention to smooth 4-dimensional manifolds.

6 Knot surgery

As I said, the list of smooth 4-manifolds is not sorted out. To see where the problems lie, I now describe a construction due to Ron Fintushel and Ron Stern that can, under certain circumstances, change a given smooth, 4-dimensional manifold into a different smooth manifold without changing the listing as a topological manifold. The Fintushel-Stern construction goes by the name of *knot surgery* [FS].

I use X in what follows to denote the given smooth 4-dimensional manifold. Knot surgery requires as input an incarnation in X of a 2-dimensional torus. To say somewhat more about this torus incarnation in X, I introduce as notation D to denote the disk of radius 1 about the origin in the 2-dimensional plane and T to denote the 2-dimensional torus. The torus in X must have a neighborhood in X that can be identified with the space of $D \times T$, the space of pairs of points (x, p) with x being a point in D and p being a point T. This neighborhood of T in X is depicted schematically in Figure 25. The figure accurately depicts the disks in $D \times T$ but it depicts the torus T as a circle.

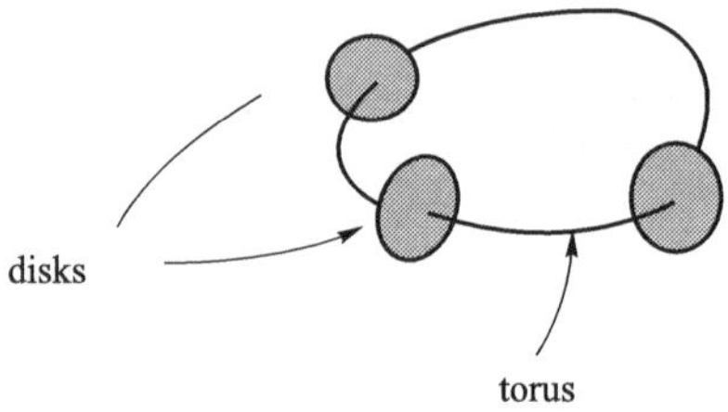

Figure 25.

There are other conditions on the incarnation of T but these I won't describe.

The boundary of $D \times T$ is the product space $S^1 \times T$, the space of pairs of the form (t, p) with t here denoting a point in the circle (the boundary of D) and with p again denoting a point in T. This is depicted schematically in the next figure.

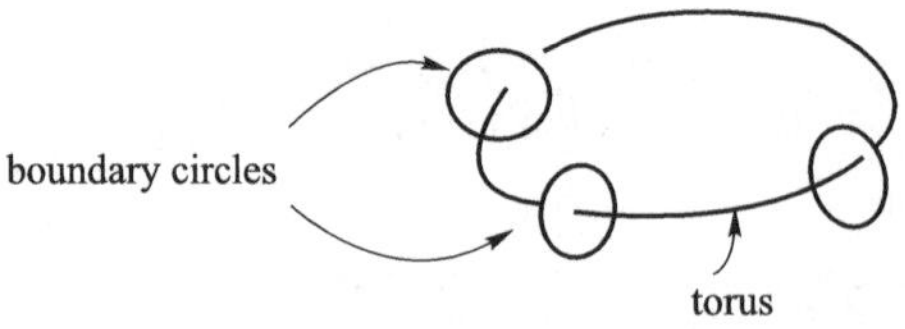

Figure 26.

Since T is $S^1 \times S^1$, the boundary of $D \times T$ is the space that is depicted in Figure 20.

The plan now is to remove the incarnation of $D \times T$ from X and replace it with a space that has the same boundary. To find such a space, depict the

3-dimensional sphere as in Figures 15 and 16 as two balls in Euclidean space with their boundaries identified. Choose a knotted circle in one of the balls and thicken the knotted circle to get a knotted tube in the 3-dimensional sphere. Two versions of this are depicted in Figures 22 and 24. Take out this solid tube around the knot to get a 3-dimensional space with boundary. The boundary is an incarnation of T in the 3-dimensional sphere. Letting K denote the chosen knot, I use M_K to denote the resulting 3-dimensional space with boundary T. The space $M_K \times S^1$ is 4-dimensional, it is the space of pairs (z, t) with z being a point in M_K and t being a point on the circle. The boundary of $M_K \times S^1$ is $T \times S^1$ and since T is $S^1 \times S^1$, this boundary is the same as that of the complement in X of the incarnation in X of $D \times T^2$.

Granted the preceding observation, remove the $D \times T$ incarnation from X and replace it by $M_K \times S^1$ by gluing along their common boundaries. The result of this gluing is the knot surgery space. I denote it by X_K. If the boundary identifications for the gluing are chosen in an appropriate way and if the incarnation of T in X has certain special properties, then X_K and X will be the same when viewed as topological manifolds. As explained momentarily, the knot K can be chosen so that X_K and X are *definitively not* the same when viewed as smooth manifolds.

7 4-manifolds and the Alexander polynomial

Two knots in Euclidean space are said to be equivalent if one can be obtained from the other via a continuous, 1-parameter family of deformations with each member of the family being a bonafide knot. Some examples of inequivalent knots are drawn below in Figure 27.

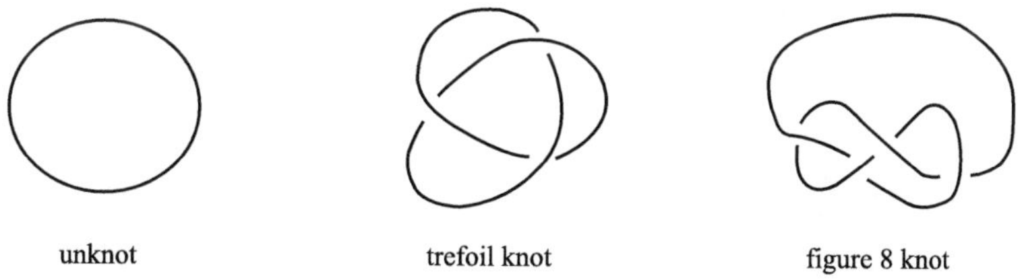

Figure 27.

The left most knot in the preceding figure is called the *unknot* since it is just a standard round circle in space. The knots in the next figure are all equivalent to the unknot.

Figure 28.

Each knot has an associated *Alexander polynomial.* This is a polynomial in the variables t and t^{-1}. The Alexander polynomial depends only on the equivalence class of the knot. The Alexander polynomials of the knots in Figure 27 are as follows: The Alexander polynomial of the unknot is the constant $\Delta_{\text{unknot}} = 1$, that of the trefoil depicted in Figure 27 has the polynoma $\Delta_{\text{trefoil}} = t-1-t^{-1}$, and that of the figure 8 knot depicted in Figure 27 has the polynomial $\Delta_{\text{fig.8}} = -t+3-t^{-1}$.

As explained in [K], a polynomial in the variables t and t^{-1} is the Alexander polynomial of some knot if and only if the sum of its coefficients is equal to either 1 or -1 and it is symmetric under the interchange of t with t^{-1}. This implies that there is an infinite set of knots with no two having the same Alexander polynomial.

Fintushel and Stern prove the following [FS]:

Suppose that T is a torus in X with the desired properties. If K is a knot, then the knot surgery manifold X_K is equivalent to X as a topological manifold; but it is not smoothly equivalent to X if the Alexander polynomial of K is not the constant 1. Moreover, if K and K' are two knots, then X_K is equivalent as a smooth manifold to $X_{K'}$ if K is equivalent to K' as a knot, but X_K is not equivalent to $X_{K'}$ as smooth manifolds if K and K' have different Alexander polynomials.

Fintushel and Stern also prove that X_K with K being the unknot in Figure 27 is the same smooth manifold as the original manifold X.

8 A candidate for knot surgery

I use $\mathbb{C}^4$ to denote the space of 4-tuples of complex numbers and I denote a given point in this space by (z_1, z_2, z_3, z_4). Set Z to be the locus in the complement of the origin in $\mathbb{C}^4$ where the coordinates obey the equation

$$z_1^4 - z_2^4 = z_3^4 - z_4^4.$$

Note that if λ is a non-zero complex number, and if (z_1, z_2, z_3, z_4) is a point in Z, then the 4-tuple $(\lambda z_1, \lambda z_2, \lambda z_3, \lambda z_4)$ is also in Z. This understood, given points (z_1, z_2, z_3, z_4) and (z_1', z_2', z_3', z_4') in Z are said to be equivalent if $(z_1', z_2', z_3', z_4') = (\lambda z_1, \lambda z_2, \lambda z_3, \lambda z_4)$ with λ being a non-zero complex number. As it turns out, the corresponding space of equivalence classes in Z is a smooth, compact connected 4-dimensional manifold. This manifold is called $K3$. A torus in $K3$ suitable for knot surgery is the space of equivalence classes of the points in the complement of the origin in $\mathbb{C}^4$ where both

$$2(z_1^2 - z_2^2) = z_3^2 + z_4^2 \quad and \quad z_1^2 + z_2^2 = 2(z_3^2 - z_4^2).$$

The locus where these equations are satisfied is contained in Z because the equation that defines Z as a subset in $\mathbb{C}^4$ can be written as

$$(z_1^2 + z_2^2)(z_1^2 - z_2^2) = (z_3^2 + z_4^2)(z_3^2 - z_4^2).$$

If K is a knot with non-constant Alexander polynomial, then knot surgery using the manifold $K3$ and the just described torus results in an $X = K3$ version of X_K that is equivalent to $K3$ as a topological manifold but inequivalent to $K3$ as a smooth manifold.

9 What we know and what we don't know

What we know is that X_K is not the same smooth manifold as X if the Alexander polynomial of K is not the constant 1. But, we don't know if X and X_K are invariably the *same* as smooth manifolds if the Alexander polynomial of K is equal to the constant 1. One of the simplest interesting knot with Alexander polynomial equal to 1 is the Kinoshita-Terasota knot pictured below.

Figure 29.

Is X_K the same smooth manifold as X in the case when K is this Kinoshita-Terasota knot and X is the $K3$ manifold? I don't think anyone knows the answer to this question. This question sits on the ignorance side of the boundary in the year 2013 between what we know about 4-dimensional manifolds and what we don't know.

References

[DK] Simon K. Donaldson and Peter B. Kronheimer, *The Geometry of 4-Manifolds*, 1997.

[FQ] Michael H. Freedman and Frank Quinn, *Topology of 4-manifolds*, Princeton University Press 1990.

[FS] Ronald Fintushel and Ronald Stern, Knots, links and 4-manifolds, *Inventiones Math.* **134** (1998) 363-400.

[K] Aiko Kawauchi, *A Survey of Knot Theory*, Birkhauser Press 1996.

[KS] Robion Kirby and Larry Siebenmann, *Foundational Essays on Topological Manifolds, Smoothings and Triangulations*, Princeton University Press 1977.

[M] John Milnor, *Lectures on the h-Cobordism Theorem*, Princeton University Press 1965.

[MF] John Morgan and Frederick Tsz-Ho Fong, *Ricci Flow and the Geometrization of 3-Manifolds*, American Mathematical Society 2010.

[P1] Grigori Perelman, *The entropy formula for the Ricci flow and its geometric applications*, arXiv:math/0211159.

[P2] Grigori Perelman, *Ricci flow with surgery on three-manifold*, arXiv:math/0303109.

[P3] Grigori Perelman, *Finite extinction time for the solutions to the Ricci flow on certain three-manifolds*, arXiv:math/0307245.

[T] William P. Thurston, Three dimensional manifolds, Kleinian groups and hyperbolic geometry, *Amer. Math. Soc. Bull.* **6** (1982) 357-381.

Introduction to
Modern Mathematics
ALM 33, pp. 409–423

Congruent Number Problem and Arithmetic of Elliptic Curves

Ye Tian*

Abstract

There are many examples of elliptic curves over $\mathbb{Q}$ that arise naturally in Diophantine problems, such as the congruent number problem. We introduce the arithmetic theory of elliptic curves via the congruent number problem, which is probably the oldest unsolved major problem in mathematics. It can be traced back at least to the 10-th century in Arab manuscripts but it is possibly much older.

1 Introduction to congruent number problem

A positive integer is called a congruent number if it is the area of a right-angled triangle whose sides are rational numbers. The congruent number problem is to determine if a given positive integer is a congruent number or not, and to find corresponding right triangles if it is. For example, $1, 2$ and 3 are not congruent, but $5, 6,$ and 7 are congruent numbers, which are areas of right triangles with rational sides

$$(3/2, 20/3, 41/6), \quad (3, 4, 5), \quad (35/12, 24/5, 337/60),$$

respectively.

The original form of the congruent number problem was stated in an Arab manuscript as follows: given an integer n, find a (rational) square γ^2 such that $\gamma^2 \pm n$ are both (rational) squares. Suppose that n is the area of a right triangle with rational sides and let c be its hypotenuse. Taking $\gamma = c/2$, then it is easy to see that $\gamma^2 \pm n$ are all rational squares. This gives the equivalence of the two formulations. This explain the origin of the name *congruent numbers*.

It was Fermat who gave a wonderful proof of that 1 is not a congruent number, settling an open question going back four centuries to Fibonacci, and also proved the case $n = 4$ of which we now call Fermat's last theorem. Fermat's proof led to the discovery of his method of infinite descend and had a profound effect on subsequent developments in arithmetic. We review Fermat's proof here.

*The author was supported by NSFC grants 11325106, 11321101, and 11031004. Academy of Mathematics and Systems Science, Morningside center of Mathematics, Chinese Academy of Sciences, Beijing 100190. Email: ytian@math.ac.cn

Proposition 1.1 (Fermat). *The integer* 1 *is not a congruent number, namely, there does not exist a right triangle with integral sides (such triangles are called integral right triangles) whose area is a square.*

Proof. Suppose it is not the case, then there exists an integral right triangle with square area and minimal hypotenuse. It is clear that the three sides of the right triangle are pairwise coprime. By Euclidean formula, there are two positive integers $p > q$ with $2 \nmid p+q$ such that the three sides of the right triangle are

$$p^2 - q^2, \quad 2pq, \quad p^2 + q^2.$$

Since its area $pq(p+q)(p-q)$ is by assumption a square,

$$p = x^2, \quad q = y^2, \quad p+q = u^2, \quad p-q = v^2.$$

for some integers x, y, u, v. It follows that

$$x^2 = \frac{u^2+v^2}{2}, \ y^2 = \frac{u^2-v^2}{2} = \frac{(u+v)(u-v)}{2}.$$

Then $\{u+v, u-v\} = \{2r^2, 4s^2\}$ for some integer r, s and therefore

$$x^2 = \frac{(r^2+2s^2)^2 + (r^2-2s^2)^2}{2} = r^4 + 4s^4.$$

Then $(r^2, 2s^2, x)$ is another integral right triangle with square area but smaller hypotenuse $x < p^2 + q^2$. It is a contradiction. □

Using Fermat's method, one can show that neither of $1, 2, 3$ is congruent and find many non-congruent numbers $\equiv 1, 2, 3 \mod 8$. To work on the congruent number problem in more generality, however, needs another prospective.

How can we check if a given integer n is congruent or not? We know that n is congruent if and only if the two equations

$$a^2 + b^2 = c^2, \quad \frac{1}{2}ab = n$$

have a solution with a, b, c all rational numbers. By Euclidean formula, the set of rational solutions of the first equation is parameterized by $x, y \in \mathbb{Q}$ with $y \neq 0$:

$$a = 2x/y, \quad b = (x^2-1)/y, \quad c = (x^2+1)/y.$$

Therefore the second equation becomes

$$ny^2 = x^3 - x. \tag{1}$$

The above equation defines an object which we call an elliptic curve. We are led to find points on this curve with rational coordinates, i.e. rational points.

Before going deep to the world of the arithmetic of elliptic curves, let us make a simple observation. By taking $x = X/Z$ with X, Z integers in (1), we have the following elementary result about the congruent number problem.

Proposition 1.2. *A positive integer n is a congruent number if and only if there exist positive integers X, Z with $X > Z$ such that*

$$XZ(X-Z)(X+Z) = n \cdot \square,$$

where $\square$ is a square of non-zero rational number.

Taking $(X, Z) = (5,4), (2,1), (16,9)$, we see easily that $5, 6, 7$ are congruent numbers with the corresponding right triangles given above. Here is a remark. If n is a congruent number, then for any non-zero integer m, so is nm^2. Thus for congruent number problem, we only need to consider square-free integers. Taking (X, Z) to be

$$(5^2 \cdot 13, 6^2), \quad (8,1), \quad (4,1), \quad (4,3), \quad (50,49), \quad (156^2, 133^2)$$

we see that

$$13, \quad 14, \quad 15, \quad 21, \quad 22, \quad 23$$

are all congruent numbers. In this way, one can obtain the following list of congruent numbers among the first square-free integers:

$$\begin{array}{l} 5, \quad 6, \quad 7, \quad 13, \quad 14, \quad 15, \quad 21, \quad 22, \quad 23, \quad 29, \quad 30, \quad 31, \\ 34, \quad 37, \quad 38, \quad 39, \quad 41, \quad 45, \quad 46, \quad 47 \cdots \end{array}$$

What can we say from this list on congruent number problem? Consider these numbers in the list modulo 8, we obtain

$$\begin{array}{l} 5, \quad 6, \quad 7, \quad 5, \quad 6, \quad 7, \quad 5, \quad 6, \quad 7, \quad 5, \quad 6, \quad 7, \\ 2, \quad 5, \quad 6, \quad 7, \quad 1, \quad 5, \quad 6, \quad 7, \ldots. \end{array}$$

The following is a basic conjecture on the congruent number problem.

Conjecture 1. *1. Any positive integer $\equiv 5, 6, 7$ modulo 8 is a congruent number;*
2. There are "very few", in the sense of density, congruent numbers which are $\equiv 1, 2, 3$ modulo 8.

Remark. Here the "very few" in the conjecture means that for $i = 1, 2, 3$,

$$\lim_{X \to +\infty} \frac{\#\{1 \le n \le X, \ n \equiv i \mod 8 \quad \text{is congruent}\}}{\#\{1 \le n \le X, \ n \equiv i \mod 8\}} = 0.$$

Note that $41, 34, 219$ are the smallest congruent numbers $\equiv 1, 2, 3 \mod 8$, respectively, by taking $(X, Z) = (25, 16)$ or $(41, 9)$ for 41; $(17, 1)$ or $(17, 8)$ for 34; $(73, 48)$ or $(169, 73)$ for 219. On the other hand, one can easily construct infinitely many square-free congruent numbers in each residue class of $1, 2, 3, 5, 6, 7$ modulo 8. In fact, taking $(X, Z) = (16k, 1)$ with k a positive integer, one can see that $n = (16k-1)k(16k+1) \equiv -k \mod 8$ is a congruent number, and the desired construction follows from suitable choice of k and Dirichlet's theorem that if $a > 0, b$ are two coprime integers then there are infinitely many primes congruent to b modulo a.

However, it is not easy to verify the conjecture even for special cases. We will know that the prime number $157 \equiv 5 \mod 8$ is actually a congruent number. The simplest rational right triangle with area 157 has side lengths

$$a = \frac{6803298487826435051217540}{411340519227716149383203}, \quad b = \frac{411340519227716149383203}{21666555693714761309610},$$

and

$$c = \frac{224403517704336969924557513090674863160948472041}{8912332268928859588025535178967163570016480830}.$$

This triangle would be difficult to find by a brute force search. There is another basic conjecture on congruent number problem

Conjecture 2. *There is an algorithm for deciding in a finite number of steps whether a given positive integer n is congruent or not.*

Any deep result towards the above conjectures comes from the study of the arithmetic of elliptic curves. We now introduce several results on the congruent number problem, obtained via arithmetic theory of elliptic curves. The first is the following celebrated result of Heegner.

Theorem 1.3 (Heegner [11], 1952). *Any positive integer $n \equiv 5, 6, 7 \mod 8$ with exactly one odd prime factor is a congruent number.*

Let p_1, p_3, p_5, p_7 denote primes $\equiv 1, 3, 5$ and $7 \mod 8$. Then Heegner's theorem says that $p_5, p_7, 2p_7, 2p_3$ are all congruent numbers. Later on, based on Heegner's construction, Monsky found more congruent numbers with exactly two odd prime factors.

Theorem 1.4 (Monsky [18], 1990). *Let p_1, p_3, p_5, p_7 denote primes $\equiv 1, 3, 5$ and $7 \mod 8$. Then the following integers are all congruent numbers:*

- *$p_3p_7, p_3p_5, 2p_3p_5$ and $2p_5p_7$;*
- *p_1p_5 provided $(\frac{p_1}{p_5}) = -1$, p_1p_7 and $2p_1p_7$ provided $(\frac{p_1}{p_7}) = -1$, $2p_1p_3$ provided $(\frac{p_1}{p_3}) = -1$.*

Here $(\frac{p}{q})$ is the usual Legendre symbol. How about (square-free) congruent numbers with arbitrary number of prime factors?

Theorem 1.5 (Tian [26] [27], 2012). *For any integer $k \geq 1$, there are infinitely many square-free congruent numbers with exactly k odd prime factors in each residue class of* $5, 6, 7 \mod 8$.

We also gave some sufficient condition of being congruent numbers which extends the results Heegner and Monsky. Recently, using the method we developed in [26] and [27], Tian-Yuan-Zhang [28] obtained a more general result. To introduce this result, we need some preparation.

Let n be a square-free positive integer and $K = \mathbb{Q}(\sqrt{-n})$ a quadratic extension over $\mathbb{Q}$. The discriminant D of K is defined to be

$$D = \begin{cases} -n, & \text{if } -n \equiv 1 \mod 4, \\ -4n, & \text{otherwise} \end{cases}$$

Note that $D \equiv 0, 1 \mod 4$. Let $\mathcal{O}$ be the integral closure of $\mathbb{Z}$ in K, consisting of elements in K satisfying a monic polynomial with integral coefficients, then

$$\mathcal{O} = \begin{cases} \mathbb{Z}[(1+\sqrt{-n})/2], & \text{if } -n \equiv 1 \mod 4, \\ \mathbb{Z}[\sqrt{-n}], & \text{otherwise.} \end{cases}$$

The quotient of all non-zero ideals of $\mathcal{O}$ by all non-zero principal ideals of $\mathcal{O}$ has a natural abelian group structure, denoted by $\mathcal{C}$ and called the ideal class group of K. It is known that $\mathcal{C}$ is a finite abelian group. An element in $\mathcal{C}$ is called an ideal class of K.

We recall some results from Gauss' genus theory. The subgroup $\mathcal{C}[2]$ of $\mathcal{C}$ of ideal classes of order dividing 2 has order 2^{r-1} where r is the number of prime factors of D. How can we determine if $\mathcal{C}$ has ideal classes of order 4, or equivalently, if the cardinality of $2\mathcal{C}$ is even? Let $p_1, \ldots, p_\mu$ be prime factors of D and $p_\mu = 2$ if $2|D$. The Rédei's matrix of K is

$$R := \begin{pmatrix} \left[\frac{D/p_1^*}{p_1}\right] & \cdots\cdots & \left[\frac{p_{\mu-1}}{p_1}\right] & \left[\frac{p_\mu}{p_1}\right] \\ \vdots & & \vdots & \vdots \\ \vdots & & \vdots & \vdots \\ \left[\frac{p_1}{p_{\mu-1}}\right] & \cdots\cdots & \left[\frac{D/p_{\mu-1}^*}{p_{\mu-1}}\right] & \left[\frac{p_\mu}{p_{\mu-1}}\right] \end{pmatrix}$$

where for odd prime p, $p^* = (-1)^{(p-1)/2}p$, and for any a prime to p, $[\frac{a}{p}]$ is the additive Legendre symbol, i.e. $[\frac{a}{p}] = 0$ if a is a square modulo p and $[\frac{a}{p}] = 1$ otherwise. Then the 4-rank of $\mathcal{C}$, i.e. the $\mathbb{F}_2$-dimension of $2\mathcal{C}/4\mathcal{C}$, is given by

$$\mu - 1 - \text{rank}_{\mathbb{F}_2} R.$$

Define $g(n) \in \mathbb{F}_2$ by

$$g(n) = \begin{cases} 1, & \text{if } \text{rank}_{\mathbb{F}_2} R = \mu - 1, \\ 0, & \text{otherwise.} \end{cases}$$

Example. Let $n \equiv 1 \mod 8$ be a square-free positive integer such that either all prime factors of n are congruent to 1 modulo 4 or all prime factors of n are congruent to ± 1 modulo 8. Then $g(n) = 0$.

Theorem 1.6 (Tian-Yuan-Zhang, [28]). *Let $n \equiv 5, 6, 7 \mod 8$ be a square-free positive integer. Assume that either*

$$\sum_{\substack{n=d_0\cdots d_k \\ d_i \equiv 1 \mod 8,\ i>0}} \prod_i g(d_i) = 1$$

or

$$\sum_{\substack{n=d_0\cdots d_k, \\ d_0 \equiv 5,6,7 \mod 8 \\ d_1 \equiv 2,3 \mod 8 \\ d_i \equiv 1 \mod 8,\ i>1}} \prod_i g(d_i) = 1.$$

Here all decompositions $n = d_0 d_1 \cdots d_k$ are non-ordered with all $d_i > 1$. Then n is a congruent number.

Examples. 1. The number $1887 = 3 \cdot 17 \cdot 37$. The first summation in the above theorem reads

$$g(3 \cdot 17 \cdot 37) + g(3 \cdot 37)g(17) = 1 + 0 = 1.$$

Thus 1887 is a congruent number.

2. Let $n \equiv 5 \mod 8$ be a square-free positive integer with all prime factors congruent to 1 modulo 4. If $\mathbb{Q}(\sqrt{-n})$ has no ideal class of order 4, i.e. $g(n) = 0$, then n is a congruent number.

Towards Conjecture 2, there is a very beautiful result of Tunnell which gives a sufficient condition to determine if a given integer n is a non-congruent number. The criterion of Tunnell is also necessary if assuming the Brich and Swinnerton-Dyer conjecture (which we will introduce in next section) for the elliptic curve $E^{(n)} : ny^2 = x^3 - x$.

Theorem 1.7 (Tunnell [29]). *For a square-free positive integer n, let $c = 1$ if n is odd and $c = 2$ if n is even, and let*

$$A_n = \#\{(x, y, z) \in \mathbb{Z}^3 : 2cx^2 + y^2 + 8z^2 = n/c, \quad z \equiv 1 \mod 2\},$$
$$B_n = \#\{(x, y, z) \in \mathbb{Z}^3 : 2cx^2 + y^2 + 8z^2 = n/c, \quad z \equiv 0 \mod 2\}.$$

If $A_n \neq B_n$, then n is a non-congruent number.

One can easily show that $A_n = B_n$ for any positive $n \equiv 5, 6, 7 \mod 8$. The book of Koblitz [13] is a very nice introduction of the above result.

Above results, even though their statements are elementary, are all obtained via the study of arithmetic of elliptic curves, including their L-series.

Many facts on arithmetic of elliptic curves can be found in the references Tate [24], Koblitz [13] and Silverman [23].

2 Elliptic curves and BSD conjecture

So what is an elliptic curve? In a high-brow language, an *elliptic curve* E over the field of rational numbers $\mathbb{Q}$ is a projective smooth curve of genus one over $\mathbb{Q}$ with a $\mathbb{Q}$-rational point O. Let $\mathbb{P}^2$ denote the projective plane over $\mathbb{Q}$, whose $\bar{\mathbb{Q}}$-points are represented by $[x, y, z]$, where $(x, y, z) \in \bar{\mathbb{Q}}^3$ with some of x, y, z non-zero, and $[x, y, z] = [x', y', z']$ if and only if there is $\lambda \in \bar{\mathbb{Q}}^\times$ such that $(x', y', z') = \lambda(x, y, z)$. In down-to-earth term (by the great Riemann-Roch formula), an elliptic curve E is a curve in the projective plane $\mathbb{P}^2$ defined by a homogeneous Weierstrass equation:

$$y^2 z = x^3 + axz^2 + bz^3, \quad \text{for some } a, b \in \mathbb{Q},$$

with discriminant $\Delta = -16(4a^3 + 27b^2) \neq 0$. Here $O = [0 : 1 : 0]$. By convention people also use the (affine) Weierstrass equation $y^2 = x^3 + ax + b$ to denote the elliptic curve. Furthermore, replacing $\mathbb{Q}$ by any field K one obtains the definition

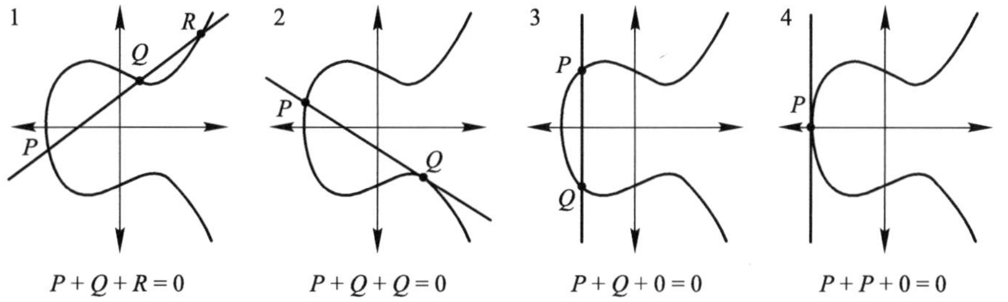

Figure 1. Addition Law of Elliptic Curves.

of an elliptic curve over K, with the general form of Weierstrass equation given by

$$y^2 + a_1xy + a_3y = x^3 + a_2x^2 + a_4x + a_6, \quad a_i \in K. \tag{2}$$

with non-zero discriminant Δ (for definition, see p.46 of [23]). An elliptic curve E over a field K has a natural abelian group structure with O as its zero element, which can be illustrated in Figure 1.. Note that $\mathbb{Z} \subset \mathrm{End}_K(E)$. An elliptic curve E over a number field K (i.e. a finite extension over $\mathbb{Q}$) is called with complex multiplication if $\mathrm{End}_{\bar{K}}(E)$ is an order of an imaginary quadratic field.

For an elliptic curve over complex numbers $\mathbb{C}$, one can view it either as a complex torus of dimension one as follows. For any lattice $\Lambda = \mathbb{Z}\omega_1 + \mathbb{Z}\omega_2$ in $\mathbb{C}$ so that ω_1/ω_2 has non-zero imaginary part, let $\mathcal{P}(z)$ be its associated Weierstrass function on $\mathbb{C}/\Lambda$ defined by

$$\mathcal{P}(z) = \frac{1}{z^2} + \sum_{0 \neq \omega \in \Lambda} \left(\frac{1}{(z-\omega)^2} - \frac{1}{\omega^2} \right),$$

and let $\mathcal{P}'(z)$ be its derivative. Then $(\mathcal{P}(z), \mathcal{P}'(z))$ satisfy the equation of an elliptic curve in $\mathbb{P}^2$:

$$E_\Lambda : \quad y^2 = 4x^3 - 60G_4(\Lambda)x - 140G_6(\Lambda),$$

where for any even $k \geq 4$, $G_k(\Lambda) = \sum_{0 \neq \omega \in \Lambda} \omega^{-k}$. The map $z \mapsto (\mathcal{P}(z), \mathcal{P}'(z))$ defines an isomorphism from $\mathbb{C}/\Lambda$ to $E_\Lambda(\mathbb{C})$.

Let K be a local field, $\mathcal{O}$ its ring of integers, $\mathfrak{p}$ the maximal ideal of $\mathcal{O}$, and k the residue field of $\mathcal{O}$. For an elliptic curve E over a local field K, a minimal Weierstrass equation of E is of form (2) with all $a_i \in \mathcal{O}$ and the additive valuation of Δ minimized. A minimal Weierstrass equation modulo $\mathfrak{p}$ defines a curve $\widetilde{E}$ over k and let $\widetilde{E}^{ns}$ denote its non-singular part. Let $E_0(K)$ be the preimage of $\widetilde{E}^{ns}(k)$ under the reduction map $E(K) \to \widetilde{E}(k)$.

For any elliptic curve E over a number field K (i.e., a finite extension of $\mathbb{Q}$), the group $E(K)$ is called the *Mordell-Weil group* of E over K. Mordell proved the following important theorem by generalizing Fermat's infinite descend method.

Theorem 1 (Mordell). *Let E be an elliptic curve defined over a number field K, then the abelian group $E(K)$ is finitely generated.*

By the structure of finitely generated abelian groups, we have that

$$E(K) \cong \mathbb{Z}^r \oplus E(K)_{\text{tor}},$$

for some integer $r \geq 0$, which is called the rank of $E(K)$ and denoted by $\text{rank}E(K)$, and $E(K)_{\text{tor}}$ a finite abelian group, which is called the torsion subgroup of $E(K)$. The torsion subgroup is easier to compute and in general described in the case $K = \mathbb{Q}$ by

Theorem 2 (Mazur). *Let E be an elliptic curve over $\mathbb{Q}$. The torsion subgroup $E(\mathbb{Q})_{\text{tor}}$ is isomorphic to one of the following finite groups*

$$\mathbb{Z}/N\mathbb{Z}, \quad 1 \leq N \leq 10, \quad N = 12 \quad \textit{and} \quad \mathbb{Z}/2\mathbb{Z} \times \mathbb{Z}/2N\mathbb{Z}, \quad N = 1, 2, 3, 4.$$

However, it is difficult to determine $\text{rank}_{\mathbb{Z}} E(\mathbb{Q})$, the rank of Mordell-Weil group $E(\mathbb{Q})$. The conjecture of Birch and Swinnerton-Dyer (for short, BSD conjecture) of elliptic curves relates the rank of Mordell-Weil group to an important analytic invariant of the elliptic curve E over $\mathbb{Q}$, called the L-function of E. We now recall its definition. Let

$$y^2 + a_1 xy + a_3 y = x^3 + a_2 x^2 + a_4 x + a_6$$

be a global minimal Weierstrass model of $E/\mathbb{Q}$ with discriminant Δ (i.e. the p-valuation of Δ is minimized for all primes p). Define the L-function of E by

$$L(s, E) = \prod_{p \mid \Delta} \frac{1}{1 - c_p p^{-s}} \prod_{p \nmid \Delta} \frac{1}{1 - c_p p^{-s} + p^{1-2s}}$$

where $c_p = p + 1 - \#\widetilde{E}(\mathbb{F}_p)$ for each prime number p, where the infinite product converges absolutely for $\text{Re}(s) > \frac{3}{2}$.

Let $N \geq 1$ be an integer. The group of matrices

$$\Gamma_0(N) = \left\{ \begin{pmatrix} a & b \\ c & d \end{pmatrix} \in \text{SL}_2(\mathbb{Z}) : c \in N\mathbb{Z} \right\}$$

acts on the complex upper half plane $\mathcal{H}$ by linear fractional transformations. Its quotient space $Y_0(N) := \Gamma_0(N) \backslash \mathcal{H}$ has a natural open Riemann surface structure and let $X_0(N) = \Gamma_0(N) \backslash (\mathcal{H} \cup \mathbb{P}^1(\mathbb{Q}))$ be its completion by adding cusps to $Y_0(N)$. There is an interpretation of $Y_0(N)$ as moduli space of elliptic curves with a cyclic subgroup of order N.

For any $\tau \in \mathcal{H}$, let Λ_τ be the lattice $\tau\mathbb{Z} \oplus \mathbb{Z}$ in $\mathbb{C}$ then $\mathbb{C}/\Lambda_\tau$ is an elliptic curves, together with a cyclic subgroup generated by $1/N \mod \Lambda_\tau$ which is of order N. This explains that $\Gamma_0(N) \backslash \mathcal{H}$ parametrizes isomorphism classes of data "elliptic curves with a cyclic subgroup of order N". It is via this interpretation that one defines algebraic curves $Y_0(N)$ and $X_0(N)$ over $\mathbb{Q}$. If $\tau \in \mathcal{H}$ satisfies a quadratic equation over $\mathbb{Q}$ so that $K := \mathbb{Q}(\tau)$ is an imaginary quadratic field, then the point $[\tau] := \Gamma_0(N)\tau \in X_0(N)$ is a point represented by an elliptic curve with complex multiplication and is defined over an abelian extension over K. Shimura's reciprocity law describes explicitly the Galois action of $\text{Gal}(\overline{K}/K)$ on such points.

Theorem 3 (Wiles, Taylor-Wiles, BCDT). *Let E be an elliptic curve over $\mathbb{Q}$ and $L(s,E)=\sum_{n=1}^{\infty}a_n n^{-s}$ its L-function. Then*

$$\varphi(z):=\sum_{n=1}^{\infty}a_n e^{2\pi i n z},\quad \mathrm{Im}(z)>0,$$

is a newform of weight 2 and level $\Gamma_0(N)$ for some integer $N\geq 1$ (called the conductor of E). Moreover, let $X_0(N)$ be the modular curve of level $\Gamma_0(N)$, then there is a non-constant morphism $f: X_0(N)\to E$ defined over $\mathbb{Q}$.

By Hecke's theory,

$$\Lambda(s,\varphi):=2\int_0^{\infty}\varphi(iy)y^s dy/y=2(2\pi)^{-s}\Gamma(s)L(s,E)$$

has an analytic continuation to the whole s-plane and satisfies a functional equation:

$$\Lambda(s,\varphi)=\epsilon\cdot N^{1-s}\cdot\Lambda(2-s,\varphi)$$

where $\epsilon=\pm 1$ is called the sign of the functional equation. It follows that $L(s,E)$ is an entire function in $s\in\mathbb{C}$. We denote by $\mathrm{ord}_{s=1}L(E,s)$ the degree of first non-vanishing term in the Taylor expansion of $L(E,s)$ at $s=1$. The first part of BSD conjecture says that

Conjecture 3 (Conjecture of Birch and Swinnerton-Dyer, Part 1). *For an elliptic curve E over $\mathbb{Q}$,*

$$\mathrm{rank}_{\mathbb{Z}}E(\mathbb{Q})=\mathrm{ord}_{s=1}L(E,s).$$

Birch and Swinnerton-Dyer also give a conjecture to express the leading coefficient of Taylor expansion of $L(E,s)$ at $s=1$ in term of several arithmetic invariants of the elliptic curves, including the Shafarevich-Tate group $\text{Ш}(E)$ of E, which is an abelian group defined as

$$\text{Ш}(E/\mathbb{Q})=\ker\left(H^1(\mathbb{Q},E)\longrightarrow\prod_p H^1(\mathbb{Q}_p,E)\right).$$

Conjecture 4 (Tate). *For any elliptic curve E over $\mathbb{Q}$, the Shafarevich-Tate group $\text{Ш}(E/\mathbb{Q})$ is a finite abelian group.*

We still need to define several invariants of E using the minimal Weierstrass equation. Let $\langle,\rangle: E(\mathbb{Q})/E(\mathbb{Q})_{\mathrm{tor}}\times E(\mathbb{Q})/E(\mathbb{Q})_{\mathrm{tor}}\longrightarrow\mathbb{R}$ be the canonical Néron-Tate height paring on $E(\mathbb{Q})$ which is the unique non-degenerate bilinear positive-definite paring on $E(\mathbb{Q})$ such that $\langle P,P\rangle-\log\max(|a|,|b|)$ is bounded for $P\in E(\mathbb{Q})$, where $P=(x,y)\in E(\mathbb{Q})$ has x-coordinate a/b with a,b coprime integers. Define the regulator $R(E/\mathbb{Q})$ for the elliptic curve $E/\mathbb{Q}$

$$R(E/\mathbb{Q})=\det(\langle P_i,P_j\rangle)$$

where $P_1,P_2,\ldots,P_r$ is a basis for $E(\mathbb{Q})/E(\mathbb{Q})_{\mathrm{tor}}$.

For each prime number p, the Tamagawa number c_p of the elliptic curve $E/\mathbb{Q}$ at p is

$$c_p = [E(\mathbb{Q}_p) : E_0(\mathbb{Q}_p)]$$

where $E_0(\mathbb{Q}_p)$ is the subgroup consists of points with non-singular reduction mod p. The real period c_∞ of E is defined to be

$$c_\infty = \int_{E(\mathbb{R})} |\omega|$$

where $\omega = dx/(2y + a_1x + a_3)$ is the Néron differential on E.

Conjecture 5 (BSD Conjecture, part 2). *Let E be an elliptic curve over $\mathbb{Q}$ and $r = \mathrm{ord}_{s=1}L(s, E)$. Then*

$$|\text{Ш}(E/\mathbb{Q})| = \lim_{s\to 1} \frac{L(E,s)}{(s-1)^r} \cdot \left(\prod_{p\le\infty} c_p \cdot \frac{|R(E/\mathbb{Q})|}{|E(\mathbb{Q})_{\mathrm{tor}}|^2} \right)^{-1}.$$

The BSD conjecture can be formulated for elliptic curves defined over any number fields. For any abelian group A and integer $n \neq 0$, we denote $A[n]$ the subgroup of A of elements annihilated by n and $A[p^\infty] = \cup_{n\ge 1} A[p^n]$ for a prime p.

Theorem 2.1 (Gross-Zagier [10], Kolyvagin [15], [16]). *If $\mathrm{ord}_{s=1}L(s, E) \le 1$, then*

$$\mathrm{rank}_{\mathbb{Z}} E(\mathbb{Q}) = \mathrm{ord}_{s=1} L(s, E)$$

and the Shafarevich-Tate group $\text{Ш}(E/\mathbb{Q})$ is finite.

The following result on BSD conjecture is for CM elliptic curves (for details see Coates' survey paper [3])

Theorem 2.2 (Coates-Wiles, Rubin, Perion-Rious, Kobayashi). *Let E be an elliptic curve over $\mathbb{Q}$ with complex multiplication by $\mathcal{O}_K$, the ring of integers in an imaginary quadratic field K. Let w be the cardinality of $\mathcal{O}_K^\times$. Then*

(1) *(i) $L(E,1) \neq 0$ if and only if $\mathrm{rank}E(\mathbb{Q}) = 0$ and $\text{Ш}(E/\mathbb{Q})$ is finite. (ii) If $\mathrm{rank}E(\mathbb{Q}) = 0$ and $\text{Ш}(E/\mathbb{Q})[p^\infty]$ is finite for some prime p with $(p, w) = 1$, then $L(E,1) \neq 0$. (iii) if $L(E,1) \neq 0$ and p is any prime with $(p,w) = 1$, then the order of $\text{Ш}(E/\mathbb{Q})[p^\infty]$ is as predicts by the BSD conjecture.*
(2) *(i) $L(E,s)$ has a simple zero at $s = 1$ if and only if $\mathrm{rank}E(\mathbb{Q}) = 1$ and $\text{Ш}(E/\mathbb{Q})$ is finite. (ii) if $\mathrm{rank}E(\mathbb{Q}) = 1$ and $\text{Ш}(E/\mathbb{Q})[p^\infty]$ is finite for some good prime p with $(p,w) = 1$, then $L(E,s)$ has a simple zero at $s = 1$. (iii) if $L(E,s)$ has a simple zero at $s = 1$ and p is any good prime with $(p,w) = 1$, then the order of $\text{Ш}(E/\mathbb{Q})[p^\infty]$ is as predicts as the BSD conjecture.*

Let p be a prime, then $T_p(E) = \varprojlim_n E[p^n]$, where the morphisms $E[p^{n+1}] \to E[p^n]$ in the limit are the multiplication by p, is a free $\mathbb{Z}_p$-module of rank 2 endowed with an action of the absolute Galois group $G_{\mathbb{Q}}$ over $\mathbb{Q}$. $T_p(E)$ is called the Tate-module of E. Let $\rho_{E,p} : G_{\mathbb{Q}} \to \mathrm{Aut}_{\mathbb{Z}_p}(T_p(E))$ be the associated Galois representation and $\overline{\rho}_{E,p} : G_{\mathbb{Q}} \to \mathrm{GL}_2(\mathbb{F}_p)$ its residue representation obtained from $\rho_{E,p}$ by modulo p. The following result on the part 2 of BSD conjecture is for elliptic curve E over $\mathbb{Q}$ without complex multiplication.

Theorem 2.3 (Kato [12], Skinner-Urban [22]). *Let E be an elliptic curve over $\mathbb{Q}$ with conductor N and p a prime. Suppose that*

- *E has good ordinary reduction at p, i.e. E has good reduction at p with $p \nmid a_p$.*
- *$\overline{\rho}_{E,p}$ is surjective and ramified at some prime $q \neq p$ such that $q \| N$.*

If $L(E, 1) \neq 0$, then the order of $\text{Ш}(E/\mathbb{Q})[p^\infty]$ is as predicted by BSD conjecture.

About the behavior of L-function, Goldfeld made the following conjecture.

Conjecture 6 (Goldfeld [6]). *Let $E : y^2 = f(x)$ be an elliptic curve over $\mathbb{Q}$. For any non-zero integer n, $E^{(n)}$ the twist of E defined by the equation $ny^2 = f(x)$ and ϵ_n the epsilon-factor in the function equation of $L(E^{(n)}, s)$. Then among all integers n with $\epsilon_n = +1$ (resp. $\epsilon_n = -1$), for almost all n, the L-fucntion $L(E^{(n)}, s)$ has vanishing order 0 (resp. 1) at $s = 1$.*

3 Congruent number problem revisited

For the congruent elliptic curve $E^{(n)} : y^2 = x^3 - n^2x$, we know that $E^{(n)}(\mathbb{Q})_{\mathrm{tor}} = E^{(n)}[2] \cong (\mathbb{Z}/2\mathbb{Z})^2$ consisting of the infinite point and points with y-coordinates 0. Thus

Proposition 3.1. *A positive integer n is a congruent number if and only if the elliptic curve $E^{(n)} : y^2z = x^3 - n^2xz^2$ has Mordell-Weil group of positive rank.*

In view of BSD conjecture, we would like to understand the L-function of $E^{(n)}$. The following proposition gives some information of $L(E^{(n)}, s)$.

Proposition 3.2. *Let n be a square-free positive integer. Then for the elliptic curve $E^{(n)} : ny^2 = x^3 - x$,*

$$L(E^{(n)}, s) = \prod_{p \nmid 2n} (1 - c_p p^{-s} + p^{1-2s})^{-1}$$

has entire continuation to the whole s-plane and satisfies the following function equation

$$\Lambda(E^{(n)}, s) := 2(2\pi)^{-s}\Gamma(s)L(E^{(n)}, s) = \epsilon_n N_n^{1-s}\Lambda(E^{(n)}, 2 - s),$$

where $N_n = 32n^2$ if $2 \nmid n$ and $N_n = 16n^2$ if $2|n$, and the epsilon-factor ϵ_n is given by

$$\epsilon_n = \begin{cases} +1, & \text{if } n \equiv 1, 2, 3 \mod 8, \\ -1, & \text{if } n \equiv 5, 6, 7 \mod 8. \end{cases}$$

Now it is easy to see that Conjecture 1 is just a corollary of Conjecture 3 of Birch and Swinnerton-Dyer and Conjecture 6 of Goldfeld.

Now we explain the proof of Theorem 1.6. The modular curve $X_0(32)$ is an algebraic curve over $\mathbb{Q}$ of genus one whose cusp $[\infty]$ is a rational point. Thus $A=(X_0(32),[\infty])$ is an elliptic curve. It turns out that A has equation $v^2=u^3+4u$ such that the cusp $[0]$ has coordinate $(2,4)$. Let $A^{(n)}$ denote the elliptic curve $nv^2=u^3+4u$. Then there is a degree 2 isogeny $\varphi_n: A^{(n)}\to E^{(n)}$. Let $\rho(n)\geq 0$ be the integer defined by

$$2^{\rho(n)}=[E^{(n)}(\mathbb{Q}):\varphi_n(A^{(n)}(\mathbb{Q}))+E^{(n)}[2]].$$

Then $0\leq \rho(n)\leq \mathrm{rank}E^{(n)}(\mathbb{Q})=\mathrm{rank}A^{(n)}(\mathbb{Q})$. Note that by Proposition 3.1, n is a congruent number if and only if $\mathrm{rank}A^{(n)}(\mathbb{Q})>0$.

For any square-free positive integer $n\equiv 5,6,7 \mod 8$, the L-series $L(E^{(n)},s)$ has odd vanishing order at $s=1$. Define a non-negative number $\mathcal{L}(n)\in\mathbb{R}_{\geq 0}$ such that if $\mathrm{ord}_{s=1}L(E^{(n)},s)=1$ then $\mathcal{L}(n)^2$ is equal to the right hand of the equality in Conjecture 5, the so-called analytic Sha of $E^{(n)}$, and such that if $\mathrm{ord}_{s=1}L(E^{(n)},s)>1$, then $\mathcal{L}(n)=0$. We have the following result stronger than Theorem 1.6:

Theorem 3.3 (Tian-Yuan-Zhang, [28]). *The number $2^{-\rho(n)}\mathcal{L}(n)$ is an integer. Let $n\equiv 5,6,7 \pmod 8$ be square-free positive integer, if*

$$\sum_{\substack{n=d_0\cdots d_\ell\\ d_i\equiv 1 \pmod 8,\ i>0}}\prod_i g(d_i)=1 \quad or \quad \sum_{\substack{n=d_0\cdots d_\ell,\\ d_0\equiv 5,6,7 \pmod 8\\ d_1\equiv 2,3 \pmod 8\\ d_i\equiv 1 \pmod 8,\ i>1}}\prod_i g(d_i)=1,$$

then $2^{-\rho(n)}\mathcal{L}(n)$ is odd. In particular,

$$\mathrm{rank}E^{(n)}(\mathbb{Q})=1=\mathrm{ord}_{s=1}L(E^{(n)},s),$$

$\text{Ш}(E^{(n)}/\mathbb{Q})$ is finite, and n is a congruent number.

We explain the idea in the simple case where n is a positive integer with $n\equiv 7 \mod 8$. Let $K=\mathbb{Q}(\sqrt{-n})$ and $\mathcal{O}$ its ring of integers. There is an ideal $\mathcal{N}$ of $\mathcal{O}$ such that $\mathcal{O}/\mathcal{N}\cong\mathbb{Z}/32\mathbb{Z}$. Let $\mathfrak{a}_1,\ldots,\mathfrak{a}_h$ be a complete ideal representatives of ideal class group of K. For each i, $\mathcal{N}^{-1}\mathfrak{a}_i/\mathfrak{a}_i$ is a cyclic subgroup of order 32 of the CM elliptic curve $\mathbb{C}/\mathfrak{a}_i$. By the moduli interpretation of $X_0(32)$, the pair $(\mathbb{C}/\mathfrak{a}_i,\mathcal{N}^{-1}\mathfrak{a}_i/\mathfrak{a}_i)$ defines a point on $X_0(32)$, denoted by x_i. By reciprocity law, x_i is defined over the so-called Hilbert class field H of K and if let c be the complex conjugation, then

$$\{x_1^c,\ldots,x_h^c\}=\{-x_1+(2,4),\ldots,-x_h+(2,4)\}.$$

We are able to show that $P_n=\sum_{i=1}^h x_i\in A(K)$ is of infinite order if

$$\sum_{\substack{n=d_0\cdots d_\ell\\ d_i\equiv 1 \pmod 8,\ i>0}}\prod_i g(d_i)=1 \quad \text{or} \quad \sum_{\substack{n=d_0\cdots d_\ell,\\ d_0\equiv 5,6,7 \pmod 8\\ d_1\equiv 2,3 \pmod 8\\ d_i\equiv 1 \pmod 8,\ i>1}}\prod_i g(d_i)=1.$$

Let $A(K)^-$ denote the subgroup of $Q \in A(K)$ with $Q^c = -Q$, then one can see easily that $A(K)^- \cong A^{(n)}(\mathbb{Q})$. Note that $2P = (P + P^c) + (P - P^c)$, and $P + P^c \in A(\mathbb{Q}) = A(\mathbb{Q})_{\text{tor}}$. If P is of infinite order then so is $P - P^c \in A(K)^-$, therefore $A^{(n)}(\mathbb{Q}) \cong A(K)^-$ has positive rank and n is a congruent number.

We now give Heegner's argument in the case that $n \equiv 7 \mod 8$ is a prime. In this case, $h = \#\mathcal{C}$ is of odd cardinality, $g(p) = 1$, and the above condition automatically holds. We have seen that $P_n^c + P_n = h \cdot (2, 4)$. If P_n is a torsion, then by ramification consideration, $P_n \in A(\mathbb{Q})$. Thus $P_n^c + P_n$ is annihilated by 2. But h is odd and $h \cdot (2, 4)$ is of order 4, a contradiction. Thus P_n is of infinite order. One can further obtain the 2-divisibility of P_n in $A(K)$ and show $2^{-\rho(n)}\mathcal{L}(n)$ is odd via Gross-Zagier formula.

For the case n with many prime factors, h maybe divisible by a higher power of 2 and the above argument does not work directly. Recall that $\mathcal{C}$ is the ideal class group of K. Define the genus point

$$Z(n) := \sum_{\mathfrak{a}_i \in 2\mathcal{C}} x_i.$$

Note that $Z(n) = P_n$ when n is a prime. To deal with composite n, we use the induction method introduced in [27]: for each character χ of $\mathcal{C}$ factoring through $\mathcal{C}/2\mathcal{C}$ (thus valued in ± 1), define

$$P_\chi = \sum_{i=1}^{h} x_i \chi([\mathfrak{a}_i]) \in A(\overline{K}).$$

Thus $P_n = P_{\chi_0}$ for the trivial character χ_0. It turns out that

$$\sum_{\chi : \mathcal{C}/2\mathcal{C} \to \pm 1} P_\chi = 2^{\mu(n)-1} Z(n), \tag{3}$$

where $\mu(n)$ is the number of prime factors of n. With the following ingredients:

- generalization of Heegner's above construction P_n and $Z(n)$ to all cases $n \equiv 5, 6, 7 \mod 8$,
- Gross-Zagier formula [10] and its generalization by Yuan-Zhang-Zhang [31],
- Kolyvagin's result [15], [16] in the above special case which says that if the Heegner points P_n is non-torsion, then $A(K)$ is of rank one,

we may express P_χ in the above equation (3) in term of $Z(d)$'s and $\mathcal{L}(d)$'s with $d|n$. Then a similar argument, as Heegner in the case that $n \equiv 7 \mod 8$ is a prime leads our Theorem 3.3 [28].

References

[1] B.J. Birch, *Elliptic curves and modular functions*, Symposia Mathematica, Indam Rome 1968/1969, vol. 4, pp27–32. London: Academic Press (1970).

[2] B.J. Birch and H.P.F. Swinnerton-Dyer, *Notes on elliptic curves (II)*, J. Reine Angew. Math, 218 (1965), 79–108.
[3] J. Coates, *Elliptic Curves with Complex Multiplication and Iwasawa Theory*, Bull. London Math. Soc. 23 (1991) 321–350.
[4] J. Coates and A. Wiles, *On the conjecture of Birch and Swinnerton-Dyer*, Invent. Math. 39 (1977), 233–251.
[5] L. E. Dickson, *History of the Theory of Numbers* Volume II. Carnegie Institute, Washington DC (1920), reprinted by Chelsea (1966).
[6] D. Goldfeld, *Conjectures on elliptic curves over quadratic fields*, in Number theory, Carbondale 1979, M. B. Nathanson, ed., Lecture Notes in Math. 751, Springer, Berlin, 1979, 108–118.
[7] B. Gross, *Heights and the special values of L-series*. Canadian Mathematical Scciety, Conference Proceedings, Volume 7 (1987).
[8] B. Gross, *Local orders, root numbers, and modular curves*, Amer. J. Math., 110(1988), 1153–1182.
[9] B.Gross, *Kolyvagin's work on modular elliptic curves*, in: L-function and Arithmetic (ed. J. Coates and M.J.Talyor) Cambridge University Press (1991).
[10] B. Gross and D. Zagier, *Heegner points an derivatives of L-series*. Invent. Math. 84 (1986), no. 2, 225–320.
[11] K. Heegner, *Diophantische analysis und modulfunktionen*. Math. Z. 56, 227–253 (1952).
[12] K. Kato, *p-adic Hodge theory and values of zeta functions of modular forms*, Cohomologies p-adiques et applications arithmetiques. III. Asterisque No. 295 (2004), ix, p. 117–290.
[13] N. Koblitz, *Introduction to elliptic curves and modular forms*, Springer 1984.
[14] Shinichi Kobayshi, *The p-adic Gross-Zagier formula for elliptic curves at supersingular primes*, Preprint.
[15] V.A.Kolyvagain, *Euler system*, The Grothendieck Festschrift. Prog. in ath., Boston, Birkhauser (1990).
[16] V.A.Kolyvagain, *Finiteness of* $E(\mathbb{Q})$ *and* $Ш(E,\mathbb{Q})$ *for a subclass of Weil curves*, Math.USSR Izvestiya,Vol. 32 (1989), No. 3.
[17] T. Miyake, *Modular Forms*, Springer Monographs in Mathematics, 1989.
[18] P. Monsky, *Mock Heegner Points and Congruent Numbers*, Math. Z. 204, 45–68 (1990).
[19] J. Nekovář, *The Euler system method for CM points on Shimura curves*, In: L-functions and Galois representations, (Durham, July 2004), LMS Lecture Note Series 320, Cambridge Univ. Press, 2007, pp. 471–547.
[20] B. Perrin-Riou, *Points de Heegner et derivees de fonctions L p-adiques*, Invent. Math. 89 (1987), no. 3, pp. 455–510.
[21] H. Saito, *On Tunnell's formula for characters of* GL(2), Compositio Math. 85 (1993), 99–108.
[22] C. Skinner and E. Urban, *The Iwasawa Main Conjecture For* GL_2, Invent. math (2014) 195: 1–277.
[23] J. Silverman, *Arithmetic of elliptic curves*, GTM 106, Springer.
[24] J. Tate, *The arithmetic of elliptic curves*, Invent. Math. 23 (1974), 179–206.

[25] J. Tate, *Number Theoretic Background* in Automorphic Forms, Representations and L-functions, Proc. Symp. in Pure Math. XXXIII Part 2(1979), 3–26.
[26] Y. Tian, *Congruent Numbers with many prime factors*, PNAS, Vol 109, no. 52. 21256–21258.
[27] Y. Tian, *Congruent Numbers and Heegner Points*, Cambridge J. of Math. Vol. 2.1. 117–161, 2014.
[28] Y. Tian, X. Yuan, and S. Zhang, *Genus Periods, Genus Points and Congruent Number Problem*, preprint.
[29] J. Tunnell, *A classical diophantine problem and modular forms*, Invent. Math. 72, 323–334 (1983).
[30] J. Tunnell, *Local ϵ-factors and characters of* GL(2), Amer. J. Math. 105 (1983), 1277–1307.
[31] X.Yuan, S. Zhang, and W. Zhang, *The Gross-Zagier formula on Shimura Curves*, Annals of Mathematics Studies Number 184, 2012.

Introduction to Modern Mathematics
ALM 33, pp. 425–432

Noncommutative Geometry of Discrete Groups

Zhizhang Xie* and Guoliang Yu†

Abstract

In this expository article, we give a friendly introduction to noncommutative geometry of discrete groups, its connections to higher index theory of Dirac operators and its applications to differential geometry.

Keywords and Phrases: Noncommutative geometry, discrete groups, K-theory, higher index theory, positive scalar curvature.

In mathematics, groups naturally arise as symmetries of geometric objects. The structure of a group is reflected in its representation theory. The set of all representations of a group forms the dual space of the group. In the case of an abelian discrete group, the dual space is a compact Hausdorff space that completely characterizes the original group and can be understood by tools from algebraic topology and classical analysis. However when the group is nonabelian, its dual space often fails to be Hausdorff and is generally ill-behaved. Noncommutative geometry is a branch of mathematics that, in particular, provides novel tools to study the space of representations for general discrete groups [12]. This is achieved by studying certain noncommutative geometric objects associated to discrete groups.

Recall that a (complex) representation of a group Γ is given by a group homomorphism $\pi : \Gamma \to GL(V)$, where $GL(V)$ is the group of all invertible linear maps on a complex vector space V. Given a finite group Γ, the collection of isomorphism classes of finite dimensional representations naturally generates the so-called representation ring of Γ, denoted by $R(\Gamma)$. For example when $\Gamma = \{e\}$ is the trivial group, then $R(\{e\})$ is just $\mathbb{Z}$. This is simply a restatement of the fact that a representation of the trivial group is completely determined by its dimension. In this sense, the ring $R(\Gamma)$ is just a generalization of the notion of dimension for finite groups.

In general, however, when Γ is infinite, the representation ring $R(\Gamma)$ as defined above is too algebraic to capture enough information about the representations of Γ. Instead, we need to introduce some analysis. More precisely, let $\ell^2(\Gamma)$ be the Hilbert space of ℓ^2-summable functions on Γ. The group Γ acts on $\ell^2(\Gamma)$

*Department of Mathematics, Texas A&M University, USA. Email: Xie@math.tamu.edu
†Department of Mathematics, Texas A&M University, USA. Email: guoliangyu@math.tamu.edu

by translations, so that $\ell^2(\Gamma)$ becomes a representation of Γ, called the regular representation. Moreover the action of Γ extends linearly to an action of the group algebra $\mathbb{C}[\Gamma]$, where $\mathbb{C}[\Gamma]$ consists of all formal sums $\sum_{\gamma\in\Gamma} a_\gamma \gamma$ such that $a_\gamma = 0$ for all but finitely many $\gamma \in \Gamma$, with multiplication

$$\left(\sum_{\gamma\in\Gamma} a_\gamma \gamma\right) \cdot \left(\sum_{\alpha\in\Gamma} b_\alpha \alpha\right) = \sum_{\beta\in\Gamma} \left(\sum_{\gamma\alpha=\beta} a_\gamma b_\alpha\right) \beta.$$

Let $B(\ell^2(\Gamma))$ be the algebra of all bounded linear operators on $\ell^2(\Gamma)$; we define the reduced group C^*-algebra of Γ, denoted by $C_r^*(\Gamma)$, to be the operator-norm closure of $\mathbb{C}[\Gamma]$ in $B(\ell^2(\Gamma))$.

Now, the reduced group C^*-algebra $C^*(\Gamma)$ is precisely the noncommutative geometric object associated to Γ that we referred to at the beginning. Intuitively speaking, if we think of an algebra as the algebra of functions on some topological space, then the maximal ideals of that algebra correspond to the points of the topological space. When Γ is abelian, $C_r^*(\Gamma)$ has many maximal ideals; indeed, in this case, the set of maximal ideals forms a nice Haudorff space, which is, in fact, just the dual space of Γ we mentioned earlier. However, for many nonabelian discrete groups such as the free group F_n with $n \geq 2$ generators, $C_r^*(\Gamma)$ turns out to be simple, that is, it does not admit any nontrivial ideals [36]. We see that the concept of points no longer makes sense for $C_r^*(\Gamma)$ in general. This is a completely new phenomenon that only exists in the noncommutative world!

Algebraic topology teaches us that cohomology theory provides some of the most fundamental invariants for understanding geometric objects. However, classical singular or cellular cohomology theory does not extend to the noncommutative geometric setting. The only cohomology theory that works for both commutative and noncommutative geometric objects is called K-theory; informally, it is a generalization of the notion of representation ring. Let us recall the definition of the K-theory group, cf. [9]. Let $A = C_r^*(\Gamma)$ and $M_\infty(A) = \bigcup_n M_n(A)$, where $M_n(A)$ is the algebra of all $n \times n$ matrices with coefficients in A. Let $V(A)$ be the set of all equivalence classes of idempotents in $M_\infty(A)$, where two idempotents p and q are said to be equivalent if there exists an invertible element $u \in M_\infty(A)$ such that $upu^{-1} = q$. Observe that $V(A)$ is an abelian semigroup with addition

$$[p_1] \oplus [p_2] = \begin{bmatrix} p_1 & 0 \\ 0 & p_2 \end{bmatrix}.$$

Recall that there is a natural abelian group, called the Grothendieck group, associated to each abelian semigroup. More precisely, if $(S, +)$ is an abelian semigroup, then the Grothendieck group of S is defined to be $S\times S/\sim$, where $(s_1, s_2) \sim (t_1, t_2)$ if $s_1 + t_2 + s = s_2 + t_1 + s$ for some $s \in S$. For example, if $S = \mathbb{N}$, then the corresponding Grothendieck group is $\mathbb{Z}$. The K-theory group $K_0(A)$ is defined to be the Grothendieck group of $V(A)$. In the case where Γ is finite, we have $C_r^*(\Gamma) = \mathbb{C}[\Gamma]$ and $K_0(\mathbb{C}[\Gamma]) \cong R(\Gamma)$.

In general, it is difficult to compute $K_0(C_r^*(\Gamma))$. In the case where Γ is abelian, we have an algorithm to compute $K_0(C_r^*(\Gamma))$, using tools from algebraic topology. This is because $C_r^*(\Gamma)$ admits many ideals in this case, which allows us to decompose $C_r^*(\Gamma)$ into tractable pieces. However, this method breaks down immediately in general, for, we have seen, $C_r^*(\Gamma)$ is often simple for nonabelian groups. An algorithm for computing $K_0(C_r^*(\Gamma))$ in the general case is envisioned by the Baum–Connes conjecture [4, 5]. In very rough terms, the Baum–Connes conjecture states that $K_0(C_r^*(\Gamma))$ is isomorphic to something that is completely topological, which can therefore be computed by tools from algebraic topology such as the Mayer–Vietoris sequence.

Let us give a more precise description of the Baum–Connes conjecture. For simplicity, let us assume that Γ is torsion-free from now on. Recall that the classifying space $B\Gamma$ of Γ is a topological space such that any Γ-covering space of a compact Hausdorff space X is uniquely determined (up to homotopy) by a continuous map $f : X \to B\Gamma$. For example, if $\Gamma = \mathbb{Z}^n$, then $B\Gamma = T^n$, the n-dimensional torus. To describe the Baum–Connes conjecture, we need to introduce the K-homology group $K_0(B\Gamma)$ of the classifying space $B\Gamma$ and the Baum–Connes assembly map $\mu : K_0(B\Gamma) \to K_0(C_r^*(\Gamma))$.

Let us first introduce the Dirac operators, cf. [2, 3, 29]. Roughly speaking, a Dirac operator on a manifold is a first order differential operator whose square is the Laplacian. For example, on $\mathbb{R}^1$, the standard Dirac operator is $D = \frac{1}{i}\frac{d}{dx}$ with its square $D^2 = -\frac{d^2}{dx^2} = \Delta$; on $\mathbb{R}^2$, the standard Dirac operator is

$$D = \begin{pmatrix} 0 & -\frac{\partial}{\partial x} + i\frac{\partial}{\partial y} \\ \frac{\partial}{\partial x} + i\frac{\partial}{\partial y} & 0 \end{pmatrix} \text{ with } D^2 = \begin{pmatrix} -\frac{\partial^2}{\partial x^2} - \frac{\partial^2}{\partial y^2} & 0 \\ 0 & -\frac{\partial^2}{\partial x^2} - \frac{\partial^2}{\partial y^2} \end{pmatrix}.$$

Observe that on $\mathbb{R}^2$, we need to use matrices in order to get a "square root" of the Laplacian. In general, there is an elegant way to use Clifford algebras to define the Dirac operator on $\mathbb{R}^n$. Now, we would like to know how to construct a Dirac operator on a manifold. Recall that a manifold is a collection of coordinate charts patched together. On each coordinate chart, which can be assumed to be $\mathbb{R}^n$, we have seen how to construct a Dirac operator. We would like to glue these locally defined Dirac operators together in a coherent way to obtain a Dirac operator on the whole manifold. It turns out that a manifold needs to satisfy a certain topological condition, called spin condition, in order to have such a coherent gluing (cf. [29, Appendix D]). Finally, the K-homology group $K_0(B\Gamma)$ is an abelian group whose elements consist of Dirac operators and their generalizations on closed manifolds, i.e., compact manifolds without boundary, subject to certain equivalence relations [6, 7, 8].

Having defined the K-homology group, let us now describe the Baum–Connes assembly map μ, which is sometimes also called the higher index map. First, we recall the classical Fredholm index theory. The Dirac operator D on a manifold M is an elliptic differential operator, which naturally induces a Fredholm operator on a Hilbert space H. By definition, a Fredholm operator F is a bounded linear operator on H with a finite dimensional kernel and a finite dimensional cokernel, where the cokernel of F is the quotient of H by the range of F. The index of F is defined to be the integer

$$\mathrm{Ind}(F) = \dim(\ker F) - \dim(\operatorname{coker} F).$$

By convention, we define the Fredholm index of D to be the index of the associated Fredholm operator F. The Fredholm index is a homotopy invariant. The celebrated Atiyah–Singer index theorem states that the index of the Dirac operator D on M can be expressed by cohomological geometric data of M [2, 3].

Observe that if we view the finite dimensional vector spaces $\ker F$ and $\operatorname{coker} F$ as representations of the trivial group $\{e\}$, then the Fredholm index is reminiscent of the notion of dimension of representations. This might sound a little ad hoc at the moment, but we shall justify it as follows. Consider the universal cover $\widetilde{M}$ of a closed manifold M and let Γ be the fundamental group of M. The Dirac operator D on M naturally lifts to a Dirac operator on $\widetilde{D}$ on $\widetilde{M}$. If Γ is finite, $\widetilde{M}$ is again closed. As we have seen, $\widetilde{D}$ naturally induces a Fredholm operator in this case. Moreover, we notice that $\widetilde{D}$ is Γ-equivariant, that is, $\widetilde{D}$ commutes with the action of Γ on $\widetilde{M}$. As a result, the kernel and cokernel of $\widetilde{D}$ are, in fact, finite dimensional representations of Γ, and their formal difference gives rise to an element in $R(\Gamma) = K_0(\mathbb{C}[\Gamma])$; we call this element the higher index of D. In general, when Γ is not finite, $\widetilde{M}$ is not closed. In this case, $\widetilde{D}$ does not define a Fredholm operator on a Hilbert space, and hence the usual Fredholm index does not make sense. Rather, we need to generalize the notion of Fredholm operator by taking Γ into account, leading to the notion of generalized Fredholm operators. Regardless of Γ being finite or not, $\widetilde{D}$ always naturally gives rise to a generalized Fredholm operator, whose higher index is an element of $K_0(C_r^*(\Gamma))$, denoted by $\operatorname{Ind}_\Gamma(D)$ [32, 23]. The higher index contains the Fredholm index as its 0-dimensional information [1], and hence is a generalization of the Fredholm index. In particular, the higher index of D is a homotopy invariant, and, moreover, an obstruction to the invertibility of $\widetilde{D}$. Finally, the Baum–Connes assembly map μ is the group homomorphism defined by mapping elements of $K_0(B\Gamma)$ to their corresponding higher index in $K_0(C_r^*(\Gamma))$. The Baum–Connes conjecture states that the assembly map μ is an isomorphism.

As one can see, the noncommutative geometry of discrete groups is tied ultimately with the study of the higher index of Dirac type operators. The latter is often called higher index theory, and it has many applications to geometry and topology. For example, it is a fundamental tool in the study of the following conjecture.

Conjecture (Gromov-Lawson Conjecture). *A closed aspherical manifold does not admit a Riemannian metric of positive scalar curvature.*

An aspherical manifold is a manifold whose universal cover is contractible. For example, the n-dimensional torus T^n is aspherical for each $n \geq 1$. In fact, the following question was one of the main motivating examples behind the above conjecture.

Question. *For $n \geq 2$, does the n-dimensional torus T^n admit a Riemannian metric of positive scalar curvature?*

The answer is "no" for all $n \geq 2$. The case where $n = 2$ follows immediately from the Gauss–Bonnet theorem. The case where $n \geq 3$, which turned out to be surprisingly involved, was solved by Schoen–Yau [37, 38] and Gromov–Lawson

[15] using different methods. We sketch a proof using higher index theory. First, we recall that the Dirac operator D associated to a given Riemannian metric on T^n satisfies the Weitzenböck–Lichnerowicz identity $D^2 = \Delta + \kappa/4$, where κ is the scalar curvature of the given metric [29, 30]. Now, this Weitzenböck–Lichnerowicz identity lifts to $\mathbb{R}^n = \widetilde{T^n}$, the universal cover of T^n; in other words, we have

$$\widetilde{D}^2 = \widetilde{\Delta} + \frac{\widetilde{\kappa}}{4},$$

where $\widetilde{D}$ (resp. $\widetilde{\Delta}$ and $\widetilde{\kappa}$) is the lift of D (resp. Δ and κ) from T^n to $\mathbb{R}^n = \widetilde{T^n}$. Now if $\kappa > 0$ everywhere on T^n, then $\widetilde{D}$ would be invertible, implying that the higher index $\mathrm{Ind}_\Gamma(D) = 0$, where $\Gamma = \mathbb{Z}^n$. However, by homotopy invariance, $\mathrm{Ind}_\Gamma(D)$ is independent of the choice of metric on T^n. Moreover, an explicit calculation shows that[1] $\mathrm{Ind}_\Gamma(D) \neq 0$ for D on T^n. Hence, by contradiction, we see that T^n does not admit a Riemannian metric of positive scalar curvature. Observe that it was essential to use the higher index instead of the Fredholm index here, since the Fredholm index $\mathrm{Ind}(D)$ on T^n is, in fact, zero.

In the above example, an essential ingredient of the proof is the nonvanishing of the higher index of the Dirac operator. This turns out to be a common theme in many applications of noncommutative geometry to classical topology and geometry. From this perspective, the strong Novikov conjecture provides an algorithm to verify whether or not a higher index vanishes.

In the last two decades, there have been major breakthroughs in the study of the noncommutative geometry of discrete groups, especially those related to the Baum–Connes conjecture [17, 28, 22, 10, 31, 33, 40, 27] and the Novikov conjecture [23, 14, 13, 39, 25, 26, 45, 46, 24, 11]. In particular, the Novikov conjecture has been proven to hold for groups that are coarsely embeddable into Hilbert space [46]. This includes all discrete subgroups of linear groups [16]. In recent years, there has also been considerable progress in the study of secondary higher index-theoretical invariants, such as the higher rho invariant [19, 18, 20, 21, 35, 34, 43, 42, 44, 41]. Secondary invariants often appear when primary invariants, such as the higher index, vanish. For example, as we have seen above, if a spin manifold M is equipped with a Riemannian metric g of positive scalar curvature, then the higher index $\mathrm{Ind}_\Gamma(D)$ of its Dirac operator D is zero. However, there is a natural way to associate a secondary higher index-theoretical invariant to the pair (D, g). This invariant is called the higher rho invariant of the pair (D, g), denoted by $\rho(D, g)$. Unlike the higher index, the higher rho invariant can very well depend on the choice of metric. For example, for two positive scalar curvature metrics g_1 and g_2 on M, if $\rho(D, g_1) \neq \rho(D, g_2)$, then g_1 and g_2 can not be connected by a path of positive scalar curvature metrics on M. This gives some nice applications of the higher rho invariant to classical geometry and topology. For example, it can be applied to study the topology of the moduli space of positive scalar curvature metrics on a given spin manifold [34, 35, 42]. Moreover, similar secondary invariants also

[1] We have restricted our discussion of the higher index to the case of $K_0(C_r^*(\Gamma))$, which corresponds to even dimensional manifolds. There is a complete analogue in the case of odd dimensional manifolds, where the higher indices lie in $K_1(C_r^*(\Gamma))$. All discussions above apply equally to the odd dimensional case as well.

lead to very interesting results in the study of the rigidity problem for topological manifolds [41].

The study of secondary invariants is ultimately related to the Baum–Connes conjecture. Secondary invariants can potentially be used to construct exotic elements in $K_0(C_r^*(\Gamma))$ that are not in the image of the Baum–Connes assembly map. For instance, given two positive scalar curvature metrics on an odd dimensional spin manifold, one can construct a secondary invariant in $K_0(C_r^*(\Gamma))$. It is an open question whether or not this secondary invariant is in the image of the Baum–Connes assembly map [44]. A better understanding of secondary invariants certainly will reveal new phenomena in the noncommutative geometry of discrete groups.

References

[1] M. F. Atiyah. Elliptic operators, discrete groups and von Neumann algebras. In *Colloque "Analyse et Topologie" en l'Honneur de Henri Cartan (Orsay, 1974)*, pages 43–72. Astérisque, No. 32–33. Soc. Math. France, Paris, 1976.

[2] M. F. Atiyah and I. M. Singer. The index of elliptic operators. I. *Ann. of Math. (2)*, 87:484–530, 1968.

[3] M. F. Atiyah and I. M. Singer. The index of elliptic operators. III. *Ann. of Math. (2)*, 87:546–604, 1968.

[4] P. Baum and A. Connes. K-theory for discrete groups. In *Operator algebras and applications, Vol. 1*, volume 135 of *London Math. Soc. Lecture Note Ser.*, pages 1–20. Cambridge Univ. Press, Cambridge, 1988.

[5] P. Baum, A. Connes, and N. Higson. Classifying space for proper actions and K-theory of group C^*-algebras. In *C^*-algebras: 1943–1993 (San Antonio, TX, 1993)*, volume 167 of *Contemp. Math.*, pages 240–291. Amer. Math. Soc., Providence, RI, 1994.

[6] P. Baum and R. G. Douglas. K homology and index theory. In *Operator algebras and applications, Part I (Kingston, Ont., 1980)*, volume 38 of *Proc. Sympos. Pure Math.*, pages 117–173. Amer. Math. Soc., Providence, R.I., 1982.

[7] P. Baum, N. Higson, and T. Schick. On the equivalence of geometric and analytic K-homology. *Pure Appl. Math. Q.*, 3(1, part 3):1–24, 2007.

[8] P. Baum, N. Higson, and T. Schick. A geometric description of equivariant K-homology for proper actions. In *Quanta of maths*, volume 11 of *Clay Math. Proc.*, pages 1–22. Amer. Math. Soc., Providence, RI, 2010.

[9] B. Blackadar. *K-theory for operator algebras*, volume 5 of *Mathematical Sciences Research Institute Publications*. Cambridge University Press, Cambridge, second edition, 1998.

[10] J. Chabert, S. Echterhoff, and R. Nest. The Connes-Kasparov conjecture for almost connected groups and for linear p-adic groups. *Publ. Math. Inst. Hautes Études Sci.*, (97):239–278, 2003.

[11] X. Chen, Q. Wang, and G. Yu. The maximal coarse Baum-Connes conjecture for spaces which admit a fibred coarse embedding into Hilbert space. *Adv. Math.*, 249:88–130, 2013.

[12] A. Connes. *Noncommutative geometry.* Academic Press Inc., San Diego, CA, 1994.

[13] A. Connes, M. Gromov, and H. Moscovici. Conjecture de Novikov et fibrés presque plats. *C. R. Acad. Sci. Paris Sér. I Math.*, 310(5):273–277, 1990.

[14] A. Connes and H. Moscovici. Cyclic cohomology, the Novikov conjecture and hyperbolic groups. *Topology*, 29(3):345–388, 1990.

[15] M. Gromov and H. B. Lawson, Jr. Positive scalar curvature and the Dirac operator on complete Riemannian manifolds. *Inst. Hautes Études Sci. Publ. Math.*, (58):83–196 (1984), 1983.

[16] E. Guentner, N. Higson, and S. Weinberger. The Novikov conjecture for linear groups. *Publ. Math. Inst. Hautes Études Sci.*, (101):243–268, 2005.

[17] N. Higson and G. Kasparov. E-theory and KK-theory for groups which act properly and isometrically on Hilbert space. *Invent. Math.*, 144(1):23–74, 2001.

[18] N. Higson and J. Roe. Mapping surgery to analysis. I. Analytic signatures. *K-Theory*, 33(4):277–299, 2005.

[19] N. Higson and J. Roe. Mapping surgery to analysis. II. Geometric signatures. *K-Theory*, 33(4):301–324, 2005.

[20] N. Higson and J. Roe. Mapping surgery to analysis. III. Exact sequences. *K-Theory*, 33(4):325–346, 2005.

[21] N. Higson and J. Roe. K-homology, assembly and rigidity theorems for relative eta invariants. *Pure Appl. Math. Q.*, 6(2, Special Issue: In honor of Michael Atiyah and Isadore Singer):555–601, 2010.

[22] P. Julg and G. Kasparov. Operator K-theory for the group $\mathrm{SU}(n,1)$. *J. Reine Angew. Math.*, 463:99–152, 1995.

[23] G. Kasparov. Equivariant KK-theory and the Novikov conjecture. *Invent. Math.*, 91(1):147–201, 1988.

[24] G. Kasparov and G. Yu. The Novikov conjecture and geometry of Banach spaces. *Geom. Topol.*, 16(3):1859–1880, 2012.

[25] G. G. Kasparov. K-theory, group C^*-algebras, and higher signatures (conspectus). In *Novikov conjectures, index theorems and rigidity, Vol. 1 (Oberwolfach, 1993)*, volume 226 of *London Math. Soc. Lecture Note Ser.*, pages 101–146. Cambridge Univ. Press, Cambridge, 1995.

[26] G. G. Kasparov and G. Skandalis. Groups acting on buildings, operator K-theory, and Novikov's conjecture. *K-Theory*, 4(4):303–337, 1991.

[27] V. Lafforgue. K-théorie bivariante pour les algèbres de Banach et conjecture de Baum-Connes. *Invent. Math.*, 149(1):1–95, 2002.

[28] V. Lafforgue. La conjecture de Baum-Connes à coefficients pour les groupes hyperboliques. *J. Noncommut. Geom.*, 6(1):1–197, 2012.

[29] H. B. Lawson, Jr. and M.-L. Michelsohn. *Spin geometry*, volume 38 of *Princeton Mathematical Series.* Princeton University Press, Princeton, NJ, 1989.

[30] A. Lichnerowicz. Spineurs harmoniques. *C. R. Acad. Sci. Paris*, 257:7–9, 1963.

[31] I. Mineyev and G. Yu. The Baum-Connes conjecture for hyperbolic groups. *Invent. Math.*, 149(1):97–122, 2002.

[32] A. S. Miščenko. Infinite-dimensional representations of discrete groups, and higher signatures. *Izv. Akad. Nauk SSSR Ser. Mat.*, 38:81–106, 1974.

[33] H. Oyono-Oyono. La conjecture de Baum-Connes pour les groupes agissant sur les arbres. *C. R. Acad. Sci. Paris Sér. I Math.*, 326(7):799–804, 1998.

[34] P. Piazza and T. Schick. Groups with torsion, bordism and rho invariants. *Pacific J. Math.*, 232(2):355–378, 2007.

[35] P. Piazza and T. Schick. Rho-classes, index theory and Stolz's positive scalar curvature sequence. preprint, 2012.

[36] R. T. Powers. Simplicity of the C^*-algebra associated with the free group on two generators. *Duke Math. J.*, 42:151–156, 1975.

[37] R. Schoen and S. T. Yau. Existence of incompressible minimal surfaces and the topology of three-dimensional manifolds with nonnegative scalar curvature. *Ann. of Math. (2)*, 110(1):127–142, 1979.

[38] R. Schoen and S. T. Yau. On the structure of manifolds with positive scalar curvature. *Manuscripta Math.*, 28(1-3):159–183, 1979.

[39] G. Skandalis, J. L. Tu, and G. Yu. The coarse Baum-Connes conjecture and groupoids. *Topology*, 41(4):807–834, 2002.

[40] J.-L. Tu. The Baum-Connes conjecture and discrete group actions on trees. *K-Theory*, 17(4):303–318, 1999.

[41] S. Weinberger and G. Yu. Finite part of operator K-theory for groups finitely embeddable into Hilbert space and the degree of non-rigidity of manifolds. arXiv:1308.4744, 2013.

[42] Z. Xie and G. Yu. Higher rho invariants and the moduli space of positive scalar curvature metrics. arXiv:1310.1136, 2013.

[43] Z. Xie and G. Yu. Positive scalar curvature, higher rho invariants and localization algebras. Advances in Mathematics 262 (2014), 823–866.

[44] Z. Xie and G. Yu. A relative higher index theorem, diffeomorphisms and positive scalar curvature. *Adv. Math.*, 250:35–73, 2014.

[45] G. Yu. The Novikov conjecture for groups with finite asymptotic dimension. *Ann. of Math. (2)*, 147(2):325–355, 1998.

[46] G. Yu. The coarse Baum-Connes conjecture for spaces which admit a uniform embedding into Hilbert space. *Invent. Math.*, 139(1):201–240, 2000.

Made in United States
Orlando, FL
04 November 2026

90219996R00243